# ENVIRONMENTAL SCIENCE

ELDON D. ENGER
J. RICHARD KORMELINK
BRADLEY F. SMITH
RODNEY J. SMITH

*Delta College*
*University Center, Michigan*

wcb
Wm. C. Brown Publishers
Dubuque, Iowa

wcb group

Wm. C. Brown *Chairman of the Board*
Mark C. Falb *President and Chief Executive Officer*

wcb

*Wm. C. Brown Publishers, College Division*

Lawrence E. Cremer *President*
James L. Romig *Vice-President, Product Development*
David A. Corona *Vice-President, Production and Design*
E. F. Jogerst *Vice-President, Cost Analyst*
Bob McLaughlin *National Sales Manager*
Catherine M. Faduska *Director of Marketing Services*
Craig S. Marty *Director of Marketing Research*
Marilyn A. Phelps *Manager of Design*
Eugenia M. Collins *Production Editorial Manager*
Mary M. Heller *Photo Research Manager*

*Book Team*

John Stout *Executive Editor*
Mary J. Porter *Assistant Editor*
Anne Magno *Designer*
Vickie Blosch *Production Editor*
Carol M. Schiessl *Photo Research Editor*
Carla D. Arnold *Permissions Editor*

Cover photograph: © Pat O'Hara



Library of Congress Catalog Card Number: 85–70908

ISBN 0–697–05103–X



Printed in the United States of America
10 9 8 7 6 5 4 3 2

# Contents

# Preface

The study of environmental science is interdisciplinary. Because it is a broad area of study, it is important to develop its many aspects simultaneously, so that the reader may acquire a unified understanding of environmental issues and concerns.

Most environmental issues are best understood when historical background is provided and political, economic, and social implications are integrated with purely scientific information. This text thoroughly explores all facets of environmental issues and thus provides a comprehensive, readable, objective treatment of this complex field. The authors have diligently avoided the inclusion of personal biases and fashionable philosophies.

*Environmental Science: The Study of Interrelationships* is designed for a one-semester, introductory course. Students from a wide range of disciplines will find it informative and interesting. The text is not a catalog of facts; appropriate facts are used to develop concepts, and concepts are carefully and effectively exemplified. The concept of interrelatedness is central to the text; understanding this concept will enable students to grow in their capacity for intelligent environmental decision making.

*Organization and content*

This book is divided into six parts and twenty-four chapters. It is organized to provide an even and logical flow of concepts and to treat this complex field in a clear and thorough manner.

Part 1 establishes the theme of the book by presenting the historical, social, political, and economic ramifications of a real-world environmental conflict. Through this detailed case study, the students are exposed to the range of issues involved in environmental decision making as it functions in reality.

Part 2 addresses the ecological principles that are basic to an understanding of environmental interactions and the flow of matter and energy in ecosystems. Population principles are developed and are then related to the human population and its impact on natural resources.

All living systems can be described in terms of the flow of energy through them. Human civilizations are based on their ability to acquire and use energy. Part 3 traces the development of energy consumption and illustrates its interrelationship with economic development and the maintenance of specific life-styles. Current energy-use patterns are stressed, and future sources of energy are discussed in the context of changing economic, political, and social climates.

In addition to energy resources, civilizations exploit material resources in order to develop economically. Part 4 examines the use and misuse of mineral, forest, wildlife, agricultural, land, and water resources.

Pollution is the result of the concentrated human use of resources. Part 5 explores the development of water- and air-pollution problems and examines techniques used to reduce the impact of human activities. Solid waste, noise, and aesthetic pollution are also considered.

Part 6 emphasizes the political realities associated with environmental decision making. The structure of the United States government is reviewed to provide a base for understanding current legislation and avenues for change. The central theme of this concluding section is the role that concerned individuals can play in influencing environmental policy. A focussed discussion of environmental ethics and a look to the future complete the text.

*Special features and aids to learning*

Each part of the text begins with an introduction that places the upcoming chapters in context for the reader. Each chapter begins with a chapter outline, a set of learning objectives, and a list of key terms. The student is encouraged to refer to these while reading and reviewing the chapter. Key terms are printed in bold type where they are defined in the text. Chapters conclude with a summary, a case study, review questions, and suggested readings. The case studies have been specifically selected to allow the reader to apply chapter concepts to actual situations. Review questions are related to the chapter objectives and thus serve to reinforce understanding of basic concepts and principles.

To dramatize and clarify text material, each chapter includes a number of tables, charts, graphs, maps, drawings, and photographs. Each illustration has been carefully chosen to reinforce and exemplify the chapter content.

Each chapter also includes one or more boxed items. These point out the far-reaching impacts of environmental forces and exemplify the interrelationships that are central to environmental issues. They are intended to interest the reader and reflect current applications of chapter concepts.

The text concludes with a list of active environmental organizations, a metric conversion chart, a thorough glossary, and an index. An instructor's manual accompanies the text. It includes chapter outlines, objectives, and key terms; a range of test and discussion questions; suggestions for demonstrations and reports; answer keys for chapter review questions; and suggestions for audiovisual materials and other additional teaching aids. A set of transparencies is also available to users of the text. The transparencies duplicate text figures that clarify essential ecological, political, economic, social, and historical concepts.

**wcb** TestPak rounds out the supplementary materials. It is a computerized testing service available to adopters of this text. It provides either a call-in/mail-in testing service or, if you have access to an Apple® *II*e or *II*c, or to an IBM® PC, you can receive a complete test item file on a microcomputer diskette.

## Acknowledgments

We wish to thank our colleagues who read all or part of this text as we developed it. Their suggestions and comments contributed greatly to the final product. They are: Gerald Moshiri, U. West Florida; Jack Schlein, CUNY—York College; Edward Kormandy, Evergreen State College; William Fox, Ventura College; Edmund Bedecarrax, City College of San Francisco; John Belshe, Central Missouri State University; Ronald Payson, Columbia-Green Community College; Michael Ells, Ferris State College; Donald Van Meter, Ball State University; Brian Myres, Cypress College; John Cunningham, Keene State College; John Harley, Eastern Kentucky University; Ernie Ramstein, Consultant (Orlando, Florida); Richard Pieper, University of Southern California; Ray Williams, Rio Hondo College; Larry Oglesby, Pomona College; George Grube, Dana College; Wayne Wendland, Illinois State Water Survey; Robert Harkrider, Cypress College; James Grosklags, Northern Illinois University; David McCalley, University of Northern Iowa; Allan Pollack, Northern Essex Community College; C. Lee Rockett, Bowling Green State University; Holt Harner, Broward Community College; Harvey Ragsdale, Emory University; T. J. Jacob, Institute for Environmental Studies (Urbana); Richard J. Wright, Valencia Community College; Susan P. Speece, Anderson College; Arthur Borror, University of New Hampshire; Fred A. Racle, Michigan State University; Terry A. Larson, College of Lake County; and Charles O. Mortensen, Ball State University.

# Part 1

# Interrelatedness

Environmental science has evolved as an interdisciplinary study that seeks to describe problems caused by our use of the natural world. In addition, it seeks some of the remedies for these problems. To deal with this complex topic, it is important to have an understanding of three major areas of information. First, it is important to understand the natural processes (both physical and biological) that operate in the world. Second, it is important to appreciate the role that technology plays in our society and its capacity to alter natural processes as well as solve problems caused by human impact. Third, the complex social processes that are characteristic of human populations must be understood and integrated with knowledge of technology and natural processes to fully appreciate the role of the human animal in the natural world.

Chapter 1 deals with a specific problem that illustrates how all of these areas are interrelated. This chapter introduces the central theme of interrelatedness with a real (although somewhat simplified) example.

## Chapter Outline

I. Interrelationships between people and resources
  A. Human impact on the environment
  B. Impacts of environmental decisions on society
II. The Teal River controversy
  A. Human influence in the forest
  B. The discovery of oil
  C. Sides develop
  D. State intervention
  E. Economic impact
    Box 1.1 The environmental impact statement
  F. The environmental impact statement
  G. A view of the alternatives
  H. Compromise or sellout
  I. Three years later

## Key Terms

benefit-cost studies

environmental impact statements

grass-roots politics

lobbying

## Objectives

Understand why environmental problems are complex and interrelated.

Realize that environmental problems are tied to social, political, and economic issues.

Understand that acceptable solutions to environmental problems are not easy to achieve.

Appreciate that individuals view environmental problems from different perspectives.

Understand that all organisms have an impact on their surroundings.

Describe the ways that benefit-cost studies, environmental impact statements, lobbies, and grass-roots politics relate to environmental decision making.

# Chapter 1
# Society, government, and the environment: A case study of interrelationships

## Interrelationships between people and resources

All organisms make some changes in their surroundings. Dogs tip over garbage cans as they search for food. Bears tear open logs when they look for ants to eat. Because trees shade the ground, some plants cannot grow close to them. People also change their surroundings. The ability of humans to cause changes in their surroundings is much greater than that of any other organism. People attempt to control their environment, and in doing so they make use of large amounts of energy and resources. Whenever changes are made, waste materials are produced. If the magnitude of change is great, then the complexity and magnitude of the problems is also large.

## *Human impact on the environment*

We are all against pollution and waste. After all, no one would want to see a favorite beach or swimming hole become so polluted that swimming could no longer be allowed. We would not like to see our neighborhood basketball court destroyed to provide the city with another parking lot, nor would we like to see a beautiful park turned into a city dump. Most of us agree that we need to protect our environment. Making decisions to protect it should be simple and straightforward. If pollution is harmful, why do we not simply put a halt to it? If forests are necessary to wildlife, why don't we prevent their destruction? In reality, making these kinds of decisions is not so easy. (See fig. 1.1.)

To reduce waste and pollution, what are we willing to change? Will we walk or ride a bike in order to help reduce air pollution and save energy? Could we make changes in the way we live that would reduce the amount of wastes produced? The questions could go on and on. The point is that even though most agree that pollution and waste are harmful, we don't always agree on the solutions to the problems. A good example of this is the solid waste disposal problem facing most of the communities in the United States. What options are open to a community when its waste disposal facility can handle no more trash? While we may spend a lot of time and effort trying to find ways of disposing of waste, we have done little to reduce its production. Everyone concerned agrees that there is a problem, but agreeing to a solution is much more difficult.

## *Impacts of environmental decisions on society*

The decisions that are made concerning what to do about trash will have impacts on society. If a city requires all of its residents to separate their trash into glass, paper, and garbage, effects will be felt in such diverse areas as the sale of garbage cans and garbage disposals, the employment of people to pick up refuse, and the inconvenience experienced by individual families. In fact, environmental decisions influence employment, inflation, energy supplies, industrial development, agricultural production, and public health. Environmental problems do not stand alone, they are interrelated and complex. They are related to almost every other problem facing us at the local, national, and international levels. Since environmental problems are complex, there is no easy way to arrive at solutions.

Air pollution and energy use are directly related. Whether we use coal, oil, gas, wood, or nuclear fuel, air pollution increases as energy consumption increases. Even though this does not have to be the case, the present state of air pollution control technology and cost make it the rule rather than the exception. As a society, we must balance our demand for energy with our need for clean air.

In fact, our use of all forms of energy has continued to increase annually. (See fig. 1.2.) So, even though many people are opposed to the idea of air pollution and its harmful effects, as a society we have not altered our life-style

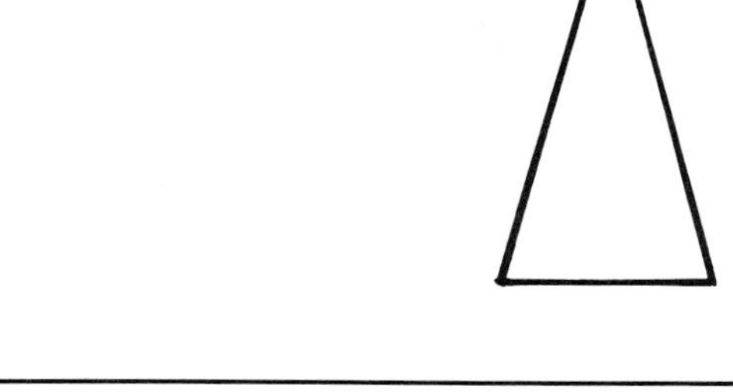

**Figure 1.1 Human impact on an area.** When people use an area, they inevitably change it. If an area is used for aesthetic value, the changes are different than if an area is used to provide housing. Determining the best use for an area is often difficult.

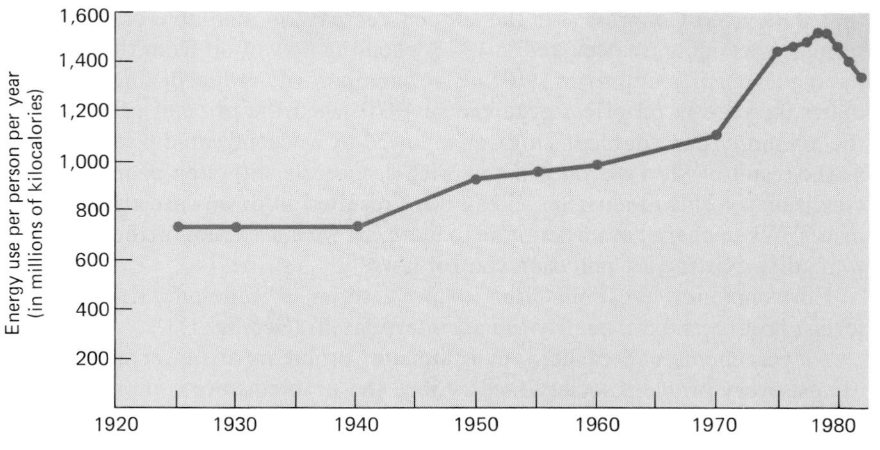


**Figure 1.2 Individual energy use in the United States.** This graph shows the amount of energy used by citizens in the United States. The standard of living in the United States is closely tied to energy consumption. Note that there was a steady increase in the amount of energy used until the increase in oil prices in 1979, and after 1979 the use of energy decreased. (From *Exploring Energy Choices*. The Ford Foundation, 1974 and *Statistical Abstract of the United States, 1984*.)

enough to significantly reduce air pollution. As the basic cost of energy continues to increase, we have become less willing to use costly air pollution control measures.

Many of the pollution control laws passed in the early 1970s were modified when the supply of oil was reduced and prices increased. For example, the Clean Air Act of 1970 specified safe concentrations of seven major air pollutants, and states were required to meet these levels by 1975. In 1977, however, this act was amended, and the deadline for the states was extended to 1982; the deadline for auto-related pollution was extended to 1987. One factor

**Figure 1.3 Pollution and progress.**
If a society is to progress, it is important that industry provide jobs for people in a community, but it is just as important that the quality of the environment is high enough so that the health and well being of the members of the community are not adversely affected. Pollution control must be a part of the industrial cost/profit equation.

that influenced Congress was the sudden decrease in available oil. The first serious price increase occurred in 1973 when the flow of oil from the Oil Producing Exporting Countries (OPEC) was temporarily reduced. The next dramatic increase in oil prices occurred in 1979 when the price of gasoline and fuel oil more than doubled. This was followed by a second rapid price increase as the result of the 1981 oil and gas price decontrols, although poor economic conditions worldwide in 1981–1984 have resulted in downward trends in oil prices. When energy costs continue to increase, we can expect further pressure to modify existing air pollution control laws.

Environmental questions often involve a series of trade-offs. Employment levels, profits, and air quality are all interrelated. (See fig. 1.3.)

As was mentioned earlier, environmental problems are interrelated with almost every problem society faces. Since the problems are complex and interrelated, arriving at solutions may be a difficult process. Often, solutions to one problem create other problems that are just as difficult.

To better understand the complexity of environmental issues, let us analyze a real-world situation. The names have been changed and the case has been somewhat simplified, but the events described actually did occur. As you read this case study, you should look for the interrelatedness and complexity of environmental issues. The situation centers on the conflict between society's increasing need for oil and its desire to preserve wilderness areas.

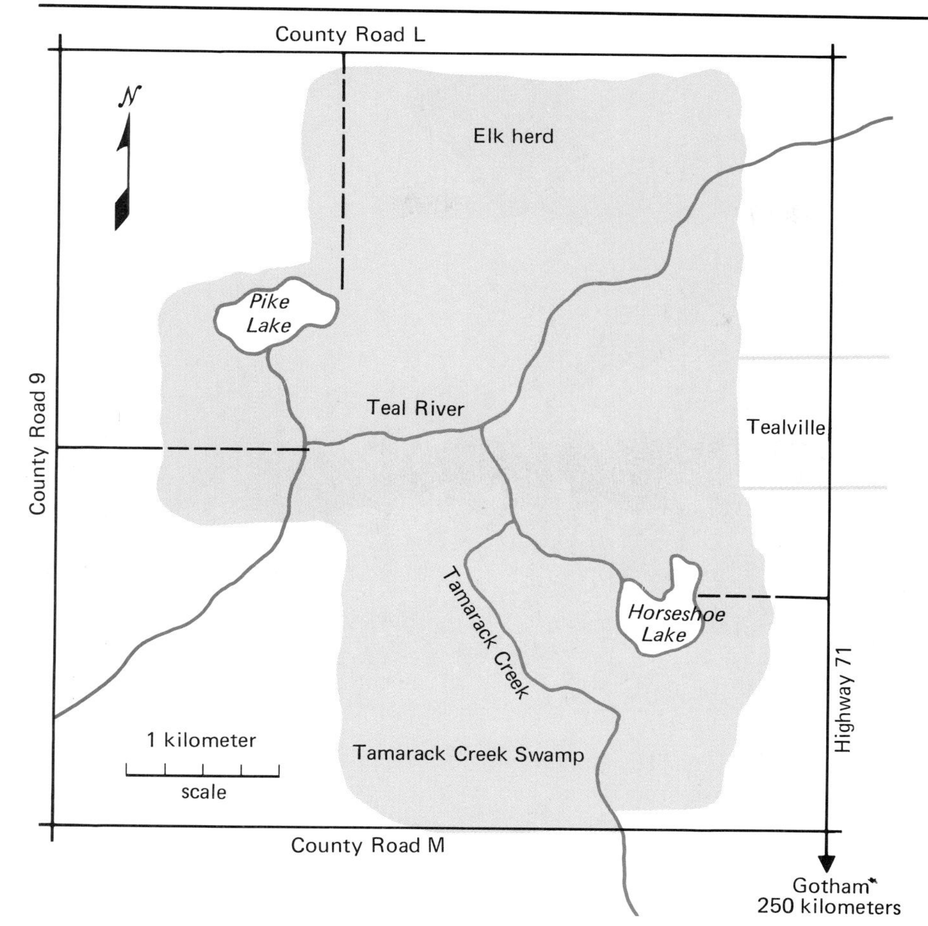


**Figure 1.4** **Map of Teal River State Forest.**

## The Teal River controversy

The Teal River State Forest is located in a north midwestern state. It is also a major industrial state. The Teal River State Forest is one of the few publicly owned large tracts of recreational land in the state. Tealville, a city of 12,000, borders the eastern edge of the forest. There are dirt roads in the forest, but no major highway passes through it. (See fig. 1.4.)

Generally speaking, the Teal River State Forest is a quiet semiwilderness area. The size of the forest and the absence of crowds make it a perfect place for visitors to escape from a fast-moving society. The area is unique, however, in that it is the home of the only large elk herd east of the Mississippi. The elk were introduced from Montana in the early 1900s and have grown to a herd of about eight hundred.

Because it has reestablished some of its original wilderness qualities, grouse, black bear, bobcats, woodcock, deer, beaver, and many other animals are also

**Figure 1.5**
**Reforestation program.**
The state's reforestation program involved the planting of trees and the control of forest fires so new forests could become established in the Teal River State Forest.

found in the region. The streams provide excellent conditions for healthy populations of native brook trout, and several of the area's lakes are well suited to game fish, such as pike and bass. In addition to hunting and fishing, other major forms of recreation in the forest include cross-country skiing, camping, and hiking.

## *Human influence in the forest*

The Teal River Forest is not virgin, untouched wilderness. Large-scale human activity began in the area in the early 1870s when the first loggers arrived. Between 1870 and 1915, the entire area was logged; first its massive white pines were removed, and later, hardwoods, such as birch and maple, were harvested.

After all the large trees had been cut, the roads and railroads the loggers had once used remained. Several forest fires swept the area in the early 1900s. At that time, the few people who still lived in the area moved out. With almost no human activity, the forest began to heal itself.

Beginning in 1925, the state gained ownership of the forest. A tree-planting program began in 1926, and soon young seedlings, both natural and planted, were growing in size and number. (See fig. 1.5.)

Over a period of time, the forest began to reestablish itself in the fields. With so few people living in or using the area, wildlife flourished. The increasing food supply and lack of human interference encouraged the growth of the elk herd. By 1939, the area was opened to deer hunting, but elk hunting was still prohibited.

During the next thirty years, the Teal River area regained some of its former wilderness character. Management during these thirty years of state ownership consisted of forest fire protection (including the building of fire breaks), protection of the elk and deer, planting of pine, and protection against timber thieves, poachers, and squatters. Some sections of the forest were developed for camping and hiking. The streams, rivers, and lakes were stocked with fish. Timber sales of pulp wood, such as aspen, increased. Finally, research programs on the elk and other wildlife began. The major objective of these programs was not only to help the animals but also to see how people could best use the resources the area had to offer.

### *The discovery of oil*

In the spring of 1968, a state agency leased land within the Teal River State Forest to four oil companies. This was important because this industrial midwestern state had to import nearly 90 percent of its energy. The rights to drill for oil contained few developmental restrictions, and these rights were sold at public auction. At that time, there was no substantial proof of oil or natural gas existing in the area, but the state received $10 million for the drilling rights. State biologists determined that exploration would have only a minimal, short-term effect on the forest's plant and animal life. A strong argument for the sale of drilling rights was that damage caused by exploration could easily be repaired with a portion of the $10 million. It seemed to be a foolproof plan at the time, and nearly everybody involved agreed with it.

However, a small group of state officials and private citizens opposed the exploration. They most strongly objected to the fact that they had little say in determining whether the drilling rights should or should not be sold. These individuals thought the forest was in danger of being destroyed; so in 1969, they attempted to put the Teal River State Forest off-limits to all oil and natural gas exploration.

However, in July 1970, two years after the initial exploration drilling rights were sold, the event that some feared and others hoped for came to pass. Oil was discovered on the edge of the Tamarack Creek Swamp, in the southern part of the Teal River State Forest.

### *Sides develop*

Within one year of the initial discovery, the oil companies realized that the amount of natural gas and oil under the forest was significant. The deposits were about three thousand feet under the surface. It would take many workers and much machinery to get the oil and gas out of the ground.

Figure 1.6 **Different opinions.**
The citizens of Tealville had different opinions about whether oil drilling should take place or be prohibited.

Many of the citizens of Tealville welcomed the discovery of oil. Many of the local landowners adjacent to the Teal River Forest were eager to lease or sell their lands to the oil companies. Unemployment was high in the area; for the unemployed individuals, the choice was clear. Oil drilling would mean additional employment opportunities and, in turn, would improve the standard of living for their families. To these individuals, the thought of steady work surpassed the desire to maintain the natural beauty of the area. The new industrial base, resulting from the oil drilling and its economic spinoffs, would definitely reduce the number of people who were out of work.

The owners of stores, restaurants, motels, and taverns were quick to support the oil companies. New money coming into the area was sure to increase the tax base for the city and county. This new revenue would improve schools, roads, and the city's hospital. New tax money would enable Tealville to improve its police and fire protection, as well as build new parks and recreational facilities. For many, the discovery of oil was seen as a great benefit to the area.

For others, the discovery of oil was not all positive. Many citizens were retirees who had left the industrial part of the state. They settled in the area primarily because of its natural beauty. To these citizens, this area represented one of the state's few remaining places where they could enjoy a natural setting. They viewed it as a sanctuary from the urban environment they had left behind. Oil drilling would create environmental conditions from which they had tried so hard to escape. (See fig. 1.6.)

Since the major industry in the area was tourism, there were many whose livelihood depended on providing the goods and services the tourists needed. The discovery of oil could pose a threat to the local bait shop, canoe livery, and sporting goods store. If the area were turned into an oil-boom town, the tourist trade would drop off greatly. Many of these small businesses would be forced either to close their doors or to relocate. What would be an economic boom for some could prove to be an economic disaster for others.

The citizens of Tealville disagreed on the issue of oil drilling in the neighboring state forest. Each side was equally convinced that its view and position was the best and that the other side's policy would surely spell doom and ultimate disaster for the area. The situation that was developing in Tealville was a smaller version of what was beginning to develop on a statewide basis.

### *State intervention*

While the citizens of Tealville were choosing sides over the issue of oil development, things began to happen at the state level. Both sides began to apply pressure on state officials. Such an attempt to influence a decision is called **lobbying.** The sides in this intensive lobbying attack did not appear evenly balanced. The pro-drilling side was made up of a powerful coalition of ten oil companies. The economic strength of the companies made it easy for them to finance a massive lobbying campaign in the state capitol. (See fig. 1.7.)

Even though the oil companies had a definite financial advantage, they were far from gaining an easy victory. A group of nine environmental organizations opposed the drilling. These organizations were supported by thousands of individuals in the state who donated their time and money. These *citizen interest groups* had little practical experience in either government or politics, but under effective leadership, these groups quickly became sophisticated in the lobbying techniques of mass meetings, visits to legislators, and letter-writing campaigns. Organized citizen action at the local level is referred to as **grass-roots politics.**

As public interest in the Teal River controversy grew, the anti-drilling group's position was strengthened through favorable newspaper and television coverage. With few exceptions, the editorial policy of the press in the state

**Figure 1.7 The impact of lobbying.** Lobbyists have a varied and significant impact on decision makers. Although some lobbying is needed and serves a purpose, other forms of lobbying are questionable, and, in some cases, illegal.

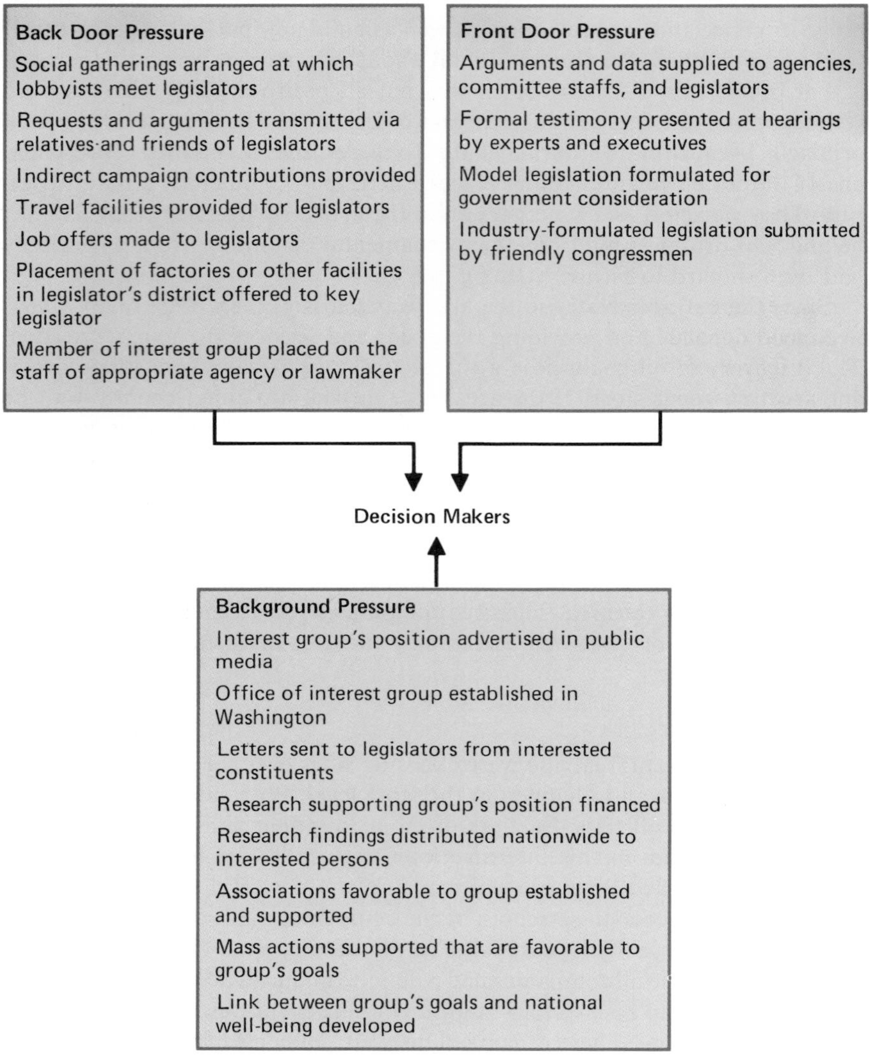

**Back Door Pressure**
- Social gatherings arranged at which lobbyists meet legislators
- Requests and arguments transmitted via relatives and friends of legislators
- Indirect campaign contributions provided
- Travel facilities provided for legislators
- Job offers made to legislators
- Placement of factories or other facilities in legislator's district offered to key legislator
- Member of interest group placed on the staff of appropriate agency or lawmaker

**Front Door Pressure**
- Arguments and data supplied to agencies, committee staffs, and legislators
- Formal testimony presented at hearings by experts and executives
- Model legislation formulated for government consideration
- Industry-formulated legislation submitted by friendly congressmen

Decision Makers

**Background Pressure**
- Interest group's position advertised in public media
- Office of interest group established in Washington
- Letters sent to legislators from interested constituents
- Research supporting group's position financed
- Research findings distributed nationwide to interested persons
- Associations favorable to group established and supported
- Mass actions supported that are favorable to group's goals
- Link between group's goals and national well-being developed

opposed the oil drilling. The media coverage was a tremendous help to these environmental groups and provided a way to offset the financial power of the oil companies.

In response to growing public involvement over the Teal River Forest controversy, the governor became directly involved. Remember that in spite of the growing strength of the anti-drilling groups, the state forest was located in an industrial state. The state's citizens were almost evenly split over the issue. As a politician, the governor had to be very careful in handling the situation. He ordered two studies to be done on the area—an economic **benefit-cost analysis** of the area and an **environmental impact statement** on the possible effects of drilling (see box 1.1).

## Box 1.1
## The environmental impact statement

In 1969, the Congress of the United States passed the National Environmental Policy Act (NEPA). NEPA is an unprecedented piece of legislation. The act declares the federal government's commitment to a broad range of environmental protection goals and measures.

The provision in NEPA that has had the greatest impact on governmental policy is its requirement of an environmental impact statement (EIS), set forth in Section 102 of the law. This section says that all federal agency plans must have a detailed report prepared that describes—

1. the environmental impact of the proposed project;
2. any adverse environmental effects that cannot be avoided should the proposal be implemented;
3. alternatives to the proposed action;
4. the relationship between local short-term use of the environment and the maintenance and enhancement of long-term productivity;
5. any irreversible or irretrievable commitments of resources that would be involved in the proposed action.

An initial EIS draft must be made available to the public and governmental agencies at least ninety days before any contemplated action is taken. Final EIS drafts must be available to the public at least thirty days before the intended action. The final document is then filed with the president's Council of Environmental Quality, where it may be used to advise the president or governmental agencies.

Requiring impact statements has a number of important purposes. It is an effort to force government administrators to make careful, thorough appraisals of the environmental effects of their activities. It also gives the president and his advisors a total view of the environmental impact of federal activities and enables the federal government to take action on the grounds of environmental impact. The exact language of the EIS gives governmental administrators clear guidelines to follow in their appraisal.

An example of how a federal EIS changed plans is the trans-Alaskan pipeline project. The EIS for this 1285 kilometer pipeline prompted many design changes; among other things, these changes minimized disruption to animal migratory routes and encouraged greater protection against oil spills.

Because there have been generally positive results from the federal EIS requirement, many states have adopted both the principles and the style of the law. At the state level, a governor could use EIS findings when making recommendations on state planning. While federal EIS is pertinent only to federal projects, state EIS regulations cover individual state actions.

**Figure 1.8 Benefit-cost analysis.**
Benefit-cost analysis is an important consideration in making decisions.

Oil available for use
State receives money for sale of drilling rights
Increased employment
Additional tax money available to community

Loss of some wildlife
Tourism would decline
Forest and river might be harmed
Drilling equipment would lower the aesthetic value of the scenery

## *Economic impact*

The economic benefit-cost study would help the governor to view the pros and cons of drilling from an *economic* standpoint. Benefit-cost analysis is one method used to arrive at an economic assessment of an issue. With a benefit-cost analysis the benefits of a proposed project are weighed against the costs of the project. If the benefits exceed the cost, then a strong argument can be made for the project. One difficulty with such an analysis is that not all benefits and costs of a project can be neatly put into columns of figures, and not all agree about the costs and the benefits. In the Teal River study, the question was whether the economic benefits of drilling for oil exceeded, were equal to, or were less than the economic costs of drilling. If the cost of drilling was greater than the benefit, then drilling would not be the proper thing to do. (See fig. 1.8.)

One of the problems of using a benefit-cost analysis in the Teal River State Forest involved the elk herd that lived in the area. Clearly the presence of elk was a benefit to the area. Since elk need large tracts of uninhabited land in order to survive, the proposed drilling could have a harmful effect on the herd and, therefore, would be a cost. How does one place an economic value on an elk? Perhaps if elk meat were to be sold at three dollars a pound, a dollar value could be assigned to the herd. However, the pleasure that people derive from elks' presence might be worth more than three dollars per pound. (See fig. 1.9.)

**Figure 1.9 The value of aesthetics.**
There really isn't any good way to place a value on some things such as scenery or solitude. We recognize that they have a value of some kind; however, we often recognize this value only after it is lost.

## *The environmental impact statement*

The second study ordered by the governor was the environmental impact statement (EIS). An EIS is an extensive study outlining the possible environmental effects of a specific project on a certain area.

The EIS for the Teal River State Forest confirmed what many had thought. Oil development would have a harmful effect on certain animals; the effect would be the greatest on those species that needed solitude and that could not live in harmony or in close proximity to humans. The elk herd would surely suffer, as would the black bear and bobcat populations; however, cutting some of the forest for drilling would actually improve the habitat, and subsequently increase populations of other animals, such as white-tailed deer.

The report showed that only a limited amount of swamp could be affected by pipelines or electric line crossings. Some bank erosion and siltation would occur due to pipeline crossings along the Teal River. If there was drilling along the streams and rivers, the damage to water quality and trout populations could be disastrous. Certain recreational uses of the forest would also suffer if development were allowed. Activities such as cross-country skiing, hiking, and camping would lose appeal near oil-drilling rigs and machinery. (See fig. 1.8.)

### *A view of the alternatives*

After reviewing the results of the two studies, the governor and other state officials met to discuss the problem. The purpose of their meeting was to consider all the information and recommend a course of action.

One possible course of action was to prohibit drilling. This would eliminate negative environmental impacts but would not provide the oil needed in this industrial state. An alternative was to delay drilling until the need for the oil would be greater. Another alternative was to allow drilling in spite of environmental objections. This was the course of action finally selected. However, strict rules regarding how drilling should proceed were laid down.

There are two ways of drilling for oil—the competitive approach and the unitized approach. When several oil companies lease or own land overlying a single oil reservoir, each has a right to drill into it to obtain its share of oil or gas. This competitive drilling causes a duplication of wells, pipelines, and other producing facilities. In this situation, the faster an oil company can get the oil out, the more money it can make; this encourages them to drill many wells in areas with oil potential.

Under a unitized drilling plan, the companies who own or lease the land above a reservoir agree to extract the oil together. A single operator is chosen to get the oil out. The costs and profits are shared by all of the companies. Because there are only a few wells drilled, the environmental impact on the surface resources is greatly minimized.

Since the decision was made to drill for the oil, the governor reviewed four separate drilling proposals—unitized drilling in only a small area in the Teal River State Forest, competitive drilling in the same area, unitized drilling in the entire forest, and competitive drilling in the entire forest.

After reviewing all of the information and after having been intensively lobbied by both sides in the conflict, the governor made his decision. (See fig. 1.10.) The decision represented a trade-off in which all sides of the conflict gave up something. The decision was to allow unitized drilling in only the southern quarter of the Teal River area. This plan included many restrictions that the oil companies had to accept before drilling would be permitted. The following is a list of these restrictions.

1. Drilling must end by the year 2000.
2. No drilling near the habitat of the elk herd.
3. Minimal noise.
4. No drilling in swamps.
5. No pipelines to cross rivers.
6. No drilling near recreation areas.
7. Minimal gas or oil odors.
8. The oil companies must install safety valves on all equipment.
9. All land used must be reforested.
10. Environmental insurance of $100,000 required per well.

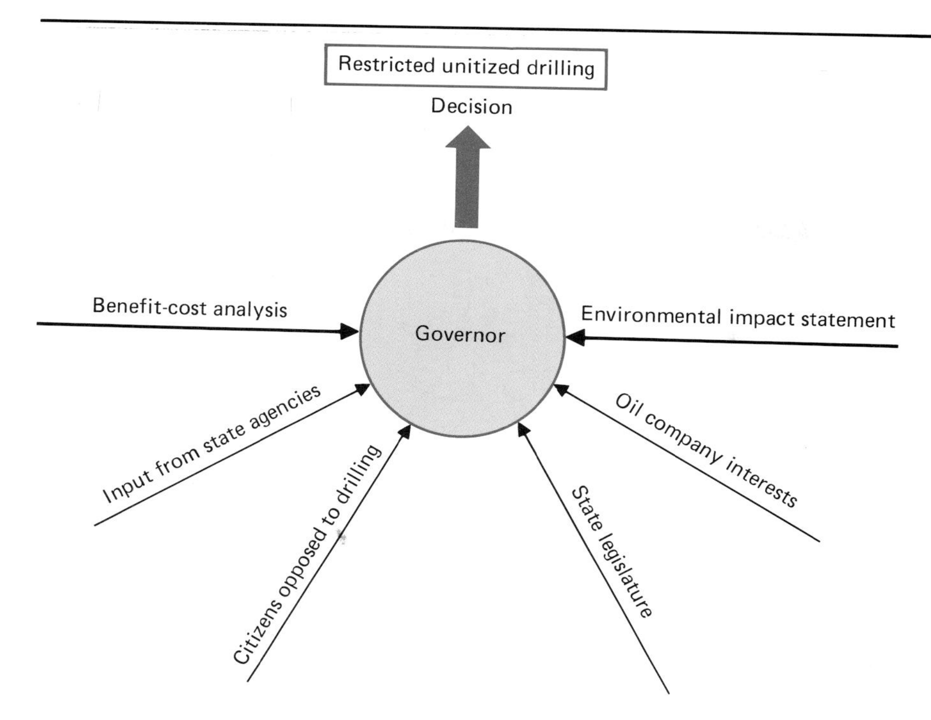


**Figure 1.10 The governor's decision.** The governor as a politician listens to all the input available and tries to make a decision that will offend the least number of people.

## *Compromise or sellout*

Was the decision to allow restricted drilling a compromise or a sellout? There were many on each side of the issue who disapproved of the decision. Some of the pro-drilling group felt that the decision was too restrictive and that "progress" had been hampered by the backward looking "environmentalists." On the other hand, some of the anti-drilling groups felt that the state had sold out to "big business" and the public would be the ultimate loser.

The majority of the participants in the issue, however, were satisfied with the final decision. Even though both sides had to yield somewhat from their original positions, they still both received a partial victory. Compromise, in this instance, was the practical answer. The needed oil would be produced but not at the total expense of the forest. The majority realized the state's need for both the oil and the forest. (See fig. 1.11.)

As was mentioned earlier, it is not easy to find answers to pressing environmental problems. Answers to environmental problems that are not carefully thought out are often as bad as no solution at all. Every environmental decision must take into account the political, economic, social, and biological considerations of the issue.

In our study of the Teal River State Forest, all of these variables were discussed and studied. In the end, no one group gained everything that it wanted, and each had to give up something.

**Figure 1.11**
**Compromise.**
When there are great differences of opinion on a matter, a compromise is usually the best solution. However, no one is completely satisfied.

## *Three years later*

After three years of drilling in the Teal River area, there have been no major problems in the region. The elk herd increased from an estimated population of 500 animals in 1980 to 800 in 1984. In fact, the State Department of Natural Resources conducted an elk hunt to reduce the size of the herd by some fifty animals. The number of elk now in the region had begun to cause damage to the vegetation in the area.

The only problem that has developed is the amount of brine resulting from the thirty-five wells drilled in the area. The oil deposits in this region are below a salt formation; therefore, to reach the oil, it is necessary to drill through this salt formation. For each well drilled, thirty-six metric tons of brine are deposited on the land surface. This material is deposited in a pit that is lined with PVC plastic and covered with soil. In the fall of 1983, it was discovered that there was some seepage at one of the pits, and there was concern that the brine might contaminate local groundwater. To correct this leakage, the oil companies have purged the spill and relined the pit. Although the drilling operations have taken place, there are still questions regarding the drilling. Exact figures are not available, but the amount of oil found is less than originally projected. Those who favored the drilling see it as doing no harm to the area. They point to the brine seepage problem and applaud the fact that it was corrected. Those people opposed to the drilling are disappointed because they claim there would have been no seepage if the pit had been constructed correctly. Thus, three years later there is still concern for the environment of the Teal River.

## Review Questions

1. Define an economic benefit-cost analysis and describe its strengths and weaknesses. Can one always be used?
2. Define lobbying and describe some of the methods involved in lobbying.
3. Describe an environmental impact statement and its uses.
4. Under what circumstances might an environmental impact statement be required in your state?
5. Describe competitive and unitized drilling.
6. Why do you think finding a solution to an environmental conflict is so complex? Explain your answer.
7. To arrive at a solution by compromise or trade-off, something has to be given up to gain something. What are the strengths and weaknesses of a compromise?
8. If you were the governor deciding the Teal River State Forest case, what would your decision have been? Why?

## Suggested Readings

Anderson, Walt, ed. *Politics and Environment.* Pacific Palisades, Calif.: Goodyear Publishing Co., 1975. Provides insight into the political problems of environmental problem solving.

Davis, David H. *Energy Politics.* 3d ed. New York: St. Martins Press, 1982. A specific look into the politics involving the oil, coal, gas, hydroelectric, and nuclear fuel options.

Louis, Theodore J. *Policide.* New York: Macmillan Publishing Co., 1976. Case study of the physical removal of a rural town to provide a nuclear power generating facility.

Michigan Department of Natural Resources, Forestry Division. "A Concept of Management: Pigeon River Country, Policies and Guidelines." Lansing, Michigan. October 1973. A description of the actual setting and conflict on which the case study in this chapter is modeled.

Ophuls, William. *Ecology and the Politics of Scarcity.* San Francisco: W. H. Freeman & Company, 1977. Resource allocation and the questions of decreasing resource supply.

Rosenbaum, Walter A. *The Politics of Environmental Concern.* New York: Holt, Rinehart & Winston, 1977. Excellent overview of public policy and governmental decision making. Deals with a wide range of environmental problems.

# PART 2

## Ecological principles

Basic information for understanding environmental problems includes explanations of how the various biological and physical processes of the world interact and are interdependent. Central to this understanding are two physical concepts—matter is composed of extremely small moving particles, and energy changes must obey certain laws. These ideas are developed in chapter 2. Other principles related to the interactions among organisms and their environment are discussed in the remaining chapters of this part. The underlying theme of these chapters illustrates the complexity and the variety of the interactions that are collectively studied as ecology. Chapters 3 and 4 describe how organisms interact with one another and the role that organisms play in the flow of matter and energy in ecosystems. The impact of human activities on natural ecosystems is the theme that is developed in chapters 4 and 5.

Since the intensity of an organism's impact on its surroundings is largely determined by the number of individuals present, chapters 6 and 7 deal with the interrelationships between population size and the resources available. Chapter 6 explains principles of population ecology, while chapter 7 emphasizes human population problems and their relationship to social forces.

## Chapter Outline

I. Scientific thinking
   Box 2.1 The scientific method
   A. Applied and theoretical science
II. Environmental science
III. The structure of matter
   A. Atomic structure
   B. Molecular structure and chemical reactions
IV. Energy principles
   A. First and second laws of thermodynamics
   B. Environmental implications of energy flow
V. Consider this case study: Four Corners Power Plant

## Objectives

Understand that science is exact because information is gathered in a manner that requires evaluation and revision.

Explain the difference between applied and theoretical science.

Understand that environmental science is a new discipline that includes both the applied and theoretical aspects of traditional science and the social, economic, and political sciences.

Understand that matter is made up of atoms that have a specific subatomic structure made up of protons, neutrons, and electrons.

Recognize that different elements have different atomic structures and that an element may have several different isotopes.

Recognize that atoms may be combined to produce molecules by forming chemical bonds.

Understand that changing chemical bonds results in a chemical reaction and that this is associated with energy changes.

Recognize that matter may be solid, liquid, or gas, depending on the amount of kinetic energy contained by the molecules.

Realize that energy can be neither created nor destroyed but that in converting energy from one form to another, some of this energy is in a form that is not useful.

Understand that energy can be of different qualities and that low-quality energy is increasing in the universe.

## Key Terms

activation energy
applied science
atom
catalyst
cellular respiration
chain reaction
chemical bond
combustion
electron
element
environmental science
first law of thermodynamics
fission
hypothesis
isotope
kinetic energy
matter
molecule
neutron
nucleus
photosynthesis
potential energy
proton
radioactive half-life
science
second law of thermodynamics
theoretical science
theory

# Chapter 2
# Interrelated scientific principles: Matter—energy—environment

## Scientific thinking

Since environmental science relies very heavily on scientific input, it is appropriate to explain how scientists gather and evaluate information. It is also necessary to understand the chemical and physical principles that underlie ecological concepts.

The word "science" creates a variety of images in the mind. Some people feel that it is a powerful word and are threatened by it. Others are baffled by topics that are scientific and, therefore, have developed an idea that scientists are brilliant individuals who can create answers to questions. Neither of these images accurately reflects what science is all about.

**Figure 2.1 Scientific method.**
The scientific method involves observations that lead to the construction of hypotheses, which are then tested to determine if the hypothesis is valid. If the hypothesis is not valid, it can be modified and tested in its modified form.

**Science** is a method of gathering and organizing information that involves observation, hypothesis formation, and experimentation. (See fig. 2.1.) Information regarding these principles is collected using whatever tools for observation are available. In some cases, the "tools" may be the observer's eyes, ears, or other unaided senses. In other cases, elaborate equipment may be used to assist the gathering of information. Often the tools used are overwhelming in complexity, which gives a somewhat erroneous and inflated impression of the actual work being done by a scientist.

Once new information has been collected, the scientist tries to fit this information into the existing framework to gain a clearer picture of the real world. It is usually more productive for a scientist to have a working hypothesis as a guide. A **hypothesis** is a logical guess about how things work. It helps to focus a person's thinking on the critical questions to be asked. If a hypothesis is used for a long period of time and is continually supported by new facts, it is often formally stated as a **theory.** Scientific theories are those unifying principles that tend to bind large areas of scientific knowledge that are not known to be true in *every* case because it would be impossible to test *every* case.

## Box 2.1
## The scientific method

The scientific method consists of observation, formulation of a hypothesis, and testing the validity of the hypothesis. Underlying these activities is constant attention to accuracy and freedom from bias. The critical steps in the scientific method involve constructing a hypothesis and testing it. A **hypothesis** is a logical guess that explains an event or answers a question. Any good hypothesis should remain simple, must account for all relevant facts, and must be testable. The testing of hypotheses requires the most scientific skill, since it is often necessary to construct artificial situations to test a hypothesis. These situations are called **experiments.** If a hypothesis is supported by the results of an experiment, it will continue to be used to explain events. If a hypothesis is unsupported, it is discarded or modified and then tested in its modified form. This constant questioning, hypothesizing, and reevaluation of hypotheses is the foundation of the scientific method, and, therefore, of science.

The method of gathering facts and the conclusions drawn from them should be free from bias. Two mechanisms that help to assure this are setting up experiments with controls and communicating one's methods and results to others in the field. This allows others to try the same experiments to determine the reliability of the results. This criterion of repeatability must be satisfied before facts are acceptable to the scientific community.

A **controlled experiment** is designed to compare two situations that differ in only one fundamental aspect. If in the experiment the two situations yield different results, it is likely that the original difference, which was designed into the experiment, was the cause. Experimental design and development of good working hypotheses are the most difficult aspects of the scientific method. It is in these two areas that average scientists are distinguished from those who exhibit brilliance.

The Nobel prize-winning scientist Karl von Frisch studied the behavior of honeybees. He thought that bees could distinguish one kind of flower from others (observation). He suspected that the bees could distinguish flowers by color (hypothesis). To determine if this was true, he designed a set of simple experiments. He first trained bees to come to a source of honey located on a piece of blue cardboard. The bees made many trips between their hive and the source of honey on the blue card.

The blue card was then removed and replaced with a clean blue card and a clean red card, neither of them with a source of food for the bees. The bees returned to the blue card and avoided the red card. The bees were able to distinguish the blue card from the red card (conclusion). After von Frisch published his results, other scientists designed similar experiments that also supported von Frisch's original hypothesis. This series of events illustrates how the scientific method is used to gather new information.

It is important to understand that knowledge and science are not synonymous and that some knowledge can be used in both a scientific and a nonscientific manner. For example, it is a fact that many birds "sing." It would be easy to infer that the birds must feel a sense of joy or that these sounds are related to the bird's desire for a mate. This is an unscientific use of knowledge. A scientist could also use the same knowledge to determine what causes the singing behavior and what reactions are produced in other birds that hear the singing. Experiments have demonstrated that the singing has nothing to do with joy or sexual desire. Although singing serves many functions, it has often been shown to cause other birds of the same species (particularly males) to avoid the area. In other words, the song was a warning for other male birds to stay away.

It is often very easy to jump to conclusions or confuse fact with hypothesis. This is particularly a problem when we generalize from personal experience. For example, it has often been thought that overweight people are happy people; actually, psychological and sociological studies indicate that most are not happy.

## *Applied and theoretical science*

It is the human perspective that is used to distinguish between applied and theoretical science. **Applied science** deals with the use of information gained through science to influence the human condition. The aim of **theoretical science** is to increase the base of scientific knowledge so that we can better understand the basic framework of the universe. Another way of looking at it is that a theoretical scientist is interested in knowledge for its own sake, regardless of its practical value to humanity. The applied scientist is more interested in the practical applications of science to everyday life. Both must use the same rules and methods of thinking, but their goals are different.

It is not always easy to categorize scientists as either theoretical or applied, since major theoretical advances in science are often made by people who have very practical goals in mind. Likewise, theoretical advances may lead to whole new areas of applied science. One goal of the United States space program was to gather information about outer space to help clarify current theories about how the universe operates (theoretical science). Before the information could be gathered, some very practical problems had to be solved. The result of all this effort has led to what is commonly called "space-age technology." Examples of this space-age technology are lightweight, miniaturized computers, new methods of communication, and advances in medicine. (See fig. 2.2.)

Because scientists tend to be methodical and follow established rules for gathering and evaluating information, there is a tendency to assume they are correct. However, scientists form and state opinions that may or may not be supportable by fact. It is common for equally reputable scientists to state opinions that are in direct contradiction. This is especially true in environmental science.

**Figure 2.2 Applied and theoretical science.** One of the objectives of the United States space program was to gain greater insight into the functioning of the universe. Many practical discoveries have also been made during this project.

## Environmental science

**Environmental science** is an interdisciplinary area of study that includes both applied and theoretical aspects of human impact on the world. It also includes many areas that are not usually considered to be scientific. Chapter 1 discussed a specific situation that included many scientific concerns but was also strongly interlaced with social, economic, and political realities. Environmental science is a field of study that is still in the process of evolving, but its beginnings are rooted in the early history of civilization. The thoughts expressed in the following quote from the American author Henry David Thoreau (1817–1862) are consistent with prevalent environmental philosophy.

> I wish to speak a word for Nature, for absolute freedom and wildness, as contrasted with a freedom and culture merely civil . . . to regard man as an inhabitant, or a part and parcel of Nature, rather than a member of society.

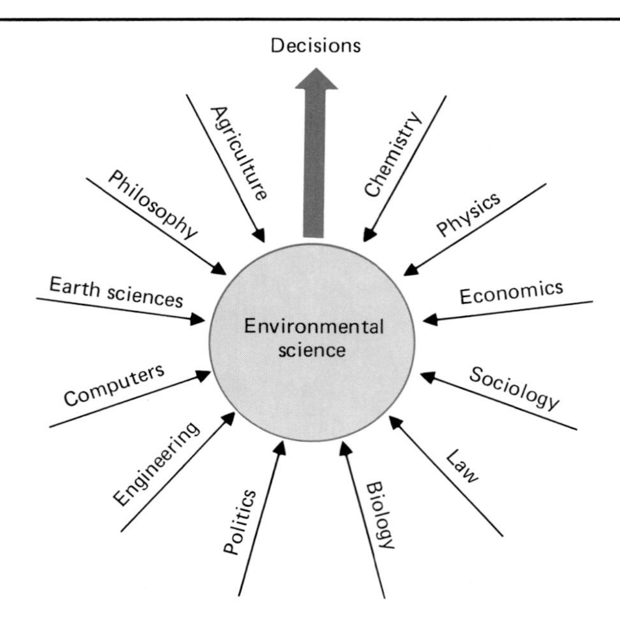


**Figure 2.3**
**Environmental science.** Environmental science draws from a variety of subject areas, which also influence decisions related to the environment. Some of these areas are shown here.

The environmental movement reached enormous proportions, and environment became a household word following Earth Day, April 22, 1970. Courses with the title "Environmental Science" were begun in high schools and colleges around Earth Day. Most of the concepts covered by environmental science courses had been taught previously in ecology classes or in conservation classes. There was also some input from the social sciences such as economics, sociology, and political science. So environmental science has evolved as an interdisciplinary field that draws information related to environmental concerns from the social sciences as well as the biological and physical sciences. (See fig. 2.3.) Since environmental science requires an understanding of the physical world, it is important to have an appreciation for the fundamental subunits of matter that make up the physical world.

## The structure of matter

The science that deals with the composition and properties of matter is chemistry. **Matter** has measurable mass and volume. The matter in our surroundings is composed of ninety-two different naturally occurring **elements.** Each element is composed of subunits called **atoms.** The structural differences between one atom and another make elements obviously different from each other. Even an untrained observer would not confuse mercury with gold or oxygen with iron. To better understand why elements differ from one another, we will look at the detailed structure of the atom.

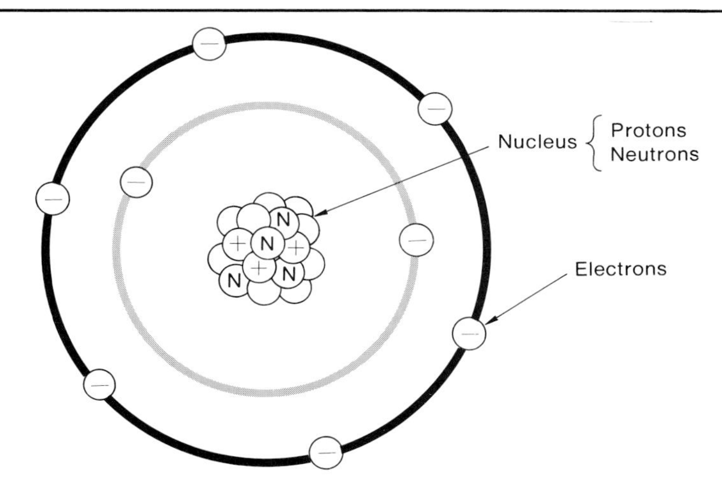


**Figure 2.4**
**Diagrammatic oxygen atom.**
Most oxygen atoms are composed of a nucleus containing eight positively charged protons and eight neutrons without a charge. In addition, there are eight negatively charged electrons that spin around the nucleus.

## *Atomic structure*

All atoms are composed of a central region called the **nucleus,** which is surrounded by smaller, negatively charged particles (**electrons**), which move around it at certain distances. The nucleus contains two types of particles, **neutrons** and **protons.** The neutrons have no charge, and the protons have a positive charge. Atoms of different elements contain specific numbers of protons in the nucleus. For instance, each atom of mercury has eighty protons, whereas each atom of gold has seventy-nine. The number of protons is constant for any one type of atom; the number of neutrons, however, may vary. For example, in oxygen there are eight protons in the nucleus of each atom. (See fig. 2.4.) Most atoms of oxygen have eight neutrons also, but some have nine neutrons, and others have ten neutrons. To be oxygen, an atom must have eight protons; but different **isotopes** of oxygen can have different numbers of neutrons. All isotopes of a particular element exhibit the same chemical properties but have different weights because the number of neutrons differs.

The neutrons and protons are held together in the nucleus by energy. Since the positively charged protons repel one another, the energy of the nucleus does the work of holding the nucleus together. Some isotopes are radioactive; that is, the nuclei of these isotopes are unstable and spontaneously decompose. As a result, a great deal of nuclear energy is released when these atoms disintegrate. The rate of decomposition is consistent for any given isotope. It is measured and expressed as its **radioactive half-life.** This is the time it takes for one half of a sample to decompose spontaneously. Table 2.1 lists the half-lives of several isotopes.

When a nucleus is hit with a very fast-moving particle, the nucleus breaks up. This is similar to what happens when you break the racked balls with the cue ball in pool. The cue ball hits the group of numbered balls, and they roll all over the table. In a nuclear reactor, the nucleus of a large atom is broken into several smaller groups of nuclear particles, and nuclear energy is released

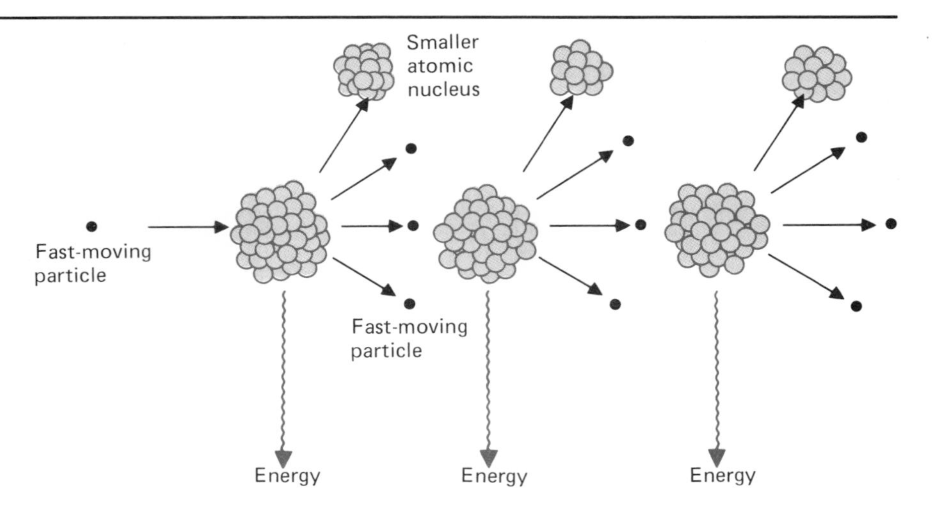


**Figure 2.5 Chain reaction.**
A chain reaction starts when a fast-moving particle hits the nucleus of a large fissionable atom. This nucleus breaks apart and releases energy and more fast-moving particles that continue the chain reaction.

**Table 2.1**
The half-lives of some radioactive isotopes

| Radioactive isotope | Half-life (years) |
|---|---|
| Carbon 14 | 5,730 |
| Protactinium 231 | 32,000 |
| Thorium 230 | 75,000 |
| Uranium 234 | 250,000 |
| Chlorine 36 | 300,000 |
| Beryllium 10 | 4.5 million |
| Potassium 40—argon 40 | 1.3 billion |
| Helium 4 | 12.5 billion |

(**fission**). When the first particle hits a nucleus, that nucleus releases a particle, which in turn hits another nucleus, and when the second nucleus releases a particle to hit a third, we have a **chain reaction.** (See fig. 2.5.) The problems associated with nuclear fission, including acquisition of fuel, control of the chain reaction, and nuclear waste disposal, will be dealt with in chapter 10. One aspect of nuclear disintegration that has intrigued chemists and physicists is the fantastic amount of potentially usable energy available from the atomic nucleus. Most traditional energy sources rely on simple chemical reactions that involve interactions between electrons rather than nuclear changes.

## *Molecular structure and chemical reactions*

When two or more atoms combine with one another, they form stable units known as **molecules.** (See fig. 2.6.) The atoms within a molecule are held together by chemical bonds. **Chemical bonds** are physical attractions between atoms resulting from the interaction of their electrons. When chemical bonds

**Figure 2.6 Some common inorganic compounds.** Table salt and water are two examples of common chemical compounds. Compounds are composed of very, very small units called molecules. In fact, one drop of water, 0.05 milliliter, contains 1,670,000,000,000,
000,000,000 molecules of water.

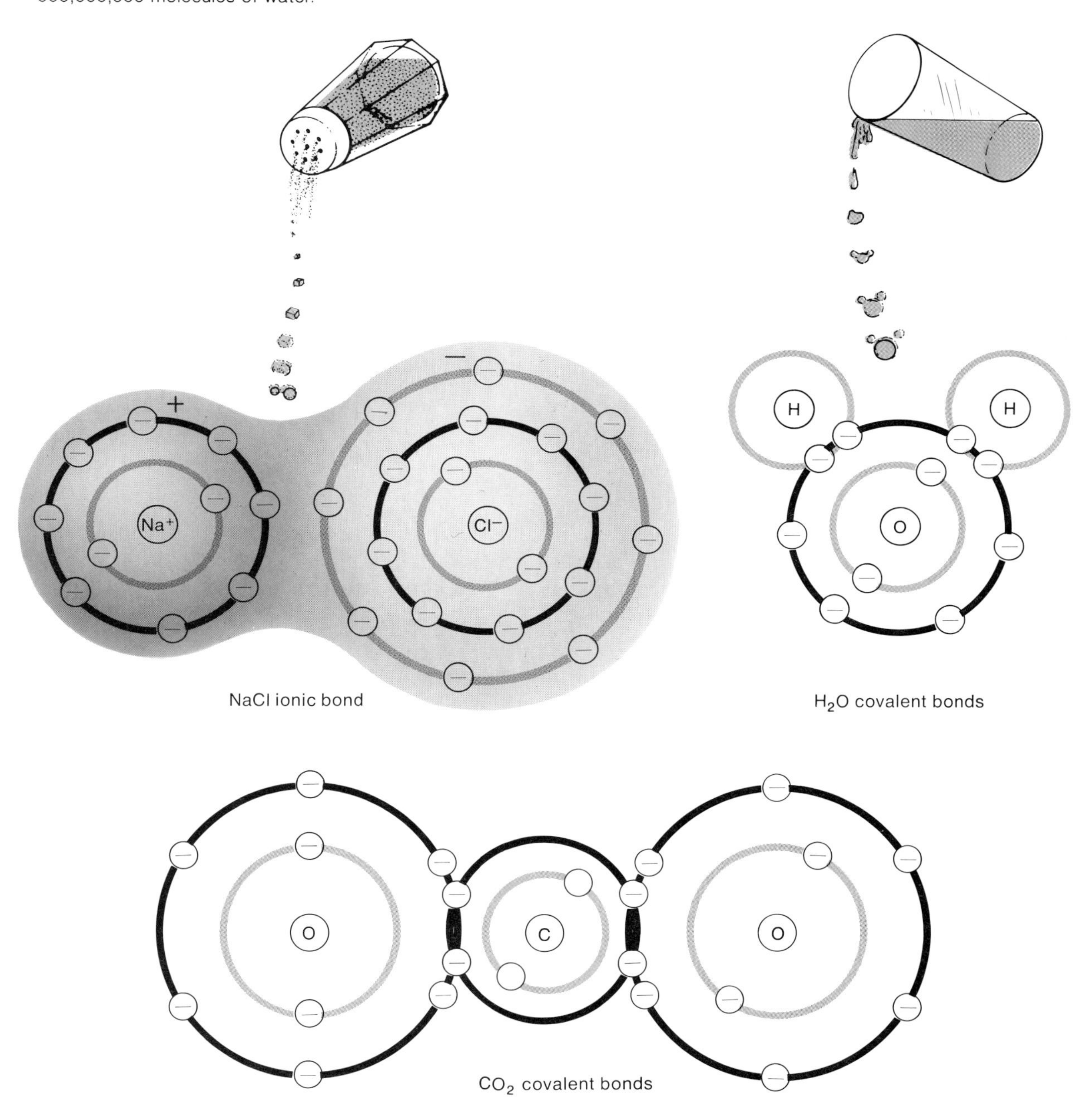

**Figure 2.7 A chemical reaction.** When methane is burned, chemical bonds are changed and the excess chemical bond energy is released as heat and light.

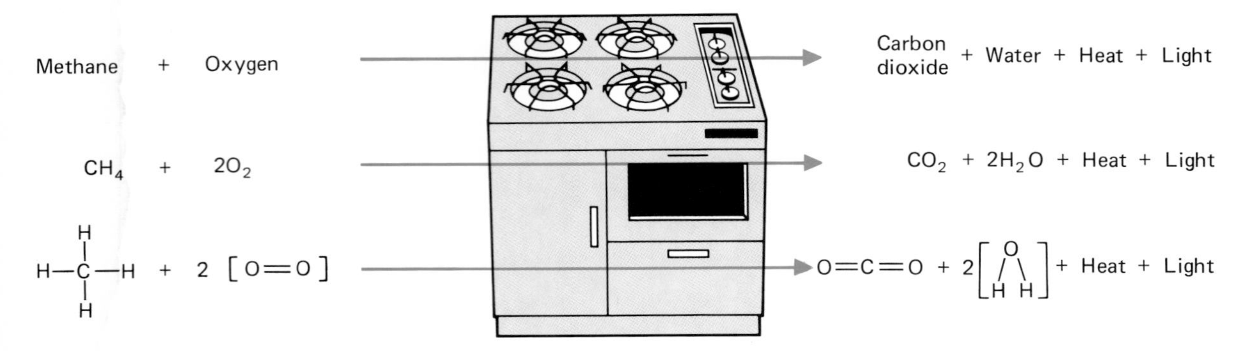


are broken or formed, a chemical reaction occurs. When a chemical reaction occurs, the amount of energy within the chemical bonds is changed. Some chemical reactions, such as burning natural gas, release energy. This reaction involves breaking the chemical bonds of methane and oxygen (reactants) and forming chemical bonds in carbon dioxide and water (products). In this kind of reaction, there is chemical bond energy left over, which is released as heat and light. (See fig. 2.7.)

Other reactions require the addition of energy to enable them to occur. **Photosynthesis** (the process by which plants manufacture food) is an example of this type of reaction. Light energy is required to enable the lower-energy reactants (carbon dioxide and water) to form higher-energy products (sugar and oxygen). (See fig. 2.8.)

In every reaction, the amount of energy in the reactants and in the products can be compared and differences accounted for. Even with energy-yielding reactions, an input of energy is required to get the reaction started. This initial input of energy is called **activation energy.** (See fig. 2.9.) In certain cases, the amount of activation energy required can be reduced by using a catalyst. A **catalyst** is a substance that alters the rate of a reaction but is not altered in the process. Catalysts are used in catalytic converters attached to automobile exhaust systems. The purpose of using the catalytic converter is to decrease the amount of air pollution by completing the combustion of unburned fuel. Large inputs of energy are required to complete the combustion of fuel, but with a catalyst, combustion can be completed at much lower temperatures.

## Energy principles

You have probably noticed that although the previous section started out with a description of matter, it was necessary to use concepts of energy to describe the nucleus of the atom, chemical bonds, and chemical reactions. Energy is very closely tied to matter. It is difficult to think of or describe one without the other.

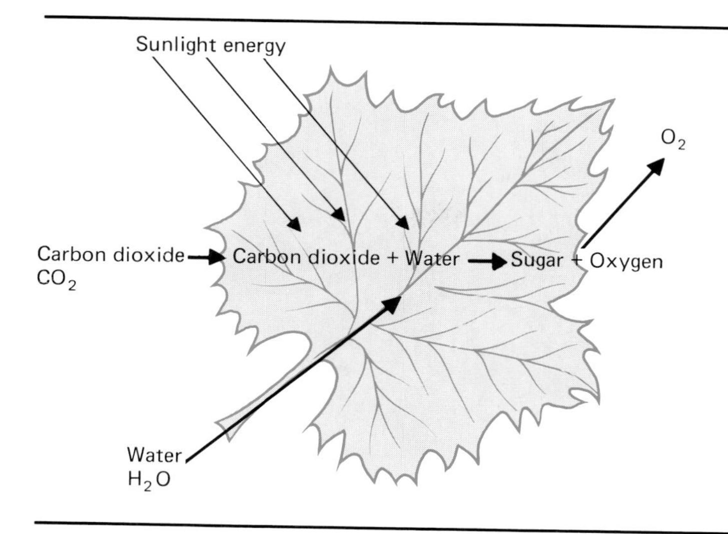


**Figure 2.8 Photosynthesis.**
This reaction is an example of one that requires an input of energy (sunlight) to combine low-energy molecules ($CO_2$ and $H_2O$) to form sugar with a greater amount of chemical bond energy.

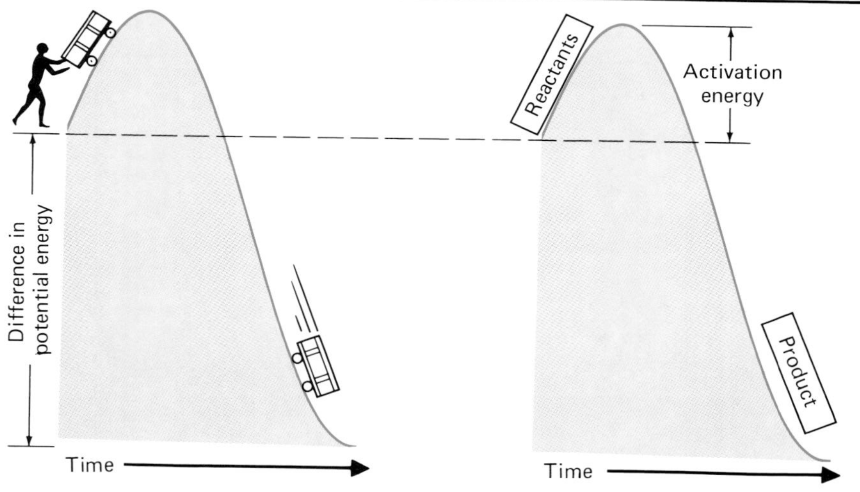


**Figure 2.9 Activation energy.**
The amount of energy needed to get the wagon rolling is the energy of activation. Chemical reactions require an initial input of activation energy to get them started.

Depending upon the amount of energy present, matter can occur in three different states—solid, liquid, or gas. The amount of energy a molecule contains determines how rapidly it moves. In solids, low-energy molecules vibrate in place and are very close together. Higher-energy molecules are farther apart and will roll, tumble, and flow over each other as in a liquid. The molecules of gases move very rapidly, collide with each other, and are very far apart. To change matter from one state to another, energy must be removed or added. (See fig. 2.10.)

When we add energy to a solid and the molecules begin to move more rapidly, they have gained kinetic energy. **Kinetic energy** is the energy contained by moving objects. Electrons moving about an atomic nucleus have kinetic energy, as does flowing water or the movement of air molecules we call wind.

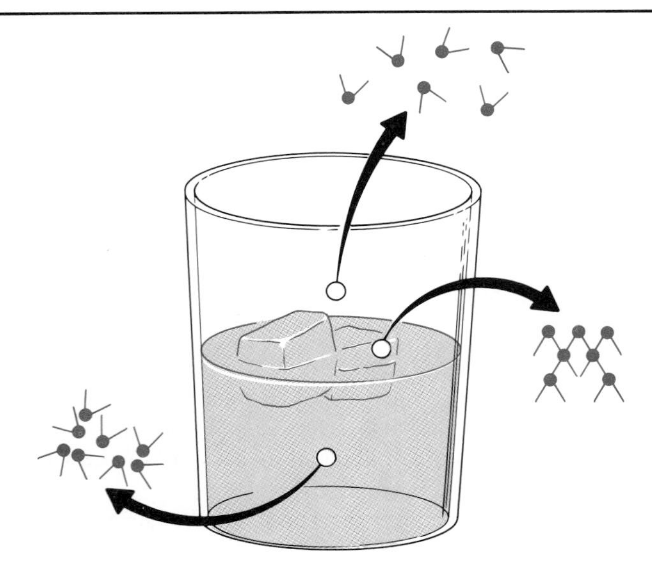

**Figure 2.10 States of matter.**
Matter exists in one of three states depending on the amount of kinetic energy the molecules have. The higher the amount of energy, the greater the distance between molecules and the greater the degree of freedom of movement of the molecules.

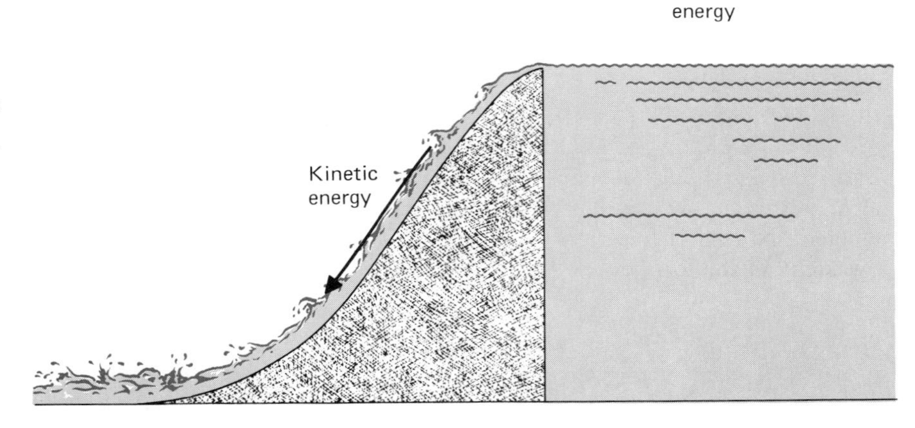


**Figure 2.11 Kinetic and potential energy.**
Kinetic and potential energy are interconvertible. The potential energy possessed by the water behind a dam is converted to kinetic energy as the water flows to a lower level.

There is another type of energy called **potential energy.** This is the energy of position. The water behind a dam has potential energy by virtue of its elevated position. (See fig. 2.11.) An electron at some distance from the nucleus has potential energy due to its position and the distance between it and the nucleus.

## *First and second laws of thermodynamics*

Kinetic and potential energy can exist in many different forms, such as chemical, nuclear, light, heat, sound, and electrical energy. It is useful to know that energy can be neither created nor destroyed. This statement is the **first law of thermodynamics.** It implies that all of the energy in the universe today was there yesterday and will be there tomorrow.

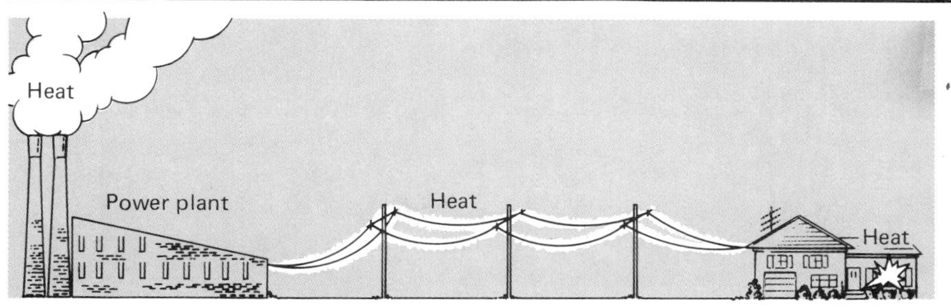


**Figure 2.12 Second law of thermodynamics.** The conversion of fuel to electricity results in the production of heat, which is lost to the atmosphere. As the electricity moves through the wire, resistance generates some additional heat. The light bulb produces not only light but also heat.

One of the problems facing modern society is that there is only a certain amount of energy readily available. Most of it is unavailable to humans, which makes the energy that is available to us precious. Because we have only a finite amount of energy, we need to learn as much about it as possible to allow us to use it most efficiently.

We use energy by converting it from one form to another. We convert gasoline (chemical-bond energy) to kinetic energy, or electricity to light, or wind to electricity, but all of these energy conversions result in the formation of some heat. This concept is known as the **second law of thermodynamics.** Specifically, this law states that whenever there is a conversion of one form of energy to another, there is a certain amount of heat produced that cannot be used. Energy is being converted from one form to another continuously within the universe. Stars are converting nuclear energy to heat and light, animals are converting food energy to movement, and plants are converting light energy into the chemical-bond energy of sugar. In each of these conversions, although no energy is destroyed, the conversion is not one-hundred percent efficient. Not all of the energy is converted to the needed form; some of it is "wasted" in the form of heat. (See fig. 2.12.)

## *Environmental implications of energy flow*

The heat produced when energy conversions occur is dissipated throughout the universe. This is a common experience. Valuable things always seem to disintegrate unless we work to prevent it. Houses require constant maintenance, automobiles rust, and appliances wear out. In reality, all of these phenomena involve the loss of heat. The organisms that decompose wood release heat. The chemical reaction that causes rust releases heat. Friction, caused by the movement of parts in a machine, generates heat and wears the parts. Orderly arrangements of matter always tend to become disordered by the constant flow of energy toward a dilute form of heat. This dissipated form of energy is low quality and, therefore, has very little value to us. Concentrated heat energy does have value and is used in steam engines and other mechanical devices. This can be compared to a hydroelectric plant. A stream flowing downhill is expending energy. The steeper the slope, the greater the amount of energy expended. The stream itself has low-quality energy, because the energy is dissipated along the entire length of the stream, if there is no point in

**Table 2.2**
Efficiency

| Energy conversion system | Efficiency* |
|---|---|
| Electric generator | 99 |
| Large electric motor | 92 |
| Dry cell battery | 90 |
| Large steam boiler | 88 |
| Home gas furnace | 85 |
| Storage battery | 72 |
| Home oil furnace | 65 |
| Small electric motor | 62 |
| Fuel cell | 60 |
| Liquid-fuel rocket | 47 |
| Steam turbine | 46 |
| Steam power plant | 40 |
| Gas laser | 39 |
| Diesel engine | 38 |
| Aircraft gas turbine | 36 |
| Industrial gas turbine | 35 |
| High-intensity lamp | 32 |
| Solid-state laser | 30 |
| Automobile engine | 25 |
| Fluorescent lamp | 20 |
| Wankel engine | 18 |
| Solar cell | 10 |
| Steam locomotive | 8 |
| Thermocouple | 7 |
| Incandescent lamp | 4 |

*Efficiency in this case means the efficiency with which energy contained in any fuel is converted to a useful form.
Source: Robert H. Romer, *Energy: An Introduction to Physics.* W. H. Freeman and Company, 1976.

the stream where the slope is very steep. To make this a high-quality (concentrated) source of energy, it is necessary to dam the water so that it will drop a long distance at one point. The amount of energy is not changed; the quality is. Likewise, when two objects differ in heat content, heat will flow to the cooler object. The greater the temperature difference, the more useful the work that can be done. Because of this, fossil fuel power plants use cold water to provide a steep heat gradient.

All organisms, humans included, produce waste heat when they convert chemical-bond energy in food into the energy needed to move, grow, or respond. The process of releasing chemical-bond energy from food by organisms is known as **cellular respiration.** The process of releasing chemical-bond energy from fuel is called **combustion.** The efficiency of these processes is relatively high. About forty percent of the energy released is in a useful form. The rest is lost as low-quality heat. (See table 2.2.)

There is a limited amount of energy in the universe. Using available energy decreases the amount of useful energy available as more heat is generated. All life and all activities are subject to these important first and second laws of thermodynamics.

## Consider This Case Study

### Four Corners Power Plant

The Four Corners Power Plant is located near Farmington, New Mexico, in the northwest corner of the state. It is a coal-fired complex of power plants operated by Arizona Public Service Company. There are currently five plants operating on the coal obtained from the second largest strip mine in the country. Utah International Corporation operates this coal mine on land leased from the Navajo nation. The power plants were built near the mines to reduce the cost of coal transportation. The mining company in turn buys back one-third of the electricity generated by the plants for its strip-mining operation. The remaining two-thirds of the electricity is sold to customers in Arizona and California. This electricity is the cheapest available, partly because little attention is given to air quality. Several groups stated that it was the dirtiest plant in western United States.

In 1977, two of the five plants that make up this complex were operating with virtually no sulfur dioxide control. Sulfur dioxide is an air pollutant. It is produced by combustion of fossil fuels that contain a variety of sulfur compounds. Sulfur dioxide causes a variety of health problems for humans and other animals. It injures many plant tissues and may combine with water to form corrosive acids that are washed from the air when it rains. At public hearings, both the New Mexico Citizens for Clean Air and Water and the Arizona Public Service Company presented information related to the problem.

The citizens' group claimed that the pollution from this plant caused the haze that is common in the southwest and also contributed to acid rain problems throughout the country. They maintained that installation and operation of pollution control devices to remove sulfur dioxide would cost consumers in Phoenix and Los Angeles less than sixty cents if their electricity bills were $100.

The utility maintained that the changes suggested were too costly, too difficult to operate, and were not needed to meet federal air quality standards. However, the utility agreed to remove 67.5% of the sulfur dioxide emissions. In 1979, they requested that this requirement be removed.

The plants continue to discharge large amounts of sulfur dioxide, but in 1984 an agreement was reached that is intended to result in the installation of equipment that will remove 72% of the $SO_2$ emissions.

List two examples that indicate the operation of the second law of thermodynamics at the Four Corners Power Plant.

Make a flow chart showing energy flow in the power plant, coal mine, and at the consumers' level.

What kind of chemical reactions occur in this situation?

## Summary

Science is a method of gathering and organizing information that involves observation, hypothesis formation, and experimentation. A hypothesis is a logical guess about how things work. If a hypothesis is used for a long period of time and is continually supported by new facts, it is often stated as a theory.

Applied science deals with the use of information gained through science to influence the human condition. Theoretical science aims to increase the base of scientific knowledge.

Environmental science is an interdisciplinary area of study that includes both applied and theoretical aspects of human impact on the world.

The fundamental unit of matter is the atom, which is made up of electrons, protons, and neutrons. The number of protons for any one type of atom is constant, but the number of neutrons may vary. The neutrons and protons are held together in the nucleus by energy. When the nucleus is hit with a fast-moving particle, the nucleus breaks up. Energy is released from the nucleus when it disintegrates.

When two or more atoms combine with one another, they form stable units known as molecules. Chemical bonds are physical attractions between atoms resulting from interaction of their electrons. When chemical bonds are broken or formed, a chemical reaction occurs, and the amount of energy within the chemical bonds is changed. Chemical reactions require activation energy to get the reaction started.

Matter can occur in three different states—solid, liquid, and gas. These three differ in the amount of energy the molecules contain and the distance between the molecules. Kinetic energy is the energy contained by moving objects. Potential energy is energy of position.

The first law of thermodynamics states that the amount of energy is constant and that energy cannot be created or destroyed. The second law of thermodynamics states that when energy is converted from one form to another, some of the useful energy is lost.

## Review Questions

1. How do scientific disciplines differ from nonscientific disciplines?
2. What is a hypothesis? Why is it an important part of the way scientists think?
3. How does "environmental science" differ from traditional science?
4. Diagram an atom of oxygen and label its parts.
5. What changes occur when an atom undergoes fission?
6. What happens to atoms during a chemical reaction?
7. State the first and second laws of thermodynamics.
8. How do solids, liquids, and gases differ from one another at the molecular level?
9. Are all kinds of energy equal in their capacity to bring about changes?
10. List five kinds of energy.
11. Why are events that happen only once difficult to analyze from a scientific point of view?

## Suggested Readings

There are many textbooks that discuss introductory chemistry and physics to which you may want to refer. The three references listed here were selected because they give a sense of how a scientist operates.

Cassidy, Harold G. *Science Restated: Physics and Chemistry for the Non-scientist*. San Francisco: Freeman, Cooper and Company, 1970. A treatment of atomic structure and thermodynamics that is both thorough and readable.

Karplus, Robert. *Physics and Man*. New York: W. A. Benjamin, Inc., 1970. Consists of a collection of articles on a variety of topics related to science in general and physics in particular. A good job of communicating what science is and how it works.

von Frisch, Karl. *Bees: Their Vision, Chemical Senses and Language*. Rev. ed. Ithaca, N.Y.: Cornell University Press, 1971. An excellent introduction to the scientific method and the way a scientist thinks. Written by a Nobel prize winner.

## Chapter Outline

I. Ecological terminology
II. Kinds of organism interactions
   A. Predation and parasitism
   B. Competition
   C. Mutualism
III. Ecosystems, communities, and individuals
   A. Energy flow
   B. Food chains
   C. Nutrient cycles
      Box 3.1 PBB
IV. Analysis of an ecosystem
V. Consider this case study: Meadowlands, New Jersey

## Objectives

Identify the abiotic and biotic factors in the ecosystem.

Describe the process of natural selection as it operates to refine the fit between organism, habitat, and niche.

Define the role of producer, herbivore, carnivore, omnivore, scavenger, parasite, and decomposer.

Describe predator-prey, parasite-host, competition, and mutualistic relationships.

Differentiate between a community and an ecosystem.

Relate the concept of food web and food chain to population size and niche.

List some components of an ecosystem.

Define niche.

Describe energy flow in an ecosystem.

Explain the cycling of nutrients such as nitrogen and carbon through an ecosystem.

## Key Terms

abiotic factors
biochemical oxygen demand
biomass
biotic factors
carbon cycle
carnivore
communities
competition
consumers
decomposer
denitrifying bacteria
detritus
ecological niche
ecology
ecosystem
ectoparasite
endoparasite
environment
food chain
food web
habitat
herbivore
host
interspecific competition
intraspecific competition
limiting factor
mutualism
natural selection
nitrogen cycle
omnivore
parasite
phytoplankton
predator
prey
primary consumers
producers
range of tolerance
secondary consumers
symbiosis
symbiotic nitrogen-fixing bacteria
trophic level
zooplankton

# Chapter 3
# Interactions: Environment and organisms

## Ecological terminology

Energy and matter are interrelated. Matter contains energy, and that energy determines the physical properties of the matter. Living systems also consist of matter and energy interrelationships. Living things require a constant flow of energy to assure their survival. The way that organisms and their surroundings interact is the basic concern of ecology.

All organisms are dependent on other organisms. This statement is a reasonably good starting point for discussions of ecology. Ecologists have developed terminology to describe the kinds of interactions that occur and the different levels at which the interactions occur.

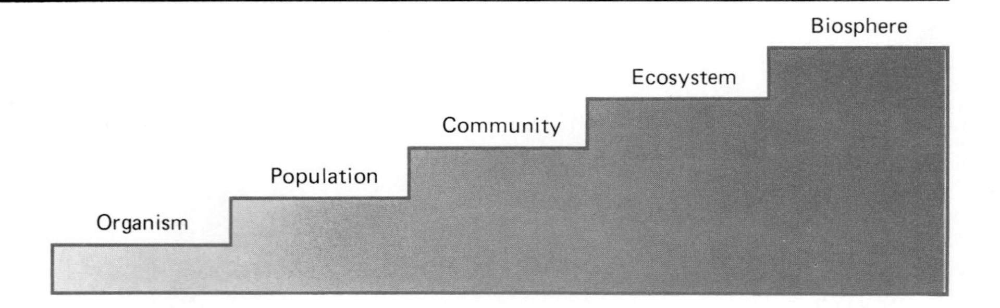


**Figure 3.1 Ecology.** The study of ecology is pursued at various levels, ranging from the individual organism to the biosphere.

**Ecology** is the branch of science that deals with the interrelationships between organisms and their environment. (See fig. 3.1.) It deals with the ways in which organisms are molded by their surroundings, how they make use of these surroundings, and how an area is altered by the presence of organisms.

Everything that affects an organism during its lifetime is collectively known as its **environment.** (See fig. 3.2.) It is convenient to classify the factors that affect an organism as **abiotic** (nonliving) or **biotic** (living) **factors.**

The abiotic factors include the flow of energy necessary to maintain the organism, the supply of chemicals required for the various life functions of the organism, and the physical factors that affect the organism.

The ultimate source of energy for any organism is the sun; in the case of plants, the sun directly supplies the energy necessary for the organisms to maintain themselves. Animals obtain their energy by eating plants or other animals. Thus the greater the supply of energy in the environment, the greater the amount of living things in that particular environment.

All forms of life require carbon, oxygen, nitrogen, phosphorus, water, and other chemicals. The organisms constantly remove these chemicals from the environment and then return them. The physical factors in the environment include the climate; temperature; type, amount, and seasonal distribution of precipitation; type of soil; and even the type and amount of space available to the organism.

The biotic factors influencing an organism include all forms of life in its environment. These include all plants, animals, and decomposers. Both the quantity and type of living forms are important in the environment of an organism. All organisms respond to the total makeup of their environment, but the one factor that limits their success is called the **limiting factor.**

For example, the limiting factor for many species of fish is the amount of dissolved oxygen present in water. In a swiftly flowing tree-lined mountain stream, the level of dissolved oxygen is high and provides a favorable condition for trout. (See fig. 3.3.) As the stream continues down the mountain, the grade decreases, which results in fewer rapids to oxygenate the water. In addition, the canopy of trees over the stream usually thins, which allows more sunlight to reach the stream. Warmer water cannot hold as much dissolved oxygen. Decreasing turbulence and increasing temperature lower the amount of oxygen in the water to a level below that required by the trout. Fish such as black

**Figure 3.2**
**Environment.**
Everything that affects an organism during its lifetime is collectively called its environment. Physical factors, such as weather, soil type, altitude, space, and amount of sunlight, have a significant effect on the kinds of organisms that can live in an area. Likewise, how the organism interacts with other organisms such as the types of plants used for food and shelter, parasites, and predators are part of the environment.

bass and walleyes replace the trout, since they have a greater **range of tolerance** and are able to withstand lower oxygen concentrations and higher water temperatures. Thus the lower level of oxygen and the higher water temperatures are the limiting factors for the distribution of trout.

In addition to the quantity of oxygen, other factors may determine the quality of the environment. Water quality determines which organisms can survive in it and, in turn, is influenced by the inhabiting organisms. For example, there may be more food available in the lower regions of the stream than in the upper fast-flowing sections of the stream. A complicated series of events is responsible for this situation.

Much of the land on the riverbanks may be devoted to agriculture, and this results in an increased amount of soil entering the stream. As the stream continues, it will become wider and slower, and most of the water will be exposed to the warming influence of the sun. Whenever the oxygen level falls, siltation increases, temperature rises, and photosynthetic activity increases, carp and other types of rough fish will inhabit the water.

**Figure 3.3 Limiting factors.** All fish require oxygen. However, the abiotic and biotic factors in a stream determine the amount of oxygen present. Cool, highly oxygenated water supports trout, but warmer, less oxygenated water is unsuited for trout. The supply of oxygen is the limiting factor for the trout. Other fish with a wider range of tolerance for oxygen occupy the lower sections of the river.

The area where an organism lives is called its **habitat.** We tend to identify habitats with particular physical environmental characteristics such as soil, water, and climate. But the living environment is equally important. For instance, the changes in the stream as it flows from the mountain to the sea are the result of a combination and interaction of abiotic and biotic factors. The increasing temperature, the leveling of the grade, the decreasing oxygen, and the changing riverbed material are examples of changing abiotic factors. The fact that trout, bass, and carp are found in different sections of the stream is an example of changes in the biotic factors.

The prime food source shifts from aquatic insects in the upper region of the stream to minnows in the lower regions, and corresponds to the alteration

**Figure 3.4 Ecological niche.** The stream provided a suitable habitat for the beaver. The activity of the beaver resulted in a pond that provided a better environment for beaver and other species, but it also destroyed the habitat for some animals.

in the quantity and type of vegetation lining the stream. It is this differing environment that determines which type of organism occupies a given area.

Just as the surroundings of an organism have an impact on the organism, so too does the organism act on its surroundings. This functional role is known as the organism's **ecological niche.** (See fig. 3.4.) The stream provides a suitable habitat for beaver, while the land on either side of the stream provides habitat for such organisms as trees, grass, squirrel, and deer. The presence of the beaver alters this habitat. With the construction of a dam, part of the terrestrial habitat becomes flooded, and various forms of aquatic vegetation replace trees and grasses. The formation of the pond and the subsequent establishment of aquatic plants creates a habitat suitable for more fish, moose, muskrats, mink, and ducks. Thus the presence of the beaver makes a drastic change in the ecosystem.

**Figure 3.5 Natural selection.** Natural selection is a process that determines which organisms survive or reproduce most effectively. As a result of this process, each generation becomes slightly better adapted to its environment. The birds that do not fly when flushed by the hunter have a better chance of survival. They will be more likely to live to reproduce, and their offspring will also be less likely to fly when flushed.

a

b

If the change in the environment is drastic and quick, such as the establishment of the beaver pond, the type of organisms in the area will change. However, if the environmental change is not as severe or is slower, the organisms may adapt to the change. **Natural selection** is the process by which a species changes in response to its environment as a result of the selection of those organisms more fit to survive and reproduce.

In the 1800s, Charles Darwin (a British naturalist) explained the process of natural selection in his book *The Origin of Species*. In this work, he suggested the following: (1) All populations of organisms show variation. No organisms have exactly the same genetic makeup. (2) Organisms produce more offspring than can survive. This means that there is not enough suitable habitat for all offspring to grow to maturity. (3) There is a struggle for survival. The organisms must compete with each other for food, space, mates, or other limited requirements. (4) Certain individuals of a species have a better chance for survival than others. It is partially the genes of an organism that determine how well it can compete and survive in its environment. (5) The result is natural selection. Those individuals with a particular combination of genes have a better chance of reproducing and passing these genes on to their offspring.

When being hunted, a quail can do one of three things: It can attempt to escape by flying away from its enemies, it can run away from the danger, or it can freeze in place. Each individual quail is capable of taking any of the three forms of action; however, some are more prone to run and others to fly. When being hunted, the birds that fly from the hunter are more easily seen and therefore more likely to be killed. (See fig. 3.5a.)

Thus the animals that run or freeze when pursued have a better chance for survival and reproduction. Their offspring are likely to respond in a similar fashion. The first time quail are hunted, the majority of the birds are likely

**Figure 3.6 Predator/prey.**
The lions are the predators, and the zebra is the prey. The quicker lions are more likely to get food, and the slower zebras are more likely to become food.

to fly. After many seasons, the percentage of birds that fly from the hunter will decrease, and more birds will run or freeze. (See fig. 3.5b.) Natural selection has changed the behavior of the birds.

This type of natural selection occurs anytime there is a change in the environment. For example, in the time of the horse and buggy, vehicles were not a threat to slow-moving animals. But with the introduction of the fast-moving automobiles on the roads, the mortality rate of slow-moving animals increased because the animals were frequently struck and killed when they attempted to cross roads. Anyone who has seen a squirrel avoid being hit by a car has observed the result of selection for a faster-reacting population of squirrels.

## Kinds of organism interactions

One important part of an organism's environment is the other organisms with which it shares its habitat. There are several different ways in which organisms interact. Organisms are not isolated and cannot exist independently. One organism may be the food for others. While this is not necessarily the only type of interaction, it is a very obvious one. The amounts and kinds of food available to an organism are central to much of its behavior. How organisms deal with getting food is a major aspect of a niche.

### *Predation and parasitism*

One kind of interaction occurs when an animal kills and eats another. The **predator** is the organism that uses another organism (**prey**) for food. (See fig. 3.6.) You are probably aware of some predator-prey relationships, such as lions and zebras, birds and worms, and frogs and insects. The predator-prey relationship is often thought to be one-sided. The benefits seem to be all with

**Figure 3.7 Parasites.** A tick is a small organism that attaches itself to the outside of a host organism. It uses the host's blood for food. It is an example of an ectoparasite. The drawing at the bottom is a tapeworm that lives inside its host's intestine. It is called an endoparasite.

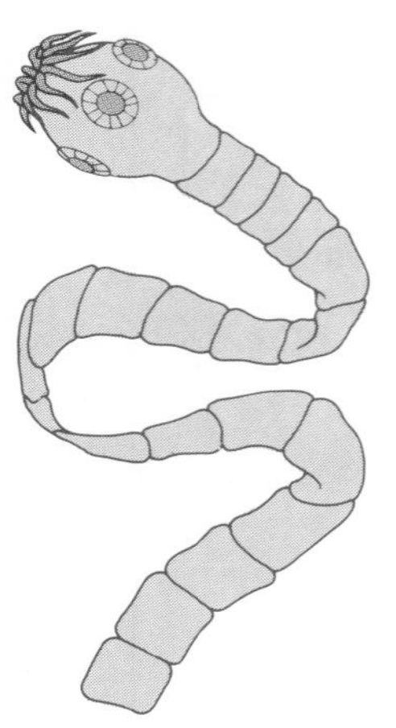

the predator. It is the predator who has the meal and goes on living after killing and eating the prey organisms. However, this interaction may be of advantage to the prey population as well. Prey species have a higher reproductive potential than predator species. Because of this higher reproductive rate, prey species can endure a higher mortality rate. Certainly the individual who has been killed and eaten did not benefit. However, the prey organisms that die are likely to be the old, the slow, the sick, and the less-fit members of the population. The healthier, quicker, and the more fit are more likely to survive. When these survivors reproduce, their offspring are more likely to be better adapted to their environment. At the same time, natural selection is occurring within the predator population, since poorly adapted individuals are less likely to capture prey. The predator is a participant in the natural selection process and so is the prey. This dynamic relationship existing between predator and prey species is a complex one that continues to intrigue ecologists.

A relationship similar to the predator-prey relationship is that of parasite-host. Any close, long-lasting physical relationship between two different species is known as **symbiosis.** The **parasite** (one species) lives in or on another living organism (the second species) from which it derives nourishment. This second organism is called the **host.** Whereas the predator kills its prey, the well-adapted parasite does not usually kill its host. It may, during its lifetime, visit several different host organisms. Some parasites live on the outside of their hosts and are called **ectoparasites.** Ticks attach themselves to a host and suck blood. (See fig. 3.7.) Other parasites are **endoparasites** and are adapted to living within their host. For example, tapeworms live in the intestines of their hosts. The tapeworm is able to resist digestion by the host. The host organism, with only minor discomfort, can live for some time supporting itself and the parasite. If you were to count all the parasites in the world, they would outnumber the free-living hosts. Parasites are not limited to the animal kingdom. In addition, viruses and many kinds of bacteria, fungi, and some plants are parasitic. Parasites are adapted to survive at the expense of the host organism, and the hosts are usually adapted to survive attack by parasites.

## *Competition*

A third type of interaction between organisms is competition. Whenever two organisms require the same limited resources, **competition** will occur. If the two organisms are unevenly matched, one eventually gets the needed resource, and the other goes without. The more similar two kinds of organisms are, the more intense the competition is between them.

Competition may occur between members of the same species as they compete for mates, food, den sites, soil nutrients, or sunlight. This is known as **intraspecific competition.** You might think that competition among members of a species only has a negative impact on the population and that cooperation between the organisms would be a better approach to the problem of limited resources. However, in many cases if the resources were equally shared, all organisms would die. Thus competition is one of the mechanisms of natural selection. The competitive pressure among individuals of the same species is

**Figure 3.8 Competition.** Whenever a needed resource is in limited supply, organisms compete for it. This competition may be between members of the same species (intraspecific) as in this photograph or between several different species (interspecific).

usually very intense, because the members of the population have similar requirements and similar ways of acquiring them. A slight advantage may mean the difference between survival or death. Natural selection is responsible for maintaining a well-adapted population.

Competition may also occur between organisms of different species. This is called **interspecific competition.** One species may be better adapted to live in an area than another. The less-fit species can migrate to new areas or through generations of selection may become adapted for a slightly different niche. The fact that competition results in organisms that are better adapted to their environment is a positive aspect of this type of interaction. The result is an organism that fits its habitat very well. (See fig. 3.8.)

## *Mutualism*

Not all interactions need to be antagonistic. Some organisms interact in a cooperative way. **Mutualism** occurs between organisms of different species in which both organisms derive benefit from the interaction. In many mutualistic relationships, the interaction is obligatory. Flowers of the yucca plant are pollinated by a specific insect, the yucca moth. The moth in turn lays its eggs in the flower, where the immature moths (larvae) feed on the developing seeds. Many other kinds of flowering plants are also completely dependent on insects for pollination. The insect benefits by getting food, while the plant is pollinated by the activities of the insect.

Many plants have nitrogen-fixing bacteria associated with their roots. The plant in turn furnishes a living site and nutrients for the bacteria. (See fig. 3.9.)

**Figure 3.9 Mutualism.** The growth on the root of this plant contains bacteria that are beneficial to the plant. The bacteria make nitrogen, a needed nutrient, available to the plant. The bacteria also benefit from the raw materials made available by the plant.

**Figure 3.10**
**Ecosystem.**
The grass population is able to survive in this area because of the climatic factors to which the grass is suited. The climate is part of the nonliving or an abiotic factor in the environment. The deer, which are part of the living or biotic environment, interact with the grass; this interaction and all other interactions between organisms constitute the community. A community in an abiotic environment is an ecosystem.

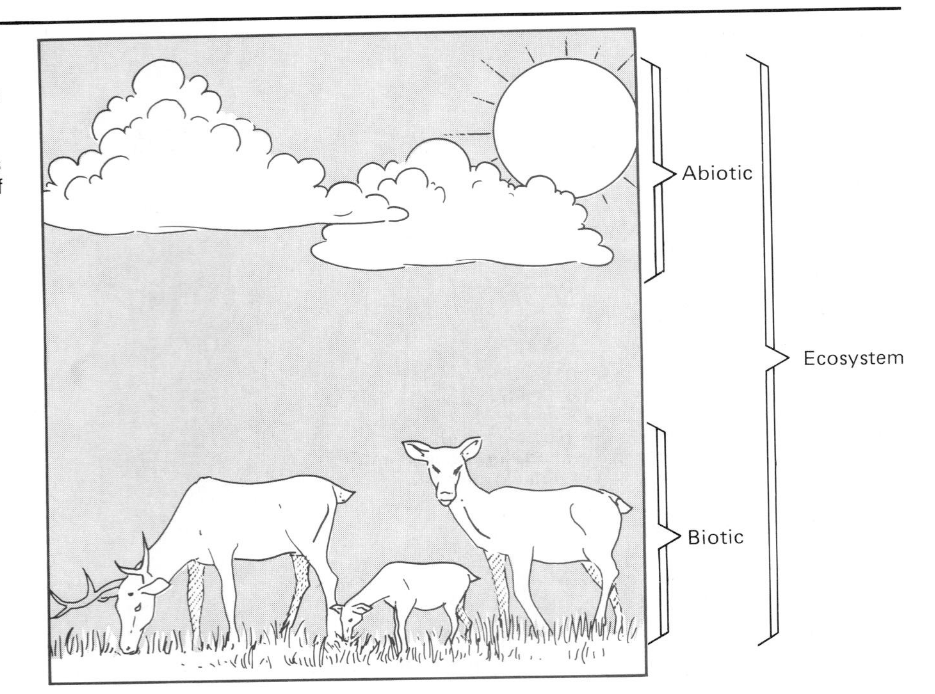


## Ecosystems, communities, and individuals

Thus far we have discussed specific ways in which individual organisms interact with their environment. This does not accurately depict the complexity of interactions that occur among groups of organisms living in the same area. Any individual is only a part of the interacting abiotic and biotic factors in the environment. An **ecosystem** is the interaction among all the living and nonliving factors in an area. It receives energy from the sun to maintain itself. (See fig. 3.10.) The energy from the sun is used by the **producers** to convert simple inorganic chemicals into complex organic compounds. Green plants are the producers in all ecosystems. The green plants in turn supply energy to the **consumers.** The consumers are classified as **herbivores, carnivores,** or **omnivores,** depending upon whether they eat plants, animals, or both. When an organism excretes waste materials or dies, the decomposers convert the complex molecules into inorganic chemicals.

Within ecosystems there may be several **communities** composed of interacting groups of organisms. For example, within the ecosystem seen in figure 3.10, the community involves only the interactions between the organisms (the deer and the plants they feed on). There may, however, be other interactions between producers, consumers, and decomposers within the same ecosystem. Within any ecosystem there are several types of consumers. **Primary consumers** eat plants directly, and **secondary consumers** eat animals who have eaten plants.

A different kind of consumer is the **decomposer,** which breaks down organic matter to inorganic matter. The primary organisms that fit into this category are bacteria and fungi. (See table 3.1.)

Table 3.1
Roles in a community

| Classification | Definition | Examples |
|---|---|---|
| **Producers** | Simple inorganic compounds converted to complex organic compounds by photosynthesis | Trees, flowers, grasses, ferns, mosses, algae |
| **Consumers** | Relies on others to produce food; animals that eat plants or other animals | |
| *Herbivores* | Eat plants directly | Deer, ducks, vegetarian humans |
| *Carnivores* | Eat meat | Wolf, pike |
| *Omnivores* | Eat both plants and meat | Rats, most humans |
| *Scavengers* | Eat meat killed by others | Coyotes, carp, vultures |
| *Parasites* | Lives in or on another individual | Tick, tapeworm, mosquito |
| **Decomposers** | Return organic compounds to inorganic compounds; completes recycling of organic material | Bacteria, fungi |

## *Energy flow*

An ecosystem is a stable, self-regulating unit. The stability of an ecosystem derives from the diversity of organisms within it, the kinds of interactions that occur, and pathways of energy flow. One of the most important characteristics of an ecosystem is the amount of energy available. The only significant source of energy is radiant energy from the sun. This light energy is captured by green plants through the process of photosynthesis and is incorporated into the structure of organic molecules. The energy that plants contain may be transferred through a series of other organisms.

Each time the food energy is passed to a different organism, it is said to enter a different **trophic level.** (See fig. 3.11.) The amount of energy contained at higher trophic levels is less than the amount of energy at lower trophic levels. Only about 10 percent of the energy is available to be passed from one trophic level to the next. The difference in energy content is partially due to the second law of thermodynamics (see chap. 2). Whenever an energy conversion occurs, some energy is dissipated to the surroundings in the form of low-quality heat. In addition to this loss of usable energy, some of the energy is used by the organisms in the lower trophic levels for their own life processes. About 90 percent of the energy at one trophic level is, therefore, unavailable to the higher trophic levels. This means that there must be ten times more available energy at one level than there is at the next higher level. Usually this comparison from one level to another is based on the weight of the organisms at each level. The weight of material tied up in organisms is called **biomass.** Since green plants are at the first trophic level, they usually have more biomass than the herbivores they support.

The plants as producers are the first organisms to trap energy from the sun, which they use to synthesize large organic molecules. All other organisms are consumers because they obtain their energy from the breakdown of the large

Figure 3.11 **Energy flow through an ecosystem.**
As energy flows through an ecosystem, it passes through several levels known as trophic levels. Each time energy moves to a new trophic level, approximately 90 percent of the useful energy is lost. Therefore, high trophic levels usually contain fewer organisms and less biomass.

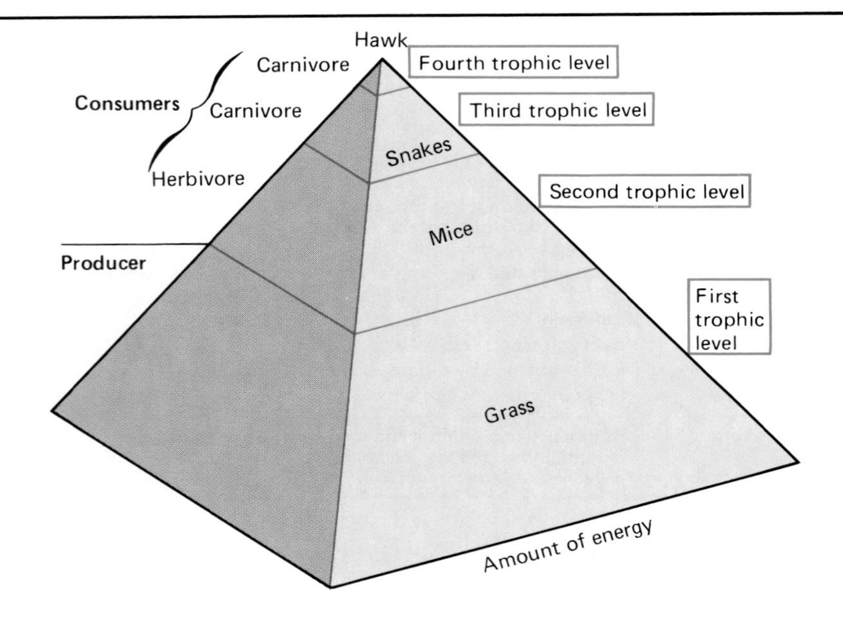


organic molecules in food. Consumers may occupy different trophic levels depending on whether they receive their energy directly or indirectly from plants. Herbivores occupy the second trophic level. The amount of energy available from the plants determines how large the herbivore population can become and the amount of biomass at this trophic level. Carnivores and omnivores occupy higher trophic levels and contain less biomass than herbivores.

When a plant or animal dies, the chemical energy contained within its body is ultimately released to the environment as heat by organisms that decompose the body into smaller molecules, such as carbon dioxide and water. The organisms of decay are the decomposers. Bacteria, fungi, some beetles, and maggots are examples of organisms that occupy this trophic level. Decomposers recycle matter within the ecosystem; and as long as the sun provides the necessary energy for photosynthesis, elements are available to be recycled over and over again.

## *Food chains*

Energy is passed from one trophic level to another in a **food chain.** For example, plants were eaten by insects, who were eaten by frogs, who were eaten by fish, who were eaten by humans. (See fig. 3.12.) In this food chain, there are five trophic levels. Several factors are important in determining an animal's place in a food chain. Each organism occupies a specific niche and has special abilities that fit it for the niche. Willow trees grow well in very moist soil that might be present near a stream, river, or pond. The leaves of the willow tree serve as the food source for insects, such as caterpillars and leaf beetles, with chewing mouth parts and a digestive system adapted to plant food. Some of these insects fall from the tree into the pond below, where they become food for resident frogs. The frog's long, sticky tongue flicks out to capture the insect. As the frog swims from one lily pad to another, a large bass consumes the frog. Occasionally a human may use an artificial frog as a lure and entice bass from their hiding places. The fish dinner is the end of a long chain of events that began with the willow tree.

**Figure 3.12 Food chains.**
As one organism feeds upon another organism, there is a flow of energy through the series. This is called a food chain. (Adapted from Ralph D. Bird, "Biotic Communities of the Aspen Parkland of Central Canada." *Ecology* 11 (April 1930): 410.)

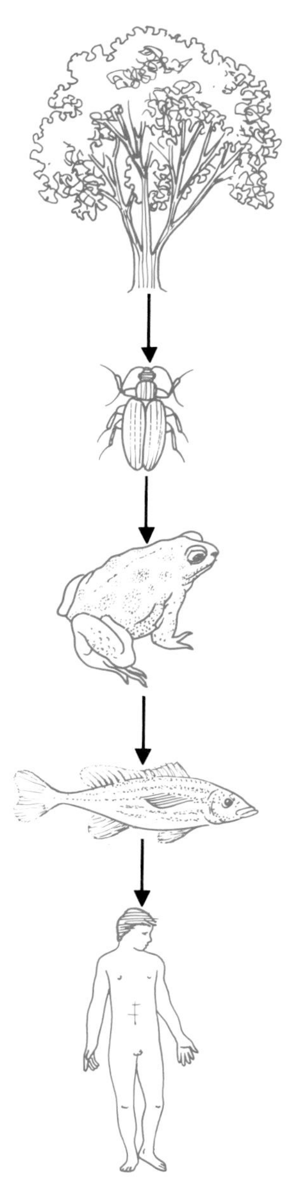

### *Detritus food chain*

We are most familiar with the producer-herbivore-carnivore food chain, but there is a second, equally important food chain: the detritus food chain. **Detritus** is the organic material that results from the decomposition of plants, animals, or fecal waste material, all of which are rich in energy and nutrients. Bacteria, fungi, and protozoans utilize the detritus as a source of energy.

Earthworms, mites, insects, small crustaceans, and other consumers also feed directly on detritus as well as the decomposers. However, when the earthworms are eaten by a bird or the small insects are eaten by a minnow, the detritus food chain becomes a part of an herbivore-carnivore food chain.

Food chains are not always as simple as the ones just described. Often, several different species may use the same item for food. For example, some insects feed on the leaves of willow trees, others bore into the wood, and some may feed on the roots. When food chains intersect and overlap, they make up a **food web.** (See fig. 3.13.) The complexity of a food web allows energy to be passed to several alternative organisms. This provides greater stability to the ecosystem.

## *Nutrient cycles*

Living systems contain many kinds of atoms. Some atoms are more common than others. Carbon, nitrogen, oxygen, hydrogen, and phosphorus are found in all living things and are recycled when an organism dies. Carbon and oxygen are combined to form the molecule carbon dioxide, which is a gas present in the atmosphere. Therefore, the atmosphere serves as a pool or reservoir for the carbon dioxide of an ecosystem. During photosynthesis, the carbon dioxide from the atmosphere is taken into a plant leaf and is combined with hydrogen from water molecules to form complex organic molecules. The water is absorbed from the soil by the roots and transported to the leaf. Oxygen molecules are released from the leaf into the atmosphere. These oxygen atoms are released when water molecules are split to provide hydrogen atoms for the organic molecules. In this total process, the light energy is converted to chemical-bond energy in organic molecules, such as sugars. These sugars are then used by the plant for growth.

Herbivores can use these organic products as food. When a herbivore eats a plant, it breaks down the complex organic molecules into simpler molecules. These are used for the herbivore's growth and activity. Much of the chemical-bond energy is released to the environment as heat. This process of releasing energy from food is called cellular respiration. When the herbivore is eaten by a carnivore, the complex molecules are again metabolized, energy is released, and the carnivore biomass increases. Ultimately, decomposers break down the organic molecules of dead organisms into carbon dioxide and water.

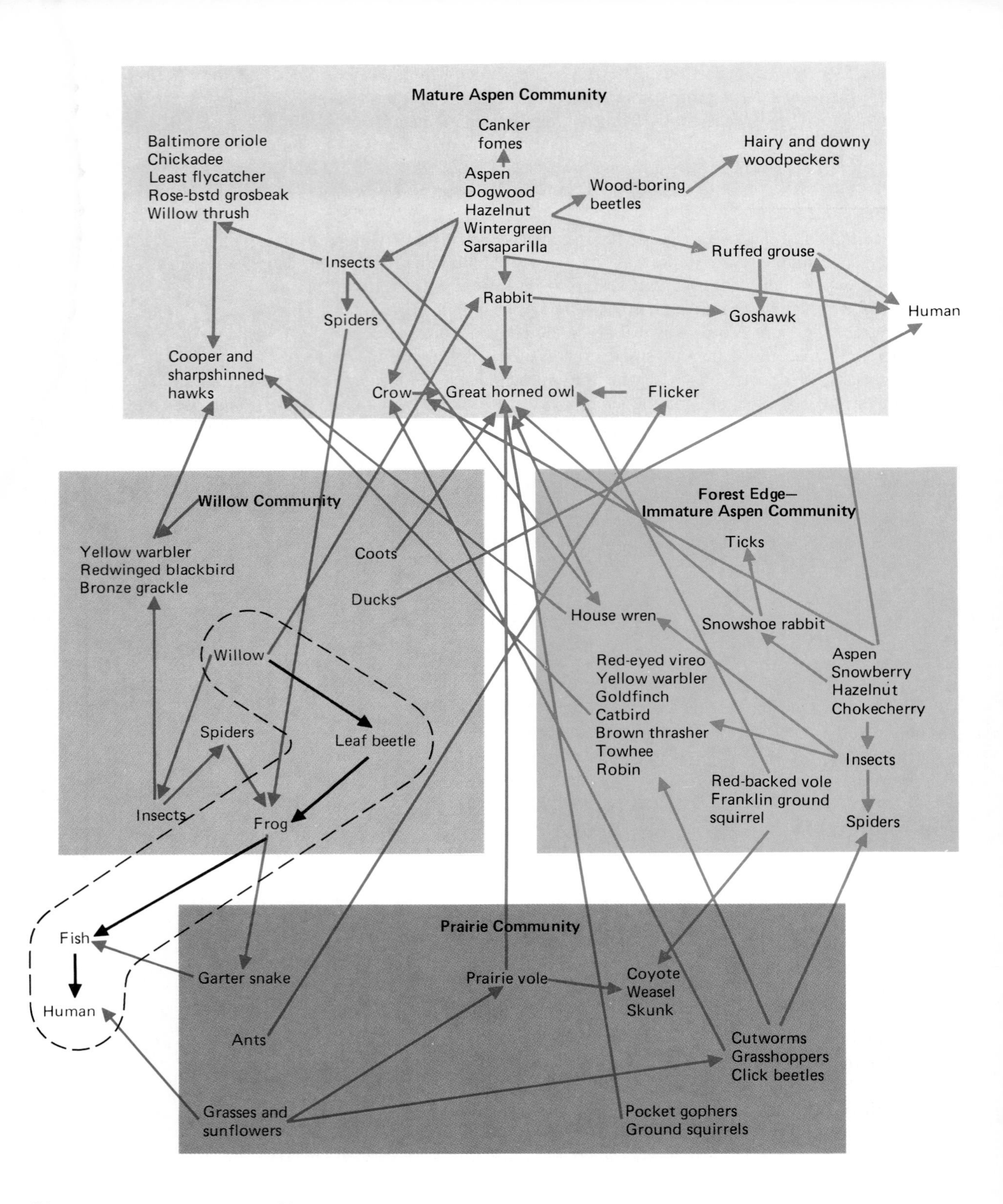

Mature Aspen Community
Canker fomes
Baltimore oriole
Chickadee
Least flycatcher
Rose-bstd grosbeak
Willow thrush
Hairy and downy woodpeckers
Aspen
Dogwood
Hazelnut
Wintergreen
Sarsaparilla
Wood-boring beetles
Insects
Ruffed grouse
Rabbit
Spiders
Goshawk
Human
Cooper and sharpshinned hawks
Crow
Great horned owl
Flicker
Willow Community
Yellow warbler
Redwinged blackbird
Bronze grackle
Coots
Ducks
Willow
Spiders
Leaf beetle
Insects
Frog
Forest Edge—Immature Aspen Community
Ticks
House wren
Snowshoe rabbit
Red-eyed vireo
Yellow warbler
Goldfinch
Catbird
Brown thrasher
Towhee
Robin
Aspen
Snowberry
Hazelnut
Chokecherry
Insects
Red-backed vole
Franklin ground squirrel
Spiders
Fish
Human
Prairie Community
Garter snake
Prairie vole
Coyote
Weasel
Skunk
Ants
Cutworms
Grasshoppers
Click beetles
Grasses and sunflowers
Pocket gophers
Ground squirrels

**Figure 3.13 Food web.** The many kinds of interactions between organisms in an ecosystem constitute a food web. In this network of interactions, several organisms may be affected if one key organism were reduced in number. Look at the rabbit in the mature aspen community and note how many organisms use it as food. (Adapted from Ralph D. Bird, "Biotic Communities of the Aspen Parkland of Central Canada." *Ecology* 11 (April 1930): 410.)

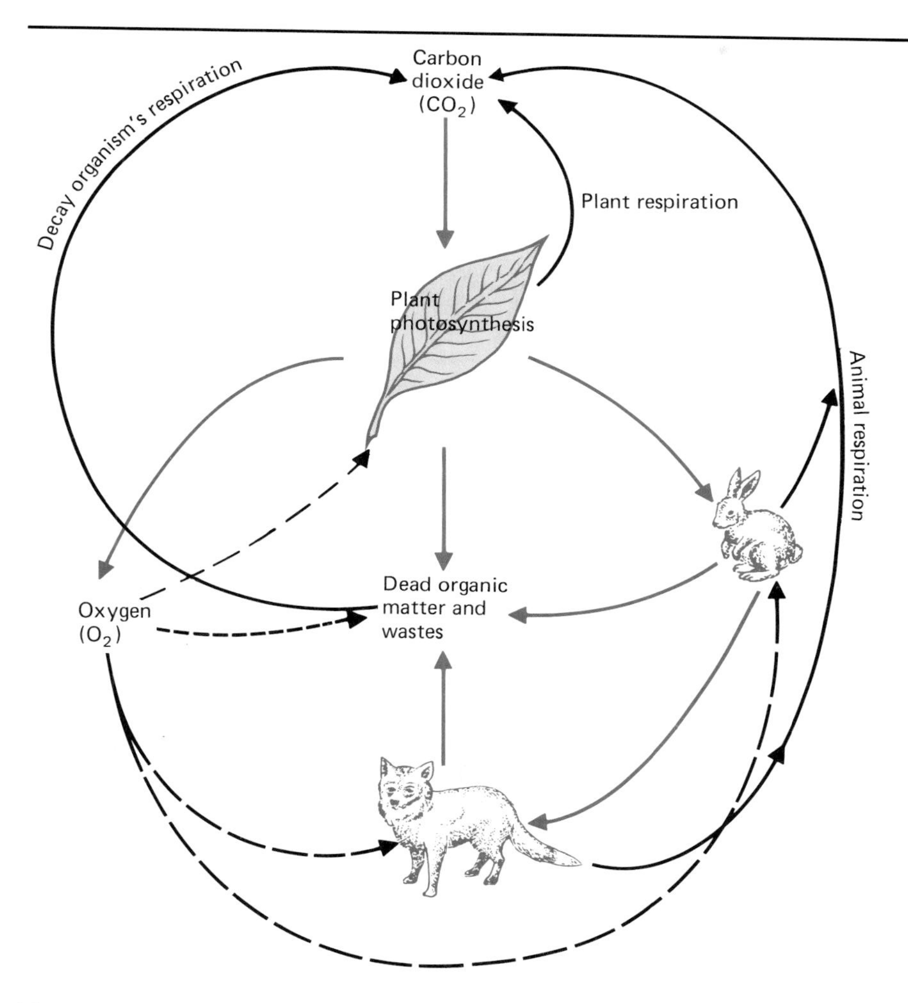


**Figure 3.14 Carbon cycle.** Carbon atoms are cycled through ecosystems. Plants can incorporate carbon atoms from carbon dioxide into organic molecules when they carry on photosynthesis. These carbon-containing organic molecules are passed to animals when they eat plants or other animals. Organic wastes or dead organisms are consumed by decay organisms. All organisms (plants, animals, and decomposers) return carbon atoms to the atmosphere when they carry on respiration. Carbon atoms are being cycled at the same time that oxygen atoms are being cycled.

The carbon atoms that were a component of atmospheric carbon dioxide passed through a series of organisms as organic compounds, such as sugars, fats, and proteins, and were finally returned to the atmosphere as carbon dioxide. This is called the **carbon cycle.** (See fig. 3.14.)

Another very important cycle, the **nitrogen cycle,** involves the flow of nitrogen. Nitrogen is a very common gas (78 percent) in the atmosphere. Plants use nitrogen if it is in the soil in the form of nitrate ($NO_3$). The amount of nitrate available to plants is an important factor that controls their growth because nitrogen is a necessary component of proteins and other plant ma-

**Figure 3.15 Nitrogen cycle.** Nitrogen atoms are cycled in ecosystems. Atmospheric nitrogen is converted by nitrogen-fixing bacteria to a form that plants can use to make proteins and other compounds. Proteins are passed to other organisms when one organism is eaten by another. Dead organisms and their waste products are acted upon by decay organisms to form ammonia, which may be reused by plants or converted to other nitrogen compounds by bacteria.

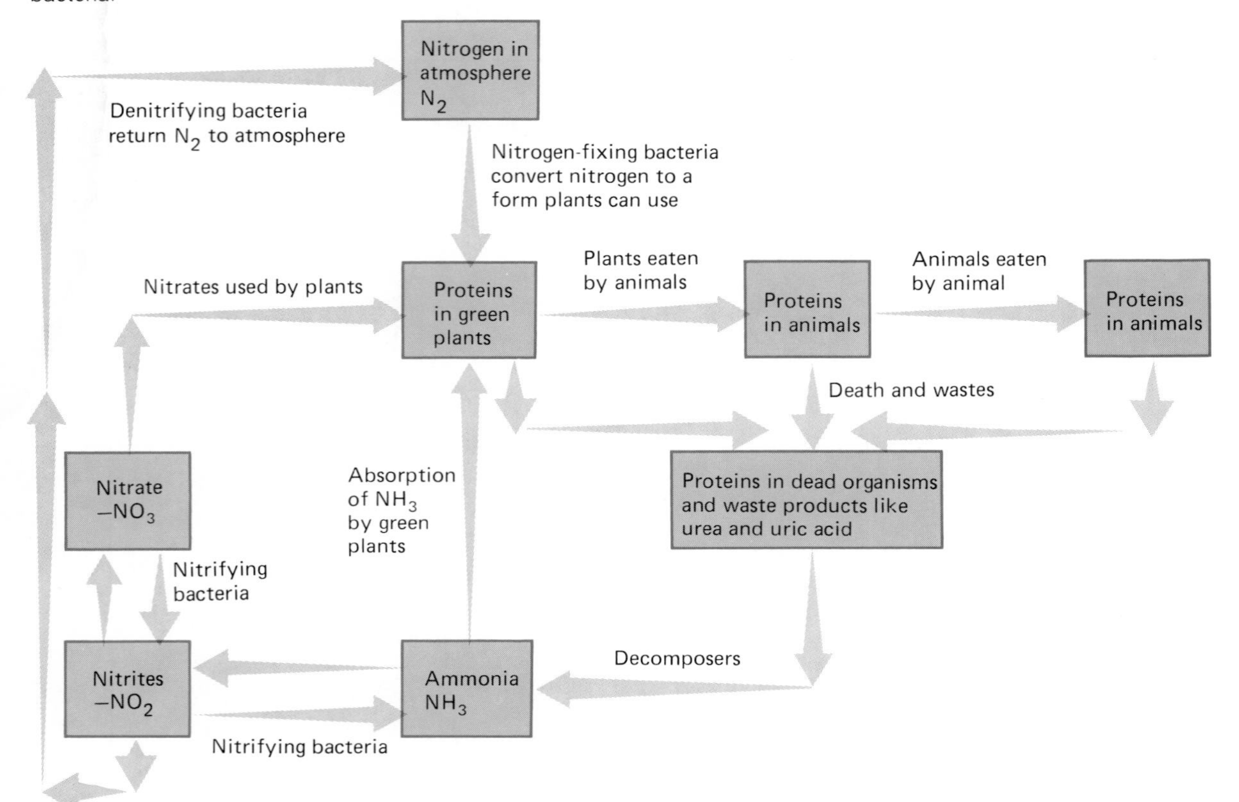


terial. There are several ways in which plants can get usable nitrogen compounds. Lightning can convert nitrogen gas to usable forms when the nitrogen and oxygen in the air are caused to react with one another.

Symbiotic bacteria in the roots of some plants can also convert nitrogen gas into a form that the host plant can use to make organic nitrogen compounds. These bacteria are called **symbiotic nitrogen-fixing bacteria.** This symbiotic relationship is an example of mutualism. Only certain plants, particularly those known as the legumes (peas, beans, and clover), can form this kind of relationship with the bacteria. Some grasses and conifers have a similar relationship with certain fungi, which seems to increase the amount of nitrogen available for growth.

Another way in which plants get usable nitrogen compounds involves certain free-living nitrogen-fixing bacteria. These bacteria are not associated with roots of plants but do convert atmospheric nitrogen to forms that plants can use.

Still another source of nitrogen for plants is the breakdown of organic compounds by a series of different bacteria. Each kind of bacterium completes one

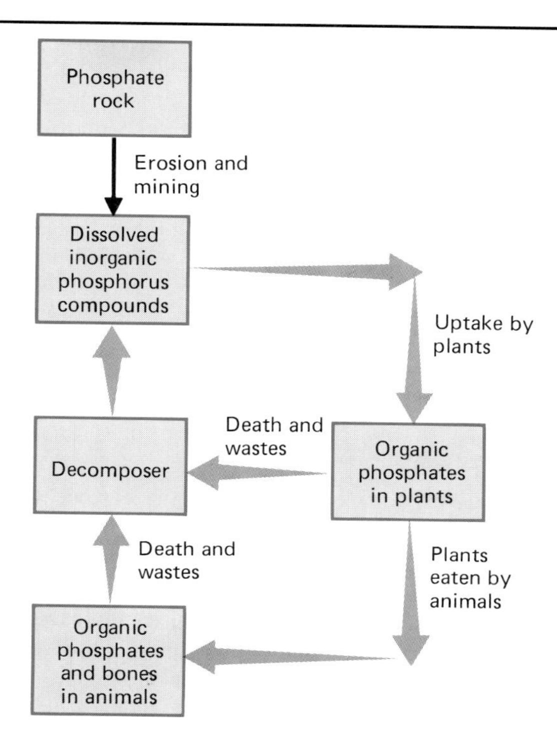


**Figure 3.16**
**Phosphorus cycle.**
The source of phosphorus is rock that, when dissolved, provides a source of phosphate to be used by plants and animals.

step in a series that leads from dead organic matter to nitrate in the soil. Some of the nitrate in soil is converted to nitrogen gas by **denitrifying bacteria.** (See fig. 3.15.)

Nitrogen is often in short supply in soil and limits the rate of plant growth. To increase yields, farmers increase soil nitrogen in a number of ways. In one year they may plant crops that have symbiotic nitrogen-fixing bacteria, and the next year they may plant a nitrogen-demanding crop in the same field. Rotating the crops in this way helps to maintain nitrogen levels in the soil. The farmer can also add nitrogen directly to the soil by spreading manure on the field and relying on the soil bacteria to convert the nitrogen in the manure into a usable form. Also, industrially produced fertilizers may be added. Some of these fertilizers still rely on soil bacteria to make the nitrogen available to plants.

Phosphorus is another kind of atom that is common in the structure of living things. It is present in many important biological molecules such as DNA and the membrane structures of cells. In addition, the bones and teeth of animals contain significant quantities of phosphorus. The source of phosphorus is rock. In nature, new phosphorus compounds are released by the erosion of rock. The plants use the dissolved phosphates to construct the molecules they need. Animals consume plants or other animals and in that way obtain phosphate compounds. When an organism dies or excretes waste products, decomposer organisms recycle the phosphates back into dissolved compounds. (See fig. 3.16.)

Fertilizer usually contains nitrogen, phosphorus, and potassium compounds. The numbers on a fertilizer bag will tell you the percentage of nitrogen, phosphorus, and potassium in the fertilizer. For example, 23–19–17 fertilizer contains 23 percent nitrogen, 19 percent phosphorus, and 17 percent potassium. In addition to carbon, nitrogen, and phosphorus, potassium and other elements are cycled within ecosystems. In an agricultural ecosystem, these elements are removed when the crop is harvested. Therefore, the farmer must return these elements to the soil as fertilizer. In a natural ecosystem, energy flows through the producers, consumers, and decomposer organisms, and the nutrients are recycled. Let us look at a simple ecosystem and analyze its components, its energetic relationships, and the interactions of the organisms that inhabit it.

## Analysis of an ecosystem

A small pond is a relatively simple ecosystem to analyze because it has definite boundaries. The abiotic features of this ecosystem include water, mineral or organic nutrients dissolved in the water or as sediments on the bottom, temperature, light, and dissolved gases.

Although water ($H_2O$) has oxygen as part of its molecular structure, this oxygen is not available to organisms. The oxygen that they need is molecular oxygen ($O_2$), which enters water from the air or as a result of photosynthesis by aquatic plants. Wave action or water flowing over waterfalls tends to mix air and water and allow more oxygen to dissolve in the water.

Dissolved oxygen in the water is an important factor since the quantity determines the kinds of organisms that can inhabit the pond. Organic materials in the pond relate to the disposal of the metabolic wastes of organisms that live in and around the pond. The breakdown of organic waste uses oxygen, and the amount of oxygen required is termed the **biochemical oxygen demand** (B.O.D.). The amount and kinds of organic wastes determine in part how much oxygen is available for organisms. Climatic features, such as amount and intensity of sunlight, average air temperature, depth of the water, geological origin of the underlying rock, latitude, and altitude, also help to determine the characteristics of the pond ecosystem.

Most likely, the producer organisms are either **phytoplankton,** which are free-floating microscopic, chlorophyll-containing organisms, or larger floating plants that often wash up on the shore of the pond. Some of these are duckweed, *Chara,* and *Elodea.* Others may be emergent forms that are rooted on the bottom of the pond but have portions that extend into the air, such as cattails, bullrushes, arrowheads, and water lilies. All animal life is dependent upon these producers for energy and nutrients.

The primary consumers of the pond are the small, swimming, microscopic animals (**zooplankton**) that strain minute phytoplankton from the water, and the larger animals that live on the vegetation. These organisms include pond snails, larval insects, other arthropods, worms, carp, tadpoles, and other similar organisms that use green plants directly for food. Secondary consumers,

## Box 3.1
## Polybrominated biphenyl (PBB)

PBB is a complex organic chemical containing bromine, which was developed as a fire retardant. It is a very stable compound with no known decomposers. This is a very important characteristic for a fire retardant, but it is also a biological problem because the compound is not biodegradable and in large quantities is toxic to living things. Because of an error at a small chemical plant in St. Louis, Michigan, a quantity of PBB was packaged and mislabeled as a cattle feed additive in March 1973. Subsequently, the mislabeled packages were distributed to feed stores where the PBB was mixed with other ingredients and sold to farmers. Two months later, some farmers began to notice reductions in milk production and symptoms among their cattle that they could not diagnose. As a result of detailed chemical analysis, it was finally proven that PBB was present in the bodies of the affected cattle.

Because time elapsed between the feed mixup and the identification of the PBB as the problem, the PBB had spread throughout the state of Michigan. People picked up PBB rather quickly because of the direct consumption of milk and meat from affected cattle. Other organisms also became contaminated by eating the tainted feed or other affected animals. PBB spread throughout the food chain.

Thousands of cattle and other domesticated animals were declared unfit for human consumption by the Michigan Department of Agriculture and were destroyed. Disposal of the affected carcasses was a serious problem. The affected carcasses had to be kept isolated from scavengers and other animals that could spread the PBB further. Since the compound was so stable, natural decomposition would not solve the problem. Incineration was considered but rejected because of the magnitude of the material needed to be burned. Therefore, it was decided to bury the carcasses in a secured landfill. This landfill was constructed with a thick layer of clay liner, and the carcasses were covered with a similar layer to prevent water from transporting the PBB from the landfill to the groundwater and surface water in the area. An intense legal and political battle surrounded the use of the designated site. The local population wanted nothing to do with the landfill and lobbied intensely in the state capitol.

This unfortunate incident cost the state of Michigan more than two million dollars. Some of the cost involved the investigation, legal maneuvering, and construction of the disposal site. In addition, many farmers had their herds condemned, so they lost their source of income. Even farmers who had clean herds were affected because the public was worried about the purity of Michigan meat and milk. Supermarkets began to advertise "western corn-fed beef" to make it clear to the buyers that the beef was not a Michigan product. Since tourism is a major industry in the state of Michigan, the national attention given to the PBB problem deterred people from vacationing there.

Although the amount of space given to the problem by the media has dwindled, the problem has not been solved. PBB is still present in the ecosystem and will remain for many years.

which are parasites or predators, are represented by a different set of arthropods, sunfish, minnows, frogs, salamanders, turtles, and snakes. Bass, pike, and other fish may represent higher-level consumers in the pond.

At each of these trophic levels there is a loss of about 90 percent of the useful energy.Therefore, do not expect many large predatory fish in such a pond ecosystem. The greatest biomass in the pond will be the plants and phytoplankton, which are the lowest trophic level.

The decomposer function in the pond is filled by bacteria and fungi, which degrade dead organisms and waste products to simpler compounds. The carbon dioxide and nitrogenous compounds they produce are in turn reused by the plants and phytoplankton to make more producer biomass. The next step in this analysis would be to characterize the specific role played by each kind of organism in the pond so that food webs, nutrient cycles, and energy flow may be quantified.

## Consider This Case Study
### Meadowlands, New Jersey

In 1972, an environmental impact study was begun prior to the construction of a sports complex in an area known as the Meadowlands in New Jersey. At that time, mercury contamination was discovered in a marshy part of this area. The mercury was probably dumped into the Meadowlands over a period of thirty-five years by mercury-processing plants formerly located there. A subsequent study has shown mercury vapor in the air, high levels of inorganic mercury compounds in the soil, and silt sediments in the marsh. The inorganic mercury is transformed by microorganisms into methyl-mercury, a more hazardous form. The methyl-mercury is absorbed by algae and other submerged vegetation. Fish and shellfish in the area also contain high levels of mercury contamination. The companies responsible for the mercury discharge contend they should not be liable, since they obeyed the law when they were in operation. The New Jersey State Department of Environmental Protection reports that the substantial amounts of mercury in the environment pose unacceptable risks to humans. They consider the site to be "an ecological time bomb." Area fish and wildlife exceed the 0.5 parts per million (ppm) mercury level considered tolerable for human consumption. Parts per million is a method of expressing very small quantities of a material found within a mixture. For example, if a fish weighs 1,000 grams (1 kg) and contains 1 milligram of mercury, the relative amount of mercury could be expressed as 1 part per million. Mercury in its methylated form is a deadly poison to the unborn, and it has been shown to cause neurological and reproductive damage to adults.

Describe as much of the food web associated with mercury contamination as you can.

Do you think the state should be responsible for the cleanup? How should the state pay for it?

If you lived in this area and had fished the creeks for years, what would your reaction to this problem be? What is the responsibility to the homeowners in this area? How will the discovery of mercury affect property values?

## Summary

Everything that affects an organism during its lifetime is collectively known as its environment. The environment of an organism can be divided into biotic (living) and abiotic (physical) components.

The area where an organism lives is called its habitat. The job an organism does, how it survives in its habitat, is called its niche. Populations have evolved so that they function well in a particular niche.

Organisms interact within a habitat in a number of ways. The predator is an organism that eats another organism (prey). A parasite lives in or on another organism from which it derives nourishment. When two organisms require the same limited resources, competition will occur. Mutualism occurs between organisms of different species in which both organisms derive benefit from interaction.

A community is a set of interacting populations of organisms. All of the organisms and physical factors of the environment are known collectively as an ecosystem. In an ecosystem, energy flows from producers through various trophic levels of consumers (herbivores, carnivores, omnivores, and decomposers). The sequence of organisms through which energy flows as food is known as a food chain. Several interconnecting food chains make up a food web. The complexity of the food web leads to its stability.

The flow of atoms through an ecosystem involves all organisms in the community. Examples of how the environment and organisms interact are demonstrated in the carbon cycle and the nitrogen cycle.

## Review Questions

1. How is an ecosystem different from a community?
2. Describe the niche of a human. Be as detailed as you can.
3. Make a list of five predators and their prey organisms.
4. When humans raise cattle for food, what type of relationship is this?
5. Define environment.
6. What are some different trophic levels in an ecosystem?
7. Give an example of an organism that is a herbivore, one that is a carnivore, and one that is an omnivore.
8. Describe the carbon cycle.
9. What organisms are involved in the nitrogen cycle? What does each of these organisms do in the cycle?
10. Analyze an aquarium as an ecosystem. Identify the major abiotic factors that influence the animals and plants living in the aquarium.

## Suggested Readings

Boughey, Arthur. *Ecology of Populations.* New York: Macmillan Publishing Co., 1973. A small book that deals well with the narrowed subject of the ecology of populations. Helpful definitions and a good bibliography.

Carson, Rachel. *Silent Spring.* New York: Fawcett Crest, 1966. A classic that is partly responsible for much of the environmental activism in the last twenty years.

Enger, E., et al., *Concepts in Biology.* 4th ed. Dubuque, Ia.: Wm. C. Brown Publishers, 1985. Easy-to-read biology text with an introduction to the concepts of natural selection and ecology.

Eshman, Robert. "The Junglei Canal: A Ditch too Big?" *Environment* 25: 16–20. Gives some insight into environmental concerns raised by the construction of a 600 kilometer long canal to divert more water into the upper region of the Nile River.

Odum, Eugene P. *Fundamentals of Ecology.* 3d ed. Philadelphia: W. B. Saunders Company, 1971. A classic text that provides an introduction to many concepts of ecological study. A must for the library of a serious student of environmental science.

Walton, Susan. "Egypt after the Aswan Dam." *Environment* 23: 31–36. Reviews the changes in an ecosystem due to the construction of the Aswan Dam.

Whittaker, Robert H. *Communities and Ecosystems.* 2d ed. London: Macmillan Publishing Co., 1975. Good background material provided in this textbook.

## Chapter Outline

I. The buffalo
II. The niche of the buffalo
  A. Prairie grazers
  B. The buffalo year
  C. Predation and competition
III. Native Americans and the buffalo
IV. European impact on the buffalo
  A. Pressure to settle the West
  B. Farms and buffalo
    Box 4.1 Barbed wire
V. Extinction
    Box 4.2 Endangered species
VI. Consider this case study: Serengeti National Park

## Objectives

List the many elements that make up an animal's niche.

Explain the profound impact humans can have on other organisms.

Describe the implications of extinction.

Explain how two different human populations (native American and European American) had different impacts on the buffalo.

List the many interlocking events that led to the buffalo's near extinction.

## Key Terms

dominant species

extinction

gradual extinction

rutting season

# Chapter 4 Competition for a niche: Humans vs. buffalo

## The buffalo

Ecologists can study the ecological relationships of an area from two different points of reference. Chapter 3 emphasized the major types of interaction by discussing how matter and energy flows through ecosystems. When studying an ecosystem in this way, it is easy to overlook the thousands of kinds of interactions and the intricate network of relationships that support an ecosystem. This chapter will look at interrelationships from a second point of view. It will use a single organism—the American bison—to develop the niche concept and will also discuss why the American bison nearly became extinct.

**Figure 4.1 Distribution of the plains buffalo.** Although buffalo were probably found throughout the North American continent at one time, they were probably most common in central North America where the grasslands were located. The area shown on the map gives the approximate area where the buffalo were most common.

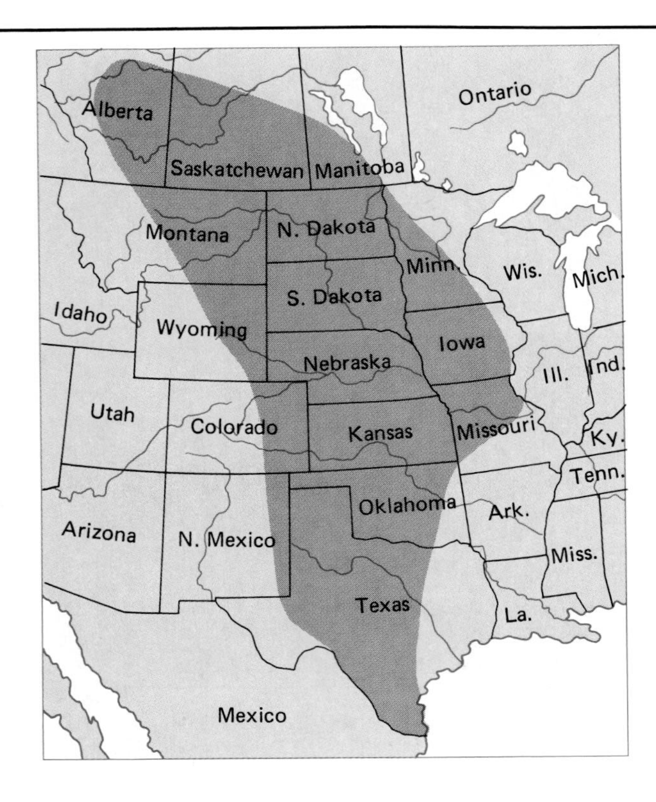


The American bison, commonly called the American buffalo, has the species name *Bison bison.* It is an animal unique to the North American continent and was originally found throughout the United States and Canada wherever grass was abundant. Since the buffalo is a grazing animal, it is dependent on grass as a source of food. Therefore, it was probably always most common in the central United States and in the prairie provinces of Canada. (See fig. 4.1.) When people became aware that prairie soil was excellent for raising crops, the stage was set for a confrontation between people and buffalo. Both people and buffalo require the same basic resource, the prairie soil, to supply them with food. Therefore, intense competition was inevitable.

The male buffalo's appearance is rather formidable. Its large front end has a hump that is accentuated by a mane of shaggy hair. A huge head with a beard hanging from the neck makes the front end of the animal appear even larger. The hind end seems quite small by comparison and ends with a short tail. The buffalo is a large animal. Full-grown bulls stand up to two meters tall at the hump and weigh over a metric ton. (See fig. 4.2.) The cows are about half that size and have a smaller hump, mane, and beard. Because of its size and population, the role the buffalo played in its environment was a

**Figure 4.2 The American buffalo.** A large bull might stand two meters tall at the shoulder and weigh a metric ton. The large head and heavy coat of fur on the front end of a buffalo make the animal appear to be even larger than it really is.

dominant one. A **dominant species** is a key organism in food chains and has a significant impact on the area in which it lives. There is no question that the buffalo was one of the most important animals in the plains and prairies of the West before it was settled.

## The niche of the buffalo

As you will recall from chapter 3, an ecological niche of an organism is the role that an organism plays in its environment. A study of the ecological niche of the buffalo shows the variety of impacts herds of buffalo had on the other organisms that shared their prairie habitat.

### *Prairie grazers*

The original prairie consisted of vast expanses of treeless country. Limited rainfall and frequent fires prohibited the establishment of forests except for narrow bands of trees along the rivers and streams that flowed through the area from the Rocky Mountains eastward to the Mississippi River. The natural plant cover was made up primarily of different species of grasses. In most

**Figure 4.3 The prairie.** The prairie was a grassland that stretched for hundreds of square kilometers. The grasses provided a good source of food for many kinds of grazing animals. The buffalo was the dominant grazing animal in this area before cattle were introduced.

areas, the grasses made up well over 90 percent of the plant cover, but interspersed with the grasses were a variety of broad-leafed plants. The eastern prairie received more rainfall and had stands of a grass called big blue-stem, which were often over two meters high. The central prairie areas were dominated by different species of shorter grasses. The western prairie, which received the least rainfall, had species of grasses that only reached a height of 10 to 40 centimeters.

Constant winds caused cold, severe winters and contributed to the drying of the prairie during the hot summers. This wind created ideal conditions for prairie fires, which swept across the land. Fires, in turn, helped to prohibit the establishment of trees. Recently burned areas with new green grass shoots were often preferred grazing areas for buffalo.

Buffalo are extremely well adapted to live on the prairie. (See fig. 4.3.) Herds consisting of thousands of animals once roamed over the prairie eating the grasses and other plants. Cowbirds accompanied the herds and ate the insects stirred up by the buffalo. Cowbirds could also be seen walking around on the buffalo's backs removing insects and ticks from their hides. In their movements, the buffalo would often frighten meadowlarks, prairie chickens, and other birds from their nests. Occasionally, nests were trampled.

If water was close by, the herd made a trip every day to a creek or stream. If it was some distance, they might wait two or three days before going to water. Since the herds moved as their needs for grass and water dictated, they frequently moved as a relatively solid mass through narrow valleys or across streams. Once they were back in open areas, the herd again dispersed.

Usually the buffalo were not evenly distributed but were in small groups within a larger herd. (See fig. 4.4.) Many of these groups probably consisted of cows and their offspring. Some years, because of drought or unusually large populations of grasshoppers, which denuded large areas of vegetation, the herds of buffalo were required to travel longer distances to get adequate food and water, but they did not migrate in any organized way.

**Figure 4.4 A herd of buffalo.**
The distribution of buffalo shown in this painting was typical of the large herds of buffalo that once roamed the prairies. Contrary to popular belief, they probably did not form solid walls of animals.

## *The buffalo year*

The buffalo shed their heavy winter coats during the early summer. As this shedding takes place, the buffalo use any convenient object to rub against, for example, trees or rocks. Many trees were destroyed, injured, or weakened and rocky outcrops or solitary rocks often had depressions around them because generations of buffalo used them as convenient rubbing and scratching sites. During the summer, when the buffalo have the least amount of fur covering, insects are quite bothersome. Also, the fruits of "needle and thread" grass often burrowed their way into the hides of buffalo and caused irritation. To ease their discomfort, the buffalo dug into moist earth and rolled in the depressions thus created, coating their hides with mud or dust. These depressions are called buffalo wallows. In the western prairie, buffalo wallows often started in prairie dog towns where the prairie dogs had cut the vegetation so short that the soil was exposed. Because the wallows were used year after year, they became permanent parts of the landscape.

The **rutting season** (breeding) occurs in the late summer and early fall. During this time, the bulls fight each other for the breeding rights to the females in the herd. Most of the offspring are probably sired by a few dominant bulls. The buffalo are in prime condition during the fall. Their winter coats return, and the animals are sleek and fat after spending the summer grazing on the abundant grass supply. During severe winter weather, buffalo turn their heads into the wind, and the heavy coats on their forequarters protect them from the cold and snow. During the following spring and early summer, reddish-colored calves are born. By midsummer, the calves are no longer nursing, although they continue to follow their mothers for a year or more. Calves are not able to defend themselves against predators; therefore, during the calving season, it is likely that wolves and other predators are able to kill many calves.

**Figure 4.5 The buffalo niche.**
Any organism's niche is a very complex concept. All of the interactions between organisms such as the native American, cowbird, grasshopper, wolves, grasses, prairie dogs, and buffalo are a part of a niche. In addition, the soil, climate, and geography contribute to the total definition of a niche.

## *Predation and competition*

The only natural predators of any significance are wolves and humans. Wherever there are buffalo, there are wolves. In most cases, the wolves simply follow the herd, picking up the less-fit organisms, such as injured, diseased, newborn, old, or otherwise weakened buffalo. At one time, humans armed with simple spears and perhaps aided by prairie fires also killed some buffalo. However, most of the buffalo probably died from other causes. During stampedes, which were relatively common, individual buffalo often fell and were trampled or crippled. The same prairie dog town that provided a dust wallow presented a hazard because running buffalo could be crippled if they stepped in prairie dog holes. Badger holes and anthills provided similar hazards. A significant number of animals were killed or injured by prairie fires. During the winter, many buffalo died during blizzards or were drowned when they fell through the ice of lakes and rivers.

Through their own activities, the buffalo helped to maintain the original prairie. The continuous grazing helped to eliminate trees and encouraged the grasses, which could grow back rapidly to replace what the buffalo had removed. The number of buffalo was controlled primarily by a few predators and by the physical conditions that maintained the prairie. The only serious competitor was the grasshopper, which fed on the same food source. Grasshoppers were only a problem during those years when their population was extremely high. Deer and pronghorn antelope did overlap with the buffalo in some of its range, but these two species are interested primarily in shrubs and twigs as food rather than grass.

It is possible to list nearly a hundred distinct items—wind, fire, grasshoppers, ticks, grass, rocks, and trees—that are a part of the niche of the buffalo. This discussion represents only a partial description of what the niche of the buffalo was before they came into extensive contact with technologically advanced humans. (See fig. 4.5.)

## Native Americans and the buffalo

The first contact between humans and buffalo probably occurred several thousand years ago when several species of *Bison* and the first humans came to North America from Asia. By the time Europeans rediscovered the Americas, several tribes of native Americans had developed rather elaborate cultures based on the buffalo as a main source of food, shelter, clothing, tools, and fuel. A variety of methods were used to hunt buffalo, including stampeding the buffalo over bluffs, approaching them disguised as wolves, and surrounding small bands of buffalo. By the 1600s, some native Americans of the plains were using horses (See fig. 4.6.) These horses were the descendants of horses brought to the Americas by early Spanish explorers. The native Americans of the plains developed much of their culture around the use of the horse to help capture the buffalo they needed. The Crow, Blackfoot, Kiowa, Sioux, Cheyenne, Comanche, and others were very proud and independent people who resented the invasion of others into their traditional hunting lands. Since the horse had become an important part of their culture, the number of horses possessed was a measure of wealth and prestige. There were frequent raids between tribes to obtain more horses.

By the early 1800s, it was estimated that about forty million buffalo roamed the plains and prairies of the central portion of North America. This population supported approximately 15 percent of the one million native Americans on the continent. At this point in the history of the buffalo, the relationship

**Figure 4.6 Native Americans and the horse.** The reintroduction of the horse to the Americas by the Spanish in the 1600s provided the native Americans of the prairie with an important new cultural element. Horses made it possible to more effectively locate and kill the buffalo they needed for food.

between the native Americans and the buffalo was a simple predator-prey relationship. The horse, lance, bow and arrow, and other technological advances certainly made these people the primary predator, far surpassing the impact that wolves and other carnivorous mammals had on the buffalo population. On the other hand, it was the buffalo that allowed these tribes to sustain their relatively large populations.

Wintertime was a time of hardship for the native Americans, just as it was for the buffalo. Food was scarce, disease was common, and the cold winter weather also contributed to keeping the native American population stable. Therefore, you would not expect the native American population to have had any significant impact on the numbers of buffalo. From what we know of predator-prey relationships, we would expect that if the number of buffalo were reduced, the number of native Americans would be reduced as well. Therefore, it is quite probable that the buffalo and the native American populations were in balance with one another before the intrusion of Europeans into the North American continent.

## European impact on the buffalo

In the early 1600s, North America was colonized by Europeans who became established on the eastern part of the continent. The prairies and plains to the west were considered a wasteland or desert by the people who had come from European farms and cities. They feared the treeless dry areas and saw little use for them. Travel through the area was made more hazardous by the native Americans, who resented the intrusion of the "white man" into their domain.

**Figure 4.7 Buffalo skinners.**
The hide hunters killed thousands of buffalo for their skins alone. By 1884, the buffalo hide business was gone because the buffalo were gone.

However, by the early 1800s European Americans were beginning to kill a significant number of buffalo for meat, for hides, and just for sport. This killing became a business by the middle of the 1800s, and by 1888 the buffalo hunters had virtually eliminated the buffalo. (See fig. 4.7.) In 1895 there were less than one thousand buffalo left.

Most people, when they listen to this story of purposeful destruction of the buffalo, are appalled and cannot understand why it was allowed to continue when it was obvious that the buffalo were being driven to extinction. The reasons behind this behavior are rather complex but had a single cause. If the prairies and plains were to be used by the European immigrants, the buffalo had to go.

## *Pressure to settle the West*

Since most European Americans had originally thought of the plains and prairies as a desert wasteland, why did they settle in this area? First of all, there were several governmental actions that encouraged them to settle in the plains. The federal government was interested in developing a rail system from the urban East to the mineral-rich West. To do this, large tracts of land were given to railroad companies to encourage building the lines. Once the railroads were built, the railroad companies encouraged people to ride them, for rather obvious reasons. The federal government also granted cheap land in the West to people such as Civil War veterans. In addition, territorial governments encouraged people to settle in their areas because once the territorial population reached a certain size it could become a state.

There were also several social reasons for settling the prairie. As families grew, it was no longer possible for all of the sons to continue to live on the family farm. Since eastern cities could not absorb the emigration from the farms, an appealing choice for young people was to move to the frontier where life was hard, but you could be your own boss. Furthermore, many farms in the East had begun to fail due to destructive farming practices. All of these social pressures were at least partially released by the safety valve of the West.

Obviously, money was also involved. Many land speculators were interested in selling land to anyone willing to buy it. Some of these land deals were financed by Europeans who encouraged their countrymen to immigrate to the United States with the enticement of cheap land. However, not all of these land dealers were totally honest.

## *Farms and buffalo*

Farming and grazing on the prairies could not occur while the buffalo still roamed. (See box 4.1.) Buffalo were just too destructive. They knocked over fences and telegraph poles and even destroyed sod houses by using them as rubbing sites. Since buffalo and cattle competed for the same grass, a reduction in the number of buffalo allowed more cattle to be raised in the area.

Converting the prairie to farmland also took away the buffalo's source of food. The native Americans of this region objected to the slaughter of their food source and retaliated by killing some of the settlers who pushed into traditional hunting areas. The United States government, in an effort to protect the settlers, went to war with the various tribes that were causing problems. Following the Civil War, the policy of the government was to exterminate the buffalo since it was the food source for the native Americans who were causing problems. The secretary of the interior in 1874 stated that ". . . the civilization of the Indian was impossible while the buffalo remained on the plains." As a further indication of this line of thought, Colonel Dodge was quoted as saying "kill every buffalo you can, every buffalo dead is an Indian gone."

Although concern about the extermination of the buffalo had been raised as early as 1870, nothing was done until 1905. In that year, sixteen men formed the American Bison Society and began to rebuild the buffalo herd to some of its former glory.

## Box 4.1
### Barbed wire

Farming the plains and prairies was inhibited by the lack of trees to build farm buildings and fences. The sod house took care of the buildings, and barbed wire took care of the fences. Fences were necessary to the farmer because they kept the farm animals contained and prevented wild animals from ruining the crops. In the eastern United States, most fences were made from stones or split rails from the abundant timber. In the treeless plains, fences were originally made of smooth wire. However, smooth wire had several disadvantages. It was heavy and it stretched. Also, animals (both wild and domestic) would lean against it and break down the fence. In 1873, Joseph Glidden invented a pattern for barbed wire and also helped to invent a machine to make it. Barbed wire had the advantage of being lighter than the smooth wire and, since it had barbs on it, animals were unlikely to lean against it. Barbed wire, therefore, became the major fencing material in the West. In 1874, five tons of barbed wire were produced. By 1907, the annual production had risen to 250,000 tons. Much of this was used to carve farms out of the prairie.

Artwork adapted from *Barbs, Prongs, Points, Prickers and Stickers, A Complete and Illustrated Catalogue of Antique Barbed Wire,* by Robert T. Clifton. Copyright © 1970 University of Oklahoma Press.

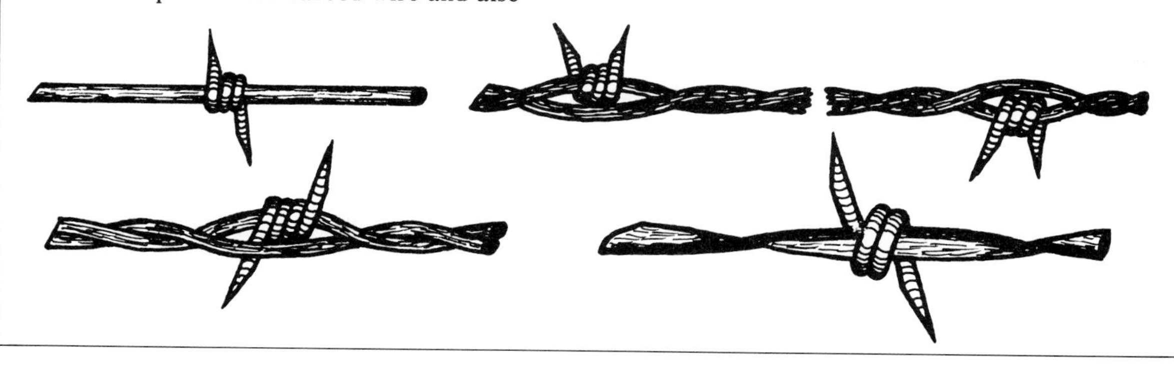

Captive herds of buffalo were maintained for years, and, eventually, in the early 1900s the federal government established the national Bison Range near Missoula, Montana. Several other sites throughout the West now have fairly sizable herds, but none approach the number prior to the settling of the prairie and plains.

Captive buffalo herds are often crossbred with cattle to produce a beefalo, with the hope that a superior animal can be produced. It is from these privately maintained herds and a few wild buffalo that our current population of approximately thirty thousand head was produced. All of these animals are protected in scattered sanctuaries, preserves, and private herds.

Regardless of the specific actions that caused the near extinction of the buffalo, the fundamental cause was the competition between the buffalo and humans for the use of the prairies and plains. The humans won.

**Figure 4.8** **Extinction.** Commercial hunting and habitat destruction are the main causes for extinction.

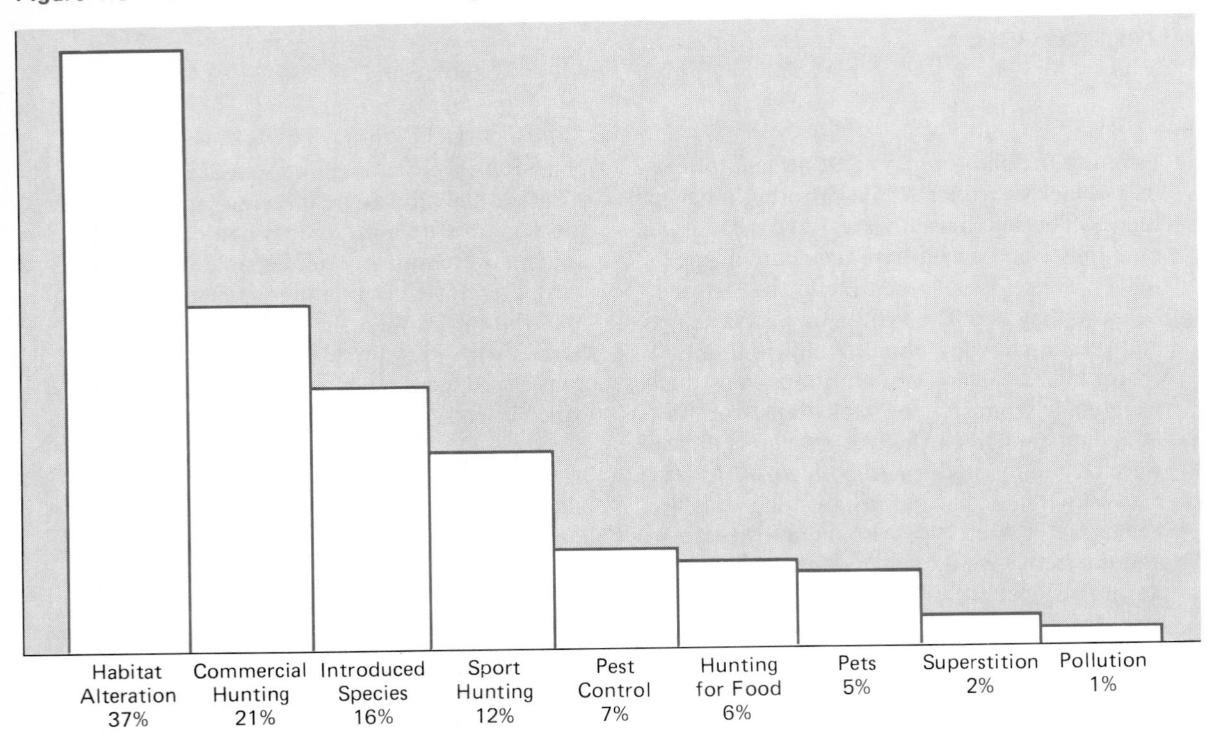


## Extinction

**Extinction** is a natural occurrence. Of the estimated 500 million species of plants and animals that have existed since life began on earth, only about 2 to 4 million are alive today. This represents about a 90 percent rate of extinction. Although biological extinction is a natural evolutionary process, it has been greatly accelerated by human activities. (See fig. 4.8.)

The International Union for the Conservation of Nature and Natural Resources (IUCN) estimates that by the year 2000 at least 500,000 species of plants and animals may be biologically exterminated, representing 13 to 25 percent of all the estimated species in the world. The IUCN classifies species in danger of extinction according to four categories: critically endangered, endangered, threatened, and rare. An endangered species is in immediate danger of extermination or extinction; a rare species is not presently in danger but is subject to risk (e.g. the Galapagos Tortoise).

The value that humans assign to a particular species plays an important role in its future. Efforts to protect species like the American Bison, elephant, polar bear, bald eagle, and blue whale are much more organized and visible than are efforts to protect rather obscure fish and plants. A case in point is the controversy that surrounded Tennessee's Tellico Dam project in 1978. The Supreme Court declared that the completion of this $116 million federal project would threaten the survival of an endangered species called the snail darter (a tiny fish that is eight centimeters long). This was a violation of the Endangered Species Act of 1973.

After a long debate between developers (who wanted to have all federally funded projects exempted from the act) and conservationists (who wanted to preserve the act as originally written), a compromise was reached. In 1978, Congress amended the act so that exemptions could be granted for federally declared major disaster areas, for national defense, or by a seven-member Endangered Species Review Committee if the committee found that the economic benefits of a project outweighed the harmful ecological effects. At their first meeting, the review committee denied the request to exempt the Tellico Dam project on the grounds that the project was economically unsound. This should have stopped the Tellico Dam Project. Nine months later, however, as a result of several political maneuvers, Congress appropriated money to complete the dam. The reservoir is now full of water. The snail darters that once dwelled there were transplanted to nearby rivers, and by 1981 they were apparently making a slow comeback.

The new amendments to the Endangered Species Act also weaken the ability of the government to add new species to the endangered and threatened lists: Before a new species can be listed, it is now necessary to determine the boundaries of its critical habitat, prepare an economic impact study, and hold public hearings—all within two years of the proposal of the listing. Some scientists have also called for an increased emphasis on evaluating the endangered status of microorganisms (such as bacteria, fungi, algae, and protozoa) and insects, earthworms, and nematodes, rather than concentrating primarily on some of the larger and aesthetically appealing animals. Invertebrates (including insects, earthworms, and nematodes) are vital for organic waste degradation, soil formation, pollination, and biological pest control.

Each individual organism is a member of a species and shows certain characteristics that are general for the species and others that are peculiar to the individual. Since a species is made up of individuals with different combinations of characteristics, some individuals are better adapted to fit the prevailing environmental conditions than others. These more successful individuals are more likely to reproduce, and, as a result, their genes are likely to become even more common in future generations.

The shift in gene frequencies as a result of different rates of reproductive success can lead to the development of new, separate, interbreeding units. These may eventually become new species if they are isolated from the main population for a long enough period of time. (See fig. 4.9.)

If some groups are better adapted than others, it follows that the less fit may eventually become extinct. This is the kind of process that has caused the extinctions of millions of species. This process is a natural consequence of environmental changes and organism interactions.

Because the environment continually changes, a species must adapt or become extinct. If this adaptation occurs over a long period, it is possible that the original species will evolve into a new species. In this situation, the original species eventually becomes extinct, but before it does, the favorable genes are passed on to the newly evolving species. This process enables the new species to be better adapted to its environment. An example of this is the dinosaur.

## Box 4.2
## Endangered species

Endangered species are organisms that are present in such small numbers that they are in immediate jeopardy of becoming extinct. This usually means that the organism will become extinct unless it has help from humans, who in most cases are also the primary cause of extinctions. Humans accelerate extinctions in two ways—hunting and habitat modification. Hunting of various species has had some impact, particularly on large mammals and birds, but the major reason for placing organisms in danger of extinction is habitat modification.

Today, the area of most concern is the modification of tropical forest ecosystems. (Most of the developed world has already caused the extinction of vulnerable species.) Tropical forests have a rich diversity of species and are presently under pressure for two reasons. Because tropical areas have very rapidly growing human populations, forest land is being cleared to provide farm and grazing land. The forests of Central America are already two-thirds gone. In the Amazon basin, about one hundred thousand square kilometers of land are cleared each year for farms or by logging companies. Logging is another pressure on tropical forests. Much of the tropical forest is being cut without regard to the long-term effects of these practices. Often the logging companies are owned by nonresident corporations that have little commitment to maintaining a continuing forest-products industry. The result of these actions will be a loss of species as a consequence of habitat destruction.

The United States Endangered Species Act was passed in 1973. This gave the federal government jurisdiction over any species that were designated as endangered. Nearly three hundred United States species and subspecies have been so designated by the Office of Endangered Species of the Department of the Interior. Among those listed are the eastern cougar, Florida panther, black-footed ferret, Sonoran pronghorn, whooping crane, bald eagle, Hawaiian crow, American alligator, Houston toad, snail darter, Warm Springs pupfish, Tan Riffle shell clam, Mission blue butterfly, Eureka dune grass, and the Furbish lousewort. In addition, nearly five hundred foreign or oceanic species, such as rhinoceros and whales, are considered endangered. The Endangered Species Act directs that no activity by a governmental agency should lead to the extinction of an endangered species and that all governmental agencies must use whatever measures are necessary to preserve these species.

As in other parts of the world, the key to preventing extinctions is preservation of the habitat for the endangered organisms. Consequently, many governmental agencies and private organizations have purchased such sensitive habitat or have managed areas to preserve suitable habitat for endangered species. Setting aside certain land areas or bodies of water forces government and private enterprise to confront one another when an animal or plant becomes the focus of a land use decision. The question becomes one of assigning a value to an endangered species. Several confrontations of this type have already occurred, and more will occur in the future.

International efforts have not been as fruitful. Although many countries have tried to preserve sensitive habitats, the pressure for food and the economic inability to adequately police the parks and preserves have left open the question of survival of endangered species.

Upper left, black-footed ferret; upper right, whooping crane; lower left, Mission blue butterfly; lower right, bald eagle.

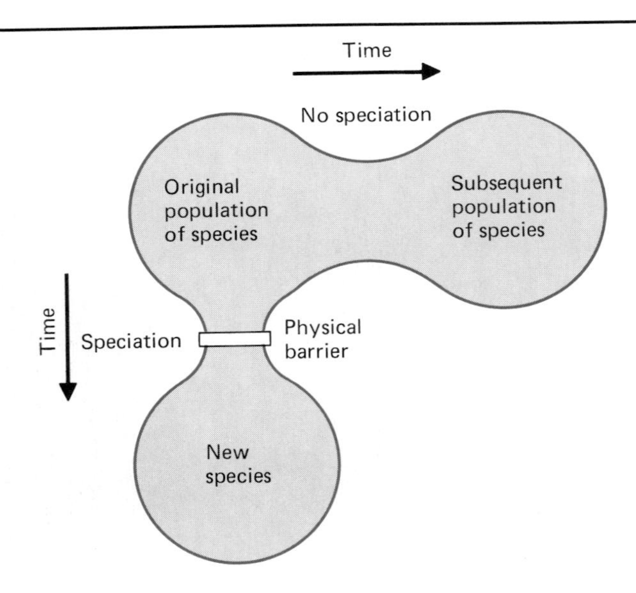


**Figure 4.9 Speciation.** Populations of a species may become isolated from one another for a long time as a result of a barrier. As their genetic makeup changes, they become separate species. As long as there is no barrier, the species may change somewhat through time but will still usually remain the same species.

As the environment on the earth changed, it became unsuitable for the dinosaur. However, before its extinction, favorable genes in the reptile line had formed the basis for newly evolving birds and mammals. Therefore, **gradual extinction** is a naturally occurring phenomenon in the formation of new species.

Examples of this kind of extinction abound. In addition to dinosaurs, many primitive plants and some bacteria have species that are only known from the fossil record.

Because humans, more than any other organism, have a tremendous capacity to alter the environment, they have a significant impact on the rate of extinction. When humans change the environment to their advantage, it is altered to other organisms' disadvantage, and the loss of certain species is inevitable. In addition to the buffalo, many other species simply could not adapt to an urban or agricultural environment.

Even though commercial hunters killed thousands of passenger pigeons, their extinction was caused primarily by increased agricultural use of the land when the forests they needed for food and nesting areas were destroyed to make way for fields. In addition, farmers sought to protect their crops from the passenger pigeons by shooting them. Similarly, populations of many endangered species such as the African elephant and the rhinoceros are being severely reduced by poachers. However, their endangered status occurred originally because habitat modification resulted from increasing human populations converting forests to farms.

Other environmental changes brought about by humans are more subtle. Diseases nearly eliminated both the American chestnut and the American elm.

The danger of human induced extinction is that it occurs in a relatively short period of time. When the genes of an extinct species are lost, the entire

**Figure 4.10 Extinction.** Many whales are in danger of becoming extinct because people use the oceans, which are the habitat for whales, for food production, waste disposal, and mineral exploitation.

global potential for adaptation and change is diminished. Aside from the biological consequences of extinction, there can also be aesthetic or emotional loss if a species becomes extinct. Even though many people will never see a whale in the ocean or a tiger in the bush, we would all suffer a little if such creatures became extinct. (See fig. 4.10.)

Even though some extinctions are inevitable, they should still serve as a warning or reminder to be careful about the kinds of environmental modifications we make. Many discussions about extinction deal with this serious environmental issue.

Other extinctions are viewed as desirable. Most people would not be disturbed about the extinction of animals like black widow spiders, mosquitos, or rats. In fact, people work hard to drive some species to extinction. For example, in the *Morbidity and Mortality Weekly Report* (October 26, 1979), the Center for Disease Control triumphantly announced that the organism that causes smallpox was extinct after many years of continuous effort to eliminate it.

There is another very selfish reason we should be concerned about extinction. Every time a species dies, its unique collection of genes dies too. We might be able to use those genes to our advantage.

For example, many of our crop plants are the result of selective breeding programs that have modified a few plants like corn, wheat, and rice so that they bear little resemblance to their wild ancestors. In some cases, the wild ancestors may even be extinct. What genes were present in those ancestors that might be valuable to us today? Because of this concern, many agricultural institutes and universities maintain "gene banks" of wild and primitive stocks of these plants.

## Consider This Case Study
### Serengeti National Park

Serengeti National Park in the east central African country of Tanzania is an area that is predominately grassland. Large herds of migratory grazing animals such as zebra, wildebeest, and Thomson's gazelle use the park and surrounding land. Together these three species number about eight hundred thousand animals. They use an area of about twenty-five thousand square kilometers. The park makes up only about one-half of this area. The animals are protected in the park but not outside the park. There are several kinds of predators that use this mixed group of grazers as a source of food. Lions, wild dogs, hyenas, and jackals are the most common. Humans are also significant predators of all of the animals in the park even though the park rangers try to prevent poaching. Tanzania is a developing country with mineral resources and agricultural potential.

What do you think will happen to the animals of the Serengeti, and why do you think it will happen?

## Summary

The concepts of habitat and niche are explored by using the buffalo as an example. The habitat of an organism is the space that it occupies, whereas the ecological niche of an organism emphasizes its interactions with both the physical and biological environment.

Historically the buffalo was the dominant organism of the western prairie of the United States and Canada. It had few competitors and became a central part of the culture of the native Americans of the prairie. The relationship between the buffalo and the native American was one of prey and predator.

Europeans, however, became competitors with both the buffalo and native Americans. Conversion of the prairie to farmland disrupted the buffalo's habitat. Furthermore, the federal government had a policy of eliminating as many buffalo as possible as a means of also eliminating the native Americans.

Extinction is a natural occurrence. If a species does not adapt to changes in its environment, it may eventually become extinct. Habitat modification resulting from human use of the environment is a major cause of extinction. The extinction of any species, whether we consider it valuable or not, has an impact on the environment through the loss of the organism's unique set of genes.

## Review Questions

1. Describe the ecological niche of the buffalo by listing ten items that contribute to the niche.
2. What were the main causes of death for buffalo before Europeans began killing them for hides and meat? How did the relationship between buffalo and humans change when Europeans came to the prairies and plains?
3. Why did the United States government allow the near extinction of the buffalo?
4. What is extinction? Why is it important to consider?
5. What is the difference between predation and competition? How do these concepts relate to the story of the buffalo?
6. What is the rutting season? When did it occur?
7. Describe how the relationship between buffalo and native Americans changed from their first contact until about 1900.

## Suggested Readings


Allen, Durward L. *The Life of Prairies and Plains.* New York: McGraw-Hill Book Company, 1967. An excellent job of introducing the prairie and the organisms that inhabit that region. Ecological principles are discussed by using examples from the prairie. Easy and fun to read.

Dary, David A. *The Buffalo Book.* Chicago: Swallow Press, 1974. A good introduction to the buffalo.

Gard, Wayne. *The Great Buffalo Hunt.* New York: Alfred A. Knopf, 1959. Deals primarily with the impact of humans on the buffalo.

Grzimek, Bernard. *Serengeti Shall Not Die.* New York: Dulton, 1961.

Haines, Francis. *The Buffalo.* New York: Thomas Y. Crowell Co., 1970. An extensive section on the relationship between the buffalo and the native American nations.

Hayes, Harold. *The Last Place on Earth.* New York: Stein & Day Publishers, 1977.

Martin, Cy. *The Saga of the Buffalo.* New York: Hart Publishing Co., 1973. An extensively illustrated, short book that presents a good overview of the life and times of the buffalo from its discovery to the present.

Roe, Frank G. *The North American Buffalo.* Toronto: University of Toronto Press, 1970. An extensively documented treatment of the whole story of the buffalo.

Schaller, George B. *Serengeti: A Kingdom of Predators.* New York: Alfred A. Knopf, 1972.

## Chapter Outline

I. Succession
  A. Pioneer community
  B. Climax community
II. Biomes as climax communities
III. Human use of terrestrial ecosystems
  A. Forests
  B. Agriculture
    Box 5.1 Farming in a tropical rain forest
  C. Philosophical approaches to the use of ecosystems
IV. Consider this case study: Indiana Dunes

## Objectives

Describe the process of succession from pioneer community to climax community.

Associate the typical plants and animals with the environmental characteristics of grassland and savannas.

Associate the typical plants and animals with the environmental characteristics of deserts.

Differentiate forests according to temperature and rainfall.

Define watershed.

Correlate species diversity and productivity with different terrestrial biomes.

Explain the underlying reasons for the exploitation philosophy.

Compare the philosophies of conservationism with preservationism.

List human uses of forests.

Explain the problems associated with using various terrestrial areas for agriculture.

List the current major uses of the lands on earth.

## Key Terms

biomes

climax community

conservation

deciduous tree

desert

ecological succession

exploitive use

grassland

northern coniferous forest

pioneer community

preservation

savanna

taiga

temperate deciduous forest

tropical rain forest

tundra

watershed

# Chapter 5
# Natural and managed ecosystems

## Succession

No matter where you live, your environment is constantly changing. Buildings are torn down to make parking lots or highways; fields become housing developments; fires destroy forests; and landslides, earthquakes, or floods physically change the skyline of cities. These represent changes in your habitat. All other organisms experience similar changes in their surroundings. As the habitat changes, populations of organisms also change. Let us look at this dynamic interplay of organisms with their environment.

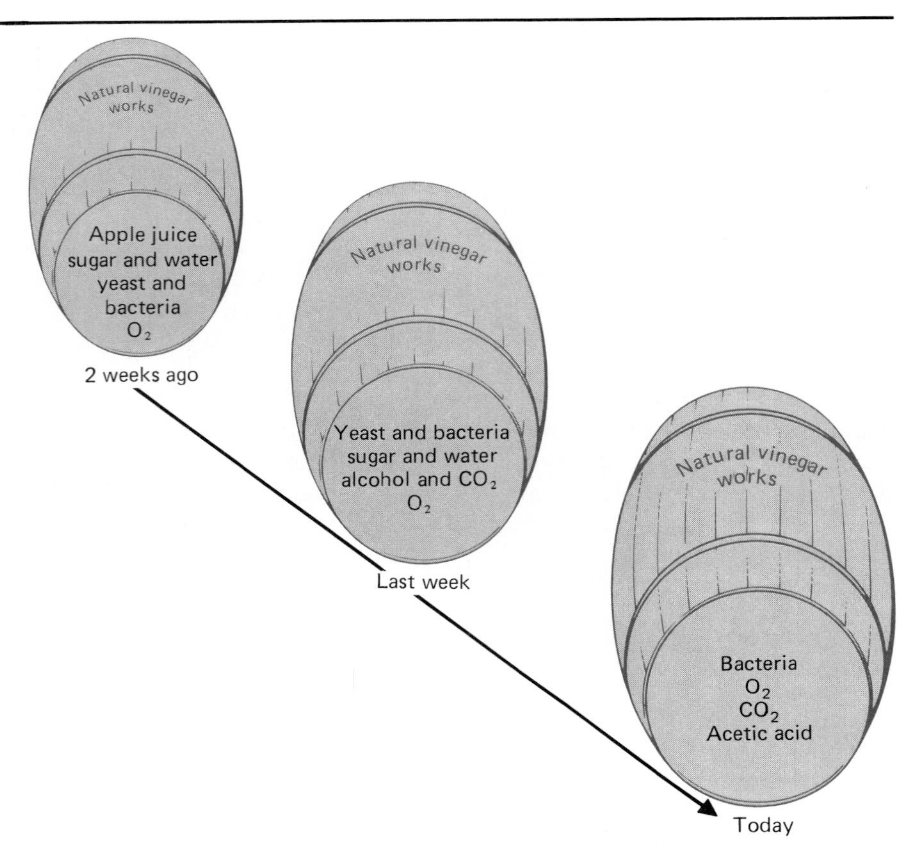


**Figure 5.1 Changing environment.**
As the yeast cells use the sugar in the apple juice, they make alcohol and carbon dioxide. The environment has changed so much that the yeasts are poisoned by the alcohol they produced. This new set of environmental conditions encourages the growth of bacteria. The bacteria convert the alcohol to acetic acid, and vinegar is produced.

Each of the changes mentioned in the opening paragraph can be related directly to human use of an ecosystem. As population increases, more houses are built. Some of these houses are built on farmlands. Because people want to live near the ocean or near sources of water, cities that are established in such areas are sometimes prone to natural disasters such as floods and hurricanes.

Ecologists who have studied changing ecosystems are convinced that organisms affect the environment just as much as the environment affects the organisms. As the environment changes, the type of community in the area also changes.

The process of one natural community replacing another is termed **ecological succession.** The kinds of organisms change and so does the physical environment. Succession occurs because organisms cause changes in their surroundings. For example, when apples are crushed, yeasts, bacteria, and other organisms that normally live on them are mixed with the juice. (See fig. 5.1.) The yeast cells begin to grow and multiply as they use the sugar for food and release carbon dioxide gas and alcohol as waste products. Eventually, the sugar supply is exhausted, and the environment of the yeast is full of alcohol wastes. So much alcohol is produced that yeast cells die. The yeast cells have changed their environment, and they can no longer survive there.

Certain oxygen-requiring bacteria use alcohol as their source of energy. Their waste product is acetic acid; the mixture produced is called vinegar. The bacteria thrive in the environment that killed the yeast cells. However, the alcohol is used up, and the accumulating acetic acid causes the bacterial population to die. The bacterial population followed the yeast population and was dependent upon the changes brought about by yeasts. This is a simple case of succession that happens when one population changes the environment in such a way that it can no longer live there but makes conditions suitable for a different population.

This simple case parallels succession in other ecosystems. Since succession is usually predictable, we can study an area and predict what the sequence of changes is going to be. For instance, the construction of a beaver dam will flood an area and cause the death of trees. Eventually, the area behind the dam will fill with silt from the stream. A wet meadow will develop, which will subsequently be replaced by shrubs and trees that grow well in moist soils. These shrubs and trees help to create a drier soil. Finally, a forest is reestablished. Thus, several different communities have succeeded one another at this beaver dam site. (See fig. 5.2.) This is secondary succession. Primary succession is different because it begins on sites that were originally uninhabited.

### *Pioneer community*

Primary succession can begin even on bare rock. The first group of organisms established in an area is called the **pioneer community.** These organisms must be hardy to survive. A bare rock environment is very harsh. There is nothing to modify the effects of wind, precipitation, and sun. Consequently, extreme temperature and moisture levels determine the kind of organisms that can survive on such a site. Typical organisms that can survive these conditions are lichens, which form a crusty growth on the surface of the rock. Small microscopic plants and animals live in and on the lichen. This combination is the pioneer community. As time passes, secretions and waste products will be produced and some organisms will die. Bits and pieces of airborne dust and debris will also be added. Over a long period of time, with the help of the physical and chemical processes called weathering, a thin layer of soil will develop on the surface of the rock.

The pioneer organisms must be firmly attached because even a drizzle is like a torrential rain to such tiny organisms. The soil they help to produce ultimately leads to the elimination of the pioneer community. The thin layer of soil is just right for mosses, fungi, certain small worms, insects, bacteria, and protozoa. These organisms encourage further soil formation and are replaced by other organisms that are better adapted. The moss community is

**Figure 5.2 Beaver pond succession.**
A colony of beaver can dam up streams and kill trees by flooding and using them for food. Once the site is abandoned, it will slowly return to the original forest community.

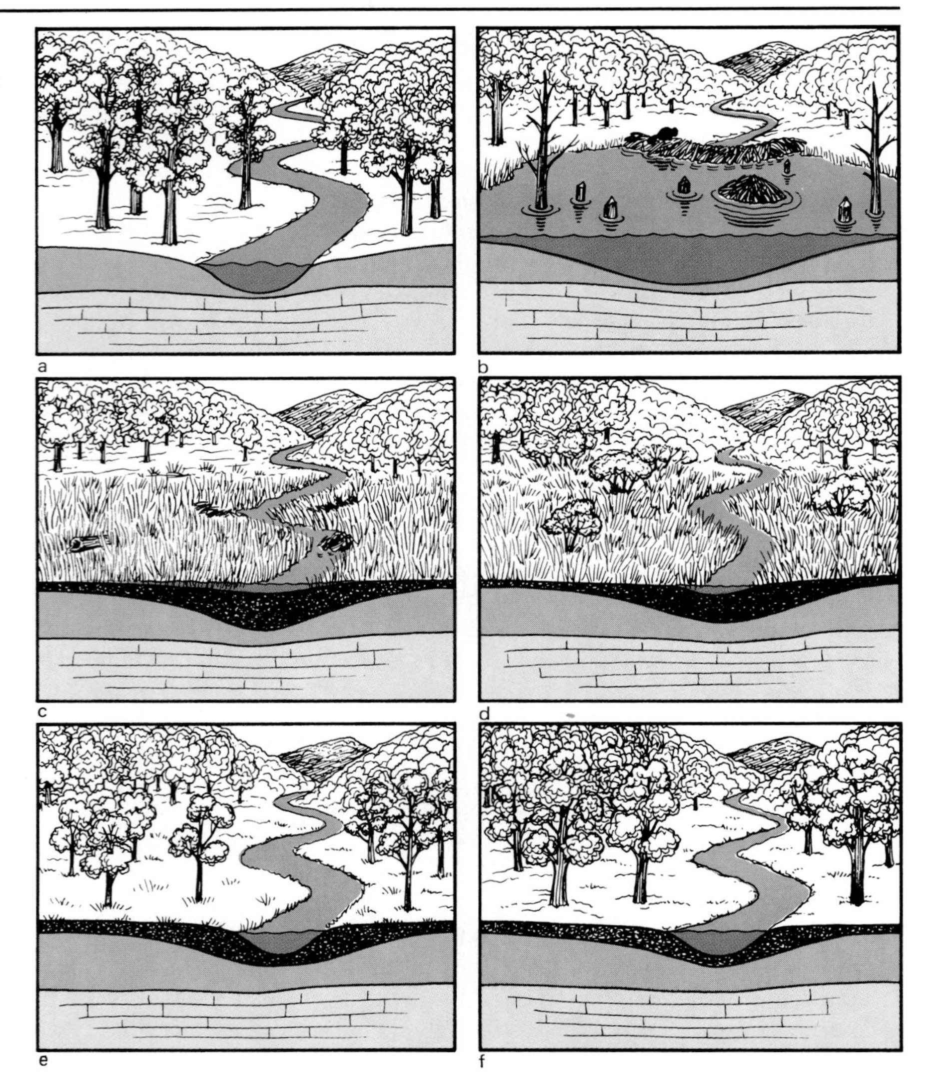


replaced by a herbaceous plant community. This community is replaced by a community dominated by grasses, followed by shrubs, and in many places, ultimately a forest.

Most examples of succession are probably of a different type, however. Secondary succession occurs when a mature ecosystem is altered. A series of changes will usually result in the establishment of the same kind of mature ecosystem. This kind of succession is often observed in abandoned farmland, forests that have burned, or other disturbed sites.

**Figure 5.3 Primary succession.** The formation of soil is a major step in primary succession. Until soil is formed, the area is unable to support large amounts of vegetation, which modify the harsh environment. Once soil formation begins, the site will proceed through a series of orderly stages to a climax community.

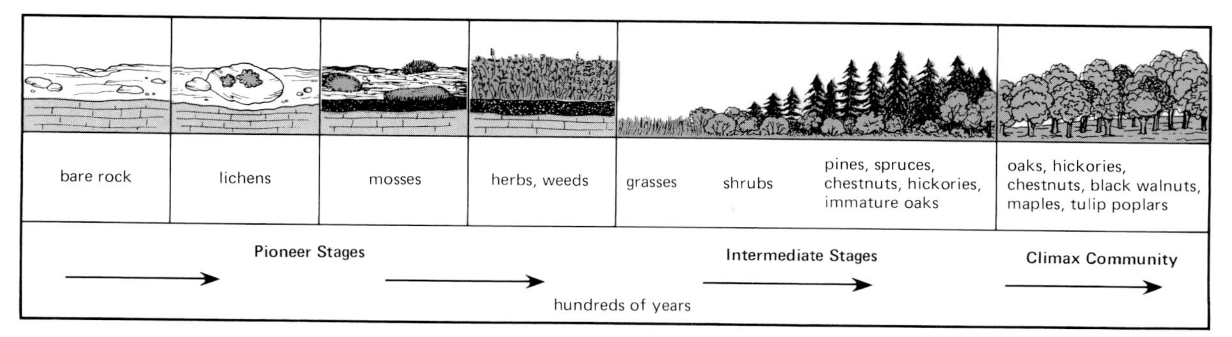


## *Climax community*

Ultimately, succession leads to a relatively stable, long-lasting interrelated group of plants and animals called the **climax community.** (See fig. 5.3.) This community and its environment are well balanced. One of the factors that contributes to this balance is the large number of different populations of organisms inhabiting the area. Each population has a small, but not necessarily equal, part to play in the whole system.

Simple ecosystems with few kinds of organisms tend to be less stable than ecosystems with many diverse populations. Simple ecosystems like agricultural ones are very unstable and require constant expenditures of energy to prevent them from being destroyed by the natural process of succession.

Almost all materials are recycled within the ecosystem. Little or no input of nutrients is required. Of course, the stability of the community is dependent upon continued input of sunlight energy. This energy is captured by a variety of producers and is distributed to the consumers. The complexity of the food web, the variety of interactions, and the cycling of nutrients within a climax community lead to its unchanging nature. Neither the populations nor the physical environment are likely to change drastically unless a major disturbance occurs. (See fig. 5.2.) The prairie community, with its native Americans, buffalo, grasses, grasshoppers, and prairie dogs, was a well-balanced climax community before farming began. Only isolated patches of this prairie remain, and these are somewhat different from the original because people have eliminated certain organisms, such as buffalo and wolves, from the community.

Humans are notorious for their disturbing influence on climax communities. Israel, Lebanon, Syria, and Iraq were at one time heavily forested. Centuries of human habitation have denuded the forested areas, and grazing animals, such as goats and sheep, have prevented the reestablishment of forests. Consequently, there has been a permanent change toward an arid desert

**Figure 5.4 Grassland and savanna.**
Both of these climax communities have grasses as their major producer organisms. The grasses are better able to withstand limited rainfall, which is characteristic of these regions. The savanna has a rainy season and a dry season, during which fires often occur. These fires kill tree seedlings and prevent the establishment of forests.

in these areas. Attempts are being made to reestablish the forests and reclaim the desert for more productive use. Today aquatic resources, such as the Monterey sardine fishery and the Peruvian anchovy fishery, are also being mismanaged. Worldwide, some whale populations are being rapidly depleted.

Even though much of the world has been altered by human activity, it is still possible to identify major regions of the world with the same types of vegetation. These regional climax communities are called **biomes.**

**Figure 5.5 Desert.** The desert gets less than twenty-five centimeters of precipitation per year, but it teems with life. Cactus, sagebrush, lichens, snakes, small mammals, birds, and insects inhabit the desert. Because daytime temperatures are high, most animals are only active at night when the air temperature drops significantly.

## Biomes as climax communities

A **grassland** or prairie is a broad region with particular environmental conditions. The major determining factor is the amount of rainfall. The grassland biome is relatively dry. The grasslands of North America, which were discussed in chapter 4, are similar to those of Europe and Asia. These kinds of grasslands receive from 25 to 75 centimeters of rain per year. The soils have a rich humus layer and can support herds of grazing animals. South America and Africa have a warmer tropical type of grassland called a **savanna.** (See fig. 5.4.) Savannas are characterized by a very wet season with 100 to 150 centimeters of rain and a very dry season during which fires commonly occur. A few scattered trees that can withstand drought and fire grow in these areas. Savannas are similar to grasslands in that they support herds of grazing animals.

**Desert** biomes are drier than grasslands. They generally receive less than 25 centimeters of precipitation per year. (See fig. 5.5.) We often think of a desert as a hot, dry wasteland, but in actuality a desert can be very cold at night and supports a wide variety of living organisms. Of course, the organisms of the desert are very specialized for life in a dry environment. Typical plants include lichens, which form crusts on surfaces; shrubs with small leaves, such as sagebrush; and succulents that store water, such as the cacti. Desert animals also exhibit several kinds of adaptations to their environment. Most have

Figure 5.6 **Biomes of the world.** Although most of the biomes are named for a type of vegetation, each includes a specialized group of animals adapted to the plants and climate of the region.

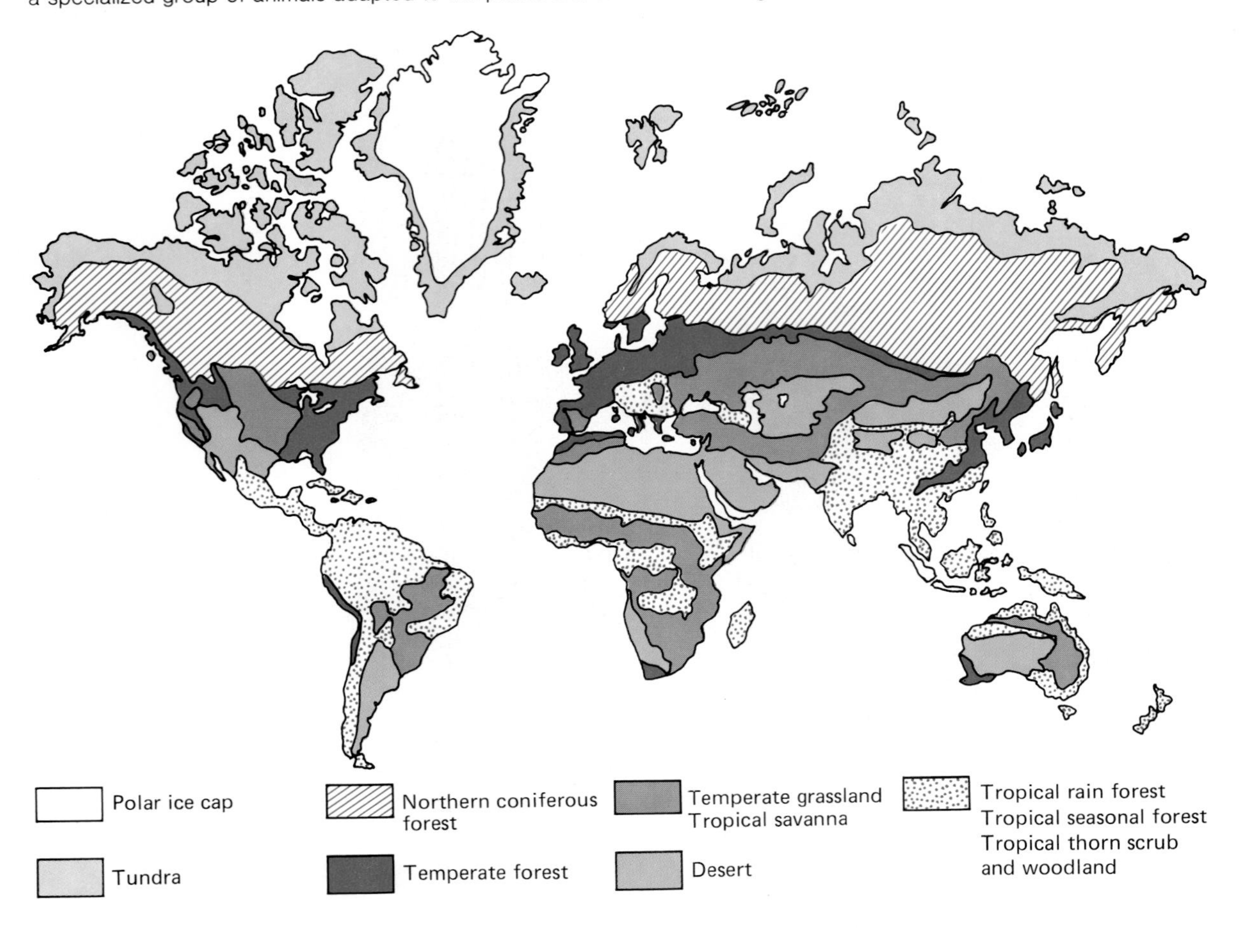


special adaptations to prevent water loss through their skin. The evening hours are cooler and more humid in the desert. Nocturnal animals avoid the hottest and driest part of the day by living in burrows where it is cool and moist. In addition, most desert animals can survive for a long time without drinking. Many kinds of lizards, snakes, rodents, camels, insects, and birds inhabit the desert.

When there is adequate precipitation, temperature becomes the major determiner of the kinds of organisms inhabiting an area. Biomes that are related to temperature as well as amount of precipitation include **tropical rain forests, temperate deciduous forests, northern coniferous forests,** and the **tundra.** The tropical rain forests are located primarily near the equator in Central and South America, Africa, Southeast Asia, and on some islands in the Caribbean and the Pacific. (See fig. 5.6.) The temperature is high and relatively constant.

**Figure 5.7 Tropical rain forest.**
A tropical rain forest is noted for the tremendous variety of plant and animal life.

Rain falls nearly every day and is excessive during certain periods of the year. The soil is a relatively poor source of nutrients since most of the nutrients in these rain forests are tied up in the biomass that exists there. There is a great variety of plant and animal life. A small area is likely to have hundreds of different species of trees. Balsa wood, teakwood, orchids, climbing vines, bromeliads, ferns, palms, and mosses are common plants of the rain forest. Snakes, lizards, birds, and climbing mammals are typical. (See fig. 5.7.) Many of these animals tend to inhabit the upper layers of the forest, rather than to dwell on the ground. There are probably more species of insects in the tropical rain forest biome than in the rest of the biomes collectively. Ants, termites, moths, butterflies, and beetles are abundant.

The biome typical of the eastern United States is the temperate deciduous forest. This is a forested region with abundant rainfall that is relatively evenly spaced. It consists largely of **deciduous trees,** which lose their leaves each fall. This characteristic is associated with a period of low temperatures when little growth occurs. Birch, oak, hickory, beech, and maple are all examples of deciduous forest trees. The number of different tree species in a temperate deciduous forest is small compared to that of a tropical rain forest. Animals of the temperate deciduous forest include birds such as turkey, thrush, ruffed

**Figure 5.8 Temperate deciduous forest.** This kind of forest once dominated the eastern half of the United States. Much of it has been cut to make farms and to build cities.

**Figure 5.9 Northern coniferous forest.** The northern coniferous forest occurs in areas with long winters and heavy snowfall. The trees have adapted to these conditions and provide food and shelter for the animals that live there.

**Figure 5.10 Tundra.** In the northern latitudes and on the tops of some mountains, the growing season is short, and plants grow very slowly. Because this growth is slow, damage to the tundra is very slow to heal.

grouse, and woodpecker, and mammals such as skunk, porcupine, deer, squirrel, and raccoon. (See fig. 5.8.) Temperate forests are also located in Europe and parts of the USSR, South America, and Asia. (See fig. 5.6.)

Through parts of southern Canada and along the mountains in the western United States, the dominant plants are evergreen (conifer) trees. (See fig. 5.9.) This area is called the northern coniferous forest or the **taiga.** Spruce and fir trees are typical members of this community, and because their dense canopy shades the ground, only a few small plants grow there. Moose, squirrels, grouse, and wolves are likely to inhabit this type of forest. This forest can receive up to 150 centimeters of precipitation per year, most of which falls as snow. The flexible branches of the pyramid-shaped conifer trees shed this snow. During its long cold season, which lasts for about six months each year, conditions are generally harsh, and the temperature is low enough to freeze the soil.

Still further north, trees are entirely absent. This area is known as the **tundra.** There are also scattered patches of tundralike communities (alpine tundra) on mountaintops throughout the world. Organisms that live in the tundra are able to tolerate the longer winters (ten months). Because the soil is frozen a few centimeters below the surface all year round, the plants of this region are small and very slow growing. Tundra organisms include reindeer moss (a lichen), grasses, slow-growing flowering plants, snowy owls, ptarmigan, lemmings, musk oxen, caribou, mosquitos, and biting flies. This is a very simple community because only a few kinds of organisms can live in this harsh environment. Because growth is slow, the tundra is easy to damage and slow to heal. Changes that occur here are, therefore, likely to have drastic and long-lasting effects. We must treat it very gently. (See fig. 5.10.)

**Figure 5.11 Habitat preservation.**
Natural habitats, such as this, must be preserved to provide niches for organisms such as moose, red squirrels, wolves, owls, mice, black bear, and warblers. If the trees are removed, the animals will be eliminated also.

## Human use of terrestrial ecosystems

Humans are not known for their gentle treatment of ecosystems. If anything, our reputation is one of exploitation and ruination. As chapter 4 stated, there are few species able to compete successfully with us. (Certainly, the buffalo was not able to do so.) A prevailing philosophy seems to be that *Homo sapiens* is the most important species and that all other species are here for our use. Because of this philosophical base, it may seem that we are generally unconcerned about the displacement of other species. Not everyone, however, shares this philosophy. Conflicts occur between those who favor total exploitation and those who favor total preservation. Disturbing the Alaskan tundra to construct an oil pipeline, systematically harvesting large whales, and attempting to eliminate most predator species were recent controversial issues. The exploitive use of forest resources is a historical example of this conflict.

### *Forests*

Almost half of the United States was originally forested. Settlers removed trees to make room for farming, to make clearings around homes, to provide fuel for heating, furnish materials for building, and also just because they were "in the way." Today it is difficult to understand why people of the past saw so little value in the forests. About one-third of the United States is still forested, and major controversies have developed about how to use those areas. Lumber companies would like to use forests only as a source of timber and timber-related products. They see forests as crops and have recently begun to plant rapidly growing high-yield species. The advantage of this is that the forest will renew itself faster. Some forest plantations will mature to harvest-

**Figure 5.12 Recreational use of forested areas.**
Camping is a very popular activity. As more people become involved in this form of recreation, additional campgrounds will need to be developed. These developments must be carefully planned so the forest will still maintain its "wild" character.

able growth within twenty years, instead of the typical one hundred years. Short-lived trees produce lower-quality timber and require a single species forest to obtain high yields.

Some people consider single species forests to be less desirable than more natural mixed forests. Single species forests are certainly less stable and more susceptible to disease. They also provide a lower-quality habitat for various animal and plant species.

Different species of wildlife have different habitat requirements. Therefore, a single species forest cannot support a wide variety of wildlife because it does not furnish a wide range of habitat requirements. The age distribution of trees within a forest is also a factor that determines the number and variety of animals present. A single species forest is usually even-aged, and an even-aged forest will not support as wide a variety of wildlife as will a mixed-age forest. These considerations emphasize the ways in which forest management is connected with wildlife management.

Forests are also important in watershed maintenance. A **watershed** is the area of land drained by a river or a stream. Management of forests is very important in maintaining the watershed because the plants and soil in forests hold the water and discharge it slowly. This protects the soil from erosion.

Forests can also be set aside as refuges or preserves. (See fig. 5.11.) For example, as discussed in chapter 1, the disturbance of a forest that is also the habitat of a large herd of elk was a major controversial topic. Many plant and animal species, which are part of forest ecosystems, are eliminated when forests are disturbed.

Finally, forests can be used for recreational purposes. Camping, hiking, hunting, and fishing are all activities of forested areas. These recreational activities will become more important as the workweek shortens and leisure time increases. Forests located near population centers are valuable as recreational areas because it takes little time and energy to get to them. Most of us will never use the Alaskan wilderness for our recreation (see fig. 5.12), but a local, state, or national forest may be a reasonable substitute.

**Figure 5.13**
**Mechanized vs. labor intensive farms.**
Large fields require a capital investment in the form of the planting and harvesting machinery. Small truck garden plots require much manual labor to operate. Because of this, it is not likely that the farmer will plant asparagus in this field next year.

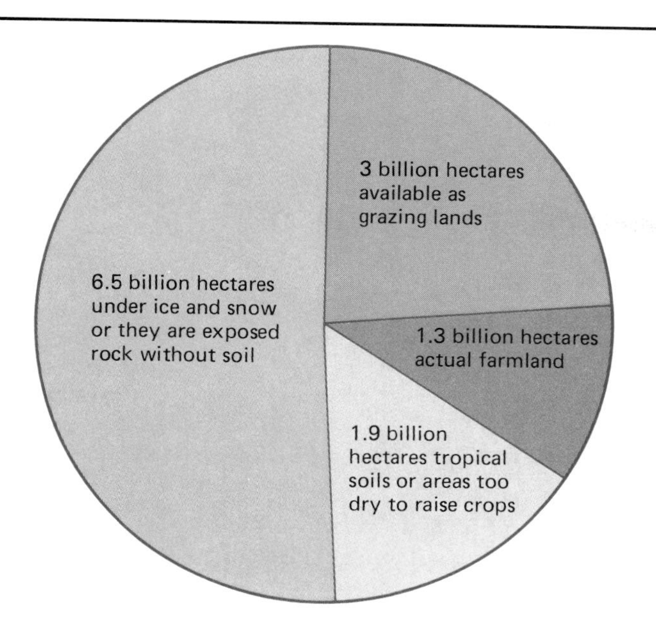


**Figure 5.14 Land areas on earth.**
There are about 12.9 billion hectares of land on the earth, but only 6.5 billion hectares have soil. Of this, only about 1.3 billion hectares are valuable as cropland. The rest is used as grazing land. It cannot be used for cropland because of location and climate.

## *Agriculture*

Another major use of terrestrial biomes is agriculture. Almost 10 percent of the total land surface of the world is devoted to agricultural food production. In the United States, approximately 57 percent of the land is used for agriculture as cropland or grazing land. As population increases, greater amounts of food are needed. Better agricultural productivity can be accomplished by increasing efficiency. A one thousand hectare farm (1 hectare = 10,000 square meters) planted with one crop, such as wheat or corn, is economically more efficient than many small farms with a variety of crops. The cost of equipment has encouraged this specialization. (See fig. 5.13.)

Another problem associated with agricultural productivity is the increasing need for pesticides, herbicides, and fertilizers. These are expensive in terms of both money and energy, but they are necessary to prevent the natural processes of succession. These additives are heavily dependent upon oil for their manufacture and application. How will decreasing oil supplies affect farming practices and productivity? Chapter 14 includes an extensive discussion about agriculture, energy, and productivity.

As world population increases, the need for food increases, but the total available land is fixed. The total land on the earth is about 12.9 billion hectares. Half of this land is not available for crops or grazing, because the land is under the permanent ice of the south pole and on mountaintops or is solid rock without soil. Of the remaining 6.5 billion hectares, half is potentially available for grazing and the other half for cropland. Of the 3.2 billion hectares available to be cultivated, only about 40 percent is actually being farmed. (See fig. 5.14.) Because these statistics show that there is a vast amount of

**Figure 5.15 Irrigation.** Many arid lands can be used to grow crops if water can be made available. Unfortunately, many areas of the world lack the necessary surface water or groundwater to make this possible.

land not being used for food production, some people believe that we should not worry about starvation and famine. However, what they do not understand is that it is not economically or ecologically practical to farm this remaining 60 percent. Two billion hectares are mostly tropical, and cultivating large tropical areas is difficult. The soil of the tropics supports tons of naturally growing vegetation (see box 5.1). The reason it is not used for agriculture is directly related to the differences between a naturally growing tropical forest and orderly rows of crop plants. If a tropical forest was used for agriculture, the extensive rainfall, which is common in many tropical areas, would erode the soil and remove nutrients. Nutrients would also be lost from the naturally growing vegetation that was removed to make room for crops. Because of this, large amounts of fertilizer would be needed to replenish the nutrients in the soil. After a few years of agricultural use, the forest soil would become extremely hard and impossible to farm.

Some areas are unacceptable for planting because they do not have the necessary precipitation for crop plants. Irrigation is not always a solution to the problem because it is possible only when groundwater or surface water is readily available nearby or can be supplied by aqueducts and channels from distant sources. This situation does not exist in most dry areas of the world. (See fig. 5.15.) Furthermore, salt buildup in the soil is a problem in many areas of the world where irrigation is used extensively. See chapter 16 for a more detailed discussion of this problem.

## *Philosophical approaches to the use of ecosystems*

This discussion of forests and croplands points out that humans must be concerned with the finiteness of their support system. There are definite and rigid limits to the use of these lands for timber, recreation, and food production. Whenever a resource is limited, judgments must be made concerning its use. Competing uses for the same resource result in controversy.

## Box 5.1
## Farming in a tropical rain forest

Because tropical rain forests are so lush and fast growing, many people have suggested that they would be excellent areas for agricultural use. In fact, primitive people in such areas have practiced *slash-and-burn* agriculture for centuries. In this method, a small plot of the forest is cut and the vegetation is burned. The burning releases the nutrients in the vegetation. If tilling and planting is done quickly, the crop will cover the soil and prevent exposure. During the first year a good crop can be harvested, but the yield will decline each succeeding year unless massive amounts of fertilizer are used. For the primitive people, this was no problem because they simply abandoned the garden and cleared a new site. The old garden was quickly repopulated by the surrounding forest. This is the only kind of farming adaptive to this kind of ecosystem.

One kind of soil found in hot, humid tropical rain forests, as in South America, is called laterite soil. Laterite soil gets its name from the Latin word for brick, because whenever it is directly exposed to weathering, it hardens into a bricklike mass. In the past, these blocks were used as building materials. A temple in the People's Republic of Kampuchea (Cambodia) is constructed entirely of blocks of laterite.

Abundant rainfall is common in the areas where laterite soil is found. As a result, the soil is subjected to a great deal of leaching, which is the removal of nutrients by water. Silica is removed, but the soil retains a relatively high concentration of oxides of iron and aluminum. The presence of these compounds contributes to the reddish color of this soil.

Because of the porous nature of the soil and the amount of rainfall, this soil is not fertile. The nutrients are quickly washed out of the soil. Why then is there such lush growth in tropical rain forests? High temperature and humidity encourage rapid plant growth so nutrients are immediately taken up and incorporated into a new plant body. Whenever a plant dies, it decays rapidly, and surrounding plants absorb the nutrients, which prevents their removal. In addition, seeds fall nearly every day of the year.

There is a mutualistic relationship involved in the cycling of nutrients in a tropical rain forest. It is estimated that 10 to 15 percent of the trees in such a forest are legumes. The leguminous trees are responsible for making nitrogen available to the other plants within the forest. Tropical forest trees bear mutualistic fungi (mycorrhiza) in association with near-surface roots. The mycorrhiza quickly penetrate each fallen leaf and take all nutrients back into the roots of the host tree. This cycle is one of the most direct forms of mineral recycling in nature. Therefore, it is very unlikely that the nutrients from decaying organisms go unused.

The nutrients of a lush tropical forest are a part of the vegetation itself. This prevents the nutrients from being leached from the soil. In addition, the soil is protected from direct exposure to sun and rain by the vegetation, so it does not harden. Slash-and-burn agriculture does not cause major disruption in the forest ecosystem because the soil is only exposed for a short period of time. This prevents hardening and reduces leaching. The garden plots are small; therefore, when the garden is abandoned, it is quickly overgrown by the surrounding forest. Large fields cannot be tolerated by this ecosystem. The poor quality of soil, cost of fertilizers, and complications with clearing and tilling the land mean that the tropical rain forest will not solve our future population/food problems.

There are three different philosophical positions on the use of resources. The first position is **preservation.** To preserve means to keep from harm or damage or to maintain. A preservationist would like to see the forests and other land maintained in their original or at least in their current state.

The opposite of a preservationist philosophy is an **exploitive use** philosophy. People with this philosophy see no value in holding anything back; they believe that technology will find replacements when we have used all of a particular commodity. For example, when copper plumbing became too expensive, plastic pipe replaced it. If plastic pipe becomes too expensive, a substitute will be found. This philosophy could also be called the technological optimist philosophy.

The middle-of-the-road philosophy is conservation. **Conservation** is the wise use of a commodity so it maintains a maximum sustained yield or a continuous supply of a resource. Conservation requires study and evaluation so that decisions can be based on an understanding of their consequences. This includes assessing the value of an item and developing a plan to use it so that maximum value is obtained over a period of time. Conservation means use without misuse or abuse.

## Summary

Succession occurs when one natural community is replaced by another. This happens because organisms cause changes in their surroundings. Some organisms make their environment an unacceptable place to live because of the changes they make. Succession progresses from a pioneer community through various intermediate stages until a stable, long-lasting climax community is established.

Regional climax communities are called biomes. The biomes that are determined by limited precipitation are grasslands, savannas, and deserts. Other biomes determined primarily by temperature are tropical forests, temperate deciduous forests, northern coniferous forests, and the tundra.

Forest and agriculture are two major uses of land. Forests provide building materials, recreational sites, wildlife habitats, and protection as watersheds. However, people convert forests for agricultural use by removing trees, but the amount of land that can be used for farming is limited.

There are three different philosophies about human use of our resources. Preservationists would like to maintain land in its original, or at least its current, state. The exploitive use philosophy believes that technology will find replacements when we have depleted a particular resource. Conservationists would use resources wisely in order to obtain a maximum sustained yield.

## Consider This Case Study
### Indiana Dunes

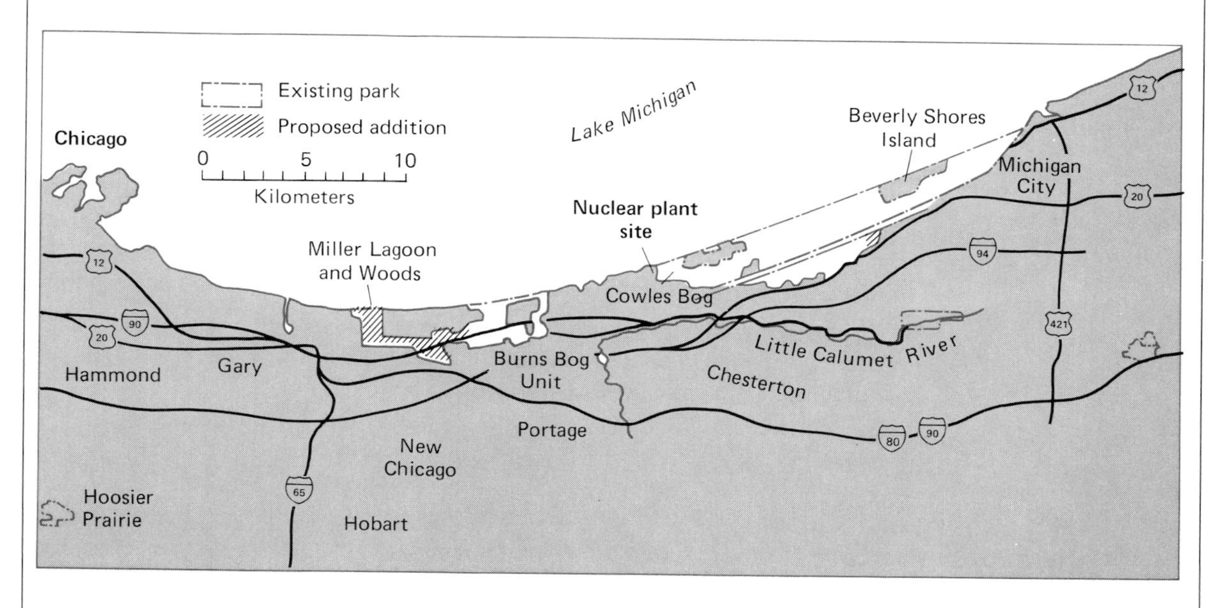


In 1966, the Indiana Dunes on southern Lake Michigan became our first national lakeshore dedicated to preserving a natural shoreline feature. This area has long been a source of controversy. This unique area is composed of slightly shifting sand dunes in the process of succession. A series of several different successional communities can be seen as one moves away from the lake. Ecologists see the area as an unusual laboratory to study plant succession and wildlife management. Therefore, conservationists and preservationists would like to keep most of the remaining shoreline on the southern tip of Lake Michigan free of industry and human-related development. The site is also well suited for public recreation.

Industries would like to develop a deep-water port. The transportation of goods to and from the nearby metropolitan area surrounding Chicago encourages deep-water port development. A nuclear power plant has also been proposed for this area. Private landowners' plans include subdivision development, railroad yards, steel fabrication plants, and industrial dump sites. Off-road vehicles are currently damaging the area so greatly that their damage is considered irreversible.

Can you justify the protection of thousands of hectares of land for public recreation and scientific investigation?

What is your response to the objectives of industry, private owners, and preservationists?

When space is needed in the Chicago area, can exploitation be prevented?

## Review Questions

1. Give detailed examples of succession.
2. What is the name of the biome in the area where you live? What are some of its physical and biological characteristics?
3. What determines whether an area will be grassland or desert?
4. What kinds of animals are common in a northern coniferous forest? In a temperate deciduous forest?
5. How is a biome different from an ecosystem?
6. What is a watershed?
7. How much of the world's land is currently being used for agriculture?
8. List three problems associated with increasing food production.
9. What problems are associated with growing crops in the tropics?
10. How does a preservationist differ from a conservationist?

## Suggested Readings

Chitty, D. *Natural Selection and the Regulation of Density in Cyclic and Non-cyclic Populations.* Baltimore: University Park Press, 1977. Excellent article on natural selection and the regulation of density.

Emmel, Thomas C. *Global Perspectives on Ecology.* Palo Alto, Calif.: Mayfield Publishing Company, 1977. Presents a series of readable essays that explain the interrelatedness of all of our ecological problems.

Milne, Lorus, and Milne, Margery. *The Arena of Life, the Dynamics of Ecology.* New York: Doubleday & Co., 1971. An attractive book, highly illustrated with photographs. Acceptable for the general reader or as an introduction to the interrelatedness in the science of ecology.

Odum, Eugene P. *Fundamentals of Ecology.* 3d ed. Philadelphia: W. B. Saunders Company, 1971. A must for the library of the serious student of ecology or environmental science.

Schofield, Edmund A., ed. *Earthcare: Global Protection of Natural Areas.* Boulder, Colo.: Westview Press, 1978. The papers presented at the Sierra Club's fourteenth biennial wilderness conference. An interesting collection of thoughts of the people actively involved in protection of our resources.

## Chapter Outline

I. Population characteristics
II. A population growth curve
III. Carrying capacity
IV. Prolific humans
  A. Humans are social animals
    Box 6.1 Control of births
  B. Ultimate size limitation
V. Consider this case study: The George Reserve

## Objectives

Understand that birthrate and deathrate together determine population growth rate.

Define the following characteristics of a population: natality, mortality, sex ratio, age distribution, reproductive potential, and spatial distribution.

Explain why the biotic potential for most populations is much greater than needed.

Describe how, as it grows, a population goes through a lag, exponential growth, and stationary growth phase.

Describe how the carrying capacity of a region for a population is determined by environmental resistance.

List the four categories under which limiting factors can be classified.

Describe how some populations enter a death phase after the stationary growth phase.

Explain why humans are subject to the same forces of environmental resistance as other organisms.

Understand the implications of overreproduction.

Understand that the human population is still growing rapidly.

Explain how human population growth is influenced by social, theological, philosophical, and political thinking.

## Key Terms

age distribution
biotic potential
birthrate (natality)
carrying capacity
death phase
deathrate (mortality)
dispersal
emigration
environmental resistance
exponential growth phase
immigration
lag phase
limiting factors
sex ratio
stationary growth phase

# Chapter 6
# Population principles

## Population characteristics

In chapter 3 population was defined as a group of individuals of the same species that inhabits an area. Just as individuals within a population are recognizable, populations themselves have specific characteristics that give them identity. Some of these characteristics are birthrate, deathrate, sex ratio, age distribution, and spatial distribution. **Birthrate (natality)** refers to the number of individuals added to the population through reproduction. It is usually expressed as the number of births per thousand individuals per year. For example, if a population of one thousand individuals had ten offspring in one year, the natality is given as ten per thousand per year.

**Figure 6.1 The effect of birthrate and deathrate on population size.** In order for a population to grow in size, it is necessary that the birthrate exceed the deathrate for a period of time. These three human populations illustrate how the combined effects of births and deaths would change population size if birthrates and deathrates were maintained for a five-year period.

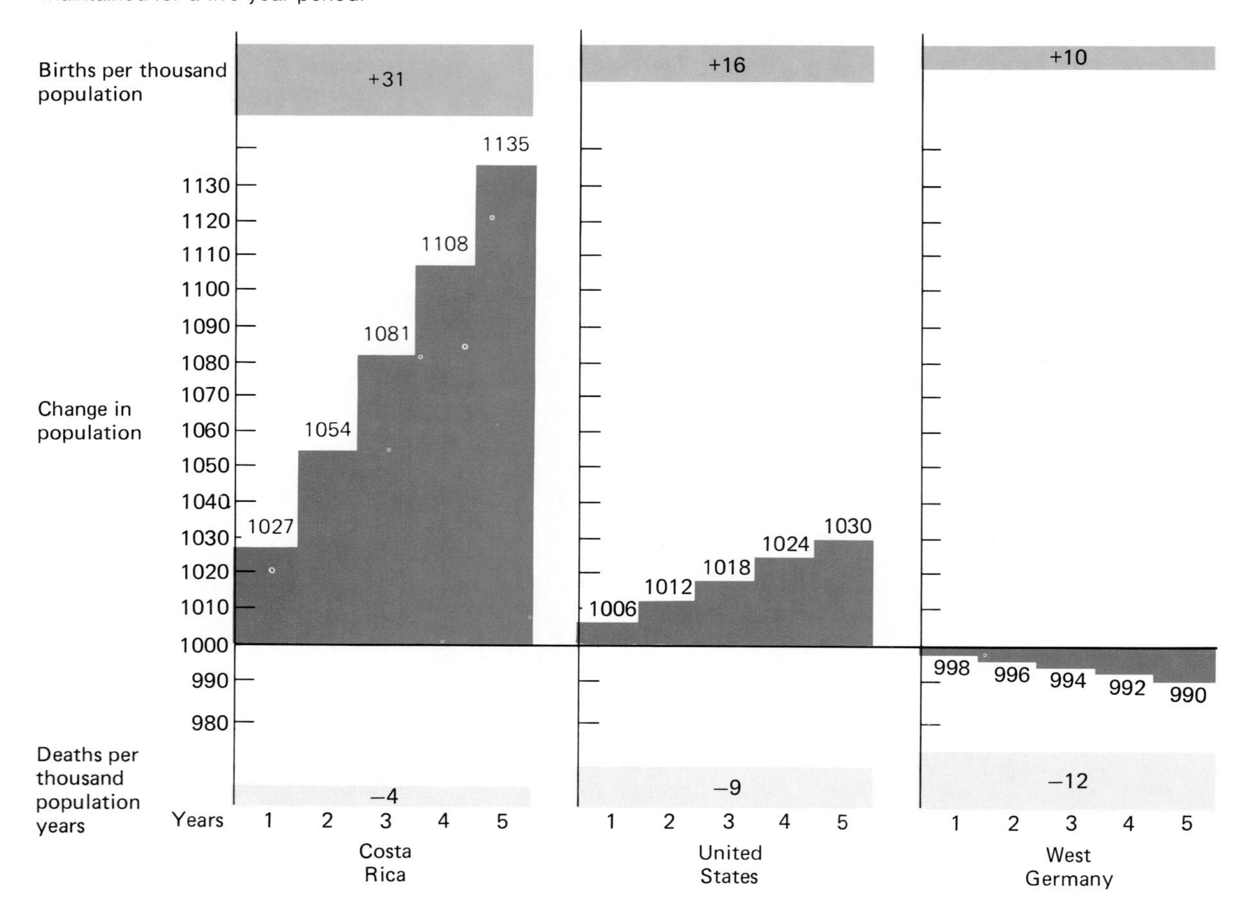


This is sometimes called the raw natality of a population. It is not the same as population growth because the growth of a population includes both natality and mortality. **Deathrate (mortality)** is the number of deaths per thousand individuals per year. If you combine natality and mortality, you can determine whether a population is increasing or decreasing and how rapidly it is changing. For a population to grow, it is necessary that the birthrate exceed the deathrate for a period of time. (See fig. 6.1.) The birthrate and deathrate of a population depends upon the sex ratio and age distribution of a population.

The **sex ratio** refers to the number of males relative to the number of females. In humans, about 106 males are born for each 100 females. However, in the United States, by the time people reach their middle twenties, a higher deathrate for males has equalized the sex ratio. The higher deathrate for males

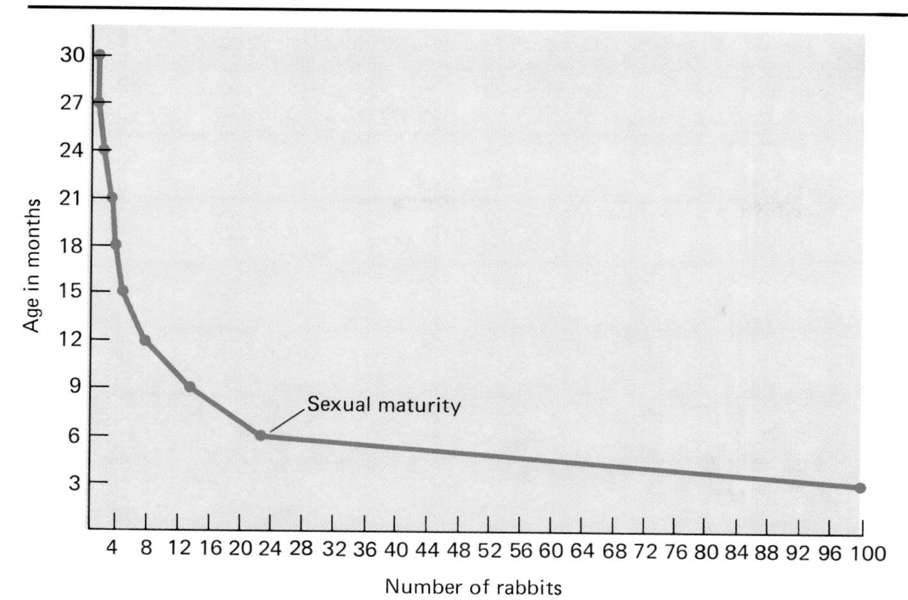


**Figure 6.2 Mortality of cottontail rabbit.**
In most natural populations, mortality rates are so high that very few individuals reach sexual maturity, and even fewer reach old age.

continues into old age when women outnumber men. In many insect populations, such as bees, ants, and wasps, the number of females greatly exceeds the number of males at all times, although most of the females are sterile.

Another feature of a population is its **age distribution.** Not all members of a population are the same age. Some are prereproductive juveniles, some are reproducing adults, and some are postreproductive adults. A stable population will have these three age classifications in a balance so the proportion of prereproductive, reproductive, and postreproductive individuals remains constant. Typically this means that there are more prereproductive individuals than reproductive individuals and more reproductive individuals than postreproductive individuals.

The age structure of a population is directly related to its mortality rate. Most species of small animals have a very high mortality rate. Therefore, such populations will have a large number of prereproductive individuals, which will decline sharply with increasing age. A good example of this type of population is the cottontail rabbit. Many young are produced, some of which reach sexual maturity, but very few live to old age. (See fig. 6.2.)

If the majority of a population is postreproductive, the population will decline. This age structure develops in many insect populations in the fall after the eggs have been laid. In human populations, cultural taboos that restrict sexual contact or selective death due to war can lead to this kind of population age structure.

If the majority of a population is made up of reproducing adults, a "baby boom" can be expected. This population age structure naturally occurs in the spring when many animal populations begin to mate and raise offspring.

**Figure 6.3 Age distribution in human populations.** The relative numbers of individuals in each of the three categories (prereproductive, reproductive, and postreproductive) can give a good clue to the future of the population. Mexico has a large number of young individuals that will become reproducing adults. Therefore, this population is likely to grow rapidly. The United States has a declining proportion of prereproductive individuals, and Sweden has a relatively balanced population. Therefore, they will probably have declining populations sometime in the future, with Sweden's population declining first.

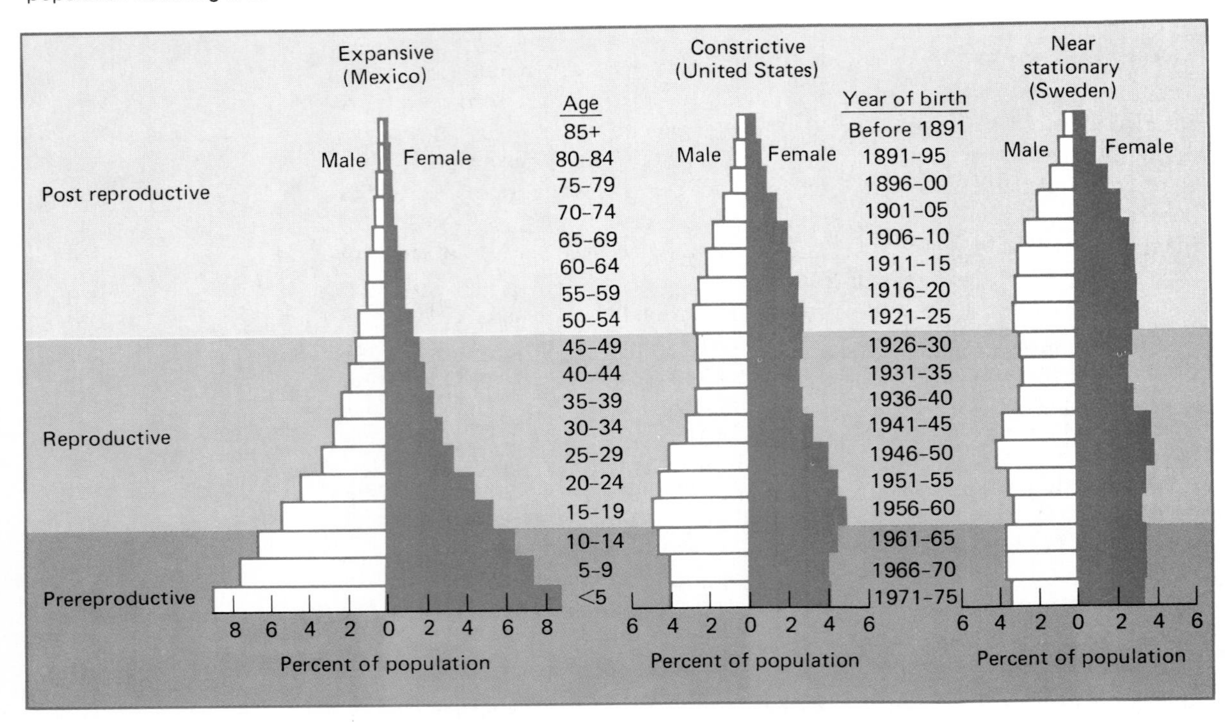


In human populations there are examples of several types of age distributions. (See fig. 6.3.) Mexico has a large prereproductive and reproductive component to its population and will continue to grow in the future. Sweden has a relatively even age distribution and will remain nearly constant in its population size. The United States has a large reproductive component with a declining number of prereproductive individuals. So eventually the United States population will begin to decline.

Just as populations are not usually divided into equal thirds according to their age distribution, organisms are not usually equally distributed spatially. Because of such factors as soil type, quality of habitat, and availability of water, there is normally an uneven or clumped distribution of organisms. When the density of organisms is too great, all organisms within the population are injured because they compete severely with each other for necessary resources. Plant populations may compete for water, soil nutrients, or sunlight. In animal populations, overcrowding might cause exploration and migration into new areas. This **dispersal** of organisms will relieve the overcrowded conditions in the home area and at the same time lead to the establishment of new populations. Often juvenile individuals relieve the overcrowding by leaving the home area. The pressure for out-migration (**emigration**) may be a result of seasonal reproduction, a rapid increase in the size of the local population, or environmental changes that cause competition to intensify among members of the same species. For example, as waterholes dry up, competition for water increases, and many desert birds will leave the area and go to areas where water is still available.

The organisms that leave one population often become members of a different population. This in-migration (**immigration**) may introduce new characteristics that were not in the existing population. When Europeans immigrated to North America, they brought with them genetic and cultural characteristics that had a tremendous impact on the existing native American population. In addition, Europeans brought diseases that were "foreign" to the native American population. These diseases increased the deathrates and lowered the birthrates of the natives.

Recent droughts, wars, and political persecution have caused people to emigrate from their native lands to other countries. The United Nations' High Commissioner for Refugees in 1981 estimated that there are ten to twelve million refugees worldwide. Often the receiving countries are unable to cope with the large influx of new inhabitants. In the 1980s, the United States accepted over 150,000 Cuban refugees as well as Haitian, Laotian, and Vietnamese refugees. Immigration to the United States accounts for about half of our annual population growth. Other countries, such as Austria, Thailand, Pakistan, and Zaire, also have refugee centers to accept and process people leaving their homeland due to political strife or famine.

## A population growth curve

Sex ratios and age distributions within a population have a direct influence on the rate of reproduction within a population. Each species has an inherent reproductive capacity, or **biotic potential,** which is the ability to produce offspring. Generally, this biotic potential is many times greater than the number of offspring that actually survive to become reproducing members of the population. (See fig. 6.4.)

Since reproduction is so important to the continuance of a population, it is logical that all species existing today would have developed mechanisms to assure adequate reproduction. Organisms have evolved two basic strategies to assure adequate reproduction. Some organisms expend their energy producing millions of possible offspring but do not provide care for them. As a result, mortality is very high and only a few individuals reach reproductive age. For example, a female oyster may produce a million eggs, but, on the average, only a few will develop into sexually mature adults. Other organisms expend much energy caring for the few young that are produced. Humans generally produce a single offspring during each pregnancy, and most of those born survive. This is even true in less-developed countries where infant mortality may be as high as two hundred infant deaths per thousand live births, which is still an 80 percent survival rate.

**Figure 6.4 Biotic potential.**
The ability of a population to reproduce greatly exceeds the numbers that are necessary to replace those who die. Here are some examples of the prodigious reproductive ability of some species.

A high reproductive potential results in a natural tendency for populations to grow in size. For example, two mice produce four offspring, which in turn produce eight, then sixteen, and so on. Figure 6.5 shows a typical population growth curve. The first portion of the graph, called the **lag phase,** indicates that initially growth occurs very slowly. During the first few weeks, the numbers do not increase very rapidly because the first two mice may have a litter of four who will not be ready to reproduce for several weeks. However, when the first litter is of reproductive age, the parents are producing a second or third litter. Now there are several pairs reproducing, not just one. When these several pairs have produced litters, the number of mice has increased greatly. Now the population is growing at an ever-increasing rate. The period during which populations increase very rapidly is called the **exponential growth phase** of population growth. However, the number of mice will not continue to increase indefinitely. Reproduction will continue within the population, but the population will stop growing so the birthrate eventually equals the deathrate. This stage is called the **stationary growth phase.** Before this stationary growth phase of population growth is discussed, the concept of carrying capacity must be introduced.

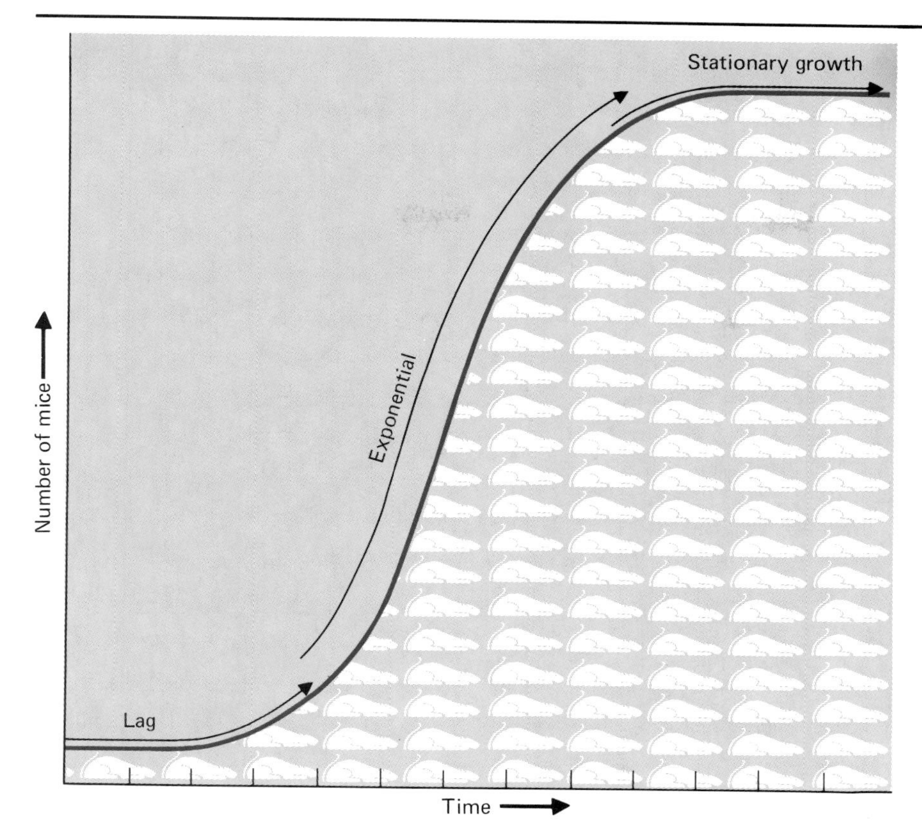


**Figure 6.5 Typical population growth curve.** In this mouse population, the period in which there is little growth is known as the lag phase. When there is a rapid increase as the offspring reach reproductive age, the curve is in the exponential phase. Eventually the population reaches a stationary growth phase, during which the birthrate equals the deathrate.

## Carrying capacity

The **carrying capacity** of an area is the optimum number of individuals of a species that can survive in that area over an extended period. In most populations, four factors interact to define the carrying capacity for a population in a given area. These four factors are the availability of raw materials, the availability of energy, the accumulation of waste products and the means of disposal, and interactions between organisms. **Environmental resistance** to continued population growth is determined by the combined influence of these four factors. In some cases, it is easy to recognize specific **limiting factors** that slow population growth and set the carrying capacity for the population. (See fig. 6.6.) For example, in grass plants, nitrogen and magnesium in the soil are necessary raw materials used in the manufacture of chlorophyll. If these minerals are not present in sufficient quantities, it is impossible to have an increase in the grass population. However, if the application of fertilizer removes this limiting factor, the grass population will increase because environmental resistance has been lessened with an increase in carrying capacity. This is one reason why garden centers sell so much fertilizer.

Plants require light energy in the form of sunlight to cause photosynthesis to occur. The amount of light can be a limiting factor for many plant species.

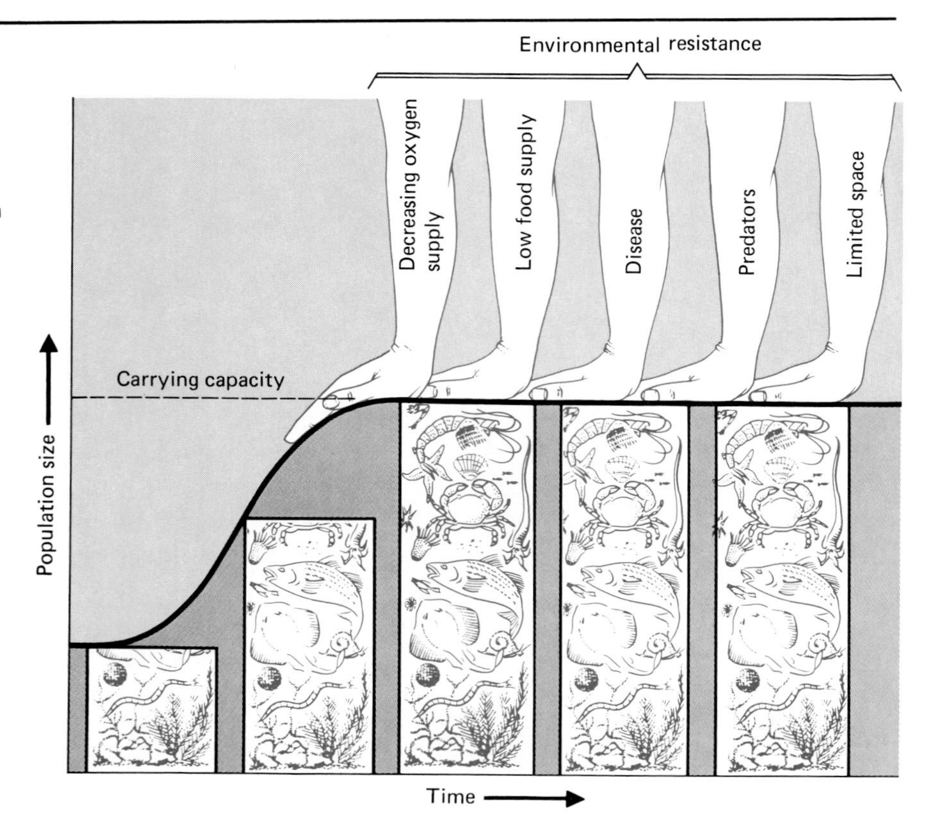


**Figure 6.6 Carrying capacity.**
A number of factors in the environment, such as food, oxygen supply, diseases, predators, and space, determine the number of organisms that can survive in a given area. This number is called the carrying capacity of that area. The environmental factors that limit populations are collectively known as environmental resistance.

In addition, the number and kinds of animals using grass for food and the number and kinds of other plants that compete with grass are involved in determining the carrying capacity of an area for grass plants.

Since there are few waste products from grass, they do not influence the carrying capacity to a great degree. In other populations, the primary determiner of population size might be the buildup of waste products from the population itself. This is common in bacterial cultures where organisms have a very limited environment. When one puts a small number of a species of bacterium in a petri plate with nutrient agar (a jellylike material containing food substance), the population growth follows the curve shown in figure 6.7. As expected, the number of bacteria increases through a lag and exponential phase of growth and eventually reaches stability during the stationary growth phase. However, as the waste products accumulate, the organisms are poisoned by their own toxic wastes. In this case, there are no other ways to handle the metabolic wastes, since space for disposal is limited and other kinds of organisms are not present to take care of these wastes. The wastes poison the environment, and the organism can no longer live; therefore, the population eventually decreases in size. This portion of the population curve is often called the **death phase.** If a population decreases rapidly, it is said to crash.

**Figure 6.7 Bacterial growth curve.** The change in population size follows a typical population growth curve until waste products become lethal. The buildup of waste products lowers the carrying capacity. When a population begins to decline, it enters the death phase.

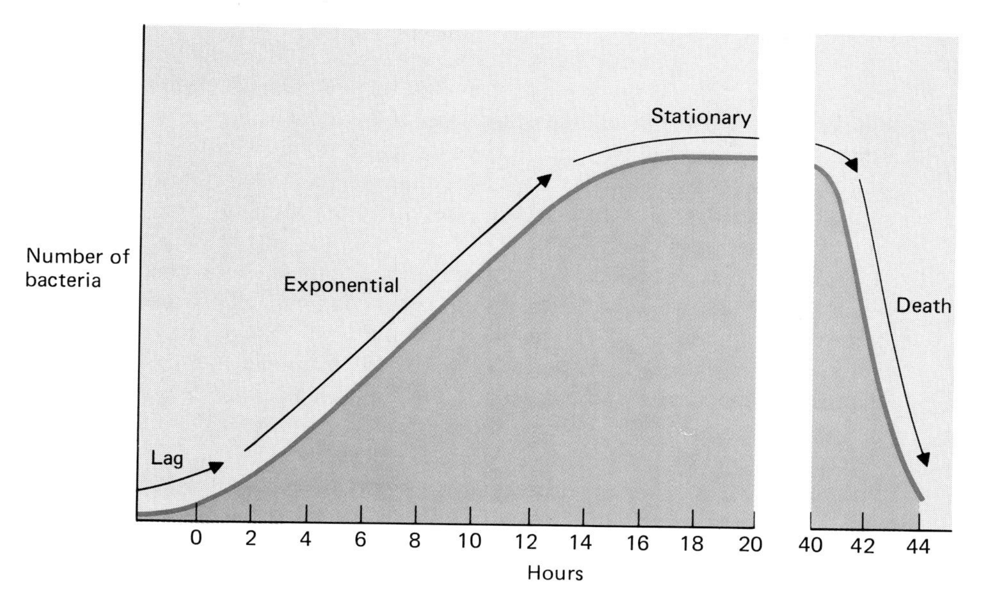


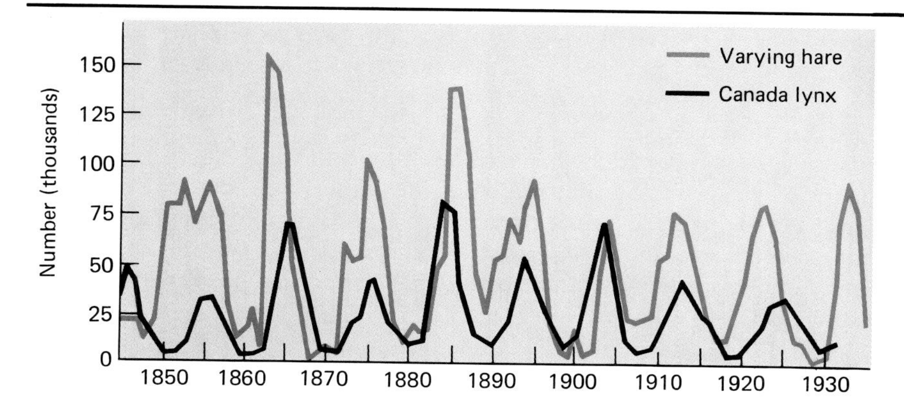


**Figure 6.8 Predator/prey interaction.** The interaction between predator and prey species is complex and often difficult to interpret. These data were collected from the records of the number of pelts purchased by The Hudson Bay Company. It shows that the two populations fluctuate with the changes in the lynx population usually following the changes in varying hare population. (Data from D. A. MacLulich, *Fluctuations in the Numbers of the Varying Hare* (*Lepus americanus*). Toronto: University of Toronto Press, 1937, reprinted 1974.)

The interaction among species of organisms is very important in controlling population size. Indirect interaction occurs when decomposer organisms help to dispose of metabolic waste products. Direct interaction occurs when organisms act as predators and kill individuals of a prey species. A good example of predator-prey interaction within a population is the interaction between Canada lynx (the predator) and varying hare (the prey) in northern Canada. (See fig. 6.8.) The varying hare has a high biotic potential that the lynx helps to control by using the hare as food. The lynx can capture and kill the weak, the old, or the diseased hares. Thus, stronger, healthier hares live to reproduce.

Some studies indicate that populations may be controlled by interactions among the individuals within a population. A study on rats shows that crowding causes a breakdown in normal social behavior, which seems to be a factor in controlling laboratory rat populations. The kinds of changes observed include abnormal mating behavior, decreased litter size, fewer litters per year, lack of maternal care, as well as increased aggression in some rats or withdrawal in others. Thus, it is possible for environmental resistance to reduce natality as well as to increase mortality. Either can influence population size or at least retard the rate of increase.

So far we have considered populations that reached a stationary growth phase because they have a relatively stable carrying capacity. However, many populations fluctuate as environmental conditions change. Environmental conditions can change in a variety of ways. For example, temperate regions change in temperature and day length; during the cold, short days of winter, environmental conditions become more stressful for organisms. Other areas experience regular seasonal variation in the amount of rainfall. With either of these periodic changes, the reproduction of organisms usually coincides with the availability of food and raw materials.

In addition to seasonal variations in population size, many organisms exhibit population cycles that are fairly regular but not annual. In figure 6.8 the size of the lynx and varying hare populations show the same general variations. The size of the lynx population may be tied to the size of the varying hare population, since the lynx uses varying hares as a source of food. However, it is unclear what causes the fluctuations in the varying hare population. It is often suggested that recurrent epidemic disease may cause many kinds of animal population fluctuations.

Although local human populations often show fluctuations, the worldwide population has increased continually for the past several hundred years. Humans have been able to reduce environmental resistance by eliminating competing organisms, increasing food production, and controlling disease organisms. This has resulted in a human population that has been increasing continuously.

## Prolific humans

The human population growth curve has a long lag phase followed by a very sharp, rapid exponential growth phase that is still increasing rapidly. (See fig. 6.9.)

A major reason for the rapid increase is that the human species has sought to reduce the deathrate. When any country has reduced environmental resistance by increasing food production or controlling disease, this technology has been shared throughout the world. Developed countries have sent health-care personnel to parts of the globe to improve the quality of life for people in the

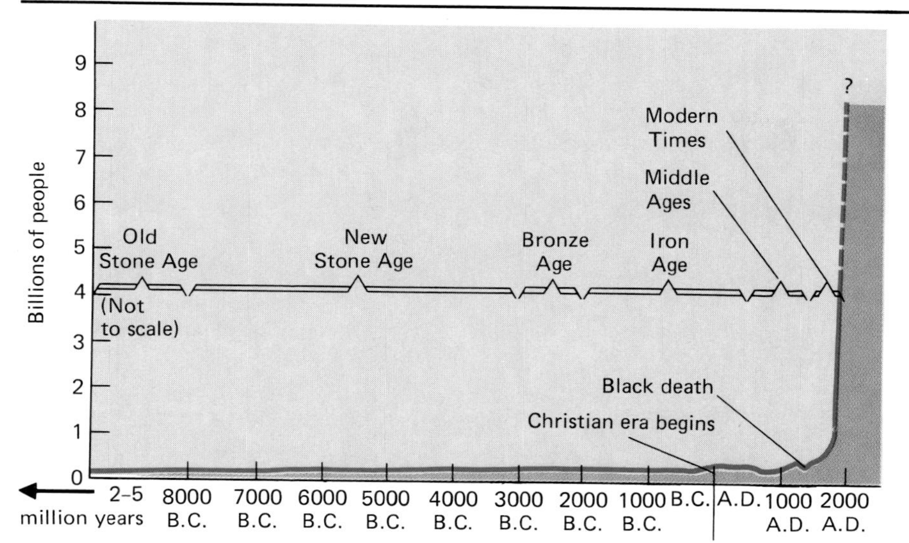


**Figure 6.9 Human population growth.** From A.D. 1850 to A.D. 1930, the number of humans doubled (from one billion to two billion) and then doubled again by 1975 (four billion) and could double again (eight billion) by the year 2010. How long can this pattern continue before the ultimate carrying capacity is reached?

less-developed countries. Physicians advise on nutrition, and engineers develop waste-water treatment systems. Improved sanitary facilities in India and Indonesia, for example, decreased mortality caused by cholera. These advancements tend to reduce mortality and directly increase the *quantity* of life because birthrates remain high. Each of these is the action of kind, humane, caring countries helping those less fortunate. At least, it appears so on the surface. Let us examine this situation from a different perspective. The world population is currently increasing at an annual rate of 1.7 percent. This is down somewhat from previous years and may indicate a slowing in the rate of human population growth. However, even at a 1.7 percent annual increase, the population is growing rapidly. It is often difficult to comprehend the impact of a 1 or 2 percent annual increase on a population. Remember that a growth rate in any population compounds itself, since the additional individuals entering a population will eventually reproduce, thus adding more individuals to the population. Another way to look at population growth is to look at how much time is needed to double the population. This is a valuable way to examine population growth because most of us can appreciate what life would be like if the number of people in our locality were to double, particularly if it were to occur within our lifetime.

Figure 6.10 shows the relationship between the rate of annual increase for the human population and the number of years it would take to double the population if that rate were to continue. At a 1 percent rate of annual increase, the population will double in approximately seventy years. At a 2 percent rate of annual increase, the population will double in about thirty-five years. The current worldwide rate of annual increase is about 1.7 percent, which will double the world population by the year 2025.

**Figure 6.10 Doubling time for the human population.**
This graph shows the relationship between the rate of annual increase in percent and doubling time. A population growth rate of 1 percent per year would result in the doubling of the population in about seventy years. A population growth rate of 3 percent per year would result in a population doubling in about twenty-three years.

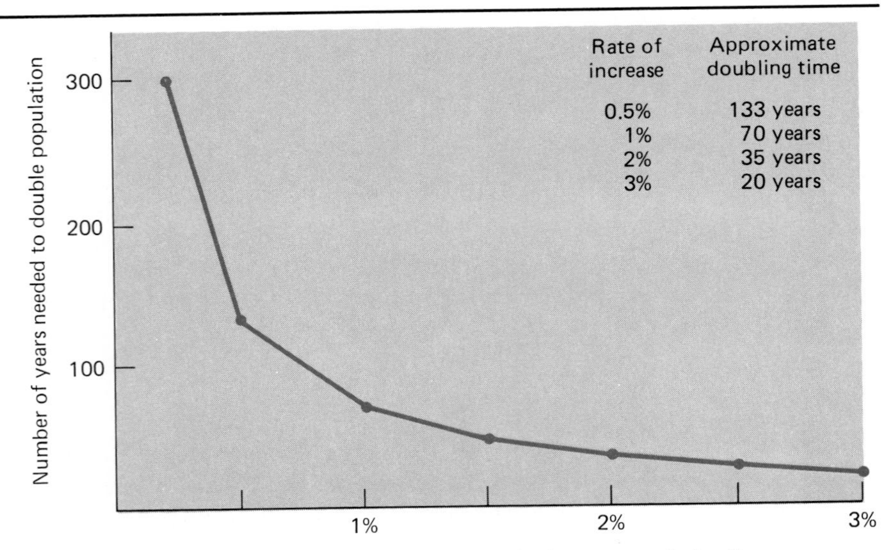


What does this very rapid rate of growth mean to the human species? First, as a species, humans are subject to the same limiting factors as all other species. We cannot increase beyond our ability to acquire raw materials, acquire energy, or safely dispose of our waste materials. We also must remember that interactions with other species and with other humans may determine our carrying capacity.

Let us look at these four factors in more detail. For many of us, raw materials simply mean the amount of food available. Do not forget, however, that we have become increasingly dependent on technology and that our life-styles are tied directly to our use of many other kinds of resources. Food production is becoming a limiting factor for some segments of the world's human population. Malnutrition is a major problem in many parts of the world because food is not available in sufficient quantities. Chapter 7 will deal with the problems of food production and distribution and their relationship to human population growth.

The second factor, available energy, involves problems similar to those of available raw materials. One important fact that should not be ignored is that all species on earth are ultimately dependent on sunlight for their energy. New, less disruptive methods of harnessing this energy must be developed to support an increasing population. Currently the world population depends upon large amounts of fossil fuels to raise food, modify the environment, and allow for movement. As energy prices rise, a large portion of the world's population is placed in jeopardy because their incomes are not sufficient to pay the increased costs for food and other essentials.

Waste disposal is the third factor determining the carrying capacity for humans. Most pollution is, in reality, the waste products of human activity. In chapter 5, the example of yeast in a fruit juice showed the importance of metabolic waste disposal. In this simple situation, the organisms produced so much waste that it poisoned them. Are humans in the same situation? Some people are convinced that disregard for the quality of our environment will be the major limiting factor. In any case, it just makes good sense to control pollution and work toward cleaning our environment.

A fourth factor that determines the carrying capacity of a species is interaction with other organisms. We need to become aware of the fact that we are not the only species of importance. When we convert land to meet our needs, we displace other species from their habitats. Many of these displaced organisms are not able to compete with us successfully and must migrate or become extinct. Unfortunately, as humans expand their domain, the areas available to these displaced organisms become more rare. Parks and natural areas have become tiny refuges for the plants and animals that once occupied vast expanses of land. If these refuges fall to the developer's bulldozer or are converted to agricultural use, some organisms will become extinct. What today seems like an unimportant organism, one that we could easily do without, may someday be seen as an important link to our very survival.

## *Humans are social animals*

Human survival depends upon interaction and cooperation with other humans. Current technology and medical knowledge are available to control human population growth (see box 6.1) and to improve the health of the people of the world. Why then does the population continue to increase, and why does poverty become greater every year? Humans are social animals that have freedom of choice and frequently do not do what is considered to be "best" from an unemotional, uninvolved biological point of view. People are "involved" and make decisions based on history, social situations, ethical and religious considerations, and personal desires. The biggest obstacle to controlling human population is not biological but falls into the province of philosophers, theologians, politicians, sociologists, and others. People in all fields need to understand the cause of the population problem if they are to deal successfully with every aspect of the problem.

## *Ultimate size limitation*

There is no logical way that the population can increase indefinitely, because eventually the weight of the human tissue would equal the weight of the earth. We can say with certainty that the population will eventually reach its ultimate carrying capacity and will level off.

There is much speculation about what the limits to human growth are. In its 1981 report, the United Nations Fund for Population Activities in its medium estimate suggests that the human population will peak at 10.5 billion people by the year 2110. Some people have suggested that a lack of food, a

## Box 6.1
## Control of births

The use of technology to control disease and famine has greatly reduced the deathrate of the human population. Technological developments can also be used to control the birthrate. A variety of contraceptive methods are available to help people regulate their fertility. Research is continuing to develop new, more effective, more acceptable, and less expensive methods of controlling conception. Because of cultural and religious differences, some forms of contraception may be more acceptable to one segment of the world's population than another.

The most common methods of contraception are the oral contraceptive pill, diaphragm and spermicidal jelly, intrauterine contraceptive device, spermicidal vaginal foam, condom, vasectomy, and tubal ligation.

The range of effectiveness of these methods, shown in the table, is the result of individual fertility differences and the degree of care employed in the use of each method.

In addition to various methods of conception control, abortion can terminate unwanted conceptions early in pregnancy. Most countries with low birthrates, like the United States, Japan, and many European countries, allow abortions.

Effectiveness of fertility control methods

| **Method** | **Average pregnancy rate per 100 women per year** |
|---|---|
| Abstinence | 0 |
| Sterilization (vasectomy or tubal ligation) | 0-.003 |
| Oral contraceptive pill | 0.1-1 |
| Intrauterine contraceptive device (IUD) | 1-6 |
| Diaphragm with jelly | 2-20 |
| Vaginal foam* | 2-29 |
| Condom (rubber)* | 3-36 |
| Withdrawal (coitus interruptus) | 18-23 |
| Rhythm (periodic abstinence) | 1-47 |
| Rhythm (temperature method) | 1-20 |
| None | 60-80 |

*Use of condom and vaginal foam together is more effective than either used singly.

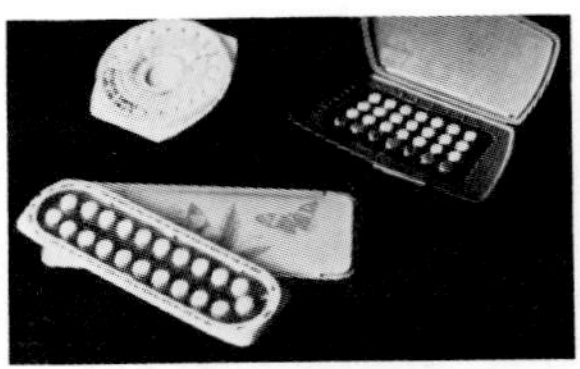
Oral contraceptive pills

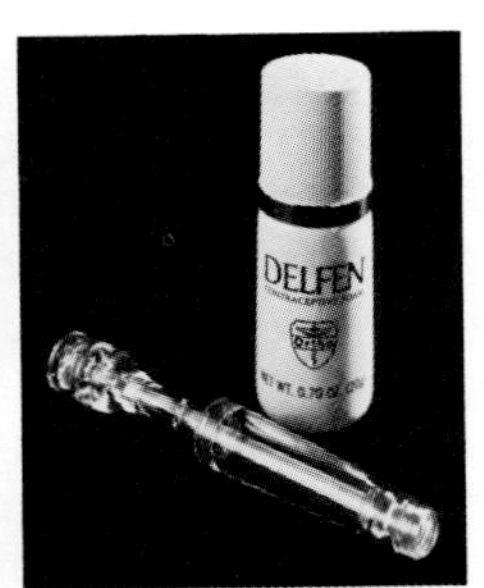

Spermicidal vaginal foam

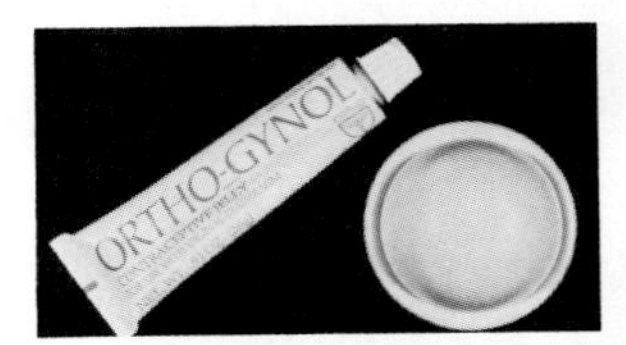

Diaphragm and spermicidal jelly

Condom

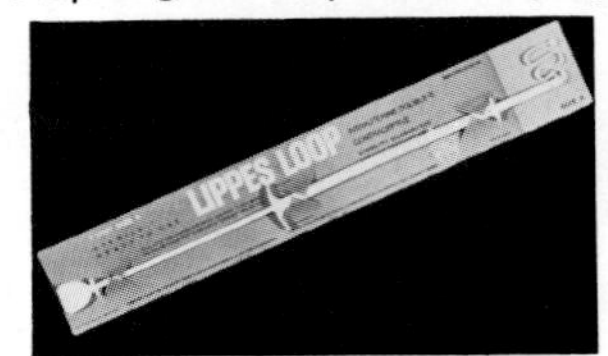

Intrauterine contraceptive device

lack of water, or increased waste heat will ultimately control our human population size. Still others have suggested that in the future we will experience social controls that will limit population growth or politically related destruction of humanity. These are only predictions—no one knows what the ultimate population size will be or what the most potent limiting factors will be.

## Summary

A population is a group of organisms of the same species that inhabits an area. The birthrate (natality) is the number of individuals entering the population by reproduction during a certain period. Deathrate (mortality) measures the number of individuals who die in a population during a certain period. Population growth is determined by the combined effects of the birthrate and deathrate.

The sex ratio of a population is a way of stating the relative number of males and females in a population. Age distribution within a population and the sex ratio have a profound impact on the future population growth. Most organisms have a biotic potential that is much greater than is needed to replace dying organisms.

Interactions between individuals in a population, such as competition, predation, and parasitism, are also important in determining population size. Organisms may migrate into (immigrate) or migrate out of (emigrate) an area as a result of competitive pressure.

A typical population growth curve shows a lag phase followed by an exponential growth phase and a stationary growth phase at the carrying capacity. The carrying capacity is determined by many limiting factors that are collectively known as environmental resistance. The four major categories of environmental resistance are available raw materials, available energy, disposal of wastes, and interactions among organisms. Some populations experience a death phase following the stationary growth phase.

The human population is still increasing at a rapid rate. The ultimate carrying capacity of the earth for humans is not known. The causes for the human population growth are not just biological, they are also social, political, philosophical, and theological.

## Review Questions

1. How are biotic potential and age distribution interrelated?
2. List three characteristics that populations might have.
3. Why do populations grow? What factors help to determine the rate of this growth?
4. Under what kind of condition might a death phase occur?
5. List four factors that could determine the carrying capacity of an animal species.
6. How do the concepts of birthrate and population growth rate differ?
7. How does the population growth curve of humans compare with that of bacteria on a petri dish?
8. What is meant by the term typical growth curve?
9. As the human population continues to increase, what might happen to other species?
10. All successful organisms overreproduce. What advantage does this provide for the species? What disadvantages may occur?

## Consider This Case Study
### The George Reserve

In 1930 the George Reserve was established as a wildlife study area through the generous gift of a Detroit industrialist, Colonel Edwin S. George. This property, about 464 hectares in southeastern Michigan, was donated to the University of Michigan to further research natural populations. It is bounded by a game fence 2.9 meters high. The reserve includes some wooded areas, both deciduous and coniferous, some wet and marshy areas, some permanent and temporary ponds, and some open, grassy meadowlike areas. Originally, six deer were imported into the reserve. There were four females (thought to be pregnant) and two males. Natural predators were excluded from the reserve, so the population of deer was expected to increase in number rapidly.

One goal of the early research was to keep an accurate record of how the population grew. Researchers decided to count the number of deer in various age classes so they could make a cross-check whenever an animal died.

Each year a census was conducted in the reserve. During this census, a group of individuals (usually graduate students) would line up and walk from one end of the reserve to the other. The people taking the census would ideally be close enough to each other that they could maintain contact. Anytime a deer passed through the line of the census takers, it was to be counted. (If it passed on the right of an individual, it was counted; if it passed on the left, the next person in the line would count it.) Any living deer or skeletal remains of deer were counted. At the end of the census the total deer population was tabulated.

The graph shows the actual population changes for almost forty years. After only six years, the number of deer had increased from six to over one hundred and sixty. Why was there such a rapid increase in just six years? To answer this question, some knowledge of the biology of deer is needed. For example, deer are sometimes sexually mature before they are one year old, and twins or triplets can occur. What else would you need to know to answer this question fully? In what phase of population growth was the deer population in 1930? In 1940? What would prevent the number of deer from increasing forever? Note that the curve is not a neat smooth one but has a series of abrupt spikes. What is the reason for this? What is the explanation for the drastic decline in population in the mid-1930s? Why was there an increase in the next few years?

Recently, one of the research activities at the George Reserve has been an attempt to manipulate the population by harvest and to reduce it to a particular level. Contrary to expectations, when the population harvest was increased, there was not an equivalent steplike decrease in population.

Can you think of any reasons why there is a gradual decline in the population, rather than a stair-step reduction?

## Suggested Readings


Andrewartha, H. C. *Introduction to the Study of Animal Populations.* 2d ed. Chicago: The University of Chicago Press, 1971. An excellent treatment of the many factors that contribute to determining population size and distribution.

Ehrlich, Paul R. *The Population Bomb.* New York: Ballantine Books, 1976. A popular, persuasive paperback designed to alert the general public.

Haupt, Arthur, and Kane, Thomas T. *Population Handbook: A Quick Guide*

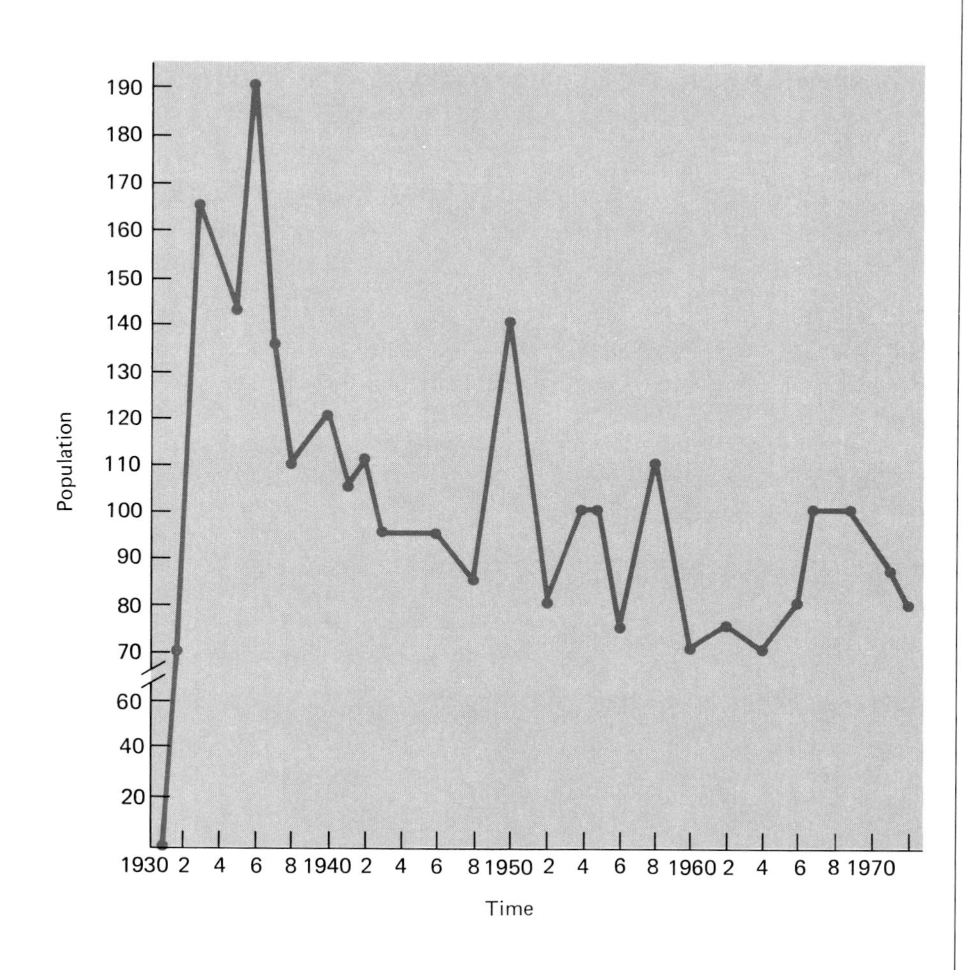


*to Population Dynamics for Journalists, Policymakers, Teachers, Students and Other People Interested in People.* Updated ed. Washington, D.C.: The Population Reference Bureau, 1982. This, along with other references from the Population Reference Bureau, provides the most up-to-date information related to human population topics.

McCullough, Dale R. *The George Reserve Deer Herd.* Ann Arbor, Mich.: The University of Michigan Press, 1979. This provides a detailed study of the George Reserve deer herd.

## Chapter Outline

I. Social implications of population growth
II. United States population changes
III. World population changes
IV. The demographic transition
V. Environmental costs
  A. Food pyramid
  B. Pollution is a cost
VI. Standard of living: A comparison
    Box 7.1 Governmental policy and population control
VII. Changes we should anticipate
VIII. Consider this case study: Costa Rica

## Objectives

Be able to apply some of the principles discussed in chapter 6 to the human population.

Differentiate between birthrate and population growth rate.

Explain the current population situation in the United States.

Explain how age distribution affects population projections.

Recognize that our society is in the process of adjusting as the average age in the United States increases.

Recognize that most of the world still has a rapidly growing population.

Describe the implications of the hypothesis that a demographic transition occurs.

Understand how increased world population will alter the worldwide ecosystem.

Understand the implications of the fact that most of the countries of the world are not able to produce enough food to feed themselves.

Explain why less-developed nations have high birthrates and why they will continue to have a low standard of living.

Recognize that the developed nations of the world will be under greater pressure to share their abundance.

## Key Terms

demographic transition

demography

gross national product (GNP)

postwar "baby boom"

replacement fertility

standard of living

zero population growth

# Chapter 7
# Human population issues

## Social implications of population growth

In chapter 6, we examined populations in general. Population characteristics, the way populations grow, and the forces that cause populations to stabilize were discussed. These forces also act on human populations. However, unlike other kinds of populations, human populations are influenced by a variety of social, political, and economic forces. Humans are able to predict the likely course of events and adjust their activities accordingly. They can make conscious decisions based on what they perceive the future political, economic, or social pressures will be. Therefore, people spend considerable time and energy trying to ascertain the future so they can plan their lives.

Decisions made by individuals can affect a whole population and cause changes at the social or political level, which may have international implications. For example, war has a basis in population pressure. During World War II, both Germany and Japan were partially motivated by a need to relieve population pressures by acquiring new land.

Recently, because of the implications of human population growth, the press has written many articles about the prospects for controlling population and the impact of population growth on our standard of living. Unfortunately, many of the news stories and discussions of population trends are misleading. Projections of population growth are logical guesses about what will happen in the future. Any discussion of future events is a guess, even though it may be a very well-informed guess. Experts will differ because they use different sets of data to make their projections. Many of these projections are based on statistics of past population trends and assume that future populations will grow at the same rate.

Another important point is that some experts have a history of being accurate while others do not. When we listen to accounts of what our future will be like, we should evaluate the credibility of the experts. Because humans are highly social organisms, we cannot look at the human population from a strictly biological point of view. How are the biological and social natures of the human species interrelated in the United States and the rest of the world?

## United States population changes

Individuals in a population tend to produce more offspring than the number required to just replace the parents when they die. Therefore, populations tend to increase until the environment cannot support any more people. At this point, the number of births will become equal to the number of deaths. This regulation may involve either a decrease in births or an increase in deaths. The result is the same whether populations are controlled by increasing deaths or decreasing births. Each mechanism operates in human populations in different countries. Some countries, such as Ethiopia and Afghanistan, have high birthrates and high deathrates. In general, a high deathrate is not considered to be a humane way of controlling population. These countries usually have an extremely high deathrate among children because of disease and malnutrition. Other countries, such as the United States and much of Europe, have low birthrates and low deathrates. (See table 7.1.) The children in these countries have much higher life expectancies.

The study of populations, their characteristics, and what happens to them is known as **demography.** Demographers have made several predictions about what will happen to the population of the United States by the year 2000. The current population of the United States is about 232 million people. The United States Commerce Department estimates that by the year 2000 the United States population will reach about 268 million people. This is about a 13 percent increase in fifteen years, or just less than 1 percent per year. Yet we continue to hear that the United States is reproducing at a rate that will lead to

Table 7.1
Population growth rates in selected countries (1985)

| Country | Births per 1,000 individuals | Deaths per 1,000 individuals | Rate of natural increase (annual %) | Time needed to double population (years) |
|---|---|---|---|---|
| Germany (West) | 10 | 11 | −0.2 | — |
| Belgium | 12 | 11 | 0.1 | 1,155 |
| Sweden | 11 | 11 | 0.0 | 6,930 |
| United Kingdom | 13 | 12 | 0.1 | 630 |
| United States | 16 | 9 | 0.7 | 100 |
| Japan | 13 | 6 | 0.6 | 110 |
| USSR | 20 | 10 | 1.0 | 71 |
| Iceland | 19 | 7 | 1.2 | 60 |
| Ethiopia | 43 | 22 | 2.1 | 33 |
| Paraguay | 35 | 7 | 2.8 | 25 |
| Afghanistan | 48 | 23 | 2.5 | 28 |
| Viet Nam | 34 | 9 | 2.5 | 28 |
| Iran | 41 | 10 | 3.0 | 23 |
| Oman | 47 | 16 | 3.1 | 22 |
| Botswana | 50 | 13 | 3.7 | 19 |

Source: 1985 World population Data Sheet, Population Reference Bureau, Inc.

**zero population growth.** A population at zero population growth has reached a stationary growth stage where births equal deaths. However, even after a country reaches a birthrate that just replaces parents, it takes many years for the population growth to stop. This condition exists because the deathrate may continue to fall as people live longer.

Another factor affecting our population growth rate is the amount of immigration. In the United States about half of the annual increase in population is due to immigrants, which number about 1 million persons per year. This number is somewhat uncertain, since the number of illegal immigrants has been estimated to be between 100,000 and 500,000 per year. There are about 700,000 legal immigrants per year.

A third reason populations continue to grow even after reaching a reproductive rate that will lead to zero population growth is the age structure of a population. If a population has a large number of young people who are in the process of raising families or who will be raising families in the near future, there will be an increase in the population even if the families limit themselves to two children. Certainly the United States population is not growing as rapidly today as it was twenty years ago, but it is still growing. (See fig. 7.1.) In the long run, if the number of children produced in the population is 2.1 children per woman, the population growth rate will become zero. This is known as **replacement fertility.** A fertility rate of 2.1 children per woman is used because some babies die soon after birth and do not contribute to the population for very long.

Depending on the number of young people in a population, it may take twenty years to a century for the population to stabilize so that there is no net growth. Regardless of what the birthrate is, there is no way, short of disaster,

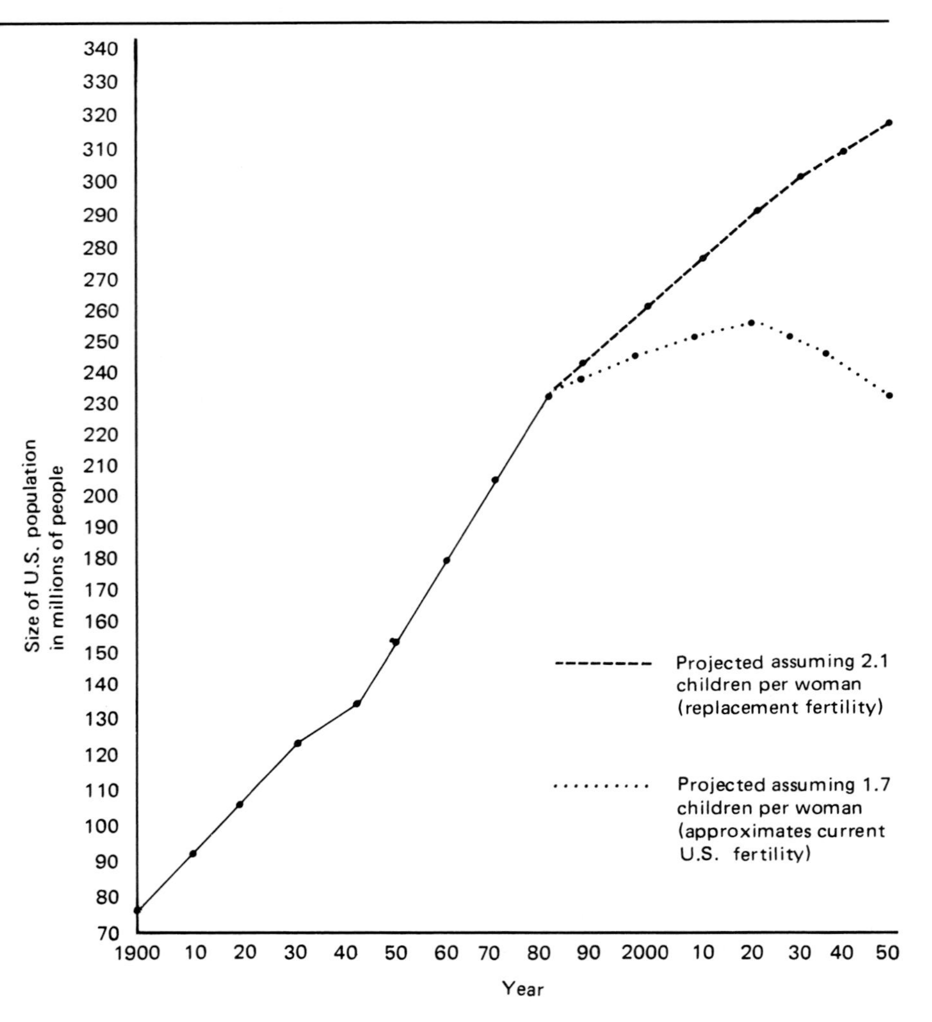


**Figure 7.1 Population growth.**
The population of the United States has grown continuously until the present. The graph indicates that the size of the population had risen to about 230 million people in 1980. The United States Census Bureau has made projections based on a birthrate of 2.1 children per woman (the rate at which the population will eventually become stable) and 1.7 children per woman (the current birthrate). Both will eventually result in a leveling off of population growth. There is a considerable difference in the ultimate size of the United States population depending on which of the estimates is used. In both cases, the population will continue to increase at least until the year 2020, and if birthrates are above 1.7 children per woman, the population will continue to increase past the year 2020. (Data from the United States Department of Commerce, Bureau of the Census.)

that the population of the United States can be prevented from growing between now and the year 2000. Demographers will be watching the United States birthrate closely for the next few years to see if we can actually expect the population size to become constant or decline.

Another item of interest to demographers is the age distribution within the population. Populations may have a majority of individuals in the young age group, or they may have some other age distribution. Figure 7.2 shows the age distribution in the United States in 1980. A decided "bulge" begins at the age of thirty-three years. These people would have been born about a year after the end of World War II and were the beginning of the **postwar "baby boom."** This "baby boom" peaked about fifteen years later, and births have continued at lower levels ever since. The women who were born during this period are now in the middle of their reproductive years. Many of these women

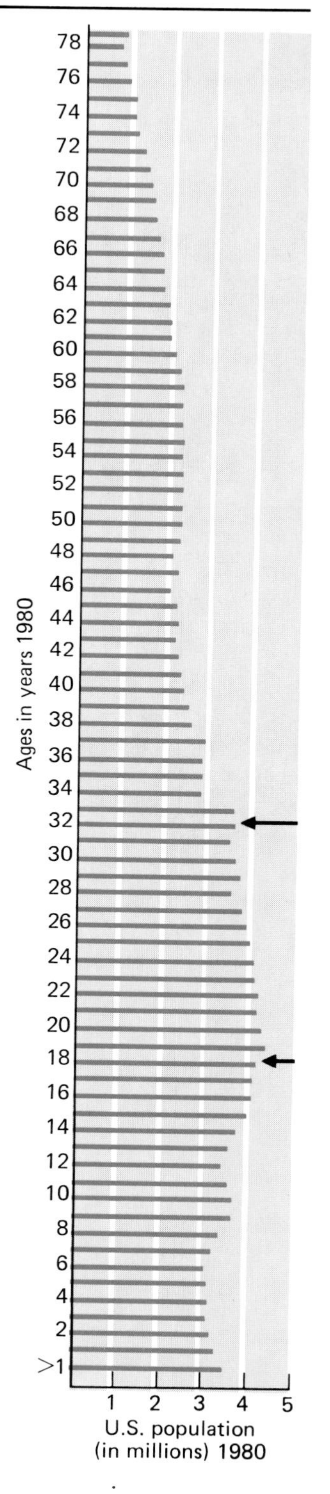


**Figure 7.2 Age distribution of United States population (1980).** This graph shows the number of people at each age level. Notice that at age thirty-three a bulge begins that peaks at about age nineteen. These people represent the "baby boom" that followed World War II. Since then, the birthrate has been declining. (Data from the United States Department of Commerce, Bureau of the Census.)

have delayed getting married and are having fewer children than their mothers did. If this trend continues, it will have a decided effect on the population for the United States beyond the year 2000. Later marriages tend to reduce fertility because there are fewer fertile years per woman in which to produce children. Data from the Population Reference Bureau indicate that 44 percent of African women in the 15- to 19-year-old age group are married and that the fertility rate is about 6.5 children per woman. In Europe only 7 percent of the 15- to 19-year-old women are married, and the fertility rate is 1.9 children per woman. Many people suggest that social changes in North America and Europe, particularly changes in the role of women, are related to the decline in birthrate. As women become better educated and obtain higher paying jobs, they become financially independent and can afford to marry later and consequently have fewer children. Changing attitudes toward divorce may also be tied to reduced fertility.

With the recent decline in the birthrate, there has been a corresponding change in the businesses that serve specific age groups. The "baby boom" of the late 1940s and 1950s resulted in growth in the service industries that are needed by young families. Maternity wards had to be expanded, schools could not be built fast enough, baby care companies saw unprecedented sales, and the toy industry flourished. Today these "babies" are in their thirties and are now buying homes, cars, and appliances. Schools are being closed because of a lack of students. Hospitals are reducing their maternity wards and are beginning to anticipate the need for geriatric services as the average age of the United States population increases. What will our social needs be in the year 2020 when many of these "baby boom" babies retire? Will they be demanding social services similar to those provided to their parents? As our population size stabilizes, the average age will increase. Young populations are growing populations, which describes much of the rest of the world.

## World population changes

Currently the world population is approximately 4.8 billion people. By the year 2000 this is expected to increase to over 6 billion people. Much of this increase is expected to occur in Africa, Asia, and Latin America, which already have about 77 percent of the world population. (See fig. 7.3.)

If world population trends continue, these three regions will increase from their current population size of about 3.6 billion people to about 5.0 billion people by the year 2000. If projections are correct, they would then constitute over 80 percent of the world's population. These regions not only have the highest population growth rates but also have the lowest per capita **gross national product (GNP).** The GNP is an index that measures the total goods and

Figure 7.3 **Population growth in the world.** The population of the world is not evenly distributed. Currently about 78 percent of the world's population is in the regions shown in color on the map. These areas also have the highest rates of increase and are generally considered to be less developed. Because of the high birthrates, they are likely to remain less developed and will constitute about 80 percent of the world's population by the year 2000.

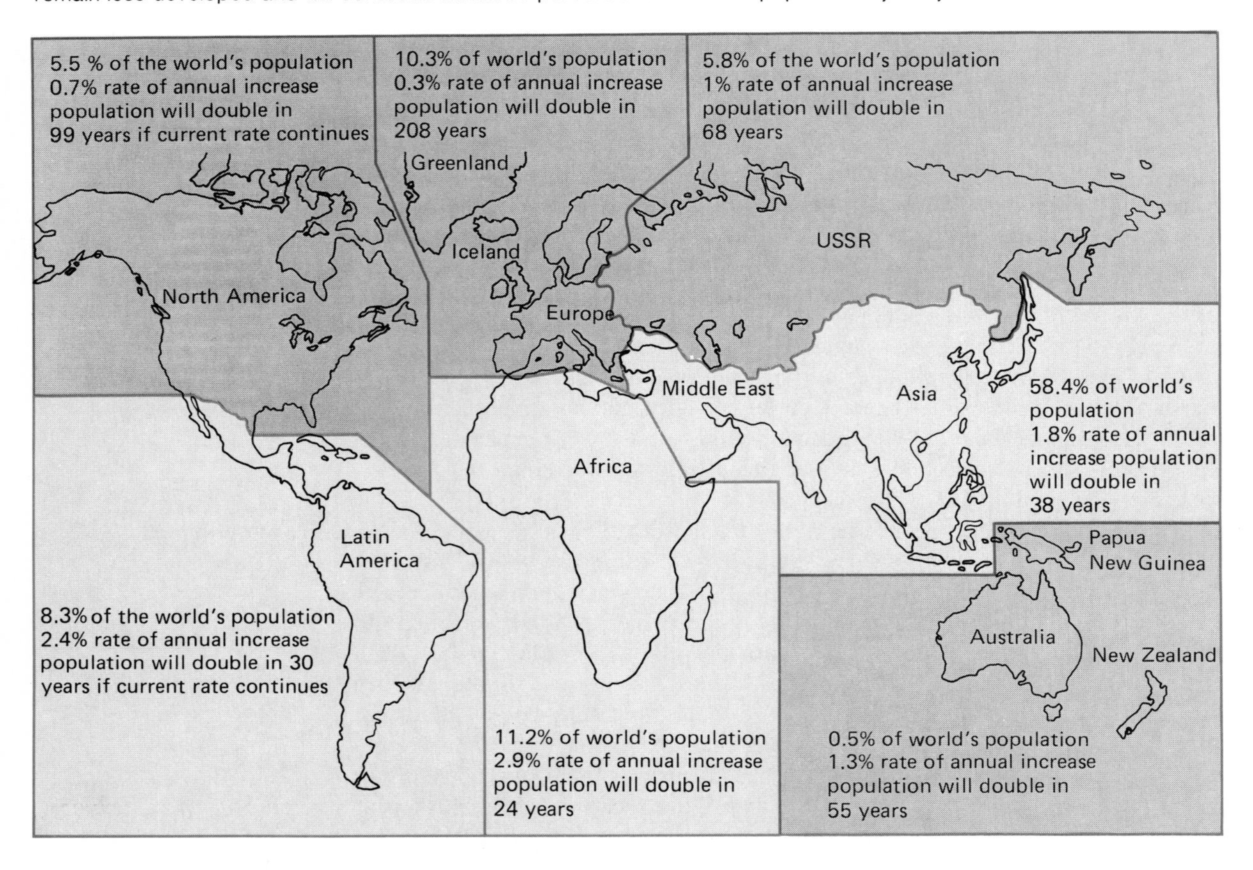


services generated within a country. (See fig. 7.4.) Obviously, there is a wide gap between these less-developed countries with a low GNP and the developed countries with a high GNP. Yet these less-developed countries aspire to the same **standard of living** exhibited by developed countries of the world.

## The demographic transition

The relationship between standard of living and birthrate seems to be one in which those countries with the highest standard of living have the lowest population growth rate, and those with the lowest standard of living have the highest population growth rate. This has led many people to suggest that countries naturally go through a series of stages called a **demographic transition.** This model is based on the historical, social, and economic development of Europe and North America. It suggests that in a demographic transition the following four stages occur. (See fig. 7.5.)

**Figure 7.4 Per capita gross national product of nations.** The world can be arbitrarily divided into developed and less-developed regions. One measure of the degree to which a country is developed is its per capita gross national product. The less-developed countries produce very little for each person in the country, while the developed countries have high productivity, and, therefore, also have goods and services available for consumption by its people.

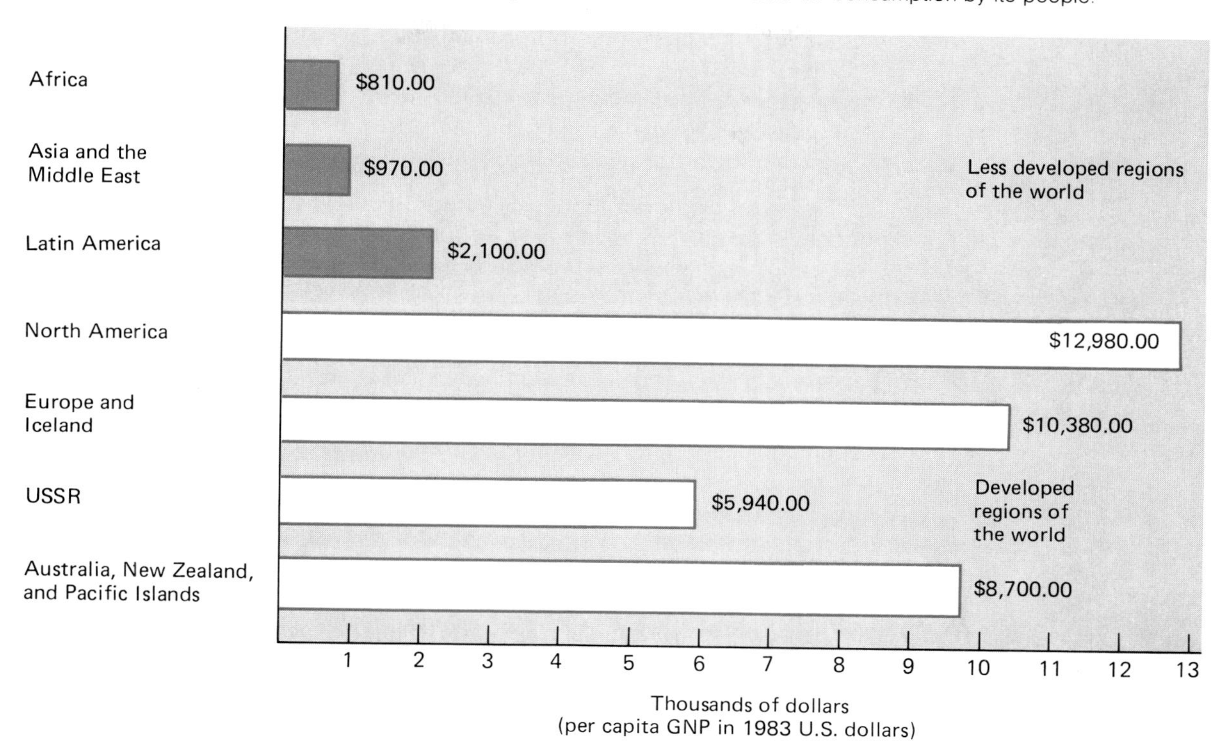


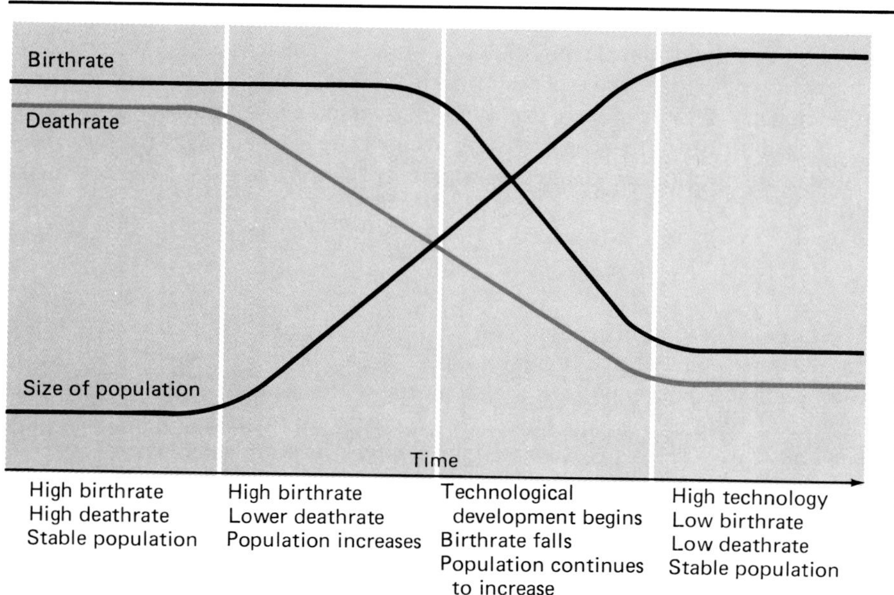


**Figure 7.5 Demographic transition.** The demographic transition model suggests that as a country develops technologically, it will automatically experience a drop in the birthrate. This is certainly what has been experienced in the developed countries of the world. However, the developed countries make up less than 25 percent of the world's population. It is doubtful whether the less-developed countries can achieve the kind of technological advances experienced in the developed world.

1. Initially, countries have a stable population with a high birthrate and a high deathrate. Deathrates are often quite variable because of famine and epidemic disease.
2. Improved economic and social conditions (control of disease and increased food availability) bring about a period of rapid population growth as deathrates fall. Birthrates remain high.
3. As countries become industrialized, the birthrates begin to drop because people make use of contraceptives.
4. Eventually, birthrates and deathrates will again become balanced.

This is a very comfortable model, because it suggests that if a country can become industrialized, then social, political, and economic processes will naturally cause its population to stabilize.

However, this leads to some serious questions. Can the historical pattern exhibited by Europe and North America be repeated in the less-developed countries? Europe, North America, Japan, and Australia passed through this transition period when energy and natural resources were still abundant. It is doubtful whether these supplies are adequate to allow for the industrialization of the major portion of the world that is now classified as less developed.

A second concern is the time element. With world population increasing as rapidly as it is, it is doubtful that industrialization can occur fast enough to have a significant impact on population growth. In many less-developed countries, the number of children produced is a form of social security. Children will take care of their parents in old age. Only people in developed countries can save money for their old age. Thus, they can make decisions about whether to have children, who are expensive to raise, or to invest money in some other way to provide for their old age. In other words, in less-developed countries, children are considered to be an economic asset, and in developed countries, they are an economic liability.

When Europe and North America passed through this transition, they had access to large expanses of unexploited lands. This provided a safety valve for expanding populations during the early stages of the transition. Without this safety valve it would have been impossible to deal adequately with the population while simultaneously encouraging economic development. Today, less-developed countries may be unable to accumulate the necessary capital to develop economically, since an ever-increasing population is a severe economic drain.

## Environmental costs

The human population can increase only at the expense of populations of other animals and plants. Each ecosystem has a finite carrying capacity and, therefore, has a maximum biomass that can exist within that ecosystem. There can be shifts within ecosystems to allow an increase in the population of one species, but this will always adversely affect certain other populations because they will be competing for the same basic resources. When the population of farmers increased in the prairie, the population of buffalo declined (see chap. 4). One basic need that all people have is food. When humans need food, they turn to agricultural practices and convert natural ecosystems to artificially maintained agricultural ecosystems. Mismanaged agricultural resources are often irreversibly destroyed. The Dust Bowl of the United States and problems in tropical rain forests are well-known examples. (See chapter 13 and box 5.1.) Humans may eventually degrade the biological productivity of a region rather than increase it. However, to a starving population, the short-term gain is all that is important.

## Food pyramid

In countries where the population is increasing, the pressures on their agricultural resources are also increasing. A consequence of this basic need for food is that people in less-developed countries generally feed at lower trophic levels than do those in the developed world. (See fig. 7.6.) Converting low-quality plant material into high-quality animal protein is an expensive process. During this conversion, 90 percent of the energy present in the plants is lost. Therefore, it is important economically and energetically for people in less-developed countries to consume the plants themselves rather than to feed the plants to animals and then consume the animals. In most cases, if they did feed the plants to animals, many people would starve to death. On the other hand, a lack of protein in the diet leads to malnutrition. In countries where food is in short supply, agricultural land is already being exploited to its limit, and there is still a need for more food. Most of Europe also needs to import food. This leaves the United States, Canada, Australia, Argentina, and New Zealand as net food exporters. The rest of the world does not grow enough food to have any left for export. Often, most of these countries do not grow enough to meet their own needs and, therefore, must import food.

## Pollution is a cost

There is little doubt that the United States is also a leader in the production and consumption of goods other than food. Although a few countries such as Switzerland, Sweden, Qatar, Kuwait, and the United Arab Emirates have a higher per capita GNP than the United States, their total population is only a small percentage of the United States population. Therefore, the impact of the United States is much greater. Our consumption of energy is especially large because it fuels our standard of living. The ultimate cost of maintaining a particular standard of living must include not only the economic costs but also the costs to the environment (e.g., air pollution, water pollution, loss of farmland, and the buildup of pollutants in the environment). It is primarily in the more developed countries with their high energy demands that environmental costs have become an issue.

## Box 7.1
## Governmental policy and population control

The actions of government can have a significant impact on the population growth pattern within a nation. Some policies are aimed at either stimulating an increase in population or attempting to control population. The Canadian government pays a bonus to the parents upon the birth of a child. Other countries give incentives to couples to encourage them to have no children or to limit their family to one child. Many countries make contraceptives available at no cost and have programs to educate the people about effective conception control.

Some governments are more subtle in their policies. Most of industrialized Europe has a low birthrate but needs a large labor force. They rely on importing labor from less-industrialized nations. Therefore, they do not need to encourage births in their own countries. They can simply import the labor when it is needed and deport the workers when there is no need for them. This also reduces the cost of production, since they do not need to pay many of the fringe benefits to alien workers.

Immigration policies of a government can have a significant effect on the population growth of a country. Some countries, such as Australia, encourage people (particularly Europeans) to immigrate. They can have their passage paid if they remain in the country. However, most countries have strict limits on the number of people who can enter the country each year.

Some countries have a profound effect on population growth within their country by taking no action at all. For example, many governments in South and Central America have taken no action to control their

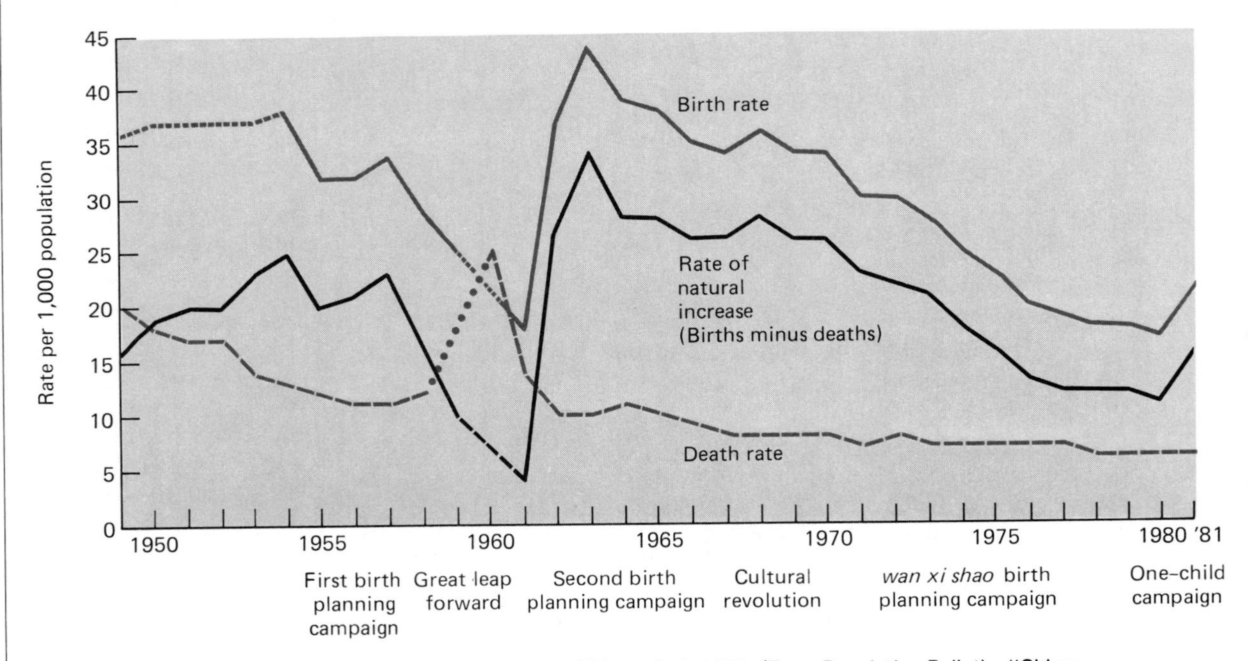


**Box figure 7.1** Birth, death, and natural increase rates: China, 1919–1982. (From Population Bulletin, "China: Demographic Billionaire," by H. Yuan Tien, 1983. Population Reference Bureau, Inc.)

populations because of the conservative Catholic prohibition on birth control. This in effect assures a continued high birthrate in these countries.

In an attempt to curtail population increases, the People's Republic of China established the *wan xi shao* family-planning program in 1971. Translated, this means "later" (marriages), "longer" (intervals between births), "fewer" (children). China has long been the most populated nation; it now contains about one-fourth of the world's population.

When the People's Republic of China was established in 1949, the population was 540 million. Because of a high birthrate, the population increased to 614 million by 1955. This rapid increase in population resulted in some changing attitudes. Abortions became legal in 1953, and the first family-planning campaign began in 1955 (see box fig. 7.1.), followed by a second family-planning program in 1962.

The forerunner of the present family-planning policy began in 1971 with the launching of the *wan xi shao* campaign. As part of this program, the legal ages for marriages were raised. For women and men in rural areas, the ages were raised to twenty-three and twenty-five, respectively; for women and men in urban areas, the ages were raised to twenty-five and twenty-eight, respectively. These steps reduced the birthrate. (See box fig. 7.1.)

Now the People's Republic of China has the one-child campaign. (See box fig. 7.2.) Initial surveys indicate that 20 percent of the married couples favor this program. To secure more participation, a series of penalties and rewards has been established for those who pledge to have no more than one child. These couples are awarded the "only-child glory certificate." Recipients of these certificates receive free medical care, cash bonuses for their work, special housing treatment, and extra old-age pensions. The child receives free school tuition and preferential treatment when entering the job market.

Those who break their pledge and have additional children forfeit all benefits after the birth of their second child. If they have a third child, their wages are reduced by 10 percent, they must live in a housing area designed for a smaller family, and they are required to pay for the food for the third child.

Since the introduction of reward and punishment, the one-child program has been more favorably received. Several large urban areas report a 95 percent adherence to the policy; however, in rural areas only 25 percent favor the program. How effective this policy will be remains to be seen.

Box figure 7.2 One-child program.

**Figure 7.6 Populations and trophic levels.**
The larger a population is, the more energy it takes to sustain the population. Every time one organism is eaten by another organism, approximately 90 percent of the energy is lost. Therefore, when populations are very large, they usually feed at the herbivore trophic level because they cannot afford the 90 percent energy loss that occurs when plants are fed to animals. The same amount of grain can support ten times more people at the herbivore level than at the carnivore level.

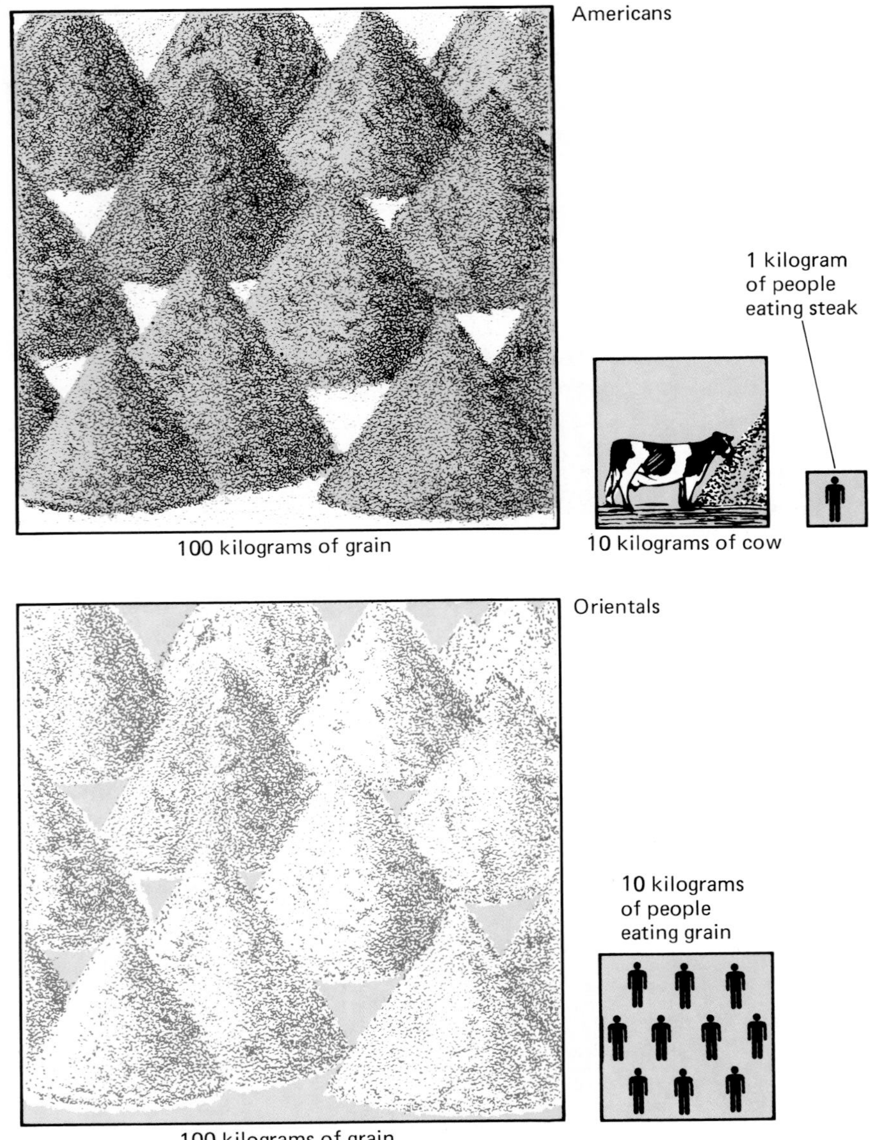


## Standard of living: A comparison

The standard of living is a difficult concept to quantify since different cultures have different attitudes and feelings about what is good and desirable. For the sake of comparison, we will look at three countries and compare them with respect to several aspects of their culture. We will look at averages. Even though most countries have both rich and poor people, we should be able to get some

**Figure 7.7 Standard of living in three countries.** Standard of living is an arbitrary measure of how well one lives. It is not possible to get a precise definition, but when we compare the United States, Argentina, and India, it is obvious that there are great differences in how the people in these countries live. India has a low life expectancy, a high deathrate, and low productivity, and the people are often starving. The United States has a high life expectancy, a low deathrate, and high productivity, and the people eat too much. Argentina is intermediate in all of these characteristics.

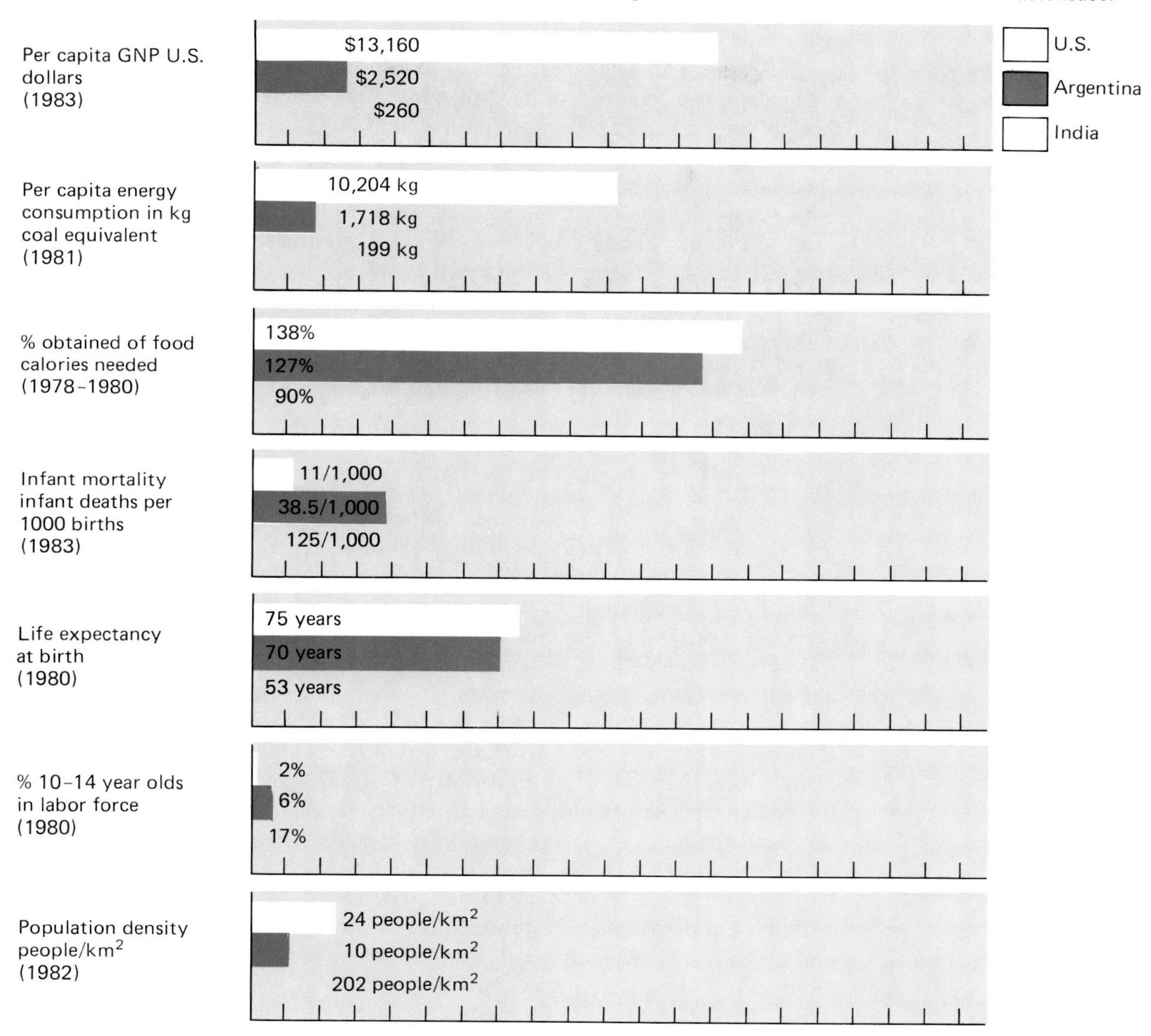


idea of what life is like for the average person in each country. The three countries chosen are the United States, which is an example of a highly developed industrialized country; Argentina, which is a moderately developed country; and India, which is less developed. Figure 7.7 shows several statistics that relate to the standard of living in these three countries. The United States produces approximately five times more goods and services per person than does

Argentina and over fifty times more goods and services than India. United States citizens consume six times more energy than the average Argentinean and fifty times more than the average Indian. Infant death in India is very high, Argentina is intermediate, and the United States infant deathrate is low. People live longer in the United States and go to school longer than in the other two countries. India is much more densely populated than either Argentina or the United States. The average Indian is starving, a factor related to high infant mortality and low life expectancy, whereas the average citizen of the United States and Argentina has more food than is needed.

Obviously, there are tremendous differences in the standard of living among these three countries. What the average United States citizen would consider to be a poverty level of existence would be considered a luxurious life for the average Indian. Standard of living seems to be rather closely tied to energy consumption. In general, the higher the per capita energy consumption, the higher the standard of living of the people. This is related to the degree of industrialization or development of the country. However, with industrialization comes pollution; huge amounts of fossil fuels are consumed to provide the goods and services that contribute to the standard of living.

## Changes we should anticipate

If the human population continues to increase (and it appears that it will), the pressures for the necessities of life will become greater. Differences in standard of living between developed and less-developed countries will remain great because most population increases will occur in less-developed countries. The supply of fuel and other resources is dwindling. The pressures for these resources will intensify as the industrialized countries seek to maintain their current standard of living. People in less-developed countries will continue to seek more land to raise the crops needed to feed themselves. Since most of these people live in tropical areas, tropical forests will be cleared for farmland. Using tropical forests as farmland often causes erosion or alteration of the soil, which can then no longer support either forests or crops. This could cause profound changes in the world ecosystem as natural ecosystems are converted to agricultural ecosystems. Developed countries will have to choose between helping the less-developed countries and maintaining their friendship or building a wall around themselves in relative splendor and isolation. Neither one of these governmental policies will be able to prevent a change in life-style as the world population increases. The resources of the world are finite. Even if the industrialized countries continue to get a disproportionate share of the world's resources, the amount of resource per person will decline as population rises. It seems that as world population increases, the less-developed areas will maintain their low standard of living. It is difficult to see how their standard of living could get much lower since some of their people are already starving to death. Other countries will probably see their life-styles become less consumption oriented. Some things that are currently necessities (television sets, microwave ovens, vacations to remote sites, and two cars per family) will probably become luxuries. Many people who enjoy the freedom of mobility associated with the automobile will have their travel limited as public transportation replaces private transportation. Recreation will probably not involve expensive, energy-demanding machines (television games, motorcycles, and electric-powered toys) but will emphasize activities such as hiking, bicycling, and reading. These changes need not be thought of as catastrophic changes. Most likely, many changes will occur slowly as economic pressures affect families. We will probably be healthier, under less pressure, and certainly more in balance with the rest of the world in the future.

## Summary

Demography is the study of human populations and the things that affect them. Demographers study the sex ratio and age distribution within a population to predict population growth.

Many social changes, such as later marriages and the role of women, affect population growth. The current birthrate of the United States population will eventually result in zero population growth, a stationary growth stage in which births equal deaths. If current trends continue, Africa, Asia, and Latin America will be the three areas of most rapid population growth. These are also the areas with the lowest GNP and lower standards of living than the industrialized nations.

The demographic transition model suggests that as a country becomes industrialized, its population begins to stabilize, but there is little hope that the earth can support the entire world in the style of industrialized nations. It is doubtful whether there is enough energy and natural resources to develop these areas or enough time to change the trends of population growth.

The United States is a leader in the production and consumption of goods. Highly developed countries should anticipate increased pressure to share their wealth with less-developed countries.

## Review Questions

1. What is demography?
2. What is demographic transition? What is it based upon?
3. What is the “baby boom” of the United States population?
4. What does age distribution of a population mean?
5. List ten differences between your standard of living and that of someone in a less-developed country.
6. Why do people who live in overpopulated countries use plants as their main source of food?

## Consider This Case Study
### Costa Rica

Costa Rica is a small Central American country located between Nicaragua and Panama. Its economy is primarily agricultural, with a per capita GNP (as of 1983) of $1,020. The 1985 population is estimated at 2.6 million.

According to the World Fertility Survey report, fertility has declined substantially during the past two decades; according to estimates, Costa Rica's crude birthrate peaked at forty-eight births per thousand population in the early 1960s and then declined steadily to about thirty in 1976. In 1960 the total fertility rate (average births per woman) was high, at 7.3 births per woman. By 1983 the total fertility rate had declined to 3.5. Altogether, the average number of births per woman declined by half in just twenty years.

On the average, women who had no living children wished to have 2.4 children; those with one living child wanted to have an average of 1.6 more children. As a general rule, the Survey shows that as the number of living children increases, the desire for additional children decreases.

Fertility was found to be lower and contraceptive use higher among the better-educated women and women living in urban areas. The Survey indicates that women with less than three years of education had twice as many children as did women who attended school for eleven years or more, regardless of the number of years since first marriage. Women in metropolitan areas were found to have the lowest fertility, averaging 3.3 children. In rural areas, the average number of children born per woman ranged from 4.6 to 5.4.

Almost three-fourths of rural women exposed to risk of pregnancy were found to be using contraception, compared to 81 percent of urban women. As might be expected, the percentage of exposed women currently using contraception was found to be highest among the better-educated women.

The Survey showed that the most common method of contraception in Costa Rica is the pill. The next most common method is female sterilization.

The Survey concludes that "Fertility in Costa Rica declined because contraceptive practices became generalized among couples. This occurred simultaneously with a rapid process of modernization across all levels of society and the economy that made the large family norm dysfunctional. In addition, the existence since 1968 of a dynamic family planning program accelerated the fertility decline, since it put within reach of the many sectors of the population the means to limit their fertility."

Does this case study make you optimistic or pessimistic about the future of the world population?

Is a demographic transition occurring in Costa Rica?

Is contraception important in controlling Costa Rica's population?

Is Costa Rica's population under control?

Adapted from INTERCOM, the International Population News Magazine of the Population Reference Bureau, Inc.

7. Although predicting the future is difficult, what do you think your life will be like in the year 2000? Why?
8. List five changes that would occur if world population were to double in the next fifty years.
9. Which three areas of the world have the highest population growth rate? Which three areas of the world have the lowest standard of living?
10. How many children per woman would lead to a stable United States population?

## Suggested Readings

Burch, William R., Jr., ed. *Readings in Ecology, Energy, and Human Society: Contemporary Perspectives.* New York: Harper & Row, Publishers, 1977. Included in this book are discussions of the Club of Rome report and Garrett Hardin's lifeboat ethics.

Coale, Ansley J. "Recent Trends in Fertility in Less Developed Countries." *Science* 221: 828–830. Presents the implications of declining birth rates in these countries.

de Souza, Anthony R., and Foust, J. Brady. *World Space-Economy.* Columbus, Ohio: Charles E. Merrill Publishing Company, 1979. An economic geography text that includes good discussions on population, less-developed versus developed countries, and international trade.

Ehrlich, Paul R., and Ehrlich, Anne H. *Population, Resources, Environment.* 3d ed. San Francisco, Calif.: W. H. Freeman & Company Publishers, 1977. Provides a good introduction to the subject of human population problems.

Gwatkin, Davidson R. "Life Expectancy and Population Growth in the Third World." *Scientific American* 246: 57–65. Considers birthrate, deathrate, and life expectancy in population projections to the year 2100.

Hardin, Garrett, ed. *Population, Evolution and Birth Control.* 2d ed. San Francisco, Calif.: W. H. Freeman & Company Publishers, 1969. Contains a series of short readings about three controversial topics.

Keely, Charles B. "Illegal Migration." *Scientific American* 246: 41–47. Presents data and social implications regarding illegal immigration into the United States.

Population Reference Bureau, Inc. 1337 Connecticut Avenue N.W. Washington, D.C. 20036. Publishes population materials of many kinds. A good source of up-to-date information on world population trends.

# Part 3

# Energy

All living systems can be described by the flow of energy through them. Energy enables simple forms of matter to be changed into more complex forms; energy is needed to maintain this complexity; and energy expenditure is also necessary to sustain the complex technical and social units typical of human populations.

All living things (including people) rely on the sun as a source of energy. Wood, coal, and oil are sources of energy that are available today because organisms in the past captured sunlight energy and stored it in the complex organic molecules that made up their bodies.

Technological development and fossil fuel exploitation are directly related to one another. Their use has allowed us to have a unique standard of living. Chapter 8 traces the development of energy consumption, its interrelationship with economic development and lifestyle, and our current supply problem. Chapter 9 discusses the sources of energy that are currently being used as well as the significance of each source and its impact. Chapter 10 discusses the use and concerns of nuclear energy. Chapter 11 discusses a variety of emerging energy technologies, their potential value, and the kinds of problems their use could cause.

## Chapter Outline

- I. History of energy consumption
  - A. Basic energy requirements
  - B. Wood
    - Box 8.1 Deforestation and wood shortages
  - C. Coal
  - D. Oil and natural gas
- II. Energy and economics
  - A. Economic growth and energy consumption
  - B. Progress: The American dream
  - C. Gross national product, energy, and life-style
- III. Approaches to our energy situation
  - A. Historical energy growth pattern
  - B. Zero energy growth pattern
  - C. Decreasing energy growth pattern
    - Box 8.2 Davis, California: Energy-efficient community
  - D. Current energy trends
- IV. Consider this case study: Shirley Highway, Washington, D.C.

## Objectives

Explain why all organisms require a constant input of energy.

Describe how the per capita energy consumption increased as civilization developed from hunting and gathering through primitive agriculture to major ancient civilizations.

Describe why the development of advanced modern civilizations occurred as new fuels were used to run machines.

Correlate the Industrial Revolution with various social and economic changes.

Explain why cheap oil and gas led to the development of a consumption-oriented society.

Explain how the use of the automobile changed the life-style of the people of the United States.

Correlate between abundant energy and the rapid growth of goods and services (gross national product) in the United States.

List three possible future patterns of energy use.

## Key Terms

coal

cottage industry

decreasing energy growth pattern

factory system

government regulation

historical energy growth pattern

Industrial Revolution

natural gas

oil

wood

zero energy growth pattern

# CHAPTER 8
# Energy and civilization: Patterns of consumption

## History of energy consumption

Because of the second law of thermodynamics, every form of life and all societies require a constant input of energy. If the flow of energy through organisms or societies ceases, they stop growing and begin to disintegrate. However, some organisms and societies are more energy efficient than others. In general, more complex industrial societies have the greatest energy needs. History shows the correlation between the amount of energy used and the complexity of the society. If societies are to survive, they must continue to expend energy; however, it may be necessary to change the pattern of energy consumption as traditional sources become limited.

**Figure 8.1 Hunter-gatherer society.**
In this type of society, people obtain nearly all of their energy from the collection of wild plants and the hunting of animals. These societies do not make large demands on fossil fuels.

## *Basic energy requirements*

An energy input is essential to maintain life. In any ecosystem, the sun provides the energy to sustain all forms of life. (See chap. 3.) The first transfer of energy occurs during the process of photosynthesis, when plants convert light energy into chemical energy in the production of food. Herbivorous animals utilize the food energy in the plants. The herbivores in turn provide the source of energy for the carnivores. Because nearly all of their energy requirements were supplied by food, primitive humans were no different from other animals in an ecosystem. In such hunter-gatherer cultures, only biological energy demands are made. (See fig. 8.1)

Early in history, people began to use additional sources of energy to make their lives more comfortable. The controlled use of fire was the first change that required energy in a form other than food. Wood provided the fuel to meet this demand. The energy provided by the wood enabled them to cook their food, heat their dwellings, and develop a primitive form of metallurgy. Such advances separated humans from other animals.

Another major human advance was in the area of agriculture. The development of domesticated plants and animals provided a more dependable supply of food. People no longer needed to depend solely upon the gathering of wild

## Box 8.1
## Deforestation and wood shortages

For centuries, humans have utilized the forests as a source of fuel; at the same time, vast expanses of forests have been cleared to provide agricultural land. This has resulted in shortages of wood for fuel. At one time, the mountainous areas of China were 90 percent forested. Now, nearly all of these forests have been cleared to provide agricultural land, and an acute shortage of firewood in the densely populated areas of China and other regions of the Far East has resulted.

The same pattern occurred in Europe. The European forest began to be cleared for farming in the eighth century. To clear fields for farms, the Emperor Charlemagne told his officials, "whenever they found capable man, give them wood to clear." The main thrust, however, in clearing European forests began four hundred years later, in the twelfth century. At that time, the quest for wealth by various orders of monks and the nobility led to the clearing of large tracts of land. This pattern of clearing land to raise cash crops resulted in the destruction of forests in all parts of Europe. Between 1639 and 1655, the ruling class of Spain deforested the remnants of the Spanish forest. They obtained royal permission to use tree shoots for animal fodder "in every dry season." The practice of clearing the forest was also widespread in the British Isles. In 1458, a visitor to England wrote about a countryside so barren of trees that people were forced to "burn a kind of rock which may contain sulfur or some rich material."

The timber shortage in England had repercussions in the New World. The British crown established the very unpopular *broad-arrow* policy in 1695. According to this policy, royal deputies were sent to "the Province of Massachusetts Bay in New England" and with a broad arrow marked all trees that were "more than twenty-five inches in diameter one foot above the ground." These trees were reserved for use by the Royal Navy. This policy later included New York and New Jersey. But even before this time there were local shortages of wood for fuel in North America. In 1720, because of a timber shortage, the French in Canada enacted severe timber trespass laws. Also in 1720, New Hampshire limited the number of trees that a citizen might cut down at one time. One village even imposed a fine for every tree taken for other than home use. The timber shortage in Massachusetts was so critical that a Timber Trespass Act and a Timber Fire Law were enacted in 1742. Later, when the country expanded westward, local lumber shortages followed the settlers in their movement across the continent.

Today the United States forests are being considered as potential sources of fuel for industrial processes. One manufacturing company in the Midwest has developed a "steam and electric co-generation" plant (SECO). This operation uses *wood* to generate steam; the steam produces electricity for use in the manufacturing process and also serves as a source of heat for the buildings and manufacturing processes. Producing electricity and heat in a single operation is more economical to the company than producing only heat and buying electricity from a local utility company.

The fuel for this Midwestern manufacturing plant will come from logging operations within a 125 kilometer radius of the plant. The trees are cut and ground into chips for transportation to the plant. Two-thirds of the wood required comes from sawmill residue, agricultural and commercial land clearing, residential and municipal tree trimming, and scrap wood. These sources are often considered to be waste material and are discarded. This plan then, will help reduce a waste disposal problem. We seem to have come full circle—back to wood as a primary source of fuel.

**Figure 8.2 Animal power.**
This bas-relief panel from an Egyptian tomb depicts an important accomplishment in the development of the human civilization. With the use of domesticated animals, people had a source of power other than their own muscles.

plants and the hunting of wild animals for their food. In addition, domesticated animals furnished a source of energy for transportation, farming, and other tasks. (See fig. 8.2.)

Early civilizations, such as the Aztecs, Greeks, Egyptians, Romans, and Chinese, were based on human muscle, animal muscle, and fire as sources of energy. Although these civilizations were culturally advanced, they were still directly dependent upon plants for their energy.

## *Wood*

Except for limited use of some wind-powered and water-powered machines, **wood** furnished most of the energy and materials needed for developing civilizations. Home construction and shipbuilding utilized wood as lumber. Firewood supplied energy for home heating and cooking. Wood as a source of charcoal was used for smelting ore and metalworking. (See fig. 8.3.) The heavy use of wood in these societies eventually resulted in a shortage. People were forced to seek alternative forms of fuel.

Because of a long history of a high population density, India and some other parts of the world experienced a shortage of wood hundreds of years before Europe and the United States did. In many of these areas, animal dung replaced wood as a source of fuel. It is still an important fuel source today.

**Figure 8.3 Early metal working.**
Using wood as fuel, people could heat metal and work it into the desired shape.

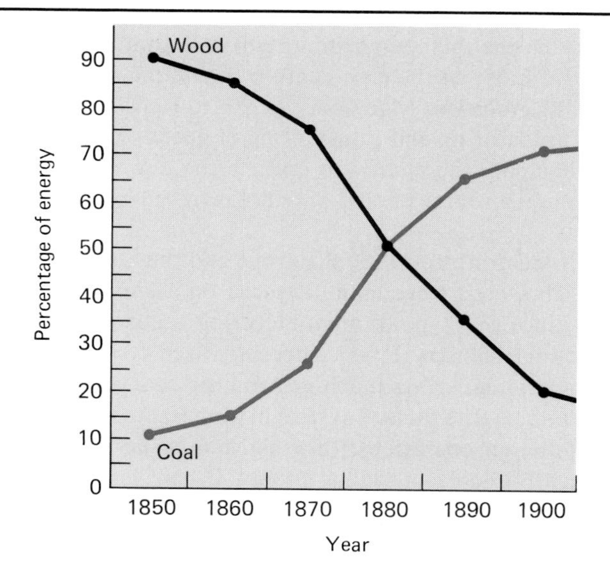


**Figure 8.4 Coal replaces wood.**
In 1850, wood furnished 90 percent of the United States energy, and coal most of the remaining 10 percent. Fifty years later, wood only supplied 20 percent of the energy and coal supplied 70 percent. The remainder was furnished by oil and natural gas. (Data from the United States Bureau of Mines.)

Western Europe and North America were able to use wood as a fuel for a longer period of time. The forests of Europe supplied sufficient fuel until the thirteenth century. In the United States, the vast expanses of virgin forests supplied adequate fuel until 1870. Fortunately, when local supplies of wood declined in Europe and the United States, coal was available as an alternative source of energy. Large deposits of easily obtainable coal were discovered, and coal replaced wood as a major source of fuel. (See fig. 8.4)

**Figure 8.5 Cottage industry.**
In this type of decentralized manufacturing, human energy was the main source of power, and the manufacturing of goods was mainly a family effort.

## *Coal*

The availability of **coal** played an important role in determining social and economic growth. Early in the eighteenth century, the world entered the era of the **Industrial Revolution.** Machinery began to replace human and animal labor in the manufacturing and transporting of goods. A large amount of fuel was required to supply the energy to operate the new industrial machinery. The nations without a source of coal were not involved in this developing technology.

Prior to the Industrial Revolution, Europe and the United States were predominately rural. Goods were manufactured on a small scale in the home. (See fig. 8.5.) Since this type of manufacturing occurred in people's homes, it was called **cottage industry.** It was a decentralized system of manufacturing that relied on human labor. As machines and the coal to power them became increasingly available, the **factory system** of manufacturing products replaced the small home-based operations. (See fig. 8.6.) The growing factories required a constantly increasing labor supply. People left the farms and congregated in the areas surrounding the factories. Villages became towns, and towns became cities. With these changes came problems typically associated with large aggregations of people. One of these was a change in the social structure. In a cottage industry, people worked for themselves, and the family was the center of the operation. In a factory, the people worked long hours for the factory owner, and contact with the family was reduced. Widespread use of coal in cities resulted in increased air pollution. In spite of these changes, the Industrial Revolution was viewed as progress. Energy consumption increased, economic growth continued, and people prospered. Within a span of two hundred years, the daily per capita energy consumption of industrialized nations increased eightfold. This energy was furnished primarily by coal.

**Figure 8.6 Industrial Revolution.**
The development of coal-powered machines resulted in factory manufacturing of goods. This consolidation of manufacturing resulted in water and air pollution. Factories such as the one shown here required a large labor force. Because people wanted to live near their places of employment, towns developed near factories.

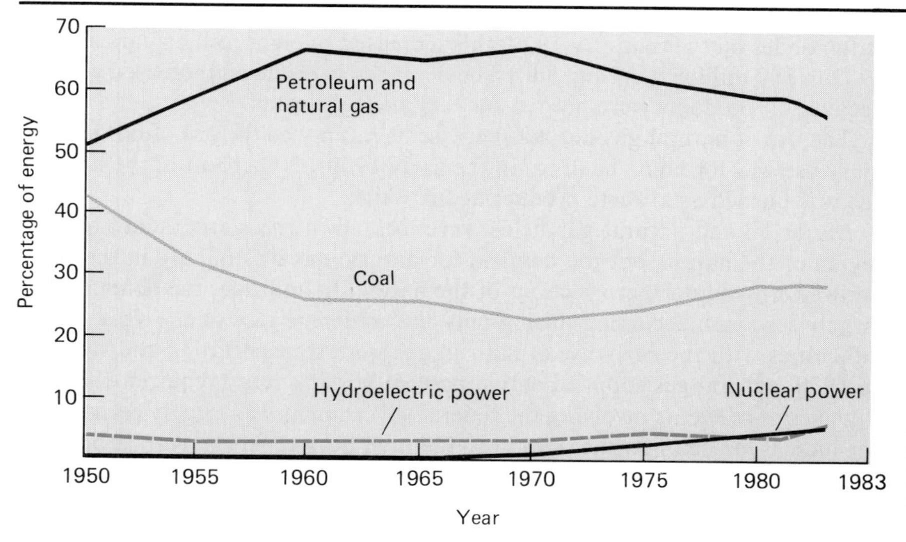


**Figure 8.7 Oil replaces coal.**
Just as wood was replaced by coal, in later years coal was replaced by petroleum products. This graph represents the production of energy by various sources. You can see that petroleum has remained a dominant energy source for the last thirty years, coal has decreased slightly, hydroelectric power has remained about the same, and nuclear power has begun to be a significant energy source in the United States. (Annual Energy Review, 1983, Energy Information Administration.)

## *Oil and natural gas*

Even though there was some use of gas and oil by the Chinese as early as 1000 B.C., these resources remained virtually untapped until fairly recently. The oil well that Edwin L. Drake, an early oil prospector, drilled in 1859 was not the world's first oil well, but it was the beginning of the modern petroleum era. By 1870, **oil** production in the United States reached over four million barrels a year and supplied 1 percent of our energy requirements. (See fig. 8.7.)

**Figure 8.8 The rise of the automobile.** The development and widespread use of the automobile caused a rapid increase in the demand for oil products.

For the first sixty years of production, the principal use of crude oil was to make kerosene, a fuel for lamps. The gasoline produced was discarded as a waste product. During this time, the supply of gasoline exceeded the demand. However, the development of the automobile caused a marked increase in the demand for crude oil. In 1900, the United States had only eight thousand automobiles on the road. By 1920, this increased to eight million cars and by 1982 to 160 million. (See fig. 8.8.) The demand for crude oil increased greatly because oil products were needed for fuel and lubricants.

The use of **natural gas** did not increase as rapidly as the use of oil. Its primary use was for home heating. In the early 1900s, 90 percent of the natural gas was burned as a waste product at the wells.

Major oil and natural gas fields were located in the warm, southwestern region of the nation, but the demand for natural gas was mainly in the cold, midwestern and northern sections of the nation. In addition, the demand was largely seasonal, occurring during only the colder periods of the year. Some difficulties with the early use of natural gas were transportation and storage. In 1920, natural gas supplied only 5 percent of our energy requirements.

A series of events involving the federal government was largely responsible for increased use of natural gas in the United States. World War II brought new energy requirements for manufacturing and transportation. Our war machinery needed energy to operate. Before this war, the majority of crude oil produced in the Southwest was transported by tankers to refineries on the East Coast. This method of transportation could not meet the huge energy requirements of our war effort. In addition, German submarines were destroying many tankers.

In 1943, a federally financed pipeline was constructed to transport crude oil and natural gas. This pipeline, 2,000 kilometers long, allowed oil to be transported more efficiently from the wells in Texas to the refineries and factories in the eastern section of the country. In 1944, a longer federally financed

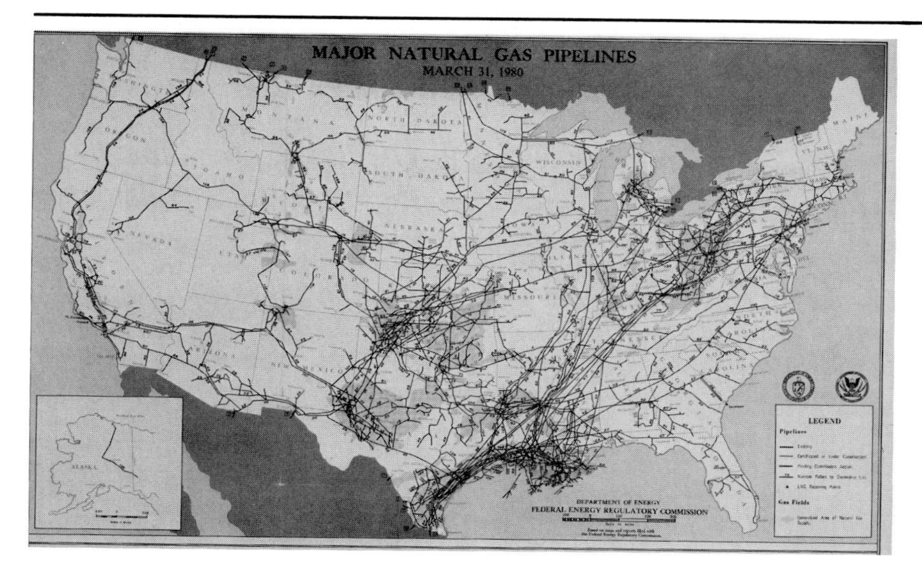


**Figure 8.9 Pipelines.** Beginning in the mid-1940s, the United States began to construct a network of pipelines to transport natural gas. Today a system of 1,613,000 kilometers of pipelines distributes gas to most communities in the United States.

pipeline (2,400 kilometers) was built to increase the flow of oil to the eastern and midwestern sections of the country. By 1971, there were 400,000 kilometers of long-range transmission pipelines and 986,000 kilometers of distribution pipelines in the United States. Approximately 1,613,000 kilometers of natural gas distribution pipelines are used. (See fig. 8.9) Improvements in transporting the natural gas were coupled with improvements in storage facilities in the form of large underground storage sites. As a result of the improvements, natural gas now supplies 31 percent of our energy.

## Energy and economics

In addition to the construction of pipelines and storage areas, another factor that provided the United States with an abundant source of cheap energy was **government regulation** and intervention. In addition to financing the early pipelines that enabled the transportation of natural gas and oil, the government regulated the price of these products at artificially low levels. The government also reinforced a policy of maintaining low prices for oil and natural gas by continuing to grant tax advantages to oil and natural gas companies. This helped companies to be profitable without raising prices. The current oil depletion allowance permits oil companies to exempt 22 percent of their gross income from taxes. In addition, all exploration, drilling, and development costs (including foreign operations) are tax deductible. Oil companies can also list royalty payments to foreign governments as "taxes," which further reduces the taxes the oil companies pay to the federal government. This policy of artificially low prices encouraged a high consumption of fuels.

As the result of the oil pricing policies of OPEC, major efforts were begun in 1973 to alter our dependency on foreign oil. This pricing by OPEC led to the oil pricing deregulation in 1980 and natural gas pricing deregulation in 1981 in the United States. Since 1981, a policy was instituted to stimulate exploration for oil and natural gas deposits in the United States. By allowing larger profits for the companies, money was made available for expanding exploration. This rapid rise in prices for oil and natural gas has stimulated people to make changes in their life-styles that have resulted in the conservation of fossil fuels.

## *Economic growth and energy consumption*

The policy of encouraging fossil fuel consumption led to the development of a technology-oriented society. The replacement of human and animal energy by fossil fuels began with the Industrial Revolution and was greatly accelerated by the supply of cheap, easy-to-handle, and highly efficient fuels. The result was unprecedented economic growth in the United States and the rest of the industrialized world.

World War II was a prime factor in ending the depression of the 1930s. The demands of the military created millions of defense jobs. Almost everyone was employed, but there was a scarcity of consumer goods. After World War II, there was a great demand for consumer goods that had been unavailable during the war years. Industry made the transition from the production of military goods to the production of consumer goods. A high employment level, a rapidly expanding population, and a supply of inexpensive energy encouraged a period of rapid economic growth. By 1984, the energy consumption in the United States was four times greater than in 1945, the year World War II ended.

In 1984, the United States, which represented approximately 5 percent of the world's population, consumed approximately 34 percent of the world's total annual energy output. As domestic natural gas and oil were consumed, the United States became dependent on foreign sources, which are more expensive. No single factor has caused the present energy problem. Since the recent history of the United States is colored by the presence of cheap energy and rapid economic growth, it is easy to understand why people are consumption oriented.

## *Progress: The American dream*

The cheap, abundant energy that fueled industries produced an ever-increasing amount and array of consumer goods. One such product was the automobile. At the beginning of this century, Americans began to drive cars. The growth of the automobile industry led to the construction of improved roadways, which required energy for their construction. Therefore, the energy costs of driving a car were greater than just the fuel consumed. As roads improved, higher speeds were possible. The demand for faster cars grew, and the automobile companies were quick to meet the demands. Bigger cars required even more fuel and better roads. So roads were continually being improved and more cars were produced. A cycle of *more chasing more* had begun.

**Figure 8.10 Energy-demanding life-style.** Building private homes on large individual lots some distance from shopping areas and places of employment is directly related to the heavy use of the automobile as a mode of transportation. Heating and cooling a large enclosed shopping mall, along with the gasoline consumed in driving to the shopping center, increases our demand on our energy resources.

Why have only one family car? Two cars are more convenient. A demand for more cars is coupled with demand for more energy. It requires energy to mine the ore, process the ore into metals, form the metals into automobile components, and transport all the materials. As our economy grew, so did our energy requirements.

More cars meant more jobs in the automobile industry, the steel industry, the glass industry, and hundreds of other related industries. The construction of thousands of kilometers of roads created additional jobs. The oil companies grew from being the suppliers of lamp oil to being the largest industry in the country. The automobile industry has had a major role in the economic development of the United States. All this wealth gave people more money for cars and other necessities of life. The car, originally a luxury, was now considered a necessity.

The car not only created new jobs in the automobile and related industries but altered people's life-styles as well. The car helped people to travel greater distances during their vacations. New resorts, chains of motels, restaurants, and other service industries developed to serve the motoring public. Thousands of new jobs were created. Because people could live farther from work, they began to move to the suburbs. (See fig. 8.10.) The growth of large shopping centers in the suburban areas hastened the decline of some of the central cities. In 1920, over 90 percent of all retail business was conducted in central business districts. Today, less than 50 percent of retail sales are made in central business districts. This loss of sales has caused a loss of jobs in these areas. In Philadelphia, 79 percent of all retail jobs were within the city in 1930; by 1970, this had declined to 43 percent.

While people were moving to the suburbs, they were also changing their buying habits. Labor-saving, energy-consuming devices became essential in the home. The vacuum cleaner, dishwasher, garbage disposal, and automatic garage door opener are only a few of the ways human power is replaced with electrical power. Eleven percent of our electrical energy is used to operate home appliances. In addition, there are other aspects of our life-style that point out our energy dependence. The small, horse-powered farm of yesterday has grown into the big, diesel-powered farm of today. Regardless of where we live, we expect Florida oranges, California lettuce, Texas beef, Hawaiian pineapples, and Maine lobsters to be readily available at all times. What we often fail to consider is the amount of energy required to process, refrigerate, and transport these items. The car, the modern home, the farm, and the variety of items on our grocery shelves are only a few indications of the fact that our life-style is based on cheap, abundant energy.

### *Gross national product, energy, and life-style*

One way to compare standards of living is based on the availability of goods and services. The total value of all goods and services produced each year is called the **gross national product** (GNP). The 1985 per capita GNP of the United States was approximately $14,090. (This means $14,090 for every man, woman, and child.) You probably find this figure hard to believe, since you have not used many of the goods and services directly. Some of the products and services are easily identifiable—cars, washing machines, repair of home appliances, food, and clothing. Other parts of our GNP are less tangible—national defense, education, health care, and governmental services.

Figure 8.11 illustrates the relationship between the per capita GNP of countries and their energy consumption. It indicates that the United States has a high GNP and a per capita energy consumption that is the highest in the world. In general, countries with high GNPs also have high energy consumption. However, the United States with its high energy consumption has a lower GNP than Switzerland. This illustrates that it is possible to be energy efficient and still enjoy a high level of consumer goods and services.

Using muscles instead of machinery to open cans, sharpen pencils, and mix our food would lower our energy demands. The elimination of air conditioning alone would reduce our electrical consumption by 9 percent. Eliminating wasteful packaging and recycling reusable containers would also save energy. Trade-offs that will reduce our energy consumption drastically are possible, but they will cause only slight changes to our life-styles.

## Approaches to our energy situation

Anyone who reads the newspaper, listens to the news on the radio, or watches the news on television is aware of the fact that the United States and the world are facing a serious energy situation. Figure 8.12 shows three possible patterns of energy use for the future.

**Figure 8.11 Energy and gross national product.** Although there is some relationship between the gross national product (GNP) and the amount of energy a nation consumes, it is not a direct relationship. Countries with a high GNP can have a lower energy consumption (Switzerland) or some countries, such as Canada and the United States, have only a moderately high GNP but a very high rate of consumption of energy.

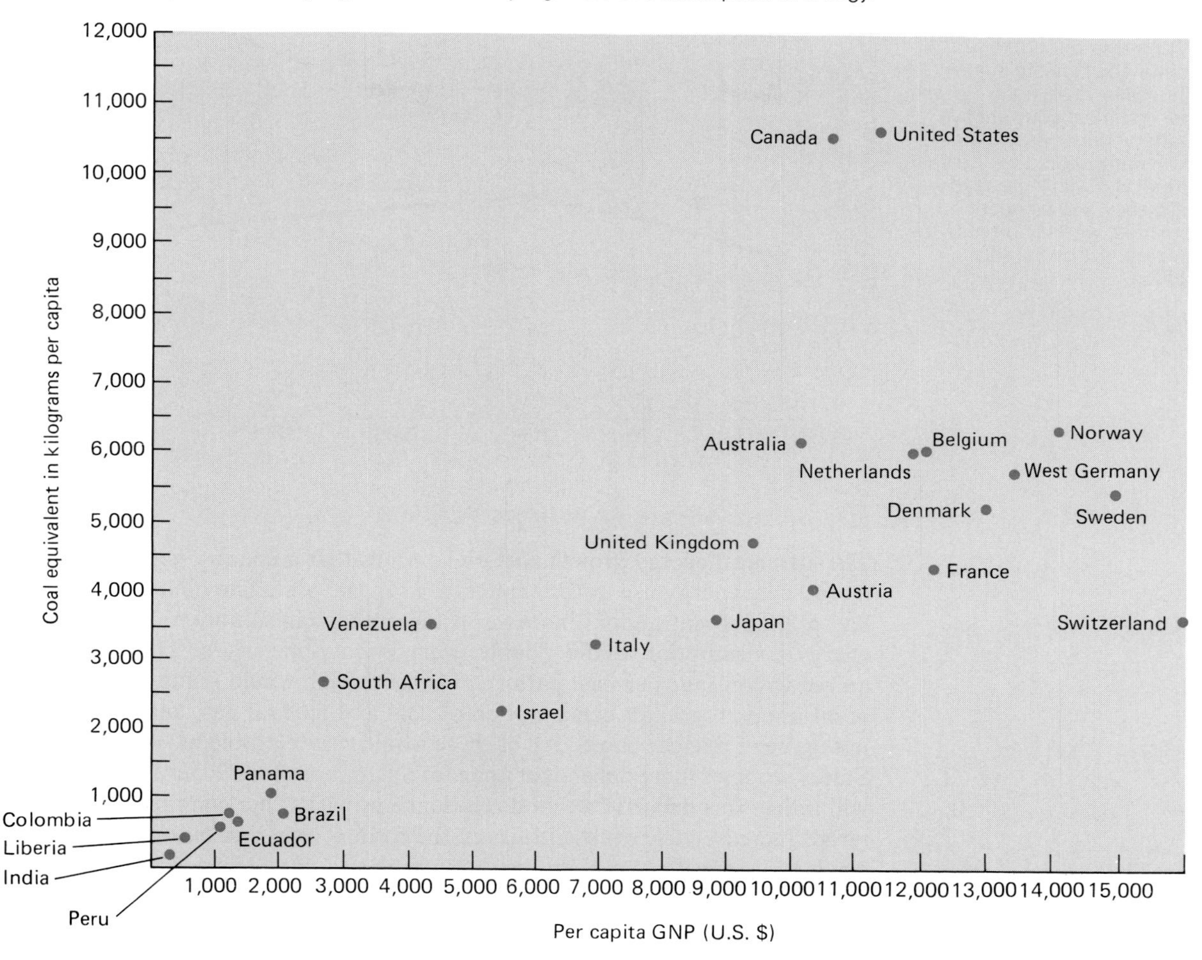

**Figure 8.12 Patterns of energy consumption.** Future energy consumption will be determined by government policy and the willingness of people to make adjustments in their life-styles. If we continue with our historical consumptive pattern, our energy usage will nearly double by the year 2000. But with increased efficiency and stringent controls, we may be able to decrease it by the year 2000.

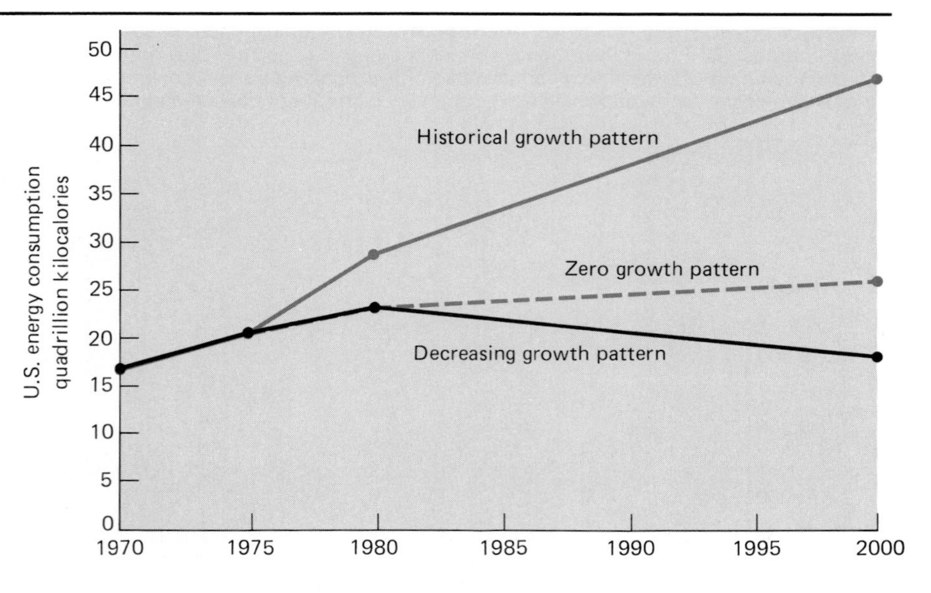


## *Historical energy growth pattern*

The **historical energy growth pattern** assumes that a country will continue to increase its energy use at the same rate as in the past. The United States has had a 3.2 percent annual increase in the energy consumption. At that rate, energy consumption would double every twenty-one years. The energy required to continue our past pattern of consumption would require an increase in oil imports, greater consumption of coal and natural gas, and an increase in the use of nuclear power. All of these would create problems. As the United States becomes more dependent upon foreign nations for oil, our foreign policy will reflect this dependence, and relationships with other countries will be altered. Increased use of fossil fuels in the form of coal and oil could cause serious environmental problems, including the risk of oil spills, soil disturbance by strip mining, and an increase in air pollution. A policy of increased energy consumption would necessitate an increase in the number of nuclear power plants, which would also add to the political, social, and environmental problems already associated with the nuclear alternative.

The cost of importing foreign oil, mining more coal, drilling deeper oil wells, or building nuclear power plants greatly increases the cost of energy. Anyone who has paid an electric or gas bill, purchased heating oil, or bought gasoline is aware of this fact. Therefore, it is impossible to continue a historical energy growth pattern that demands more energy each year.

## *Zero energy growth pattern*

A **zero energy growth pattern** would require governmental policy aimed at gradually reducing the annual rate of increase in energy consumption over the next fifteen to twenty years. This would result in no annual growth in energy

**Figure 8.13 Energy-efficient community.** The use of multiple-family dwellings reduces the fuel requirements for heating and cooling. A cluster of dwellings near a central transit system enables the commuters to travel by rail instead of car. This type of living greatly reduces an individual's energy demands.

consumption after that time. (See fig. 8.12.) From a material viewpoint, we would not need to lower our standard of living significantly. (See chap. 24.) In addition, any decrease in our population or significant technological advances could result in more energy being available per person.

Governments can encourage energy conservation by granting tax advantages for the use of energy-saving techniques; monies presently used for highway construction could be diverted to the development of mass transportation systems. This would cause great political consternation in the Congress and in states where the highway lobbies are well entrenched.

Suburbs with multiple-family dwellings could replace the traditional one-family homes located on large lots. A multiple-family dwelling saves energy and places people within walking distance of public transit systems. (See fig. 8.13.) Living in cities could become more popular.

Greater emphasis on durability and less emphasis on disposable goods would also be needed to achieve a zero energy growth pattern. Because of higher

## Box 8.2
## Davis, California: Energy-efficient community

Davis, California, is a unique community. The city council has passed several ordinances and modified several policies in order to meet the goal of being energy efficient. These actions mandate some changes in historical practices and strongly encourage others. Nearly everyone in the community will be influenced by these actions.

The city's energy conserving building code, originally enacted in 1968 and revised in 1982, mandates energy efficient homes by requiring certain design specifications. These specifications include the color of the exterior, the type of outside doors and windows, the slope of the roof, the amount of insulation on all steam heating and hot water pipes, the amount of insulation in walls, ceilings, and floors, the percentage of glazing (glass) in relation to the amount of floor space, and the use of appropriate overhangs and shades to reduce heat loss in the winter and heat gain in the summer.

All existing homes offered for sale are inspected. If the home is not up to code, the buyer is required to make the necessary improvements, including certain energy conservation measures.

There is a master shade-tree program for the city. The location of shade is part of the formula used in establishing allowable heat gain in the summer. Deciduous trees are planted to provide shade in the summer and sun in the winter.

At one time, Davis passed an ordinance banning the use of clotheslines within the city because clotheslines were considered unattractive. However, in April 1977 the city passed an ordinance that not only allowed but encouraged clotheslines as a means of saving energy. Part of the ordinance reads:

prices, people would be encouraged to keep their cars, television sets, refrigerators, and other material goods rather than replacing them often. (As a result of higher prices in the early 1980s, people bought fewer new homes, automobiles, home appliances, and other forms of consumer goods.). This approach would require less use of high-energy manufacturing industries (plastics, steel, aluminum, and copper) and an increase in service-related industries.

Goods would be designed to last longer and to be easier to repair. There are great social and economic implications with such a shift in emphasis. It

Section 29–169.1 Clotheslines: It shall be unlawful and a nullity to establish any private covenant or restriction which prohibits the use of a clothesline in any residential zone, except that all multi-family developments (three-family and greater densities) requiring Design Review Commission approval shall require suitable space for facilities except where such space would preclude good project design, to enable residents to dry their clothes using the sun. Such clotheslines shall be convenient to washing facilities and oriented so as to receive sufficient sun to dry clothes through the year.

The city has also adopted several policies that discourage the use of automobiles because they are not energy efficient. These include a city-subsidized bus route for commuters to Sacramento, reestablishment of passenger rail services for commuters to Oakland, and double-decker buses for in-city transportation.

In 1966, even before the Energy Code of 1968, energy was a topic in Davis. The city council election of that year had as its main issue the encouragement of the use of bicycles within the community. The pro-bicycle faction won. Today, 40 percent of in-city travel is by bicycle. Bicycles are provided to city employees, who are encouraged, though not required, to ride them. All new housing tracts must construct a separate bicycle lane. An added advantage of using bicycles is that less land is required for parking spaces.

As part of the policy of saving energy and land, Davis has reduced the width requirements for new streets from ten meters to eight meters. The narrower streets use less land, require less asphalt, cost less to construct, are cheaper to maintain, and reduce speed limits.

In still another attempt to save energy by reducing car travel, Davis has passed an ordinance encouraging people to work at home. Within limits, certain types of businesses are permitted within private residences.

The results of the Energy Code have been favorable. There has been a 12 percent drop in electrical consumption per customer. Although most houses are built with air conditioning, awnings, shade screens, arbors, and other devices are widely used to reduce air-conditioner usage (the summer daytime average temperature is around 32°C).

In many ways Davis is unique. It is a city concerned about its total environment. The rich farmland surrounding the city is regarded as an asset to the community, and its development for residential or commercial use is restricted. The city has also limited the rate and type of growth. Each year, all new building requests are reviewed at the same time. In approving a request, the review board maintains a mix of homes for people of low, medium, and high income. No construction is allowed without the certainty that the existing city services to that area are adequate.

It is indeed unusual for a city to adopt a policy that regulates growth, protects farmland from development, encourages bicycle travel, allows for narrower streets, and promotes businesses in the home. Many communities are doing exactly the opposite. The people of Davis believe that the life-style of a community can shift to energy efficiency and still maintain high quality.

would reduce the number of jobs in production fields and increase the number of jobs available in service-oriented areas. During the transition period from a production industry to a more service-oriented industry, some governmental assistance in retaining workers may be required. Even though major changes are necessary for a zero energy growth pattern, many people feel that this pattern is too optimistic, and we will need to consider decreasing energy consumption.

### *Decreasing energy growth pattern*

Since there is a limit to the amount of fossil and nuclear fuels in the world, a time will come when these energy resources are not adequate to meet the demand. At that time, a decrease in energy consumption must occur. A **decreasing energy growth pattern** necessitates changes in our life-styles. Unless some unforeseen energy technology is developed, we must anticipate a lower-energy society. This lower-energy society would increase the use of muscle power. It would be decentralized and would rely on renewable sources of energy such as the sun. Fossil fuels would become more valuable and would be used for the production of high-technology devices, such as solar collectors, wind generators, or communication devices.

There would also be a transition to a more service-oriented economy (fix it, don't replace it). For many people, it could mean less job specialization. With greater durability of goods, less energy is needed, and air and water pollution are lessened. However, many of our traditional "freedoms" would be limited, such as the "freedom" to commute long distances to work, the "freedom" to travel for pleasure, and the "freedom" to enjoy a wide choice of foods. There would also be less emphasis on advertising, because buying would be more related to needs than to wants. More people would grow their own food.

These changes would not occur overnight. They might take a generation or more to be realized. Although the changes may be subtle, they represent actual energy conserving activities. While it is true that the energy-saving activities of one individual may seem insignificant, it is also true that the activities of several individuals could significantly reduce energy consumption. (An example is the dramatic decrease in gasoline consumption in the United States from 1979 to 1981.)

Here are some suggestions *you* can follow without drastically interfering with your life-style. There are many others that you might add to the list.

**Transportation**

Reduce kilometers driven by planning travel. Car pool or use public transportation when possible.

**Reduce heat loss from homes during winter**

Weather stripping.
Increase insulation.
Install storm windows.
Adjust the thermostat downward and wear a sweater.
Set thermostat at a lower temperature at night (some thermostats will do this automatically).

**Reduce heat gain during the summer**

Increase insulation.
Leave storm windows on as many windows as possible.
Adjust thermostat upward and dress accordingly.

**Appliances**

Consider whether an appliance is really necessary. (Although many gadgets use very little energy to operate, it does take energy to manufacture them.)

Reduce hot water temperatures.

Reduce the use of hot water.

- Use flow restriction devices on showers.
- Wash with cold or warm water.
- Fix dripping faucets.

Insulate the hot water heater.

When buying appliances, look at the energy efficiency tag as well as the price tag.

Turn off lights, radios, air conditioners, television sets, etc., when not needed.

**Personal energy conserving behaviors**

Turn off all lights when leaving the room.

If you are going to be absent for a long time, cut back on the use of the air conditioning or heating system while you are out.

During the winter, close all open windows and doors to reduce heat loss.

Reduce the amount of light used.

Encourage others to conserve energy.

## Current energy trends

Recent energy consumption data indicates a decrease in the rate of energy consumption from 1979 to 1983, with the beginning of an upward trend in 1984. (See fig. 8.14.) Industrial growth increased 17 percent from 1973 to 1979, but energy use by industry increased only 2 percent. During that period the standard of living did not drop. United States gas and oil consumption peaked in 1978 and declined each year until 1984 when it increased by about 3 percent. Electric utilities are seeing little change in demand and are adjusting their projections for future demands accordingly.

Several forces have contributed to this decrease in energy consumption. Increases in imported oil prices (due to OPEC) and domestic oil and natural gas prices (due to deregulation) have forced businesses and individuals to become energy conscious. Government tax incentives have encouraged changes in existing and new building construction that reduce energy consumption. By insisting that energy consumption data be made available to the consumer, government has forced the design of automobiles and appliances to be more energy efficient. Consumers can make purchases based on energy efficiency. In addition to financial considerations, many people are altering their energy consumption for philosophical reasons. They are finding that they can do so without significantly altering their standard of living.

In the past, whenever we have experienced changes in energy requirements, greater consumption has been the result. Although a few people still predict an increase in energy consumption, most believe that our energy situation will lie somewhere between that of a zero energy growth pattern and a decreasing energy growth pattern.

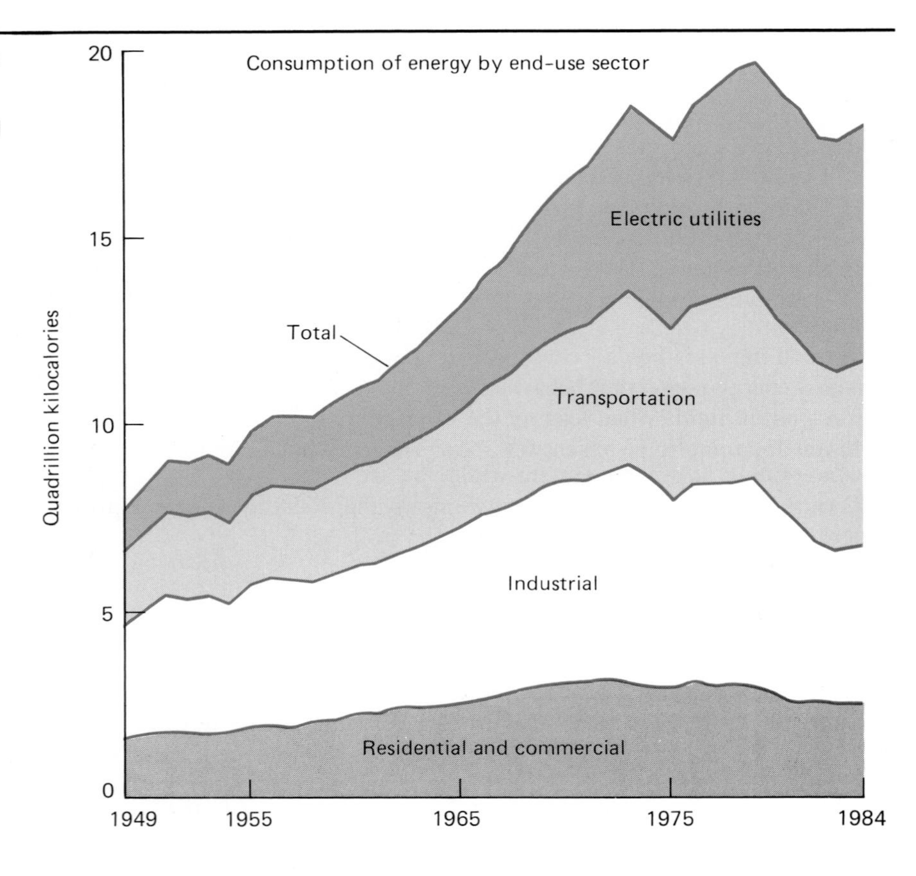


**Figure 8.14 Changes in energy consumption.** From 1979 to 1983, total energy consumption declined each year. Part of this decline was tied to an economic slow down. With the improvement in economic conditions in 1984, energy consumption began to increase. (Source: Annual Energy Review, 1984.)

## Consider This Case Study
## Shirley Highway, Washington, D.C.

In the United States, commuting to and from work consumes large amounts of energy. If more energy efficient means of transportation could be found, there would be a substantial energy savings.

Shirley Highway is a heavily traveled urban freeway connecting residential communities in Virginia with Washington, D.C. In an attempt to conserve gasoline, separate lanes for regular traffic and for buses and carpools were designated. During peak travel times, certain lanes were designated for exclusive use by buses and carpools of four or more persons. The Urban Mass Transportation Administration conducted a study to determine the amount of energy required to travel during the rush hour on the Shirley Highway. The results are shown in the table.

How much more efficient is carpooling than regular car travel? How much more efficient are buses than regular car travel? What is a disadvantage to carpooling and travel by bus?

## Summary

Energy is a prerequisite for all life and any type of civilization. There is a direct correlation between the amount of energy used and the complexity of the civilization.

Wood furnished most of the energy and construction materials for early civilizations. Heavy use of wood in densely populated areas eventually resulted in a shortage of fuel, which was replaced by coal, oil, and natural gas. However, these fuels are also limited in supply.

The United States and Western Europe, with their abundant supply of cheap energy, had experienced tremendous economic growth until the recent energy problem. For example, a large segment of our economy is based upon the production and daily use of the automobile. The car not only created new jobs in the automobile and related industries but also altered our life-styles by influencing where people live, work, and go on vacations. The energy supply problem will necessitate changes in our life-styles.

There are three patterns of future energy use. The historical energy growth pattern emphasizes obtaining ever-increasing amounts of energy. The zero energy growth pattern allows no increase in our energy consumption but does not significantly alter the standard of living. The decreasing energy growth pattern reduces individual energy consumption and changes the life-style.

Can carpooling save anything besides energy? What is the advantage of using specific lanes for carpools and buses? Can you think of any disadvantages of carpools?

Whenever a community studies transportation, various alternatives such as carpools, buses, and rapid rail transportation are suggested. Which method can be most quickly initiated? Which can be initiated with the smallest capital investment? Which would save the most energy? What can be done in your community to make urban transportation more energy efficient?

| Mode of transportation | Total kilocalories expended per kilometer | Average number of passengers | Kilocalories expended per passenger per kilometer |
|---|---|---|---|
| Non-carpool auto | 1,620 | 1.29 | 1,255 |
| Carpool auto | 1,900 | 4.59 | 414 |
| Bus | 10,375 | 35.9 | 289 |

## Review Questions

1. Why was the sun able to provide all the energy requirements for human needs before the Industrial Revolution?
2. In addition to food, what other energy requirements does a civilization have?
3. Why were some countries unable to use the technology that began to develop with the Industrial Revolution?
4. What factors caused a shift from wood to coal as a source of energy?
5. How were the needs for energy in World War II responsible for the subsequent increased consumption of natural gas?
6. What part does government regulation play in changing the consumption of gas and oil?
7. Why was much of the natural gas that was first produced wasted?
8. What was the initial use of crude oil? What single factor was responsible for a rapid increase in crude oil consumption?
9. What are three patterns of energy growth, and what changes will they cause?
10. In addition to gasoline, what are some other uses for crude oil?

## Suggested Readings

Braiterman, Marta; Fabos, Julius Gy; and Foster, John H. "Energy Saving Landscapes." *Environment* 20: 30–37. Decisions about how land is to be used can affect the amount of energy a community requires to supply its needs for food, water, and other goods.

Carter, Luther J. "Energy Standards for Buildings Face Delay." *Science* 209: 784–87. New conservation standards are criticized by builders and electric utilities.

Meadows, Dennis L., ed. *Alternatives to Growth I, A Search for Sustainable Future.* Cambridge, Mass.: Ballinger Publishing Company, 1978. A series of seventeen papers by leading environmentalists, economists, policymakers, and philosphers concerning the possibilities of a steady-state society. To survive, such a society would not demand a constant increase of raw materials and energy. Excellent presentations that address the future from the viewpoints of several different disciplines.

Revelle, R. "Energy Dilemmas in Asia: The Need for Research and Development." *Science* 209: 164–74. Not including China, the countries of southern and eastern Asia contain 30 percent of the world's population but only 2 percent of known fossil fuel reserves. Research and development in energy production, conversion, and conservation should eventually allow local energy sources, of which the most promising is biomass energy, to be substituted for imported fuels.

Rosenbaum, Walter A. "Energy, Politics and Public Policy." *Congr. Quarterly,* 1981. An excellent presentation of politics and the use of coal. It also presents questions about increased coal production, environmental and regulatory policies, and the future cost of coal.

Sant, Roger W. "Coming Markets for Energy Services." *Harvard Business Review* 58: 6–24. Presents "least-cost" marketing approaches for the energy problem by suggesting new products and technological opportunities to reduce the cost of energy.

Sant, Roger W. "Cutting Energy Cost." *Environment* 22: 14–20, 42–43. Presents methods of lowering the United States energy consumption by 17 percent without lowering our energy services.

Steinhart, John, S., et al. *Pathway to Energy Sufficiency: The 2050 Study.* San Francisco: Friends of the Earth, 1979. An informative, short, easy-to-read book that presents an energy scenario. This book includes all segments concerned with energy conservation, including personal, business, and government.

Stobaugh, Robert, and Yergin, Daniel. "The Energy Outlook: Combining the Options." *Harvard Business Review* 58: 57–73. An excellent presentation of the need for conservation and solar energy as a means of reducing our dependency upon fossil fuels as a source of energy.

Teller, Edward. *Energy from Heaven and Earth.* San Francisco: W. H. Freeman and Company, 1979. In the introduction, the author states, "My purpose is not to be right. My purpose is not to convince anybody. My purpose is to give information from my particular viewpoint." The book begins with the role of energy in the origin of the earth and closes with models for the future. The author presents the political, social, economic, and scientific aspects of our use of energy.

## Chapter Outline

I. Major fossil fuels
  A. Fossil fuel formation
  B. Resources and reserves
  C. Mining and the use of coal
  D. Production and use of oil
  E. Production and use of natural gas
II. Other conventional energy sources
  A. Water Power
  B. Wood
    Box 9.1 Pumped storage
III. Consider this case study: The Alaskan pipeline

## Objectives

Explain how millions of years were involved in the process of converting plant remains from peat to coal.

List peat, lignite, bituminous coal, and anthracite coal as steps in a process of coal formation.

Recognize that coal deposits are not uniformly distributed and that the United States has thirty percent of the world coal reserves.

Explain how natural gas and oil were formed from ancient marine deposits.

Differentiate between resources and reserves.

Explain how some methods of mining coal can have negative environmental impacts and why strip mining is the most common type of coal mining in the United States.

Explain why more expensive processes and exploration in less accessible areas are necessary.

Explain why storage of natural gas and transportation from its production site to its use site is a major problem.

Recognize that water power and wood fuel are currently minor sources of energy compared to fossil fuels.

Recognize that most of the available water power sites are already being used.

Understand that wood is a limited source of energy.

## Key Terms

acid mine drainage

black lung disease

heavy oil

liquified natural gas

overburden

reserves

resource

secondary recovery

shaft mining

strip mining

# CHAPTER 9
# Current energy sources

## Major fossil fuels

Fossil fuels are the remains of plants, animals, and microorganisms that lived millions of years ago. During the Carboniferous period, 275 to 350 million years ago, conditions in the world were conducive to the formation of large deposits of fossil fuels. (See fig. 9.1.) Since the Industrial Revolution, the major sources of energy for the world have been fossil remains from the distant past.

**Figure 9.1**
**Carboniferous period.**
Approximately three hundred million years ago, this kind of ecosystem was common throughout the world. Plant material accumulated in these swamps and was ultimately converted to coal.

## *Fossil fuel formation*

Coal is the result of plant material that grew in freshwater swamps approximately three hundred million years ago. As this plant material died and accumulated, peat was formed. Because the plant material accumulated under water, decay was inhibited. Many areas of peat were inundated by oceans, and sediments from the sea were deposited over the peat. The weight of these sediments and the heat of the earth gradually changed the composition of the peat bog, and coal was formed. Although peat is used as a source of fuel in some parts of the world, its high water content makes it a low-grade fuel.

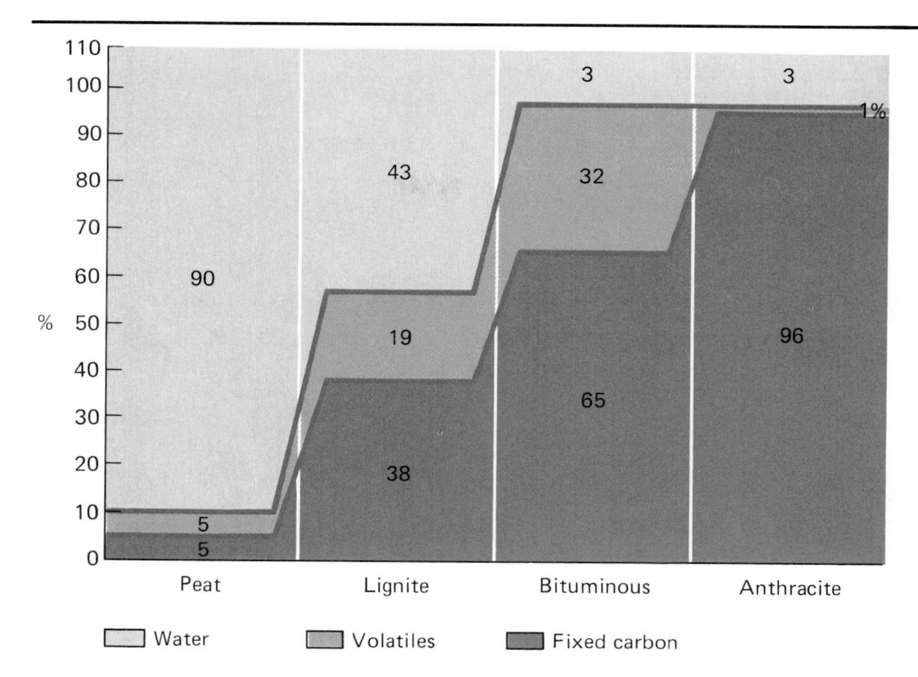


**Figure 9.2**
**Composition of coal.**
The high percentage of water in peat and lignite make them low-energy-yielding fuels. Increased pressure and heat decreases the moisture content and increases the percentage of carbon during formation of bituminous or anthracite coal. Bituminous and anthracite coal are, therefore, better sources of energy.

An estimated nine meters of peat were required to produce and form a 0.3 meter vein of coal, and it required three hundred years to accumulate that much peat. Due to a series of changes in the water level, as many as forty veins of coal could have been formed in an area. The longer the vegetation was allowed to accumulate, the thicker the vein of coal. In the United States, these deposits range from a few millimeters to fifteen meters thick. It required fifteen thousand years of plant growth to produce the material in a 15 meter vein. After many centuries of being compressed by the weight of sediment, the peat is changed into a low-grade coal known as lignite. The percentage of carbon in the lignite is higher than peat. Continued pressure and the heat from the earth changed the lignite into bituminous coal. Bituminous coal (soft coal) is the most common. In changing from lignite to bituminous coal, the carbon concentration is increased. If the heat and pressure were great enough, anthracite (hard coal) would be formed. The components of these three types of coal are shown in figure 9.2.

The location of these coal deposits is extremely important in the world economy. As mentioned in chapter 8, coal deposits were vital to developing nations at the time of the Industrial Revolution. Figure 9.3 indicates the general location of the world's coal deposits.

**Figure 9.3 Recoverable coal reserves of the world.** The percentages indicate the coal reserves that can be recovered under present local economic conditions using available technology.

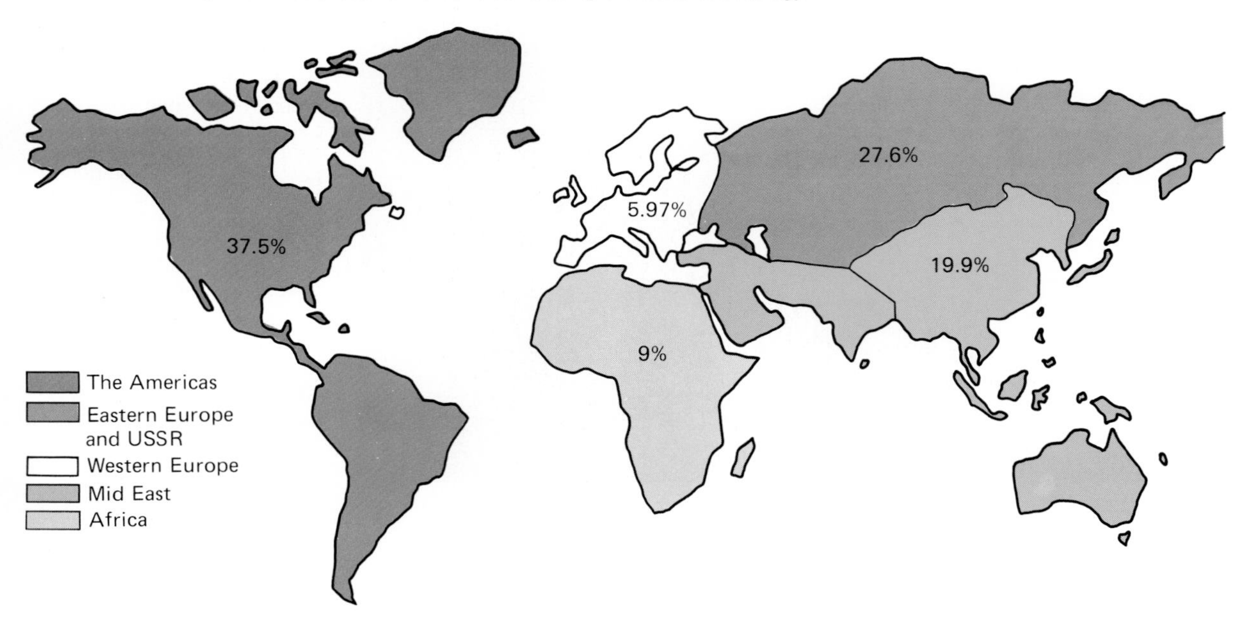


Oil is also a product from the past. Crude oil probably originated within microscopic marine organisms. When these organisms died and accumulated on the ocean bottom, their decay released oil droplets. Gradually, the muddy sediment formed rock containing dispersed oil droplets. Although rock of this type is very common and contains a great deal of oil, it is very difficult to extract the oil because it is not concentrated. Usually a pool of trapped oil does not exist as a liquid mass but rather as a concentration of oil within the pores of sandstone. (See fig. 9.4.) When a layer of sandstone was laid on the oil-containing rock and an additional impermeable layer of rock on the sandstone, conditions were suitable for the formation of oil pools. This was particularly true if the rock layers were folded by geologic forces.

Natural gas, like coal and crude oil, was also formed from fossil remains. In fact, the geological conditions conducive for crude oil formation are the same as those for natural gas. The two fuels are often found together. However, in the formation of natural gas, the organic material changed to lighter, more volatile hydrocarbons than those found in oil, the most common one being methane gas ($CH_4$). Water, liquid hydrocarbons, and other gases may be present in natural gas as it is pumped from a well.

The conditions that led to the formation of oil and gas pools were not evenly distributed throughout the world. Figure 9.5 illustrates this uneven distribution. Some of these deposits are easy to exploit, while others are not.

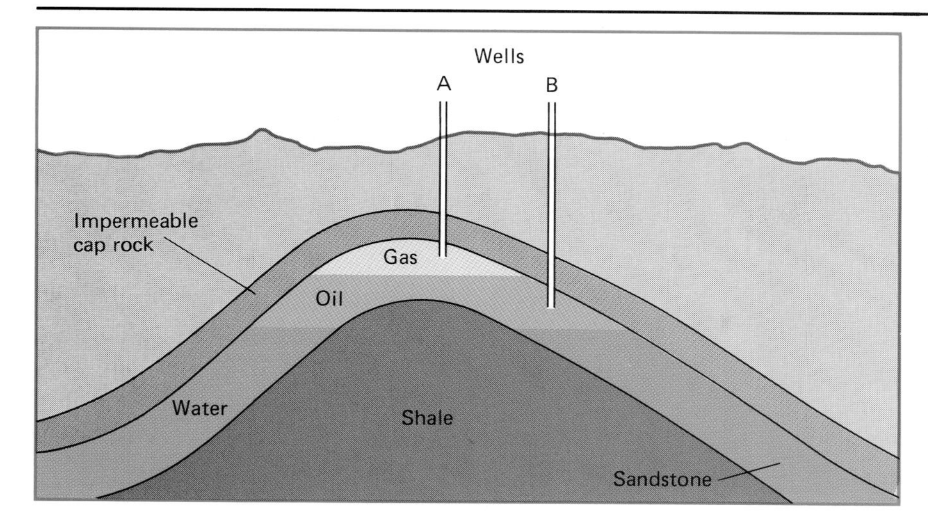


**Figure 9.4 Crude oil and natural gas pool.** The water and gas pressure forces the oil and gas out of the shale, and they collect in a pool beneath the impermeable rock.

**Figure 9.5 World's oil deposits.** The world's supply of oil is not distributed equally. Certain areas of the world enjoy an economic advantage because they control the vast amounts of oil.

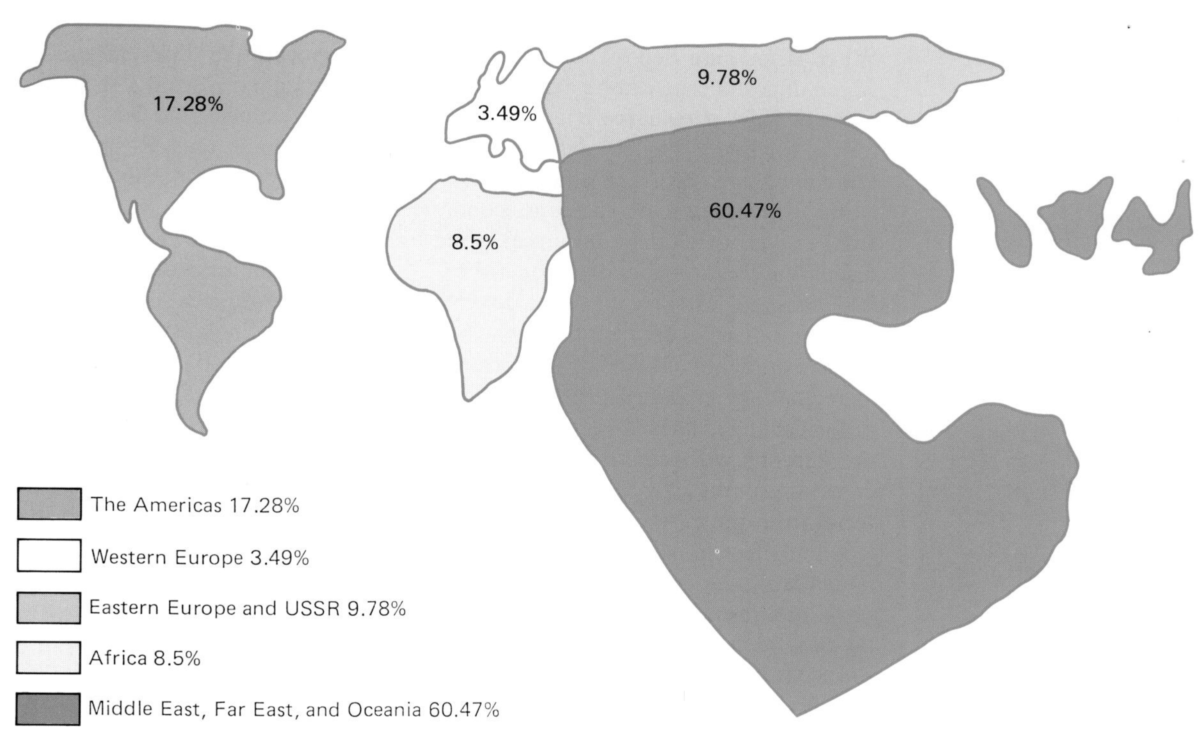

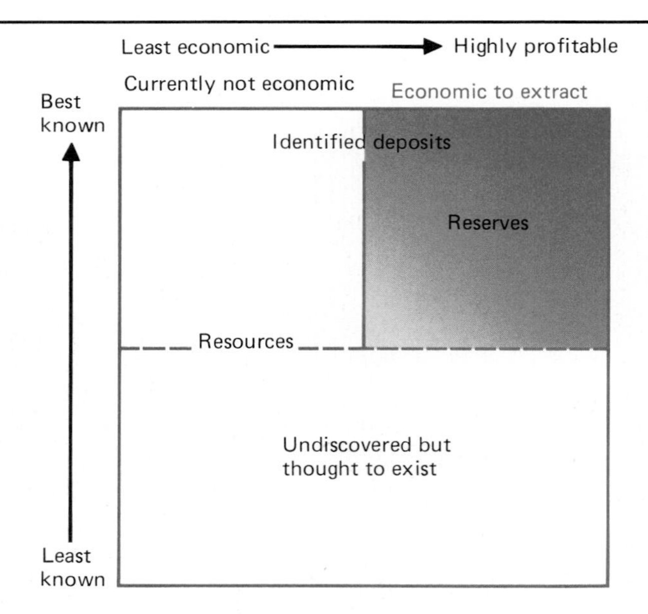


**Figure 9.6 Resources and reserves.**
Each term describes the amount of a natural resource that is present. Reserves are those known deposits that can be profitably obtained using current technology under current economic conditions. Reserves are shown in color in this diagram. The darker the color, the more valuable the reserve. Resources are a much larger quantity that includes undiscovered deposits and deposits that currently cannot be profitably used, although it is feasible to do so if technology or market conditions change. (Adapted from the United States Bureau of Mines.)

### *Resources and reserves*

When discussing deposits of fossil fuels, it is important to differentiate between deposits that can be exploited and those that cannot. From a technical point of view, a **resource** is a naturally occurring substance that has the potential for being feasibly extracted under prevailing conditions. **Reserves** are the known deposits from which materials can be extracted profitably with existing technology under present economic conditions. Therefore, reserves are smaller quantities than resources. (See fig. 9.6.) Both terms are used when discussing the amounts of a mineral or a fossil fuel resource a country has at its disposal. This can cause considerable confusion because most people do not understand the difference between these two concepts. Look at the definition of reserves again. Obviously, the reserves of a mineral or fossil fuel will change as technology changes, as new deposits are discovered, and as economic conditions change. Therefore, there can be changes in the amount of reserves, but the deposits will remain constant. When you read about the availability of certain resources, remember that energy is needed to extract the wanted material from the earth. The more dispersed the material is, the more expensive it is to collect in one place. So, when people talk about how much fossil fuel is available, be careful to determine what they mean. Much of the world's fossil fuel resources is located in deposits that we are unable to use. This may be due to economic reasons. If the cost of removing a fuel and processing it for use is greater than the market value of the fuel, no one is going to produce that fuel. Also, if the amount of energy used to produce, refine, and transport a fuel is greater than the potential energy of that fuel, it will not be produced.

**Figure 9.7 Strip mining.**
Large power shovels or draglines are used to remove the overburden, which is piled to the side. The vein of coal can then be loaded into trucks. When the coal has been removed, the overburden is placed back in the trench.

There must be a net useful energy yield in order to exploit the source. However, new technology or changing price structures may permit the profitable removal of some resources in the future. If so, those resources will be reclassified as reserves.

## *Mining and the use of coal*

Coal is the most abundant fossil fuel in the United States. We have about 30 percent of the world's coal reserves. The conditions responsible for the formation of coal deposits determine the type of mining operation used to obtain a particular vein of coal. There are two common methods of mining coal, **strip mining** and underground or deep **shaft mining.** Strip mining is used when the **overburden** (the amount of material on top of a vein of coal) is less than one hundred meters thick. This type of mining operation removes 100 percent of the coal in a vein and can be profitably used for a seam of coal as thin as half a meter. (See fig. 9.7.) Because it removes all the coal and can be used to obtain coal from thin veins, strip mining makes the best utilization of the coal reserves. Today, 60 percent of the coal mined in the United States is obtained from strip mines. If the overburden is too thick, strip mining becomes so expensive that operators must use underground mines.

In underground mining, a vertical pit or shaft is dug down to the vein of coal. Lateral shafts are extended so the coal can be removed and taken to the

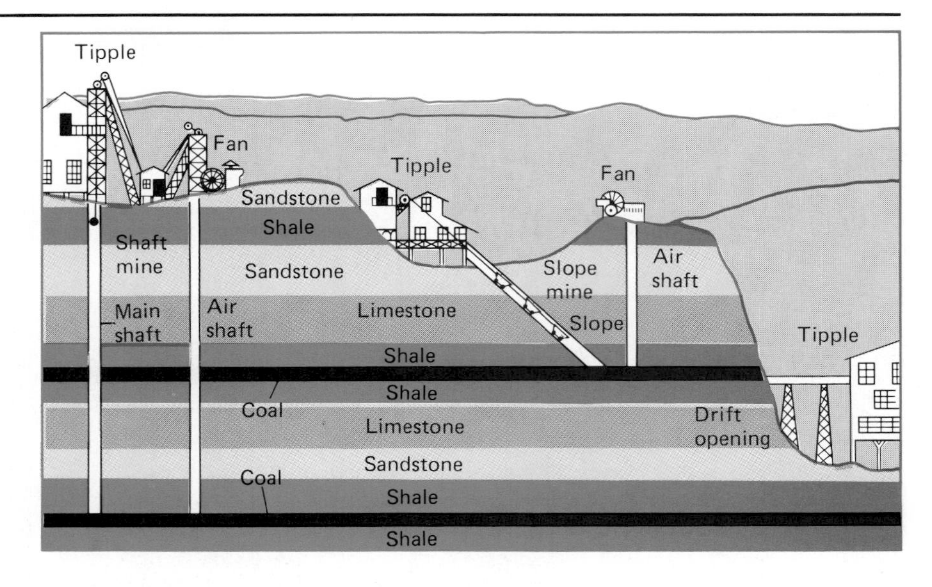


**Figure 9.8 Underground mining.** Three types of underground mine are shown here. They are shaft mines, slope mines, and drift mines. The method used depends in part upon the location of the coal and the type of overburden.

surface. (See fig. 9.8.) Underground mining is done to a depth of more than one thousand meters. The problems associated with underground mining are the cost and the amount of coal that must be left unmined. Some of the coal must be left as pillars to support the shafts. A vein of coal must be at least half a meter thick to make underground mining profitable.

The major use of coal is as a fuel. The type of coal determines its use. The high moisture content and crumbly nature of lignite makes it the least desirable form of coal. Bituminous coal is the most widely used coal in the United States because it is the easiest to mine and the most abundant. It supplies 17 percent of our energy requirements. For most purposes, anthracite coal is the most desirable, but it is not as common and is usually much more expensive because it must be mined by underground mining techniques.

Coal is an abundant resource, but there are problems associated with its mining, transportation, and use. Strip mining disrupts the landscape. However, it is possible to minimize this disturbance by reclaiming the area after the mining operations are completed. (See fig. 9.9.) Reclamation rarely if ever returns the land to its previous level of productivity. The cost of such reclamation is passed on to the consumer in the form of higher coal prices. Although underground mining does not disrupt the surface environment as much as strip mining, there are the problems of large waste heaps, water pollution, and miner safety.

With underground mining, there is also the problem of **black lung disease,** which is a respiratory condition that results from the accumulation of large amounts of fine coal dust particles in the miner's lungs. This accumulation of dust in the lungs inhibits the exchange of gases between the lungs and the blood. Health care costs and death benefits related to black lung disease are

**Figure 9.9 Strip-mine reclamation.** Photograph *a* shows a large area that has been strip mined with little effort to reclaim the land. The windrows created by past mining activity are clearly evident, and little effort has been made to reforest the land. By contrast, *b* is an example of proper strip-mine reclamation. The sides of the cut have been graded and planted with trees. The topsoil has been returned, and the level land is now productive farmland.

a

b

**Table 9.1**
Coal as a source of air pollution

| | Thousands of tons of pollutants per year | | | | |
|---|---|---|---|---|---|
| Sources of Pollution | Particulates | Sulfur oxides | Nitrogen oxides | Hydrocarbons | Carbon monoxide |
| Residential | 222 | 193 | 360 | 40 | 1,527 |
| Electric generation | 1,412 | 17,899 | 7,877 | 58 | 296 |
| Industrial | 777 | 3,907 | 3,798 | 821 | 603 |
| Commercial and institutional | 106 | 765 | 385 | 14 | 79 |
| *Totals* | 2,517 | 22,764 | 12,420 | 933 | 2,505 |

Source: EPA 1980 National Emissions Report.

actually an indirect cost of coal mining. Since these are partially paid by the federal government, their full price is not reflected in the price of the coal.

The problem of coal dust is also a concern when transporting coal. Because coal is bulky, rail shipment is the most economic way to transport it. Rail shipment costs include the expense of constructing and maintaining the tracks as well as the energy required by the engines to move the long unit trains. Therefore, coal can only be used if the user is located near a railroad. Each year the burning of coal releases over twenty-eight million metric tons of material into the atmosphere and is responsible for millions of dollars of damage to the environment. Coal is a major source of air pollution, and the burning of coal for electric generation is the prime source of this type of pollution. (See table 9.1.)

Another problem often associated with coal mining is **acid mine drainage.** Even before the first coal mine was opened in this country in 1761, natural

**Figure 9.10 Off-shore drilling.**
Once the drilling platform is secured to the ocean floor, a number of wells can be sunk to obtain the gas or oil. (Source: American Petroleum Institute.)

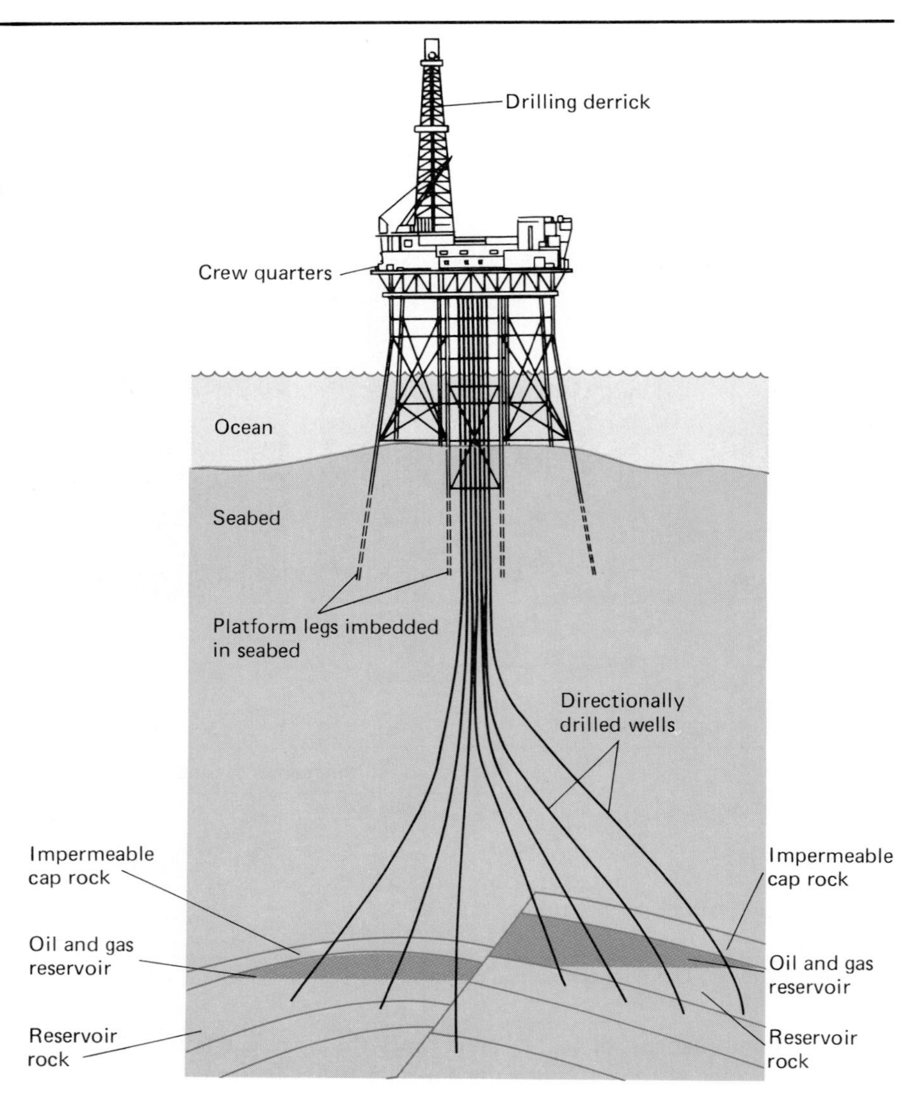


acid seepage from coal deposits was a topic of concern. This type of pollution results from bacterial action converting the sulfur in the coal into compounds that form sulfuric acid.

This problem becomes most acute when the overburden is disturbed, allowing heavy rains to wash the sulfuric acid from the mines into the streams. Streams may become so acid they will not support most forms of life. Only certain species of bacteria and algae can survive in these acid streams. Today, many states have regulations to limit the amount of runoff from the mines, but underground mines and strip-mined areas abandoned before these regulations were enacted continue to pollute the water.

Although there are environmental problems associated with the use of coal, the cost of coal was the motivation for changing to an alternative source of fuel—a more convenient and less expensive source of energy. That alternative was crude oil.

## *Production and use of oil*

In the 1920s, the use of oil was encouraged because it was more convenient and often less expensive than coal. This stimulated the oil exploration industry. Geologists use a series of tests to locate underground formations that may contain oil. A test well is then drilled to determine if oil is actually present. If found, the oil is pumped to the surface. Since the fields where the oil is the easiest to reach have already been tapped, drilling for oil is becoming more expensive. As the oil deposits become more difficult to locate, geologists have widened the search for possible sites to include the ocean floor. The cost of building an offshore drilling platform can be millions of dollars. To reduce the cost of drilling, as many as seventy wells may be sunk from a single offshore platform. (See fig. 9.10.)

If the water or gas pressure associated with a pool of oil is great enough, the oil will be forced to the surface when a well is drilled. This is called a gusher. When the natural pressure is not great enough, the oil must be pumped to the surface. Present technology allows only about one-third of the oil in a pool to be removed. This means that two barrels of oil are left in the ground for every barrel produced. In some oil fields, **secondary recovery** is used to recover a greater yield from the pool. One secondary recovery method is to inject gas or water into the pool to force the oil to the surface. (See fig. 9.11.)

As the supply of crude oil decreases, the price increases. It then becomes profitable to use expensive secondary methods to obtain a higher percentage of oil from a pool. The increased price of oil also encourages the development of new technology to exploit resources that are currently uneconomic.

One example of a resource that is becoming a reserve is heavy oil. **Heavy oil** has the consistency of asphalt. Only 10 percent of it can be pumped to the surface with ordinary pumping operations. Heavy oil also contains less energy per barrel than ordinary crude oil and yields less gasoline per barrel when refined. For these reasons, heavy oil has not had a market in the past.

However, the increased price of oil has made pumping heavy oil profitable. At present, oil companies are forcing pressurized steam into heavy oil wells. This heat causes the heavy oil to flow, so it can be pumped to the surface. This process requires the equivalent of one barrel of oil in the form of heat for every three barrels pumped to the surface. Thus, the production cost and the energy cost of pumping heavy oil is higher than that of ordinary crude oil. The low-energy value of this oil makes it more expensive to refine into gasoline. In an attempt to increase the production of gasoline from this oil, a new type of oil refinery is scheduled to go into production in California in the 1980s. The use of heavy oil has increased the world's oil reserves, but the consumer must be prepared to pay higher prices for the production of this oil.

**Figure 9.11 Secondary recovery methods.** This illustration contrasts three methods of secondary recovery. They are flank water flood, fire flood, and gas injection. In the flank water flood method, water is pumped into an injection well to force the oil into the producing well. With the fire flood method, some of the underground oil is burned to provide heat that thins the heavier oil and allows it to be pumped to the surface. Natural gas is forced into the oil reservoir to force the oil upward when the gas injection method is used. Then both the gas and oil are obtained from the well. The gas is separated and reused to force more oil upward. (Source: American Petroleum Institute.)

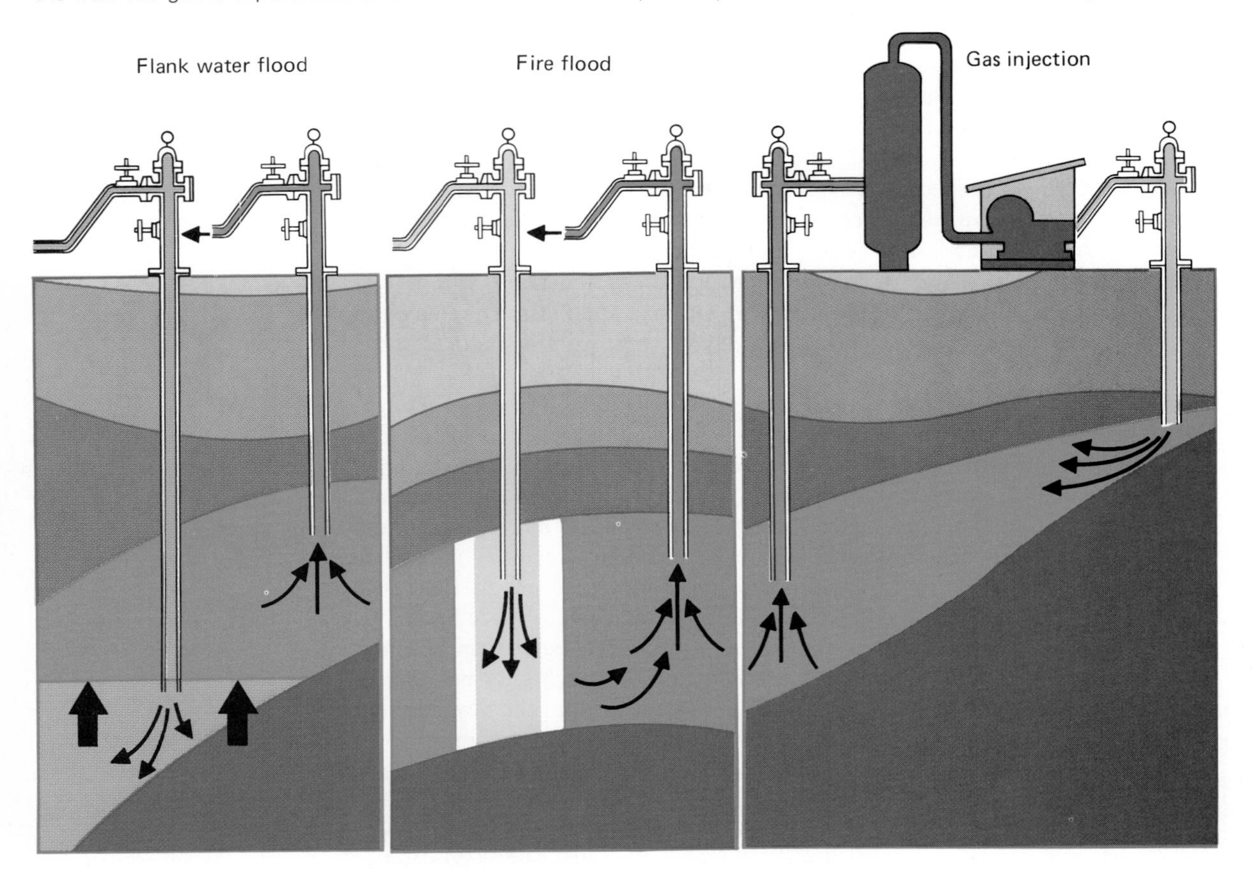


Oil as it comes from the ground is not in a form suitable for use. It must be refined in order to get the products wanted. By heating the oil in a distillation tower, the various components of the crude oil may be separated and collected. (See fig. 9.12.) After distillation, the products may be further refined by the "cracking" process. In this process, heat, pressure, and catalysts are used to produce a higher percentage of volatile chemicals such as gasoline from less-volatile liquids such as diesel fuel and furnace oils. Therefore, it is possible, within limits, to obtain different products from a barrel of oil. In addition to its use as a fuel, petrochemicals from oil serve as raw materials for a variety of synthetic compounds. (See fig. 9.13.)

**Figure 9.12 Uses of crude oil.** A great variety of products may be obtained from the distilling and refining of crude oil. A barrel of crude oil will produce slightly less than half a barrel of gasoline. This figure shows the many steps in the refining process and the variety of products that can be obtained from crude oil.

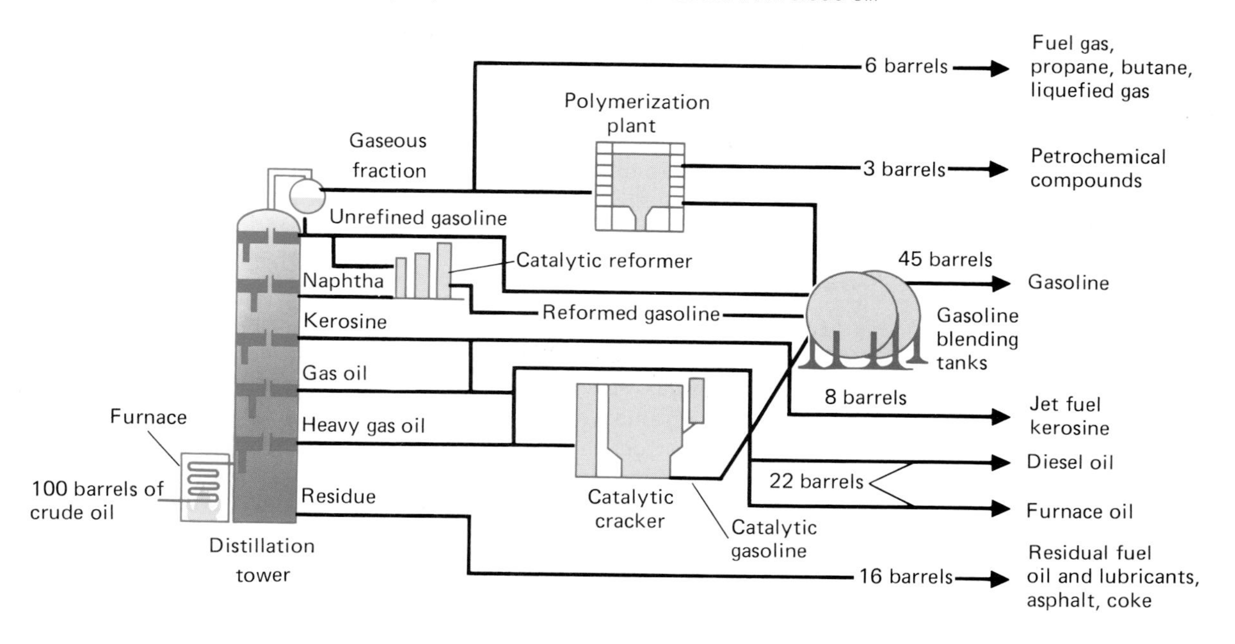

**Figure 9.13 Oil-based synthetic materials.** This array of common household items is produced from chemicals derived from oil. Although petrochemicals represent only about 3 percent of each barrel of oil, they are extremely profitable for the oil companies.

The United States has 5 percent of the world's oil reserves, but by the late 1970s it was importing 40 percent of its total oil requirements. As a result of economic changes and conservation practices, this was reduced to about 10 percent by 1982. We are still very dependent upon foreign nations for much of our energy. Changes in the attitudes of these countries can affect the supply and price of the oil we import.

Because of the international trade in oil and the increased drilling of oil wells in the oceans, oil spills are a constant concern. If a tanker is damaged or oil escapes from an offshore well, it may cause considerable injury to various life forms in the area of the spill. (See fig. 9.14.)

### *Production and use of natural gas*

Natural gas is the third major source of fossil fuels. The drilling operations for obtaining natural gas are similar to those used for oil. In fact, a well may yield both oil and gas. After processing, the gas is piped to the consumer for use. Although the primary use of gas is for its heat energy, it does have other uses, such as for petrochemicals and fertilizers. Approximately 7 percent of the natural gas produced in the United States is used to produce fertilizers.

Transportation of natural gas still presents a problem in some parts of the world. The wells are too far from the consumers to make pipelines practical. Because of this, much of the natural gas is burned as a waste product of the wells in the Middle East, Mexico, Venezuela, and Nigeria. However, several new methods of transporting gas and converting it to other products are being explored. At $-162°C$, natural gas becomes a liquid that has only 1/600 of the volume of the gas. Tankers have been designed to transport **liquified natural gas** from the area of production to the area of demand. Many people are concerned about possible accidents that would cause the tankers to explode. There is also a process for converting the gas to methanol, a liquid alcohol, and transporting it in that form. As the demand for gas increases, the amount of gas wasted will decrease, and new methods of transportation will be employed. Higher prices will make it profitable to transport the gas greater distances from the wells to the consumer.

Even though the process of forming fossil fuels is continuing today, the rate of production is very slight. For all practical purposes, the world's supply of fossil fuels is limited to what was formed some 300 million years ago. When this supply is exhausted, there will be no more. As a result, people have started to seek alternative sources of energy. Three alternatives that are already providing significant portions of our current energy requirements are water, wood, and nuclear power. Nuclear power will be discussed in chapter 10.

| Year | 1,000 tons of oil spilled |
|---|---|
| 1973 | 84 |
| 1974 | 67 |
| 1975 | 188 |
| 1976 | 204 |
| 1977 | 213 |
| 1978 | 260 |
| 1979 | 723 |
| 1980 | 135 |
| 1981 | 45 |
| 1982 | 2 |
| 1983 | 55 |

**Figure 9.14 World-wide oil spills.**
Note the decrease in the amount of oil released since 1980. This probably represents some increase in conservation and environmental awareness, but it also represents the change in the amount of oil being transported by tankers since 1980, which is a result of economics and the unstable conditions in the Middle East.

**Figure 9.15 Hydroelectric power plant.**
The water impounded in this reservoir is used to produce electricity. In addition, this reservoir serves as a means of flood control and provides an area for recreation.

## Other conventional energy sources

Although water and wood are not major energy sources, this does not mean that they are not important sources of energy. In the United States, the current energy crunch makes any type of energy important. These two sources produce approximately 5 percent of the United States energy, and they are vital economically. As the supply of fossil fuels declines, these minor sources will become even more significant.

### *Water power*

People have used water to power a variety of machines for a long time. Some early uses of water power were to mill grain, saw wood, and power machinery for the textile industry. However, today water power is used almost exclusively to generate electricity. Flowing water supplies the energy to turn a generator and produce electricity. In this country, hydroelectric power plants are commonly located on artificial reservoirs. (See fig. 9.15.) The impounded water represents a potential source of energy. In some areas where the streams have steep gradients and constant flows of water, hydroelectricity may be generated without the construction of a reservoir. However, such sites are usually found in mountainous regions and operate only small power-generating stations. At the present, hydroelectricity produces 2 percent of the world's electrical needs and 13 percent of the United States electrical requirements.

Slightly over 5 percent of the potential hydroelectric sites of the world have been developed. Many of the undeveloped areas are so remote that the construction of generating stations and transmission lines would not be profitable. In the United States, about 50 percent of our hydroelectric capacity has been developed. This does not reveal the entire picture. For example, the Wild and Scenic Rivers Act (1968) prevents the construction of dams on certain designated streams. Thirty-seven potential hydroelectric sites are on streams protected by this act. The hydroelectric generating potential often quoted for the United States includes the areas even though, at the present, it is not possible to construct a generating plant on them.

As long as the political climate of the United States favors the protection of certain rivers, it will be impossible to develop hydroelectric power plants on them. This attitude might be modified if the demand for more energy becomes acute.

In the United States, the development of many of the potential hydroelectric sites would mean constructing reservoirs that would flood fertile farmland, displace homes and communities, or flood some extremely scenic areas. As a result, farmers, homeowners, and environmental groups often oppose the construction of new reservoirs. The movement to stop the building of the Tellico Dam in Tennessee is an example of such opposition. One of the issues involved was the destruction of the habitat of an endangered species of fish known as the snail darter.

When a dam is built on a stream, it can affect the native fish population adversely. The dam disrupts the normal movements of fish in the stream. A fish ladder is often built to provide a way for the fish to migrate around the dam. However, the ladder can only be used by certain species of fish, such as salmon and some varieties of trout.

Retarding the natural flow of the stream in a reservoir causes another problem. Moving water transports soil. When the movement is stopped, the soil is no longer carried by the water, and it settles to the bottom. The greater the amount of silt within a reservoir, the less capacity for water behind the dam, and the more difficult it is to regulate flow. Along with water power, attention is being focused on another energy source commonly used in the past—wood.

## Box 9.1
## Pumped storage

When designing units to produce electricity, the utility companies face the problem of fluctuating demand. The demand for electricity during the day may be 40 percent greater than at night. If a utility company builds a generator to meet the peak day demand, there will be a surplus of electricity produced at night. This surplus would be wasted. In order to adjust to the high and low demand periods, companies construct large coal-fired, nuclear, or hydroelectric plants that produce the "base load" of electricity. Such plants operate continuously at maximum efficiency and produce 75 percent of our electricity. This amount corresponds to our low demands. When the demand for electricity exceeds the production possibility of the "base load" generators, small generating units are used to produce the additional power required during peak demand. These smaller units are often gas or oil-fired systems, which are less efficient and more costly to operate than the larger coal or nuclear "base load" systems. Therefore, the use of these generators during peak demand increases the cost of electricity.

If people would make an effort to use less electricity during the peak demand period and shift this use to low demand periods, the "base load" plants could supply more of our needs. Adjusting the thermostat on the home air conditioner or heating system in the daytime when no one is home or baking and washing at night are ways that each of us can help to level these peaks. If people adopted this pattern, the cost of electricity would not rise as rapidly.

One way that utility companies try to reduce the need for expensive peak generating capacity is through pumped storage. Pumped storage stores the energy produced by generating stations during periods of low demand and uses it during times of peak electrical demand. This energy is not stored as electricity but as potential energy. In a pumped storage site, an electrical motor pumps water up into a

### *Wood*

Wood, one of our earliest fuel sources, is being reexamined as a potential energy source. The most common use of wood as an energy source is for cooking and heating in private dwellings. The amount of wood used for this purpose now supplies about 1 percent of the home heating needs in the United States. It is not profitable from an economic or energy standpoint to transport wood over long distances. Therefore, the homeowner or industry using the wood must be relatively close to the supply.

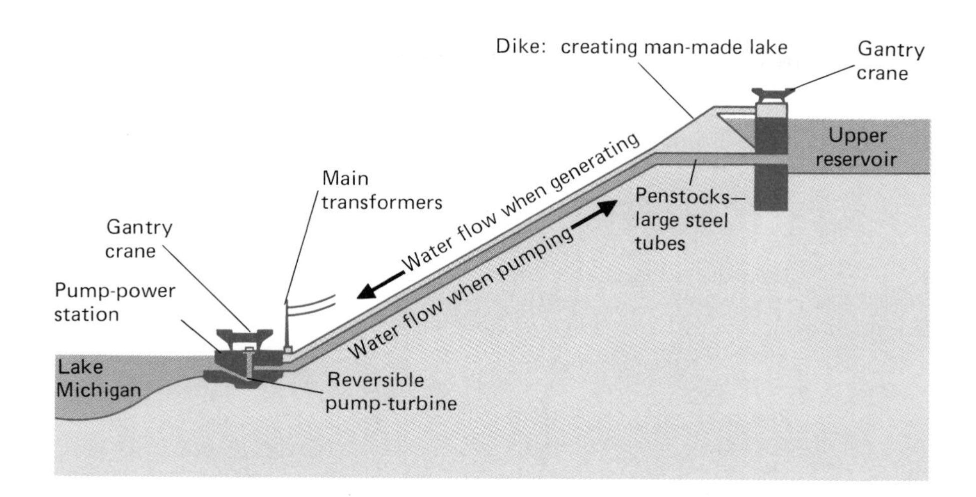


reservoir. When the flow of water is reversed, this same motor becomes an electrical generator, as shown in this figure. During periods of peak demand, the water flows down the pipe and produces electricity. Pumped storage is about 66 percent efficient; for every three hundred kilowatt hours of electricity used to pump the water into the reservoir, two hundred kilowatt hours of electricity is produced when the water flows from the reservoir. This efficiency does not account for line losses in transmitting the power. The closer the "base load" generator is to the pumped storage site, the smaller these losses are.

Indiscriminate cutting of wood will destroy some wildlife habitats. The greater our use of wood as a fuel, the greater this potential problem. Another problem associated with the use of wood as a fuel is the large amount of **fly ash** (particles in smoke) produced when wood is burned, which could result in increased air pollution. In 1971, Aspen, Colorado, limited the number of fireplaces to one per dwelling because of air pollution problems.

In a country with a large population, such as the United States, wood can supply little of our energy. We must continue to seek alternate sources of energy.

## CONSIDER THIS CASE STUDY
### The Alaskan pipeline

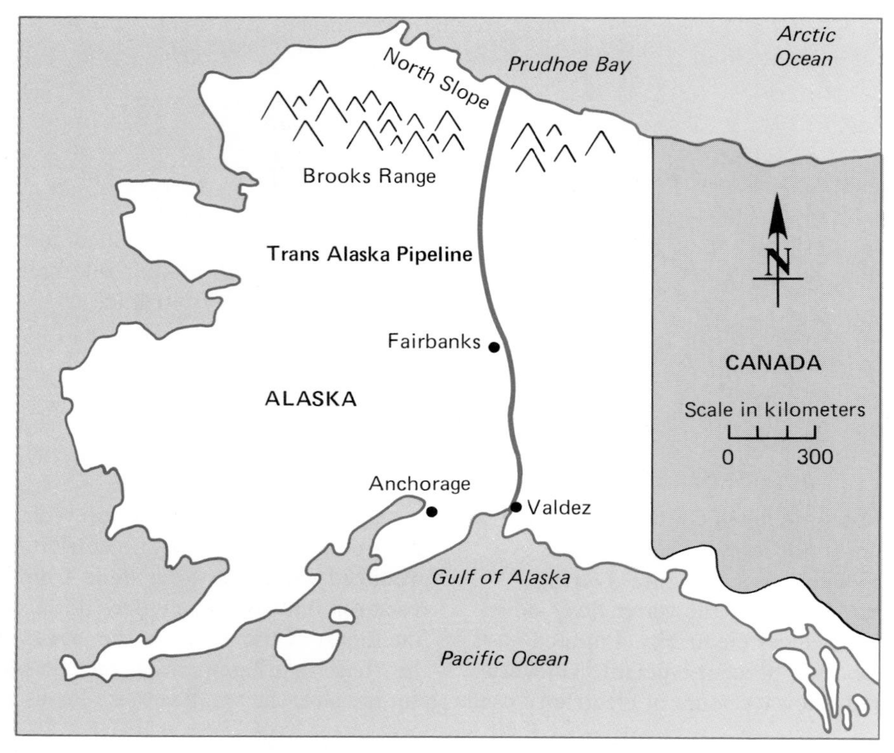


Reprinted from *U.S. News and World Report* issue of June 20, 1977. Copyright 1977, U.S. News and World Report, Inc.

"In the conflict between nature and resources development, nature invariably loses." These words by an environmentalist have been applied to the Alaskan pipeline. This $7.7 billion project was the largest construction job ever undertaken by private developers and the biggest project since the digging of the Panama Canal. People first began to think about a pipeline in 1968, when oil was discovered on the Alaskan north slope. For nearly seven years, the construction of the pipeline to transport this oil was delayed. First, the Alaskan Natives (Eskimos, Aleuts, and Native Americans) delayed construction by demanding a share of the oil's wealth. In addition, environmentalists demanded design changes. As a result, the plans were changed from a buried heated line that would have destroyed sections of the permafrost to an elevated pipeline.

Construction was begun in 1974 and completed three years later. A 1,285 kilometer pipeline now extends from Prudhoe Bay in the north to the port of Valdez in the south. This pipeline can carry 1.6 million barrels of oil per day. This is 7 percent of the oil consumption of the lower forty-eight states and represents an annual savings in excess of $6 billion over the cost of importing the same amount of foreign oil. But there are conflicting reports about the amount of environmental damage.

To safeguard against any disruption in animal migration, four hundred animal crossings have been constructed. These involved burying the pipe or elevating it 4.5 meters above the surface. Thus far, the wildlife seem to have adjusted to the pipeline, and there has been no disruption in the migration of caribou or other wildlife. In fact, the caribou and moose will cross under the pipeline anywhere it is 1.5 meters above the ground. In addition, much of the pipe was built above ground so the 35°C oil would not thaw the permafrost. Where the pipe is buried, it is insulated, and refrigerated brine prevents thawing of the permafrost. As a protection against earthquakes, the pipe has joints at three locations that allow it to move six meters vertically and six meters horizontally. Alyeska, the corporation that constructed the pipeline, states that all is well with the environment, but not everyone agrees.

During the construction process, several issues surfaced. Some pipeline workers have reported that the supports elevating the pipeline are sinking. (Alyeska denies this.) The agreement that bulldozers were only to operate in the winter when they would do less damage has not been honored. During construction, 2.1 million liters of oil were spilled on the right-of-way. This is approximately one liter of oil per sixty centimeters of pipeline.

There was also concern about the surveillance of the pipeline during its construction and, also, now that it is in operation. The surveillance is still being done largely by engineers. Ecologists, who can identify damage to the environment, are in advisory roles. The biggest problem seems to be summed up by Guy Martin, Assistant Secretary of Interior for Land and Water Resources at the time the pipeline was being built: "The major environmental effect of this pipeline and the road that parallels it is that they're there."

The delays in construction supposedly have tripled the cost of the pipeline. Were the delays worth the added cost?

What additional environmental dangers are present in the port of Valdez as a result of the pipeline?

Should Alyeska be held responsible for the material spilled during the construction?

Who should be responsible for the surveillance of the pipeline? Why?

Is the oil more important than the protection of a fragile arctic ecosystem?

## Summary

Fossil fuels were formed from the remains of plants, animals, and microorganisms that lived about 300 million years ago. Peat changed to lignite (low-grade) coal, which changed to bituminous (soft) coal, which changed to anthracite (hard) coal. Natural gas and oil were formed from ancient marine deposits.

A resource is the naturally occurring substance that is potentially feasible to extract under prevailing conditions. Reserves are known deposits from which materials can be extracted profitably with existing technology under present economic conditions.

Coal is the United States's most abundant fossil fuel. (The United States has 30 percent of the world's supply.) Coal is mined either by strip mining or shaft mining. Some problems associated with coal mining are disruptions in the landscape due to strip mining, black lung disease, waste heaps, water pollution, and acid mine drainage.

Oil was originally chosen as an alternative to coal because it was more convenient and less expensive. However, the supply of oil is also limited. As oil becomes less readily available, advanced technology is needed to acquire oil. These include multiple offshore wells, secondary recovery methods, and heavy oil exploitation.

Natural gas is another source of fossil fuel. Problems associated with natural gas are storage and transportation.

Other conventional sources of energy are water power and wood.

## Review Questions

1. How were fossil fuels formed?
2. What are the advantages of strip mining as compared to underground mining? What disadvantages are associated with strip mining?
3. Distinguish between reserves and resources.
4. List three techniques being used to exploit oil sources that were unavailable twenty years ago.
5. What single factor was responsible for the increased consumption of natural gas?
6. What are some limiting factors in the development of new hydroelectric generating sites?
7. What is likely to prevent wood from becoming a major source of energy?
8. Compare the environmental impact of the use of coal with the use of oil.

## Suggested Readings


Chironis, Nicholas P., ed. *Coal Age Operating Handbook of Coal Surface Mining and Reclamation.* New York: McGraw-Hill Book Company, 1978. Although this book is designed for coal mine operators, it can be easily understood by the average reader. It is not a highly technical book, but it presents the various methods of strip mining and reclamation. There are many excellent photographs of the machinery used in strip mining, as well as drawings of the various mining and reclamation operations.

Curran, Samuel C., and Curran, John S. *Energy and Human Needs.* New York: Halsted Press, 1980. Presents a detailed account of all forms of energy. This book has a good section on nuclear energy. It deals with energy transmission and storage as well as social, environmental, and international aspects of energy.

Dick, Richard A., and Wimpfen, Sheldon P. "Oil Mining." *Scientific American* 243: 182–88. Conventional drilling and pumping removes less than half of the oil in an average petroleum reservoir. As the price of oil rises, mining it underground or at the surface becomes economically more attractive.

Erskine, George S. "A Future for Hydropower." *Environment* 20: 33–38. Addresses upgrading and retrofitting of existing small dams to obtain more electrical power as well as the use of nondam hydrogenerators in small streams.

Kass, Ron E., et al. *Energy under the Ocean.* Folkestone, England: Bailey Brothers and Swinfen Ltd., copyright University of Oklahoma Press, 1974. A somewhat technical book concerning oil and gas operations on the outer continental shelf. Of particular interest to the student is part 2 on exploration and drilling operations. Part 3 deals with environmental quality, governmental control, and jurisdiction.

## Chapter Outline

I. Nuclear fission
  A. Nuclear power generation
  B. Nuclear power concerns
  C. Three Mile Island
II. Breeder reactors
III. Nuclear fusion
IV. Consider this case study: Consumers Power Midland Nuclear Power Plant

## Objectives

Explain how nuclear fission has the potential to provide large amounts of energy.

Describe the biological effects of "low" and "high" levels of radiation.

List the concerns that may limit the construction of new nuclear power plants.

Describe the incident at Three Mile Island.

Explain the ways in which a breeder reactor differs from other nuclear reactors.

List the technical problems associated with the design and operation of a liquid metal fast breeder reactor or a fission reactor.

Explain the process of fusion.

## Key Terms

atomic fission

decommissioning costs

fusion

isotope

liquid metal fast breeder reactor

moderator

nuclear breeder reactor

nuclear reactor

plutonium–239

pressurized water reactor

radiation

radioactive

thermal pollution

uranium–235

uranium–238

# Chapter 10 Nuclear power: Solution or problem

## Nuclear fission

**Radioactive** atoms contain nuclei that are unstable and split apart. When nuclear disintegration occurs, particles from the nucleus fly out and may cause other atomic nuclei to break apart. The process of **atomic fission** occurs when an atom is struck by a slowly moving neutron and caused to split. When this occurs, large amounts of energy radiate from the splitting nuclei of the atoms. This energy is important in the nuclear fission industry. **Uranium–235** (U–235) is a radioactive element used to fuel nuclear power plants. When atoms of U–235 undergo fission, they release neutrons from their nuclei; these nuclei strike the nuclei of other atoms of U–235 and cause them to undergo fission, which in turn releases more energy and more neutrons. Once begun, this chain reaction continues to release energy until the fuel is spent or the neutrons are prevented from striking other nuclei. (See fig. 10.1.)

**Figure 10.1 Nuclear fission.**
Whenever a neutron strikes a nucleus of U-235, energy is released, krypton and barium are produced, and several neutrons are released. These new neutrons may strike other atoms of U-235 to produce a chain reaction.

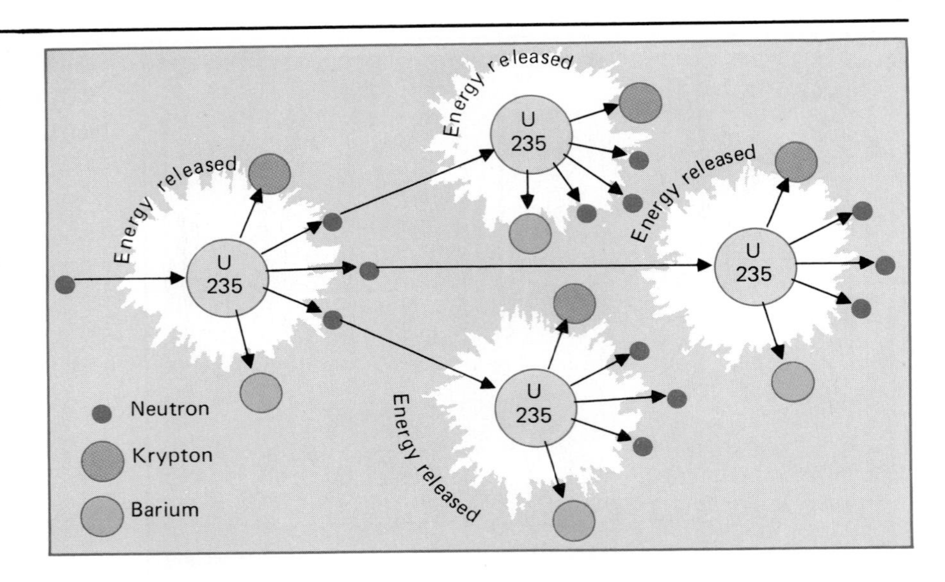


## *Nuclear power generation*

Naturally occurring uranium contains only 0.7 percent of the **isotope** U–235. This naturally occuring uranium is not suitable to power the reactors used in the United States. The U–235 isotopes must be four times as concentrated to be used as a suitable fuel for these reactors. In a **nuclear reactor** (such as the **pressurized water reactor** used in the United States) pellets of enriched uranium oxide are housed within metal fuel rods, which are surrounded by water. As the U–235 undergoes fission, the energy of the fast-moving neutrons is transferred to the water; the neutrons slow down, and the water is heated. The water that absorbs the energy is called a **moderator** because it slows neutrons and enables them to split the nuclei of fuel atoms. New fuel rods must be placed in the reactor after a certain proportion of the U–235 has disintegrated. Although the spent fuel rods still contain radioactive materials, they do not contain enough to power the reactor. These spent rods are a source of radioactive waste material produced in a nuclear reactor.

Control rods of cadmium, boron, and other nonfissionable materials are used to control the rate of fission. When these control rods are placed in the reactor, they absorb the particles produced by the fissioning uranium. Therefore, there are fewer particles to trigger fission in the other uranium atoms, and the rate of fission decreases. To produce more fission and more heat, the control rods are removed.

In a nuclear power plant, it is the energy released from the process of fission that heats the water to an extremely high temperature. This heated water travels under high pressure through the primary loop to a heat exchanger. In the exchanger, the heat from the water in the primary loop is transferred to

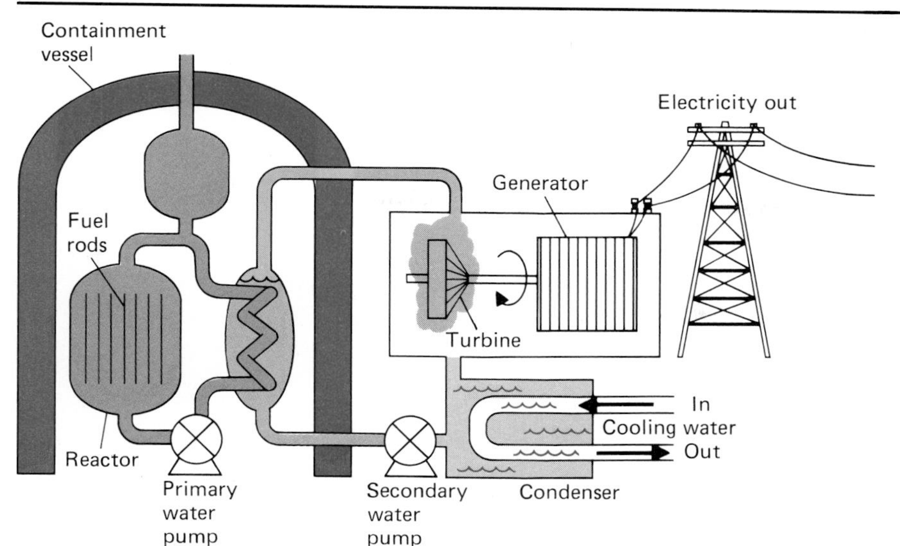


**Figure 10.2 Pressurized water reactor.**
The energy from fission within the reactor heats the water in the primary loop. In the heat exchanger, the heat from this primary loop is transferred to a secondary loop. The steam from this loop powers a turbine that produces electricity. Cooling water is used to condense the steam after it has gone through the turbine.

a secondary loop. This secondary loop conducts steam from the heat exchanger to turn a generator that produces electricity. (See fig. 10.2.) After powering the generator, the heated water in the secondary loop is cooled by an external water source, recirculated to the heat exchanger, and converted again into steam. Thus, the primary function of a nuclear reactor is to produce the heat required to make steam to turn the turbines that turn the generator.

Early in their development, nuclear power generating plants seemed to be an ideal alternative source to coal-fired plants. The nuclear plants did not pollute the air significantly, and the cost of producing electricity was less than that from a coal-fired plant. But eventually, some people began to question the safety of the nuclear power plants and the dangers associated with radioactivity. Even though studies in 1940 revealed a high percentage of cancer in European uranium miners, the United States Department of Labor did not establish radiation exposure levels for workers until 1967.

## *Nuclear power concerns*

In addition to the danger to the miners, there is a problem with the mine tailings that still contain some radioactivity. There are over 150 million metric tons of radioactive tailings in the United States. Although these tailings have a very low radioactivity level, they can be carried into the environment and create a problem. (See fig. 10.3.) For example, in some sections of the country, radioactive tailings were used for fill prior to the construction of schools, homes, and other buildings. There is concern about the effects of **radiation** on living things.

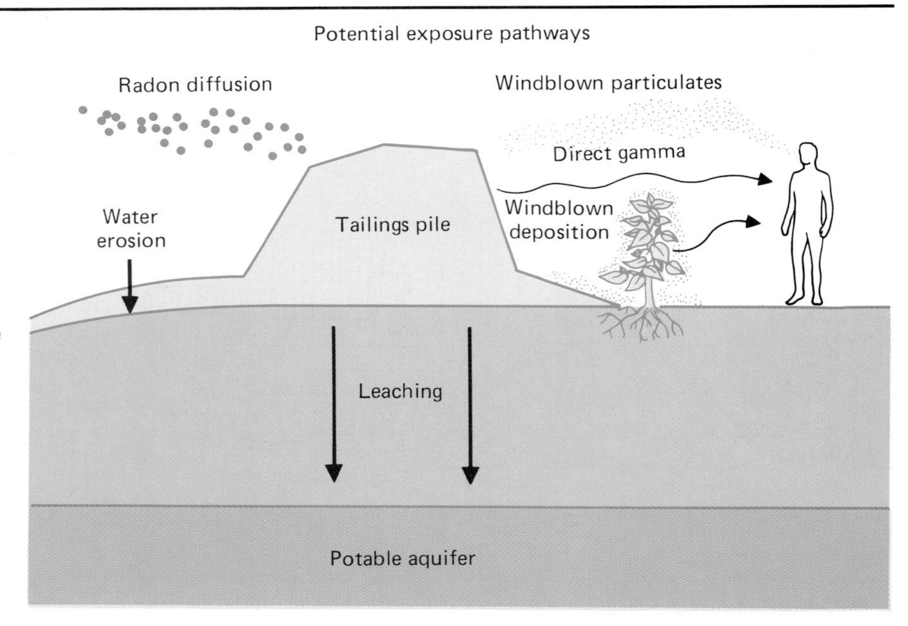


**Figure 10.3 Uranium mine tailings.**
Even though the amount of radiation in the tailings is low, it still represents a threat. There are several ways that the radioactivity is dispersed through the environment. These include seepage into the water supply, dispersal of radioactive particles by wind, and diffusion of radioactive gases. In addition, the mine tailings are a source of gamma rays, which can cause negative health effects.

Radiant energy is converted to other forms when it passes through matter. Because of this energy conversion, when organisms are irradiated, damage to them may occur at a cellular, tissue, organ, or organism level. The degree and kind of damage will vary with the amount of radiation, the duration of the exposure, and the particular type of cells irradiated.

A massive radiation dose like that produced in a nuclear blast or the radiation produced in the immediate vicinity of a nuclear reactor can result in the immediate death of cells and organisms. Lesser doses may produce changes within certain cell structures. For example, moderate doses destroy membranous structures in cells. This physical damage will eventually result in the death of the cell.

Another aspect of radiation that causes concern is its ability to cause mutations, which are changes in the genetic messages within cells. These mutations can cause two quite different problems. Mutations that occur in the ovaries or testes can form mutated eggs or sperm, which can lead to abnormal children. Therefore, care is usually taken to shield these organs from unnecessary radiation. Mutations that occur in other tissues of the body may manifest themselves as abnormal tissue growths known as cancer. Two common cancers that are strongly linked to increased radiation exposure are leukemia and breast cancer. In addition, ultraviolet light radiation is strongly linked to increased incidence of skin cancer.

Because mutations are essentially permanent, they may accumulate over an extended period of time. Therefore, the accumulated effects of radiation over a person's lifetime may result in the development of cancer late in a person's life.

Table 10.1
Radiation effects

| Source | Dose | Biological effect |
|---|---|---|
| Nuclear bomb blast or exposure in a nuclear facility | 100,000 rads/incident | Immediate death |
| | 10,000 rads/incident | Coma, death within 1-2 days |
| X-rays for cancer patients | 1,000 rads/incident | Nausea, lining of intestine damaged, death in 1-2 weeks |
| | 100 rads/incident | Increase probability of leukemia |
| | 10 rads/incident | Early embryos may show abnormalities |
| Upper limit for occupationally exposed people | 5 rems/year | Effects difficult to demonstrate |
| X-ray of the intestine | 1 rad/procedure | Effects difficult to demonstrate |
| Upper limit for release from nuclear installations (except nuclear power plants) | 0.5 rems/year | Effects difficult to demonstrate |
| Natural background radiation | 0.1-0.2 rem/year | Effects difficult to demonstrate |
| Upper limit for release by nuclear power plants | 0.005 rem/year | Effects difficult to demonstrate |

1 rad = 100 ergs absorbed per gram of tissue

1 rem = energy of the radiation (rad) × some quality factor based on the biological effectiveness of the radiation source. (i.e., quality factor for X-rays = ~ 1.0 and for neutron, proton and alpha particles = ~ 10.)

Because of the concern for radiation safety, it is necessary to quantify amounts of radiation. Two different ways of measuring radiation have been developed. The *rad* and the *rem* are both used and are roughly equivalent. After large doses (1,000 to 1,000,000 rads), effects are easily seen and can be quantified, because there is a high incidence of death at these levels.

Smaller doses make it much more difficult to demonstrate known harmful biological effects. Moderate doses (10 to 1,000 rads) are known to increase the likelihood of cancer and birth defects. (The higher the dose, the higher the incidence of abnormality.) Lower doses may cause temporary cellular changes, but it is difficult to demonstrate long-term effects. This low-level chronic radiation is the most controversial. Some people feel that all radiation is harmful and there is no safe level. Others feel that the increased risk of low-level radiation is extremely small and that current radiation standards are adequate to protect the public, especially in light of the benefits that radiation can give in terms of medical diagnoses and electrical energy. Research is currently being done to try to better assess the risks associated with repeated exposure to low-level radiation. (See table 10.1.)

In addition, there is still a question about nuclear reactor safety. One of the concerns is a possible core meltdown. A nuclear generating plant will not explode like an atomic bomb, but, if the fission rate is too rapid, the amount of heat released will melt the reactor. This hot, molten mass will release large amounts of radioactive material into the environment.

The Atomic Energy Commission sponsored a study to predict just how much damage would be done by a core meltdown. The findings, released in the Rasmussen Report, estimated that a complete core meltdown could cause as many as thirty-five hundred immediate deaths from exposure to the radiation. Another forty-five thousand people might die later as a result of cancer caused by the radioactive exposure. The study also estimated property damage from such an accident could be as much as $25 billion. People opposed to nuclear power plants use these figures to support their stand.

However, the report also predicts that the possibility of a meltdown is very unlikely. In fact, according to the report, if there were a hundred nuclear power plants operating in the United States, your likelihood of dying from an atomic accident would be one in five million. The odds of dying because of a nuclear meltdown are much less than the odds of dying because of an auto accident (one in four thousand) or being killed by lightning (one in two million). People who favor nuclear plants use these figures in their stand.

The big question is who do we believe. This question has become even more complicated. The Nuclear Regulatory Commission, which succeeded the Atomic Energy Commission, has rejected the findings of the Rasmussen Report as being too low in its risk assessment.

However, even if a nuclear power plant were completely safe, there is still the question about what to do with the radioactive waste. For over twenty years, people have been attempting to find an answer to this problem. Thus far, the only solution has been to store over five thousand metric tons of this waste in "temporary storage" sites. The federal government has suggested several sites suitable for the long-term storage of nuclear wastes. In every instance, the state involved has resisted the establishment of these storage facilities. Most of these facilities have been suggested for geologically stable salt deposits or granite rock. (See chap. 18.) Because of the problems concerning nuclear power plants, California, Iowa, Maine, and Wisconsin have banned the building of any new nuclear power plants.

Gradually, other concerns not related to radioactive material also surfaced. These include large construction costs, reliability, **thermal pollution,** and **decommissioning costs.** At first, nuclear power was hailed as a safe, cheap source of electricity. However, as construction costs of a nuclear power plant escalated, the price of producing electricity from such a plant has increased. Since 1969, the construction costs of building a nuclear power plant have increased seven times the original cost. In addition, nuclear plants are becoming more expensive to operate. By 1990, it is projected that nuclear power plants will generate electricity for 6.5 cents per kilowatt hour and coal-fired plants for 6.4 cents per kilowatt hour. Thus, nuclear-generated electricity will no longer be cheaper than electricity generated from a coal-powered plant.

Another reason for increasing the cost of nuclear-generated electricity is the loss of generating time. Nuclear power plants are most efficient when they are operating at full capacity for long periods of time. If the generating capacity must be reduced or interrupted, the costs rise rapidly. The time when the plant is not operating is called downtime. Many of the incidents that cause

**Figure 10.4 Nuclear cooling pond.**
This 350 hectare reservoir was constructed to serve as a cooling pond for a nuclear facility. The 12.5 billion liters of water in the pond will provide the coolant for the two reactors under construction.

plants to be shut down today were considered unimportant a few years ago. An increase in public awareness and concern for safety demands that these kinds of mechanical failures be corrected before the plant is allowed to continue operation.

Another reason for the increased cost of producing nuclear-generated electricity is the problem of thermal pollution. (For a more complete discussion of thermal pollution, see chap. 19. Both fossil fuel and nuclear plants generate steam to produce electricity. This produces a great deal of waste heat. In a fossil fuel plant, half of the heat energy produces electricity and half is lost as waste heat. In a nulear plant, only one-third of the heat generates electricity and two-thirds is waste heat. Therefore, the nuclear plant is less efficient. The nuclear plant also results in increased thermal pollution of the environment. To prevent this, costly cooling facilities are constructed and operated. (See fig. 10.4.)

It would seem that nuclear power plants have a history of becoming more costly. A newly appreciated cost is that of decommissioning a nuclear power plant. It is estimated that a nuclear power plant has an operational life of thirty to forty years. At the end of that time, it is taken out of operation. Recently, a midwest utility company requested permission to increase its electrical rates to raise one billion dollars over the next twenty years. This rate increase is designed to close down two nuclear power plants, one in year 2000 and another in year 2007. The original costs of constructing these two power plants was only $212 million. Groups opposed to nuclear power plants cite high decommissioning costs as another reason to stop any new construction of nuclear facilities.

**Figure 10.5 Three Mile Island nuclear reactor.** This is the site of the most publicized incident associated with a nuclear power plant. In 1979, a series of errors led to the shutdown and ultimate abandonment of the unit two reactor.

## *Three Mile Island*

One of the most publicized problems associated with a nuclear power generating facility was the incident at Three Mile Island in 1979. (See fig. 10.5.) At 4 A.M. there was a minor pump failure at the Unit Two reactor. (See fig. 10.6.) If the built-in preprogrammed safety feature of the reactor would have corrected this, the result would have been a very minor mishap. It would have posed no danger to the plant or the community.

When the pump failed, the electrical-generating turbine stopped. Automatically, the emergency core cooling system should have flooded coolant into the reactor and stabilized the temperature. However, due to an instrument error, an operator overrode the preprogrammed safety system and prevented the coolant from entering the reactor. As a result, the temperature of the core increased to the point of a near meltdown and severely damaged the facility. The events that followed not only damaged the facility but changed the attitude of the public regarding the role of nuclear reactors.

Figure 10.6 Chronology of events at Three Mile Island.

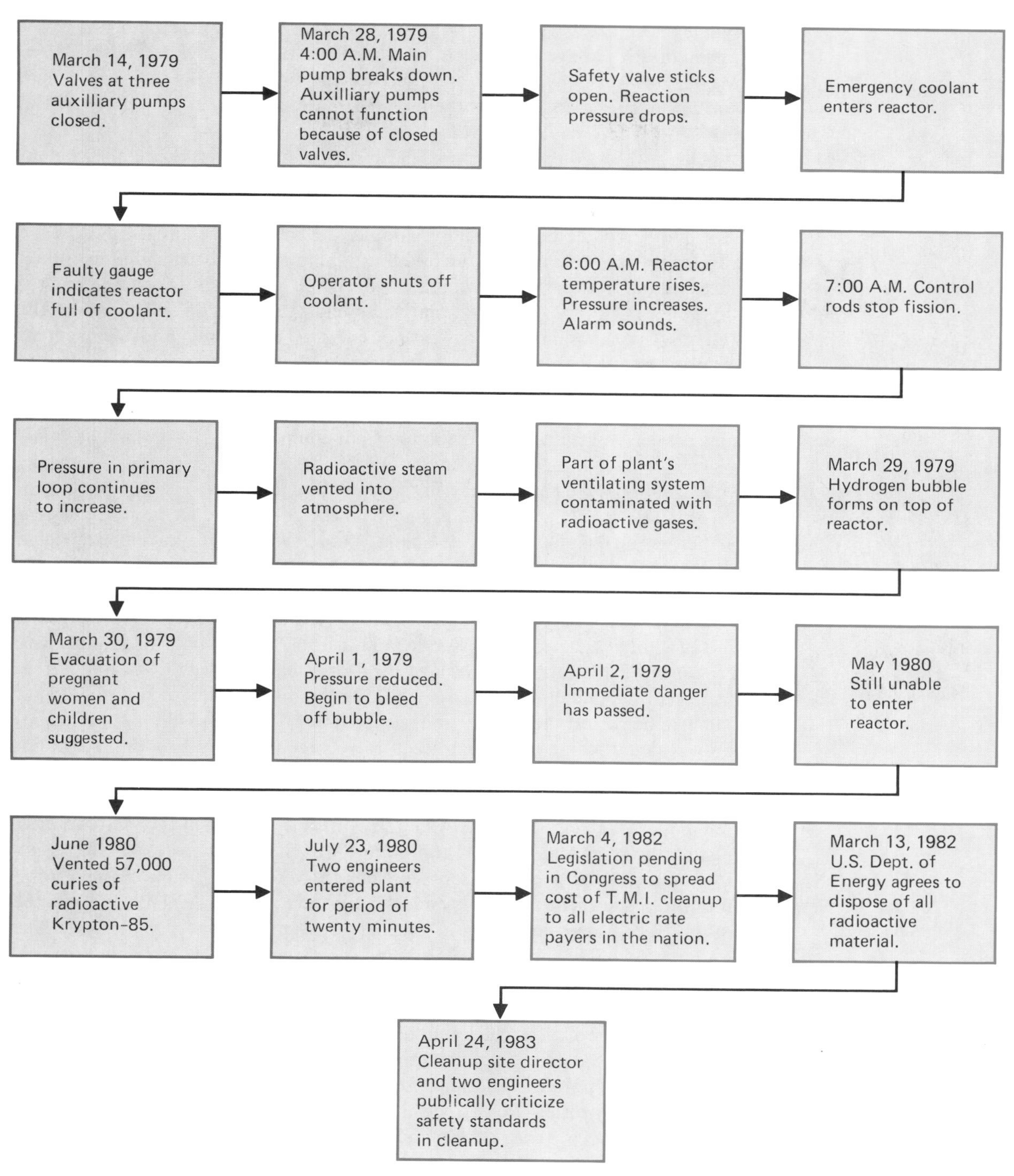

From the outset of the accident, a mass of misinformation was released about what was happening at Three Mile Island. Metropolitan Edison and the Nuclear Regulatory Commission (NRC) either were not aware of the exact extent of the damage or withheld information. So, the first victim of the Three Mile Island accident was the credibility of the entire nuclear industry. People will no longer be as likely to believe any statements by the industry or the NRC regarding nuclear power.

There was also the problem of cleaning up the reactor. Since an accident of this magnitude had never occurred before, no one was exactly certain how it should be handled. Several years after the incident, a number of problems still hampered the cleanup. For example, the radioactive gas trapped in the reactor prevented entry into the building. There were also nearly three million liters of water to be decontaminated, over half of which was in the reactor building. The NRC estimated that there had been a partial meltdown of the thirty-seven thousand fuel rods in the reactor. The cleanup and the eventual startup of Unit Two hinged on the removal of these rods. The NRC estimated that the cleanup would take five to six years to complete.

Thus far, Metropolitan Edison has paid over $106 million in cleanup costs, and they have estimated that the total bill could reach $1 billion. The NRC places the cost of cleanup at one billion dollars. Whatever the cost, someone will have to pay. The company has already applied to the public utility commission to raise its electric rates and pass the cost of the cleanup to the consumer, even though Three Mile Island is not currently generating electricity.

Regardless of what happens at Three Mile Island, the accident has made a tremendous impact on the nuclear industry. Even before Three Mile Island, nuclear power was in trouble. In 1973, forty-one nuclear reactors were ordered in the United States. In 1978, one year before Three Mile Island, only two reactors were ordered. None have been ordered since 1979. A Wall Street analyst, who called nuclear power a "dying industry," advises his clients not to put money into the nuclear industry. Three Mile Island did nothing to improve this outlook. As nuclear reactors become more costly to build, the industry will require more investors, not less.

Besides the difficulty of securing funds, the nuclear power plants face a much more concerned public. A Harris poll on the future of the United States nuclear power plants states that a majority of the public favors the construction of these plants but opposes the construction of the plants in their communities. Since Three Mile Island, the question of nuclear plant safety is the main issue. Today, no one wants nuclear plants built in their community. The high costs would probably prevent construction even if this were not the case. This places an unofficial moratorium on the construction of new nuclear power plants.

Thus, in the United States, nuclear power plants are receiving opposition because of the danger of radioactive material, possible damage to the environment, and high construction and decommissioning costs. However, in other regions of the world, nuclear power plants are still considered the best source of energy. The countries building nuclear plants are mainly those that do not have fossil fuel deposits.

A national referendum in Sweden indicated that 58 percent of the people favored the development of nuclear power plants, while 39 percent were opposed. Sweden is the world's largest per capita importer of oil. It has no oil or coal deposits but does have the largest uranium deposits in western Europe. More than 25 percent of its electricity comes from nuclear power plants.

France also lacks large deposits of fossil fuel and is building nuclear power plants. Second only to the United States in the number of nuclear power plants, it has sixteen now in operation. France has scheduled a new nuclear facility to begin operation every two months for the next five years. By 1985, 55 percent of its electricity and 20 percent of its power would have come from nuclear power plants. Nuclear power plant construction has been limited to the completion of those already started and no new construction has been allowed with the election of President Francois Mitterrand. Other countries with low fossil fuel reserves, such as West Germany, Belgium, Japan, Brazil, and Denmark, are planning major nuclear developments.

Today's energy concerns are mainly the result of the limited amount of fossil fuels and the problems resulting from the nuclear alternative. In addition, the amount of uranium suitable for nuclear fuel is limited.

## Breeder reactors

A new technology suggests a way of manufacturing nuclear fuel in a safe and economic way. This process furnishes energy for the production of electricity and also produces new radioactive material for use in other reactors. At first glance, it seems impossible to use a fuel for energy and at the same time obtain fuel for use in other reactors. A regular fission reactor produces heat to generate electricity but does not form fission products useful as fuel. A **nuclear breeder reactor** is a nuclear fission type of reactor that produces heat to be converted to steam and then to electricity; it also forms a new supply of radioactive isotopes. If a fast-moving neutron hits a nonradioactive **uranium 238 (U–238)** nucleus and is absorbed, an atom of radioactive **plutonium 239 (Pu–239)** is produced. (See fig. 10.7.) In a breeder reactor, water is not used as a moderator because it would slow the neutrons needed to produce the new fuel. What is required, however, is a material that allows the neutrons to move at a rapid rate and have good heat transfer properties.

The **liquid metal fast breeder reactor** (LMFBR) seems to be the most promising model. In this type of reactor, the fuel rods in the core are surrounded by rods of U–238 and liquid sodium. (See fig. 10.8.) The energy of the neutrons released from U–235 in the fuel rods heats the sodium to a temperature of 620°C. In addition to furnishing heat, these fast-moving neutrons are absorbed by the rods containing U–238, and some of these atoms are converted into Pu–239. After approximately ten years of operation in which electricity is produced, the LMFBR will have also produced enough radioactive material to operate a second reactor. The Pu–239 can be used in another reactor as a replacement for U–235.

**Figure 10.7 Formation of Pu-239 in a breeder reactor.** A fast-moving neutron (N) furnishes the energy for a series of reactions that results in the formation of Pu-239 from U-238. Two intermediate atoms are U-239 and Np-239, which release beta particles (electrons) as they disintegrate. This is an important reaction because, while the U-238 does not disintegrate readily and, therefore, is not a nuclear fuel, the Pu-239 is fissionable and can serve as a nuclear fuel.

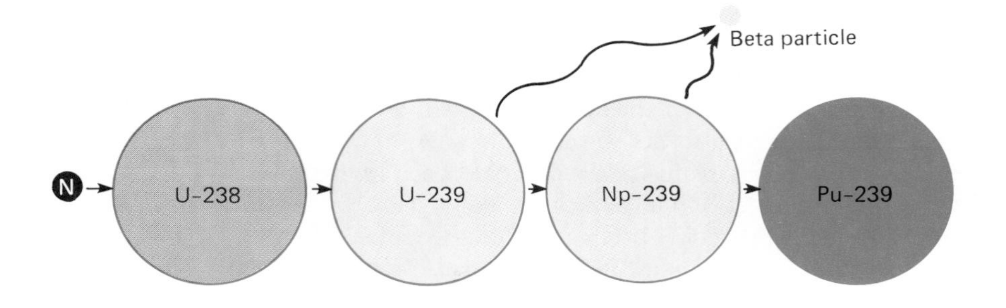


**Figure 10.8 Liquid metal-fast breeder reactor.** In the reactor unit, U-238 is converted into Pu-239. The heat from the fission reaction within the reactor also heats sodium metal in the core. This heat is transferred in a heat exchanger to the sodium in a secondary system. In a second heat exchanger, steam is generated to turn the turbine to generate electricity. Thus, a breeder reactor produces more radioactive fuel as well as electricity.

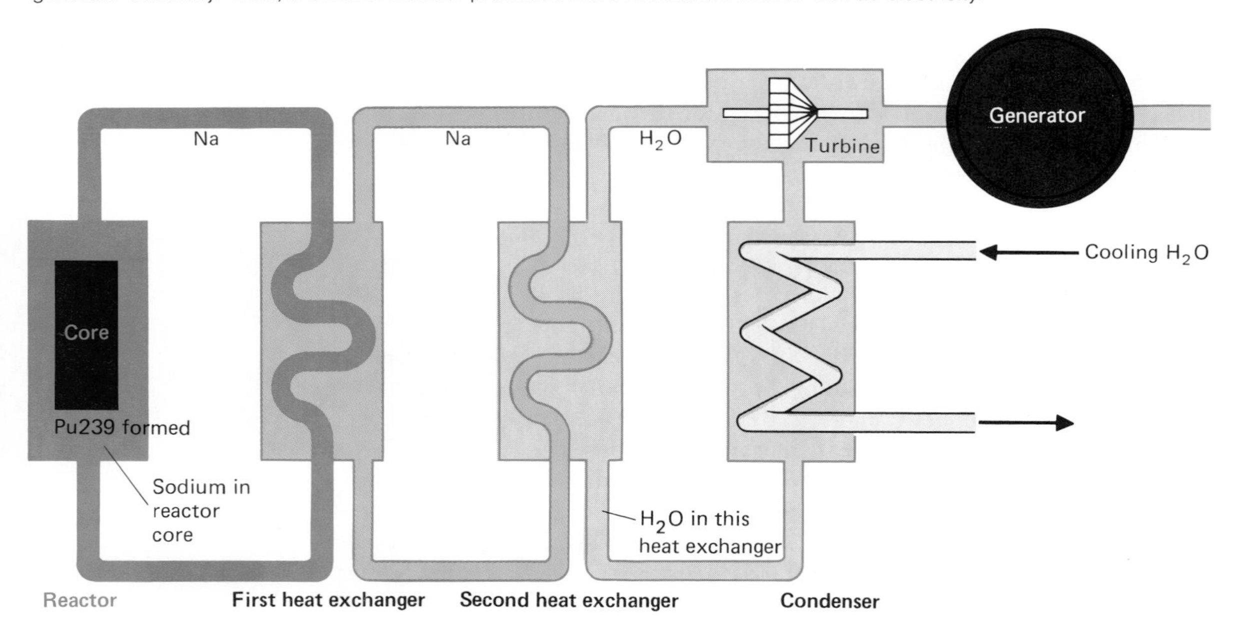

This would seem to solve our energy problem. We merely continue to breed new fuel. However, breeder reactors are not used on a large scale because of several unsolved problems.

The use of sodium for heat transfer has some drawbacks. It reacts explosively if allowed to come in contact with water or air. Therefore, costly, highly specialized equipment is required to contain and pump the sodium. If these systems fail, the sodium would boil, which would cause the chain reaction to proceed at a faster rate and could damage the reactor. There are also problems in the startup of a LMFBR. The sodium is a liquid at a high temperature. When a new breeder is starting or going on line after a shutdown for repair, the solid sodium cannot be moved by the pumps until it becomes a liquid. This presents an additional technical problem in the management of a LMFBR.

Another problem is that reaction rates are very difficult to regulate in a LMFBR. The instrumentation needed to monitor a LMFBR must be of extremely high quality because control of the reaction involves precise adjustments over a very short period of time.

The amount of fissionable material generated by a breeder reactor increases the potential amount of radioactive waste. The production of Pu–239 in a breeder would increase the amount of fuel available and allow for the proliferation of fission-type reactors. A breeder reactor also produces other types of nuclear waste, including iodine–129 (I–129). This iodine isotope is harmful to the thyroid gland and has a half-life of 17 million years. The half-life is the length of time required to reduce the number of radioactive atoms by one-half. Thus, if one kilogram of I–129 were stored as waste, in 17 million years, there would be a half kilogram of this material still present. Because we are unable to solve the problem of radioactive waste disposal from current fission reactors, the additional radioactive material made available from breeder reactors would only compound this dilemma.

Another cause for concern is the potential use of Pu–239 in the production of nuclear weapons, which could threaten world peace. Nations without the capability of producing bombs from U–235 may be capable of developing bombs from Pu–239. Technically, this is a relatively easy task compared to making weapons from U–235.

Because of their problems, no breeder reactors are scheduled for commercial use in the United States. However, the Soviet Union is operating a small-scale breeder. France's President Francois Mitterrand made a pre-election promise to halt construction of the Super-Phenix fast breeder reactor, and the commitment by the French government to an additional five breeder reactors has been reduced to only one.

Figure 10.9 Nuclear fusion.
In nuclear fusion, small atomic nuclei are combined to form heavier nuclei. Large amounts of energy are released when this occurs. Different isotopes can be used in the process of fusion. There are three possible types of fusion: (1) Two deuterium isotopes can combine to form helium-4 and release energy. (2) Two deuterium isotopes may combine to form helium-3, a free neutron, and release energy. (3) A deuterium and a tritium isotope can combine to form helium-4, a free neutron, and release energy.

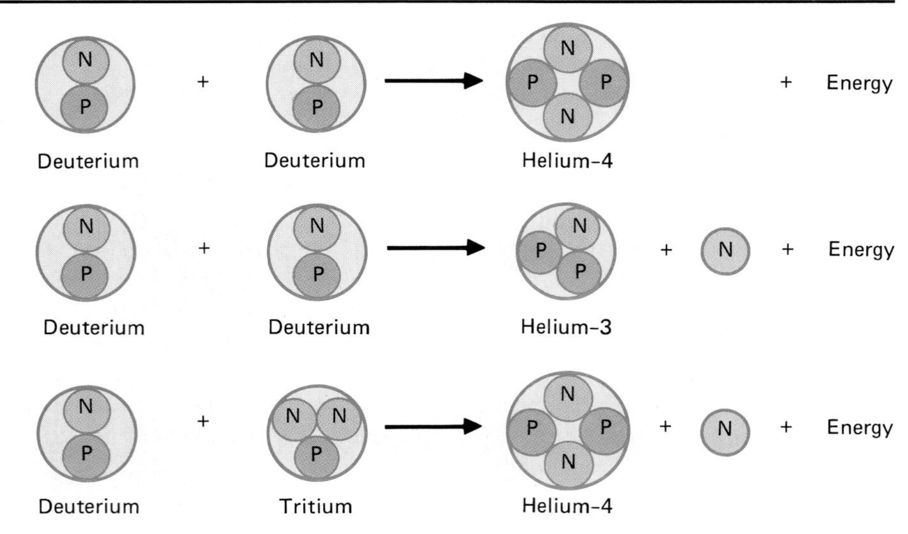


## Nuclear fusion

When two lightweight atomic nuclei combine to form a heavier nucleus, a large amount of energy is released. This process is called **fusion.** The energy produced by the sun is the result of fusion. Most studies of fusion have involved small atoms like hydrogen. Most hydrogen atoms have one proton in the nucleus. The hydrogen isotope deuterium ($H^2$) has a proton and a neutron in the nucleus. Tritium ($H^3$) has a proton and two neutrons. When deuterium and tritium isotopes are combined to form heavier atoms, large amounts of energy are released. (See fig. 10.9.) The energy that could be released by combining the deuterium in one cubic kilometer of ocean water would release an amount of energy greater than that contained in the world's entire supply of fossil fuels.

Although fusion could solve our energy problem, technology must answer several questions before fusion power can become a reality. Three factors must be met simultaneously if fusion is to occur. These are high temperature, adequate density, and confinement. If heat is used to provide the energy necessary for fusion, the temperature must approach that of the center of the sun. At the same time, the walls of the chamber must be protected from the heat, or they will vaporize. The main problem is containment of the nuclei. Because they have a positive electric charge, they repel one another.

**Figure 10.10 Tokamak.** The Soviet Tokamak project has developed magnetic fields to confine the hot plasmas used in the process of fusion.

One solution to this problem is to use a magnetic field to contain the material. The deuterium would be decomposed into a gas of charged particles known as plasma. Magnetic fields would then be used to contain the plasma and concentrate the positively charged nuclei. This is a difficult task because the nuclei repel one another. Electricity would provide energy to initiate the reaction. Experimentally, the problems of extremely high temperature, high plasma density, and confinement of the plasma have each been solved on an individual basis. (See fig. 10.10.) However, these various technologies have not been integrated to allow temperature, density, and containment to be maintained long enough to achieve a sustained fusion reaction.

Within the past twenty-five years, over two billion dollars have been spent in the development of a fusion reactor. While some progress has been made, a workable model has yet to be developed. Because of technical problems, fusion does not offer an immediate solution to the world's energy problems.

## Consider This Case Study
### Consumers Power Midland Nuclear Power Plant

In 1967, Consumers Power Company (a public utility in Michigan) began plans for a pair of nuclear reactors to be located in the city of Midland. These twin reactors were to have a capacity of 1300 megawatts. The reason two reactors were planned was associated with the possible sale of steam to Dow Chemical Company, which has manufacturing processes located in that community. To enable a continuous supply of steam, a second reactor would be in operation when the first reactor was in the process of refueling or in the event of an incident that would cause temporary shutdown. The original cost of the project was estimated at $267 million. It was to be completed in the early 1970s by Bechtel Power Company, the main contractor for the plant. This nuclear facility was to be one of several plants that Consumers Power Company was expecting to build to meet the projected energy needs of this industrial midwestern state.

Almost immediately there were delays. A local citizens group was opposed to the construction of the nuclear facility so close to the city, and they sought legal recourse to halt construction. The Nuclear Regulatory Commission scrutinized the planned facility with extra care because of its uniqueness. Subsequently, Consumers Power Company discovered that parts of the plant had settled excessively, and a massive underpinning project had to be undertaken.

Because of the problems experienced, Consumers Power Company announced that the project would be completed in the early 1980s rather than mid–1970s.

Because of additional construction problems, continued intervention by citizen groups, and changes proposed by the Nuclear Regulatory Commission, the plant was still not on line in 1983. Dow Chemical Company then decided to abandon its agreement to purchase electricity and steam from Consumers Power Company. Thus, one of the largest potential customers for the utility had been lost.

In July 1984, construction on the Midland Nuclear Power Plant was halted. It had already cost $3.9 billion, and if completed, it was estimated that the final cost would approach $6 billion. Construction of the plant was more than ten years behind the scheduled completion date. Consumers Power Company was in serious financial difficulty because of cost overruns and construction delays.

Who do you think should be held accountable for the cost of the twin reactors, which are almost 85 percent completed? Should the rate payers, the stockholders, the state, or the federal government be responsible for helping the company financially?

Should the citizens intervention group bear some of the burden for the cost?

What is your reaction to the construction of a nuclear power plant near a metropolitan area?

Because the project was so near completion, should Consumers Power Company have been helped to complete the project so that they could recoup some of their financial losses?

What part do you think the Three Mile Island incident played in the ultimate decision to abandon the project?

How will the abandonment of a power facility affect the economy of the state of Michigan, which is already suffering as a result of the changes in the automotive industry?

## Summary

The splitting of nuclei of particular isotopes of uranium is called nuclear fission. The heat energy released in a nuclear reactor is used to heat water to produce steam which generates electricity.

One of the problems associated with the release of this nuclear energy is the production and release of radioactive materials into the environment. These radioactive particles can cause mutations and cancers. Wastes from nuclear reactors are also radioactive and pose the same dangers. These wastes must be stored and monitored for many years. The potential for a core meltdown, thermal pollution, and high decommissioning costs also make nuclear energy a controversial issue.

The only significant nuclear incident associated with a power generation facility occurred in 1979 at Three Mile Island. A minor pump failure and subsequent errors and failures resulted in a great deal of adverse publicity and the ultimate abandonment of reactor unit two at Three Mile Island.

The lack of uranium–235 for nuclear fuel may not be as great a problem as it once was, because of a new technology known as breeder reactors. In this type of reactor, the fuel is used to produce electricity, and a product that can be used as fuel for other nuclear facilities is produced.

Nuclear fusion, the technology of the future, uses hydrogen as its fuel and produces helium as a product. A major problem associated with the fusion process is that a method is needed to contain the reaction.

## Review Questions

1. How does a nuclear power plant generate electricity?
2. What is a rad and a rem?
3. Why is there no specific amount of allowable radiation to which we may be exposed without endangering our health?
4. How will nuclear fuel supplies, costs of decommissioning facilities, and storage and ultimate disposal of nuclear wastes influence the nuclear power industry?
5. What happened at Three Mile Island, and why did it happen?
6. Describe a liquid metal fast breeder reactor, and tell how it works.
7. How is plutonium–239 produced in a breeder reactor?
8. Why is fusion not being used as a source of energy?

## Suggested Readings

Curran, Samuel C., and Curran, John S. *Energy and Human Needs*. New York: John Wiley & Sons, 1979. Presents a detailed account of all forms of energy. This book has a good section on nuclear energy. It deals with energy transmission and storage as well as social, environmental, and international aspects of energy.

Deffeyes, Kenneth, and MacGregor, Ian D. "World Uranium Resources." *Scientific American* 242: 66–76. This estimate of mineral resources, supported by United States mining records, indicates that the supply of uranium will not be a limiting factor in the development of nuclear power.

Lewis, Harold W. "The Safety of Fission Reactors." *Scientific American* 242: 53–65. By most standards, fission reactors are quite safe. Nonetheless, Three Mile Island confirmed that there should be greater reliance on quantitative methods of assessing risks presented by such systems.

Wade, Nicholas. "France's All-Out Nuclear Program Takes Shape." *Science* 209: 884–86, 888–89. If the experimental breeder reactor being built by France is successful, France may become a showcase for nuclear power.

## Chapter Outline

I. Solar energy
   A. Passive solar systems
   B. Active solar systems
   C. Solar-generated electricity
   D. Solar limitations
   E. Wind power
II. Other energy sources
   Box 11.1 Energy conservation
   A. Geothermal
   B. Tidal
   C. Solid waste
   D. Biomass conversion
   E. Oil shales and tar sands
   Box 11.2 Synthetic fuels and politics
III. Fuel and the future
IV. Consider this case study: Ames, Iowa

## Key Terms

active solar system

biomass

gasohol

geothermal energy

oil shales

passive solar system

photovoltaic cell

tar sands

## Objectives

Explain the use of solar energy in a passive heating system, in an active heating system, and in the generation of electricity.

Describe how wind-generated electricity is produced.

Recognize that there are limitations to the use of solar and wind energy.

Describe how geothermal and tidal energy produce electricity.

Recognize that geothermal and tidal energy can only be used in areas with the proper geologic features.

Recognize that wastes represent a source of energy.

Describe the potential and limitations of biomass as a source of energy.

List the problems associated with obtaining energy from oil shale and tar sand.

Recognize that coal will probably become more important as a source of energy.

# Chapter 11
# Energy and the future: Sources and problems

## Solar energy

The sun is often mentioned as the ultimate answer to our energy problems. It provides a continuous supply of energy that exceeds by far our energy demands. In fact, if it could be trapped, the solar energy that the United States receives is six hundred times more energy than we require. We can utilize the energy from the sun in three ways. In a passive heating system, the sun's energy is converted directly into heat for use at the site where it is collected. In an active heating system, the energy is converted into heat, but the heat must be transferred from the collection area to the place of use. The sun's energy can also be used to generate electricity.

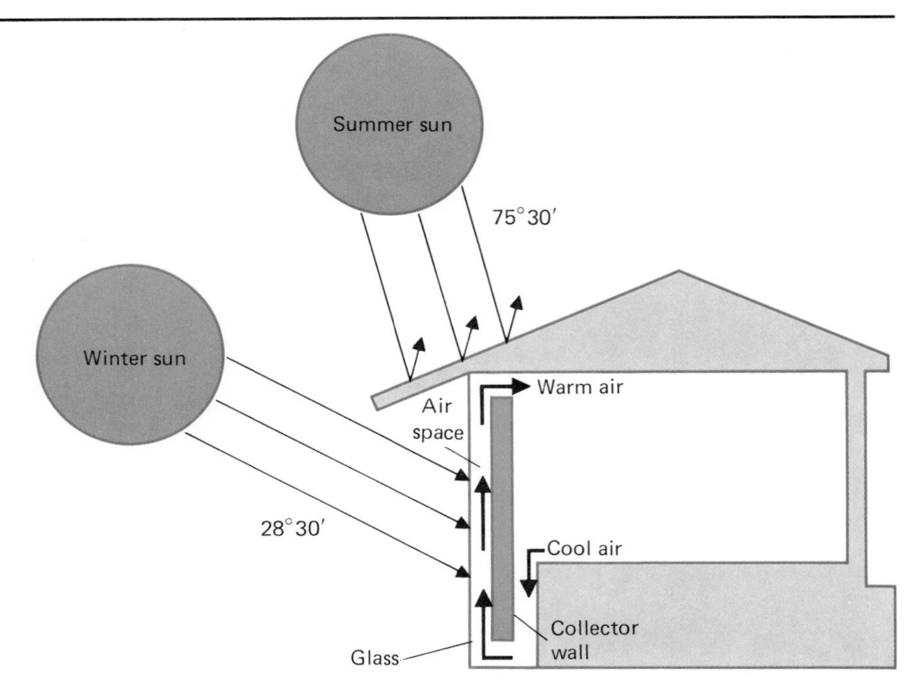


**Figure 11.1 Passive solar heating.** The length of overhang in this home is designed for solar heating at the latitude of St. Louis, Missouri (38°N). In this design, a wall thirty to forty centimeters thick is used to collect and store heat. The collecting wall is located behind a glass wall and faces south. During a mid-winter day when the sun's angle is 28°, light energy is collected by the wall and stored as heat. At night, the heat stored in the wall is used to warm the house. Natural convection causes the air to circulate past the wall, and the house is heated. During a mid-summer day when the sun's angle is 75°, the overhang shades the collector from the sun.

## *Passive solar systems*

Anyone who has walked barefooted on a sidewalk or a blacktopped surface on a sunny day has experienced the effects of a collector in a **passive solar system.** The light energy is transformed to heat energy when it is absorbed by a surface.

Some of the earliest uses of passive solar energy were to evaporate sea water to produce salt, to dry food, and to dry clothes. In fact, solar energy is still being used for these purposes. Homes and buildings are currently being designed to use passive solar energy for heating. (See fig. 11.1.)

In a passive solar design, the heat is utilized where it is produced. Some object receives the light and converts it into heat that can be reradiated into the space that surrounds the object. Darker objects are most efficient in converting light to heat. If the dark object is also massive, it can store the heat and reradiate it at a later time when the air temperature around the object falls. In many passive solar designs, thick walls, thick floors, or pools of water are used to store the heat.

A passive system is maintenance free. There are no moving parts. Therefore, none of the energy is used to transfer the heat within the system, and there are no operating costs. But passive solar design is usually only practical in new construction.

**Figure 11.2 Active solar system.** During a test period from March 14 to May 15, 1974, an active solar system supplied 91 percent of the heating requirements for this school wing. The photograph shows the solar collectors on the roof of the Timonium, Maryland, Elementary School. The heat is collected on the roof of the building, and the heated water is pumped to a storage tank. Hot water from this tank can be pumped to the classrooms when heat is needed.

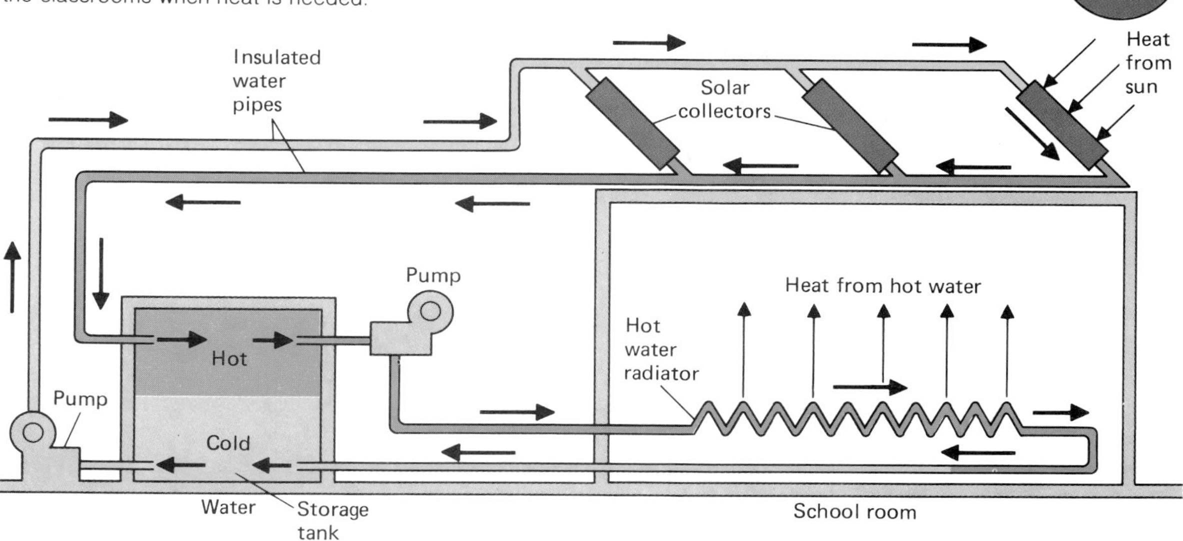

## *Active solar systems*

Another method of using solar energy for heat is in active solar systems. In addition to requiring a collector, an **active solar system** requires a pump and a system of pipes to transfer the heat from the site of its production to the area to be heated.

Active systems work best in new buildings. (See fig. 11.2.) However, in some cases, they can be installed in existing structures. The major obstacle in the use of active systems is the cost. An active system requires a more complex

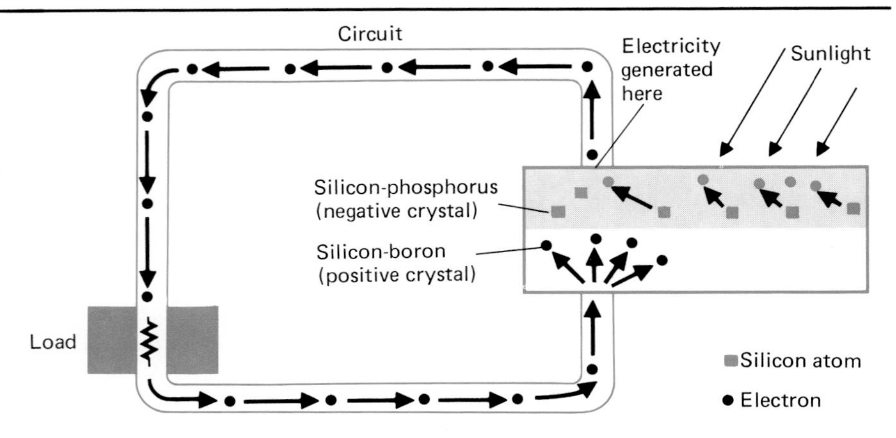


**Figure 11.3 Photovoltaic cell.**
The energy from the sunlight causes a flow of electrons from the silicon-phosphorus crystal to the silicon-boron crystal. Thus, sunlight energy is used directly to produce electricity.

collector than a passive system. In an active system, the collector consists of a series of liquid-filled tubes and a system of pipes to transfer this warm liquid. Added to the cost of the collectors is the expense of the pumps and pipes needed to transfer the heat. Since an active system has moving parts, it has operation and maintenance costs. A Saskatchewan-based company has recently perfected a storage system using sodium sulfate. The sodium sulfate filled polyethylene trays have 160 kilocalories storage capacity, ten to fifteen times the amount of an equivalent volume of rock or water. The crystals of sodium sulfate absorb the heat and change into a liquid. When the liquid sodium sulfate recrystallizes, most of this stored heat is released. The efficiency of this storage system makes it possible and economically sound to retrofit older homes. Savings in heating costs are estimated at more than $400 per year. At this rate, construction costs could be recouped in three to ten years.

## *Solar-generated electricity*

In addition to being used as a method of heating, the sun can supply energy for the production of electricity. The sun's energy can be collected and used to heat water that would turn turbines to generate electricity. Under optimum conditions, five square kilometers of collectors would be required to generate the electricity needed by a city of one hundred thousand people. The cost of constructing such a facility is several times greater than the cost of constructing a coal or nuclear power plant.

Another method of converting solar energy into electrical energy is by using photovoltaic conversion units. The **photovoltaic cell** was first developed in 1954 by Bell Laboratories. But sunpower didn't come into its own until the late 1950s, when the sun was used to power satellites in space. In the solar cell, the light causes a flow of electrons between two layers of material, and this produces an electric current. (See fig. 11.3.) Although this type of conversion is possible, large-scale use is not currently practical. Again the reason is cost; the cost of producing electricity from photovoltaic cells is about a thousand

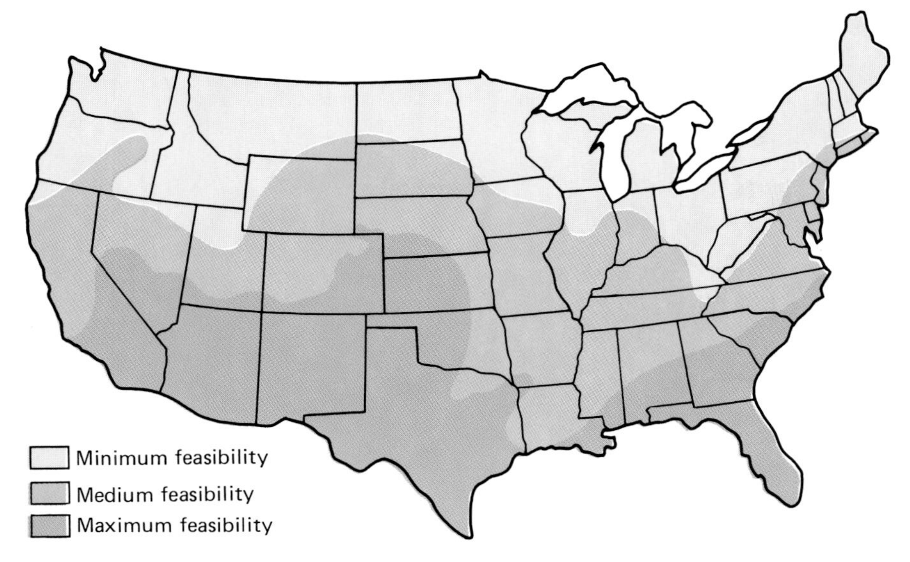


**Figure 11.4 Solar heating.**
The areas of maximum feasibility (dark color) receive enough solar energy to totally heat the home and provide a supply of hot water. In the middle portion of the United States, solar heat must be supplemented by another source of energy. Solar heating can only be an auxiliary source of energy in the northern United States. Is there enough solar energy in your area to totally heat your home? (Solar Energy Research, Staff Report of the Committee on Science and Astronautics, United States House of Representatives.)

times greater than conventional power plants. However, technological advances in the last ten years have increased the efficiency of the photovoltaic cell and increased its reliability, while at the same time the costs of production have steadily decreased.

## *Solar limitations*

In addition to cost, there are several other problems associated with the use of solar energy. In a passive system, it is necessary to orient the building so the maximum advantage can be obtained from the sun. In either a passive or an active system, there can be nothing blocking the sun's light. The topography of the area, location of trees, and proximity of other buildings must all be considered in the location and construction of a solar-heated building. In fact, the interest in solar energy has resulted in a number of new laws being drafted to insure and protect a person's "solar rights."

Only a portion of the United States obtains enough solar energy to supply all the requirements for home heating. (See fig. 11.4.) In the remainder of the nation, an additional heating system must be employed whenever there is not enough sunlight. Even though the homeowner could save on fuel costs, the savings would be partly offset by the added costs of installing two heating systems. The same concern applies to solar-powered generating stations, which can only operate at rated capacity about 25 percent of the time. As a result, they cannot replace existing base load generating plants. The main use of a solar electric generating plant would be to provide part of the electrical needs during the peak demand time. Even though the use of solar energy in housing will increase, solar energy will probably never be able to supply significant amounts of energy for industries and transportation.

**Figure 11.5 Wind energy.**
For many centuries, people have used wind as a source of energy.

## *Wind power*

As the sun's radiant energy strikes the earth, it is converted into heat. The atmosphere gains heat from the earth. The earth is unequally heated, because various portions receive different amounts of sunlight. Warm air is less dense and rises; cooler, denser air flows in to take its place. This flow of air is wind. Wind has been used to power our ships, grind our grains, pump water, and do other forms of work. (See fig. 11.5.) Windmills can be used to drive generators to produce electricity.

In the 1920s, wind generated a limited amount of electricity for farms in the United States. Small wind-powered generators provided power to run the family radio and maybe one or two light bulbs. The generator also charged batteries to provide electricity during windless periods. In 1930, when Congress established the Rural Electrification Administration (REA), a move was begun to provide farms with better electrical service. Within twenty years, this federally backed program was providing full service to most farms, and wind generators were eliminated.

Today there is a renewed interest in wind as a source of energy. The federal government proposed spending $80 million in 1981 on designing and testing large wind generators.

However, in the United States, winds of sufficient speed to operate a generator to its rated capacity occur only on about two days of each week. This means a wind generator may operate at only 20 to 25 percent of its potential.

There is also the problem of storing wind energy and transmitting the electricity produced by the generators. Wind energy does not have the immediate potential to supply a large portion of our requirements, but it can augment the energy that is furnished by fossil fuel. Anything that extends the supply of fossil fuels is beneficial.

## Other energy sources

In addition to wind and solar energy, there are several sources that may supply a small portion of what we need. Some of these have a greater potential than others, but all can reduce our pressure on the available fossil fuels.

The ultimate development and use of other energy sources seems to be in a constant state of flux. During the Carter administration, there was much talk about developing new sources of energy, and various programs were begun. However, the Reagan administration has cut back on these programs. One casualty was the budget for the Solar Energy Research Institute (SERI), which was trimmed from $120 million to $50 million. As a result, the SERI director and over half of the research staff lost their jobs. Reagan also proposed the elimination of $2.9 billion that was allotted for the development of five synthetic fuel plants.

In addition to the changing political climate, reduced use of conventional fuels has decreased the enthusiasm to develop new sources of energy. A 50 percent reduction in the amount of imported crude oil and a slight increase in the production of domestic crude oil have reduced the need for new sources of energy. Thus, sources that were being considered for development have become less economic and have fallen from political favor. The Solar Energy Research Institute is an example. Other sources have continued to provide significant amounts of energy in locally favorable situations.

### *Geothermal*

The earth's core is a molten mass of material possessing vast amounts of energy. In some regions, this material sometimes breaks through the earth's crust and produces volcanoes. In other regions, the hot material is close enough to the surface to heat the underground water and form steam. Geysers and hot springs are natural areas where this hot steam and water come to the surface. In areas where the steam is trapped underground, **geothermal energy** is tapped by drilling wells to obtain the steam. This steam is then used to power electrical generators. At present, geothermal energy is only practical in areas where this hot mass is near the surface. (See fig. 11.6.)

In the United States, the Pacific Gas and Electric Company (PG&E) has been producing electricity from geothermal energy since 1960. PG&E's complex of generating units is located in northern California. The power produced by these generators equals the hydroelectric generating capacity of Hoover

**Figure 11.6**
**Geothermal power plant.** In this plant, the steam obtained from geothermal wells is used in the production of electricity.

## Box 11.1
## Energy conservation

During the long era of inexpensive energy, many people were not overconcerned about the cost of energy; however, that has changed. In 1972, 2.2 percent of the United States GNP ($26.2 billion) was spent for energy. By 1980, this had increased to 8.5 percent of the GNP ($226 billion), an increase of 900 percent. One of the first segments of the country to begin conserving energy was industry. Industry consumes about one-third of the energy used in the United States. In an effort to reduce energy consumption and lower costs, many corporations developed energy-conservation programs. In 1979, these efforts had saved the equivalent of 850 million barrels of oil. This represents more than the amount of oil imported in three months.

In 1973, International Business Machines (IBM) undertook a conservation program to reduce its energy consumption by 10 percent. By 1977, energy consumption had been reduced by 39 percent. This was done by developing a company-wide energy education program. Posters, brochures, films, company magazines, and other media were used to encourage energy conservation on the job and at home.

By 1980, IBM's United States facilities had saved enough energy to heat 160,000 northern homes for a year. The dollar savings to the company was $160 million. Two-thirds of the energy savings involved such simple things as turning off lights and machines when not in use, adjusting thermostats upward in the summer and downward in the winter, and fine tuning heating and cooling systems.

Another approach at IBM involved retrofitting existing buildings and systems (altering buildings to make them more energy efficient). This was done by reclaiming waste heat by heat pumps and other methods, modifying combustion techniques, and managing energy use through computers. Energy was saved by recapturing heat from computers, lights, machinery, and people, and reusing it to warm buildings and water. At the plant in Poughkeepsie, New York, a waste-burning boiler is being planned that will burn about 3,500 metric tons of wastes from the plant and provide 6 percent of the steam requirements. The use of computers has reduced the consumption of energy by 10 to 15

Dam. In fact, these plants, which constitute the world's largest geothermal complex, produce the electrical power for the city of San Francisco. Other geothermal generating facilities are operating in the Soviet Union, Italy, Iceland, Mexico, and New Zealand.

Although geothermal energy does not seem to be an environmental concern, it does present some problems. The steam contains hydrogen sulfide gas, which has the odor of rotten eggs. This gas is an unpleasant form of air pollution. The minerals in the steam are also toxic to fish, and they are corrosive to pipes and equipment, causing constant maintenance problems.

## *Tidal*

Another source of energy that is related to local geologic conditions involves tidal flow. The gravitational pull of the sun and the moon, along with the earth's rotation, causes tides. The tidal movement of water represents a great deal of energy. For years, engineers have suggested that this moving water could be

percent through the control of electrical, heating, and cooling equipment.

To insure that all new IBM buildings are energy efficient, computers simulate the operation and performance of buildings and their energy usage before construction begins. This has resulted in new buildings using 50 percent less energy than those designed before the 1973 oil embargo.

An example of an energy efficient building is at Southfield, Michigan. A unique outward slanted window design reduces the heat load from the sun, but stainless steel reflectors let light in. The south and west walls of the building are painted a light color to reflect the heat and help these walls stay cooler in the summer, while the east and north walls are painted gray to absorb the sun's heat, making the building warmer in the winter.

At a new building in Atlanta, Georgia, six hundred sensors located throughout the building record the weather, temperature, and humidity within the building. This information is fed into a computer that controls heating and cooling and allows the building to operate at minimum energy levels. This system of sensors and the computer, coupled with the design of the building, saves energy equivalent to that used by seven hundred homes in the Atlanta area.

At a new building in Iikura, Japan, heat from computers, lights, people, and the outdoors is reclaimed by heat pumps. A heat wheel is used to transfer building heat from the exhaust systems to incoming outside air. As a result, the building operates at about 50 percent of the energy level of a typical office building.

As you can see, conservation can have a significant impact on energy use without altering working conditions.

**Figure 11.7 Tidal generating station.** The Rance River Estuary Power Plant is the world's largest tidal electrical generating station.

used to produce electricity. The principle is the same as that employed in a hydroelectric plant. France was the first country to put this idea into large-scale use. In 1967, France constructed a commercial tidal generating station. (See fig. 11.7.) Despite the energy potential, it is doubtful whether more tidal power plants will be constructed. The largest obstacle to tidal plants is cost.

At present, three alternative sources of energy are being investigated as having commercial potential. If dollar costs and energy costs can be reduced, more of our energy may be obtained from solid waste, biomass conversion, or oil shales and tar sands.

## *Solid waste*

In the United States, each person generates approximately two kilograms of municipal solid waste per day. About 80 percent of this is combustible and, therefore, represents a potential source of energy. In addition, we produce other forms of solid waste. However, not all of this solid waste has an energy potential. For example, agricultural waste is expensive to collect. Also, this material is naturally returned to the soil and recycled. If it were to be removed, some other material would have to be used to replace the minerals and organic material removed from the soil.

"Trash power," the use of municipal waste, depends upon several factors. First, the material must be sorted. The burnable organic material must be separated from the inorganic material. This is accomplished most economically by the homeowner who produces the waste. Second, the feasibility of waste as a source of fuel depends upon the amount of trash and the dependability of the supply. If a community constructs a facility to burn two hundred metric tons of waste a day, the community must generate two hundred metric tons of waste each day. If it generates less, the facility operates at less than full capacity. The city of Paris uses municipal waste to produce energy. This means daily collections, 365 days a year. Burning trash is not a method of making a profit; at best, it is a matter of reducing costs. With some federal funds to construct a plant, a community may be able to reduce the cost of disposing of trash by burning it to produce electricity.

Monsanto Company has developed a method for converting solid wastes to burnable gas. Baltimore produces gas from a thousand metric tons of waste per day. This gas generates steam for the Baltimore Gas and Electric Company. The daily waste also yields eighty metric tons of a charcoal-like material (char), seventy metric tons of ferrous metals, and one hundred and seventy metric tons of glass. These products all have potential uses. The char can be burned as an additional source of energy. Selling the ferrous metals reduces the cost of operating the plant, and recycling these metals conserves our mineral resources. The glass also represents an item that can be sold and recycled, further reducing energy and economic costs.

**Figure 11.8 Gasohol.** In some sections of the country, a mixture of 10 percent alcohol and 90 percent gasoline is being used in automobiles.

Organic wastes have the potential for furnishing about 15 percent of the United States total energy, but because of technological and economic factors, they will probably only supply between 1 to 3 percent. Even though trash represents a potential source of energy, more energy could be saved by reusing material instead of throwing it out. A cardboard box, a Styrofoam cup, or a paper bag all furnish energy when burned. However, the energy required to replace these items when they are burned is greater than the energy yield in their burning. For example, if all the metal, glass, and paper products were recycled just once, we would save 3.2 percent of our energy. With multiple recycling (recycling a product more than once), the energy savings could approach 10 percent of our total energy budget. Therefore, if we want to make wise use of our energy resources, we should not throw out so many items; we should reuse them.

## *Biomass conversion*

**Biomass** is an accumulation of living material. Biomass conversion is the process of obtaining energy or fuel from the chemical energy stored in the biomass. This is not a new idea; burning of wood is a form of biomass conversion. In the 1930s, studies were conducted to determine whether alcohol could be substituted for gasoline. During World War II, charcoal-fired engines were used in trucks and cars in Europe. Although we live in a world where many suffer from hunger and there is an ethical question about using plant material for fuel instead of food, there is a renewed interest in biomass conversion as a source of fuel.

Alcohol is one type of fuel obtained from plants. The plants contain carbohydrates, which can be converted to sugar. The sugar can then be fermented to produce alcohol, which is distilled from a dilute mixture. A 10 percent mixture of alcohol and gasoline, **gasohol,** is being marketed as an alternative fuel for gasoline. (See fig. 11.8.) However, the production of alcohol is limited by

**Figure 11.9 Methane digester.** In the digester unit, anaerobic bacteria convert animal waste into methane gas. This gas is then used as a source of energy. The sludge from this process serves as a fertilizer. In many less-developed countries, this type of digester has the advantages of providing a source of energy, providing a supply of fertilizer, and managing animal wastes, which helps to reduce disease.

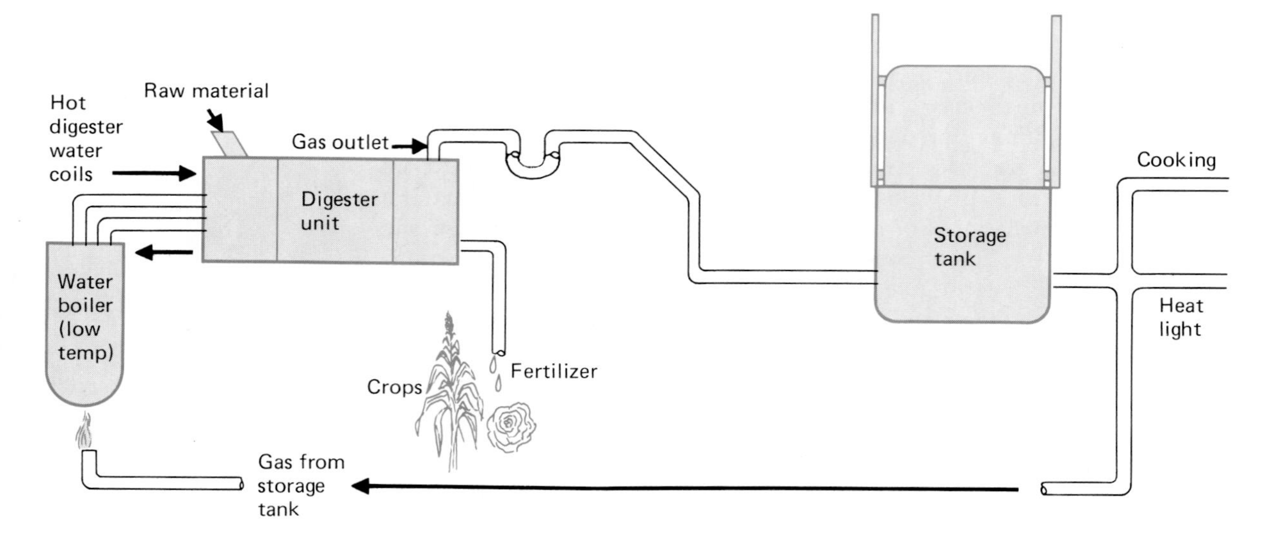


economic, environmental, and energy cost factors. At present, the energy required to produce alcohol is greater than the energy derived from using the alcohol as a fuel. This is a net loss of energy.

Alcohol may be produced cheaply if muscle power is used to shovel, grind, and handle the material to be fermented, but this conversion places no dollar value on a person's time. To determine whether there is an energy gain in producing a fuel, all factors must be considered. These include the energy used to plant, cultivate, harvest, and transport the crop, the energy used to produce, store, transport, and apply herbicides, fertilizers, and insecticides, and the energy used to distill the alcohol. When all these are considered, the use of sugar cane for alcohol production in Brazil has resulted in only a slight energy profit.

As new technology develops, the production of alcohol from biomass may not result in a net energy loss. The use of solar energy may make biomass conversion energy effective. Brazil, a country with an established alcohol program, is investigating the use of the swill from alcohol production as a source of fish food. Thirteen liters of swill are generated for each liter of alcohol produced. Brazil is experimenting with pumping this material directly into the ocean. Diluting the swill with ocean water would eliminate the oxygen depletion problem. The organic swill would be eaten by the fish. Such an undertaking could increase the fish yield, which would offset the overall energy loss in the production of alcohol.

In other parts of the world, human and animal wastes are used to generate methane. In a methane digester, organic material is decomposed by bacteria under anaerobic conditions. The Peoples Republic of China has 500,000 small

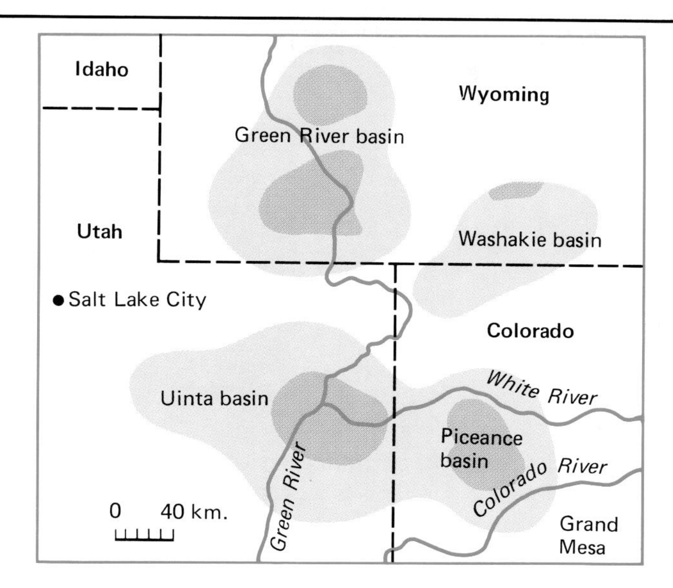


**Figure 11.10 Oil shale deposits.**
Much of the oil shale in the United States is located in the West. The best areas (dark color) yield about a barrel of oil per 1.5 metric tons of shale. (From United States Office of Oil and Gas.)

methane digesters in homes and on farms, India has 100,000, and Korea has 50,000 units. (See fig. 11.9.) For example, a large-scale operation in New York has been using this process for thirty years as a source of fuel. It is possible that methane digestion may find a wider use in the United States.

## *Oil shales and tar sands*

The industrialized world depends heavily on fossil fuels, particularly oil. The best quality sources and the most easily obtained supplies have already been used. Therefore, efforts are being made to develop technology that will allow for the use of poorer quality fossil fuels, such as **oil shales** and **tar sands.**

Oil shale does not contain oil; oil shale contains small amounts of a tarlike organic substance known as kerogen. The kerogen in the shale is heated to yield shale oil. The exact origin of shale oil is not known. It is thought that the kerogen is the remains of algae, pollens, and waxy spores that collected in large, shallow bodies of water. Sizable deposits of oil shale have been found in ten countries. In the United States, major deposits are in the Green River Basin in Wyoming, the Uinta Basin in Utah, and the Piceance Basin in Colorado. (See fig. 11.10.)

Even though oil shale is regarded as a new source of fuel, it has been used in parts of the world for over a hundred years. In 1850, shale oil was widely used in Scotland. Later it was replaced by cheaper coal. In parts of the Soviet Union, the richest forms of oil shale are burned directly as fuel.

Although oil shale has been touted by some as the fuel of the future, its potential is doubtful. The energy potential in the Green River Basin shale is greater than the oil fields of the Middle East, but there are problems involved

**Figure 11.11 Obtaining oil from shale.** Energy is required to dig, transport, crush, and heat the shale. The amount of energy obtained is less than that invested. Therefore, the cost of oil from shale is high. In addition, the strip-mining methods used to obtain the shale have serious environmental effects.

in the mining and processing of oil from the shale. Most of the shale deposits are located in semiarid regions. This is a problem because nearly two barrels of water are needed to process each barrel of oil. Also, approximately 1.5 metric tons of shale must be processed to yield one barrel of oil.

The oil shale must be mined, crushed, transported, and heated to convert the kerogen into a liquid. (See fig. 11.11.) At our present stage of technological development, it requires more energy to obtain this oil than we would get from burning the product. Therefore, this source of oil cannot compete with cheaper sources of liquid fuels.

Tar sands are closely related to oil shale. The tar sands contain a tarlike material located in deposits of sand instead of shale. The same problems associated with obtaining oil from shale are applicable to obtaining oil from tar sands. In addition, the sand is very damaging to the mining equipment. Its abrasive characteristics rapidly destroy the machinery used to mine, handle, and transport it. Tar sands are no more promising than oil shale.

## Fuel and the future

> With the downward trend of production from the present oil fields plainly in sight, the nation is turning its attention to synthetic fuel sources that within a few years will produce hundreds of millions of barrels of oil. This new industry will assure us a supply of gasoline for many generations to come.

Does this message sound familiar? How often do we read such glowing promises? But this quote does not refer to our present situation. It appeared in a 1918 *National Geographic* story entitled "Billions of Barrels of Oil Locked Up in Rock." We still hear the same promises for shale and other forms of new fuels. Almost every edition of a paper or magazine seems to carry some story about the solution to our energy situation. Satellite solar collectors in space, laserbeam energy, ocean-wave power, hydrogen gas, and others are mentioned.

There is a major difference between demonstrating that a new process works and developing a practical model that is economical and can be widely used. In 1903, the Wright brothers demonstrated that a machine could fly by staying

## Box 11.2
### Synthetic fuels and politics

Synthetic fuels are usually considered to be liquid fuels or gaseous fuels that come from nonpetroleum sources. The five major sources are coal gasification (converting coal to natural gas), coal liquefaction (converting coal to liquid fuels), oil shale retorting (removing oil from shale oil deposits), tar sands conversion (extracting oil from tar sands), and biomass conversion (making liquid or gaseous fuels from organic matter).

All of these methods were encouraged by the Carter administration. Direct subsidies from the federal government, loan guarantees, and tax incentives to both business and consumers encouraged the development of at least twenty-five different synthetic fuel installations. Most of these plants were built to demonstrate the feasibility of such technology and to determine the extent of the problems. The goal was to produce the equivalent of two million barrels of oil per day by 1992. This would be approximately 10 percent of the United States daily oil consumption.

However, the Reagan administration has withdrawn some support for a synthetic fuel program in favor of increasing incentives to the oil companies to drill for more domestic oil. Deregulation of oil and natural gas prices, while retaining the tax advantages that oil companies have long enjoyed, has sparked a large increase in the exploration and drilling efforts. In any case, the same companies have benefitted, because most of the money devoted to the synthetic fuel program was going to oil companies. Oil companies invested heavily in alternative energy technologies when it became obvious that the energy future would include a large component that was not based on oil. With lower oil prices, Congress appears to be ready to eliminate the majority of synthetic fuel plants, but oil companies have been lobbying to keep some of them active.

aloft for fifty-nine seconds. However, the first commercial use of aircraft was not accomplished until 1924, when cross-country airmail service was established. Twenty years passed between Wright's demonstration that flight was possible and its practical use. Air passenger service was not established until 1926. As time passes, new developments occur that increase efficiency. After eighty years of technological development, supersonic air travel is possible.

The suggested technology for a solution to our energy problem must follow the same pattern of development. It is reasonable to assume that at least twenty years are necessary to bring some of these ideas to practical use. Others may never become practical.

Furthermore, even if solar heating is an economical, practical method of heating at this time, it would take many years to replace our present buildings with buildings that use solar energy. While looking for new sources of energy, we may return to a source from the past.

In 1980, the World Coal Study released the results of an eighteen-month study conducted by sixteen nations. This study concluded that coal could best meet the world's future energy needs. (See fig. 11.12.) The report also stated that the increasing costs of oil have given coal a cost advantage. The industrialized western nations will still need to invest one trillion dollars within the next twenty years to make the switch to coal. This includes opening new mines, upgrading railroads, and constructing new ships designed to carry large loads of coal.

How do we know whether this report is more valid than any other? There are several reasons why it has merit. The results of this study do not involve the discovery and development of new technology. It only means spending more money to upgrade systems that have already been developed. We also know there is a large reserve of coal. However, one thing should be evident—the cost of energy is going to continue to increase in the future. The days of plentiful, cheap fuel are past.

## Consider This Case Study
### Ames, Iowa

In 1971, the city of Ames, Iowa, learned that its sanitary landfill would be obsolete by 1976. The city council decided to do more than just find a new dumping site. They established a task force to investigate alternative methods of waste disposal. As a result of the study, the city developed three systems to solve the waste disposal problem. They decided to operate a Solid Waste Recovery System (SWRS) and a Refuse Derived Fuel System (RDFS), as well as a sanitary landfill.

To collect the required volume of material, Ames entered into a twenty-five-year agreement with thirteen other communities within the county. Under this agreement, the costs of handling the waste are shared. The SWRS and RDFS receive the waste generated by sixty-five thousand people—forty-six thousand from Ames and nineteen thousand from the remainder of the county. For a variety of reasons, the costs have been higher than estimated. As a result, some of the cooperating communities have attempted to withdraw from the agreement. This is critical because the systems are designed to function with a certain volume of waste. If the machinery operates at less than capacity, the cost rises.

The problems began with the SWRS. It recovers glass, aluminum, oil, paper, and ferrous metals from the city's daily collection of 150 metric tons of waste. Here the waste is shredded into fifteen-centimeter pieces. A

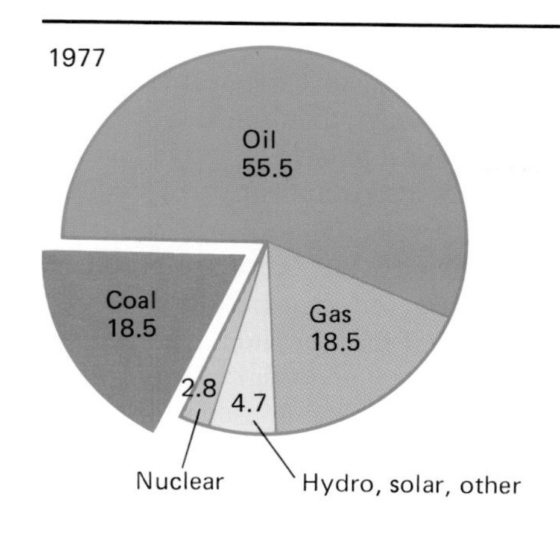


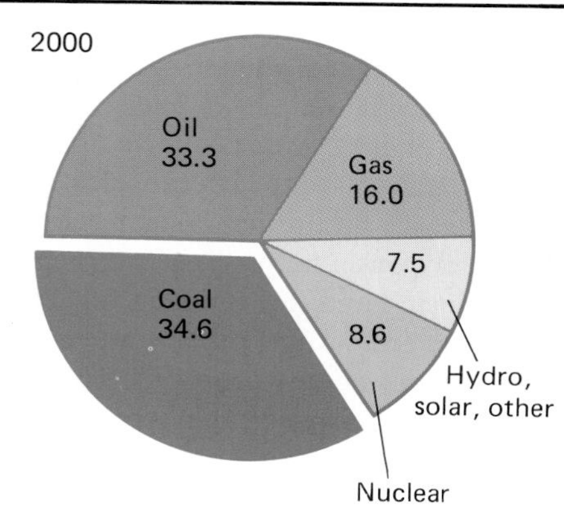


**Figure 11.12 Changing energy sources.** As the price of oil increases and the supply decreases, the industrialized nations will become more dependent upon coal as a source of energy. There will also be increases in the amount of energy furnished by nuclear, solar, and other sources.

degritter removes glass and sand to reduce wear on the machinery. The sand is separated and put in the landfill. Three magnets recover 97 percent of the ferrous metal. The sale of this material was to have been a prime source of income, but the nearest buyer for the material is in Gary, Indiana. As a result, over half of the anticipated profits are lost in shipping costs. In addition, the ferrous metal recovery equipment experienced 50 percent downtime during the first three years of operation. The low demand for wood chips and paper also contributed to the operating loss of this facility.

After the material has passed through the SWRS, it is shredded into five-centimeter pieces. Jets of air separate the lighter burnable material from the heavier nonburnable material. This burnable material is fed as an 80 percent coal and 20 percent refuse derived fuel (RDF) into boilers used to generate electricity. Costs for the RDF are comparable to those of coal. After passing through the SWRS and RDFS, only 7 percent of the waste has not been recycled or burned. This material is placed in the sanitary landfill. Thus far, the cost of processing the waste has been double the originally projected figure. But is the Ames system a failure? Will it prevent other communities from attempting SWRS and RDFS processing as a way to solve their waste problems?

The loss of revenues from the sale of ferrous metals is a result of a transportation problem. How can this problem be solved?

What can be done to increase the demand and price for scrap paper?

Ames was the first community to attempt both SWRS and RDFS as a combined operation. Regardless of the high costs of recycling, can we afford to bury our waste metals and other resources?

Should other communities, as did Ames, invest money and effort into trying to promote a better method of waste disposal?

## Summary

The long-range solution to our energy problem must involve renewable sources of energy, new sources of energy, and reducing energy requirements.

Solar energy can be collected and used in either a passive or active system. It can also be used to generate electricity. Lack of a constant supply of sunlight is the major limitation of solar energy.

Wind power may be used to generate electricity, but there are problems with storing and transmitting the electricity produced by generators.

There are several other sources of energy that may provide a partial solution to our energy problems. The use of geothermal and tidal energy will be limited by their geographical locations. Solid waste is being used as a fuel in some communities and has a potential for further development. However, there is a question about whether we could benefit more from recycling or burning our trash. Biomass conversion is a process that releases energy in a useful form from plant material. Producing this energy may result in a net energy loss. This is also true for oil shale and tar sands.

The use of coal will probably increase in the near future because of its availability.

## Review Questions

1. What two factors limit the development of tidal power as a source of electricity?
2. Compare a passive solar heating system with an active solar heating system.
3. Why is solar technology more feasible in the southwestern United States than in the northeast?
4. How can wind be considered as a form of solar energy?
5. Give two processes that generate useful fuel from organic matter.
6. In what parts of the world is geothermal energy available?
7. Why are oil shales and tar sands not widely used as a source of fuel?
8. Why may coal be more widely used to meet our future fuel needs?
9. What problems are associated with the use of solid wastes as a source of energy?

## Suggested Readings

Bergman, Elihu; Bethe, Hans A.; Marshak, Robert E. *American Energy Choices before the Year 2000*. Lexington, Mass.: Lexington Books, 1978. A series of seventeen papers presented at a conference on energy at City College of New York. Discusses the subject of energy choices before the year 2000.

Brown, Lester. "Food versus Fuel." *Environment* 22: 32–40. Presents the problems associated with the use of cropland to grow plants for biomass instead of plants for food.

DeRenza, D. J., ed. *Wind Power: Recent Developments.* Energy Technology Review no. 46. Noyes Data Corporation, Park Ridge, N.J., 1979. Presents a brief history of wind power. Technical data given on wind-energy assessments, rotor construction, and turbine construction. Presents information about wind use for electric utilities, rural use, and legal, social, and environmental issues. Good list of references at the end of each chapter.

Detweller, Raymond F. *Waste Can Produce Cheap Energy and Save Vitally Needed Material Resources.* Souderton, Penn.: E. & E. Publishing Co., 1977. Eleven types of waste material (such as used motor oil, lumber waste, and industrial waste) are cited as sources of cheap energy. The author does not detail how wastes can be used or the cost of converting wastes to energy. However, it raises some interesting questions about why we should look to wastes as a source of energy.

de Winter, F., and de Winter, J. W. *Description of the Solar Energy R & D Programs in Many Nations.* National Technical Information Service, Springfield, Virginia, 1976. A brief description of solar energy R & D programs in thirty-two countries. There is no unified format, and the reports from the individual countries vary in style and detail. Most reports list present and proposed projects. The book also includes a number of contacts for those who may desire specific information from a particular country.

Ewers, William L. *Solar Energy: A Biased Guide.* Northbrook, Ill.: Domus Books, 1979. A basic introduction into solar energy. The main emphasis is on solar heating in private homes. However, there is a section about solar energy use in public buildings and a section on some advanced methods for using solar energy.

Greenbaum, Rolf. "Alternative Energy in the USSR." *Environment* 20:26–30. Surveys the development of wind, sun, tides, and geothermal heat as alternative sources of energy in the USSR.

Hartline, Beverly K. "Trapping Sun-Warmed Ocean Water for Power." *Science* 209: 794–96. Articles concerning an experiment to use the temperature differences in the ocean as a source of energy for generating electricity.

Inglis, Rittenhouse D. *Wind Power and Other Energy Options.* Ann Arbor, Mich.: The University of Michigan Press, 1978. A well-written text covering wind, solar, geophysical, and nuclear sources of energy. Presents practical as well as highly theoretical sources of energy.

Jamison, Andrew. "Democratizing Technology." *Environment* 20: 25–28. Looks at the increase in "do-it-yourself" windmills in Denmark.

Kalhammer, Fritz R. "Energy-Storage Systems." *Scientific American* 241: 56–65. Energy reservoirs consisting of pumped water, compressed air, and batteries, and the storing of heat and "cold" can do much to help coal, nuclear, and solar energy replace substantial quantities of oil.

McDaniels, David K. *The Sun: Our Future Energy Source.* New York: John Wiley & Sons, 1979. A well-written textbook concerning all aspects of solar energy. An excellent source for the student who desires more specific information on any segment of solar energy.

Meador, Roy. *Future Energy Alternatives: Long Range Prospects for America and the World*. Ann Arbor, Mich.: Ann Arbor Science Publishers, 1978. A good presentation of proposed energy sources. Includes a variety of future fuels, including the use of coal, nuclear, and some highly theoretical solutions. Each chapter has an extensive list of references for those who desire more detailed information.

Mears, Leon G. "Energy From Alcohol." *Environment* 20: 17–20. Presents a report on Brazil's goal to substitute homegrown fuel for twenty percent of the gasoline consumed.

Paul, J. K., ed. *Passive Solar Energy: Design and Materials*. Energy Technology Review no. 41, Noyes Data corporation, Park Ridge, N.J., 1979. Technically written book on the various components of passive solar systems. Includes a number of case studies concerning passive solar systems now in use. Good list of references at the end of each chapter.

Scott, Robert. "An Economic Comparison of Three Technologies." *Environment* 20: 11–15. Compares the cost of using photovoltaic cells, cogeneration, and nuclear power plants to produce electricity.

Weisz, Paul B., and Marshall, John F. *Fuel from Biomass: A Critical Analysis of Technology and Economics*. New York: Macel Dekker, 1980. Assesses the potential and limitations of creating high-grade fuel from agricultural products (biomass) grown for the purpose of fuel production.

# Part 4

# Natural resource use

Natural resources are those items in nature that humans can use for their own purposes. Even though we use some of these daily, we do not even think of them as being resources. The air we breathe, the water that provides us with a cool drink or flushes away our wastes, and the sunlight that powers our ecosystem are seldom thought of as natural resources because the supply is so large and apparently inexhaustible.

The chapters in this part will deal with various natural resources, including mineral, forest, wildlife, agricultural, land, and water resources. Energy resources have already been covered in chapters 8, 9, 10, and 11.

Natural resources are important in determining the economic, political, and social growth of countries. The United States is a powerful country today in part because it has an abundant supply of resources available within its borders. In part 4, we will look at different types of resources and try to develop an understanding of the United States policy regarding resource utilization.

## Chapter Outline

I. Renewable and nonrenewable resources
II. Mineral resources
  A. Costs associated with mineral exploitation
  B. Steps in mineral exploitation
  C. Recycling as an alternative
III. Forest resources
  A. Costs associated with forest exploitation
  B. Reforestation and processing costs
IV. Wildlife resources
  A. The restoration of wildlife in the United States
  B. Wildlife management techniques
  C. Management of waterfowl
  D. Fish management
    Box 12.1 Native American fishing rights
V. Consider this case study: Knox County, Illinois

## Objectives

Differentiate between renewable and nonrenewable resources.

Recognize the fact that mineral resources are unevenly distributed, which creates international trade in these commodities.

List several types of costs associated with the steps involved in mineral exploitation.

Explain why energy costs are high when minerals are extracted from ores.

Describe how environmental costs and energy costs can be converted to economic costs.

Recognize that current forest resources are being used heavily and that new sources are being sought.

Explain how forests can be managed like any crop.

Explain why traditional forest practices may not work in tropical forests.

Explain why recycling reduces the need for new sources of minerals and trees.

Explain how wildlife can be managed.

Recognize that managing for wildlife involves interference with natural ecosystems.

Recognize that humans have replaced natural predators for most game animals.

## Key Terms

clear-cutting

cover

economic costs

energy costs

environmental costs

habitat management

migratory birds

natural resources

nonrenewable resources

patchwork clear-cutting

recycling

reforestation

renewable resources

selective harvest

# Chapter 12
# Mineral, forest, and wildlife resources

## Renewable and nonrenewable resources

**Natural resources** are those structures and processes that can be used by humans for their own purposes but cannot be created by them. If the supply of a resource is very large and the demand for it is low, the resource is often thought of as free. You don't pay for the air you breathe. If a resource has been consumed or if it was rare initially, it is expensive. Pearls and gold and other precious metals fall into this category. Things like sunlight, the oceans, and air are often not even thought of as natural resources, because their supply is so large. These resources will become more valuable as we begin to exploit them more intensively. There have already been lawsuits filed to seek to preserve a person's right to have sunlight for solar collectors and clouds for rainfall. The landscape itself is a natural resource. This is probably most easily seen in countries with the proper combination of mountainous terrain and high rainfall, which can be used to generate hydroelectric power. Rivers, forests, scenery, climate, and wildlife populations are also examples of natural resources.

Natural resources are usually thought of as being in one of two categories. Either a resource is renewable or it is nonrenewable. **Renewable resources** are those resources that can be regenerated by natural processes. Soil, vegetation, animal life, air, and water are renewable resources; they are replaced as we use them. Air and water are renewable primarily because they naturally repair and cleanse themselves.

Even though a natural resource is renewable, it is not inexhaustible. There are limits to how fast we can make forests or cornfields grow. In some areas, we have polluted air and water to such an extent that they are unable to cleanse themselves. The major concern with renewable resources is to match the rate of use with the rate of production or repair. One needs to understand that renewable resources must be managed properly if they are to continue to provide for the needs of future generations. In Europe and Japan, the clearing of large areas for agriculture has resulted in a shortage of wood products. Once all the trees are removed to make farmland, it will take fifty to one hundred years for that same piece of land to naturally return to a forest. Those forests that remain are intensively managed. If the demand for forest products increases, the price for such products will rise.

Because forests can be used for a variety of purposes, forest management is complex. Watershed protection, wildlife habitat, and scenery must be balanced with the production of lumber, paper, and other forest products. It is not always possible to satisfy all of these needs at the same time. If the trees are cut for lumber, the watershed will lose much of its protection, and some animals will lose their habitat. Other animals will benefit from an improved food supply, but stumps do not make very attractive scenery. Even with renewable resources, careful decisions need to be made because it is possible to harvest too much or to use a resource too intensively. (See fig. 12.1.)

**Nonrenewable resources** are those resources that are not replaced by natural processes or those whose rate of replacement is so slow as to be noneffective. Therefore, when nonrenewable resources are used up, they are gone and a substitute must be found or we must do without. Some examples of nonrenewable resources are land (as distinct from soil, which is renewable), fossil fuels, and mineral resources. Fossil fuels were covered in detail in chapters 8, 9, and 10. Land will be discussed in chapter 15. This chapter will discuss mineral resources as examples of nonrenewable resources.

**Figure 12.1 Mismanagement of a renewable resource.**
Although soil is a renewable resource, it can be used so extensively that it is permanently damaged. Many of the deserts of the world were formed or extended by the unwise use of farmland.

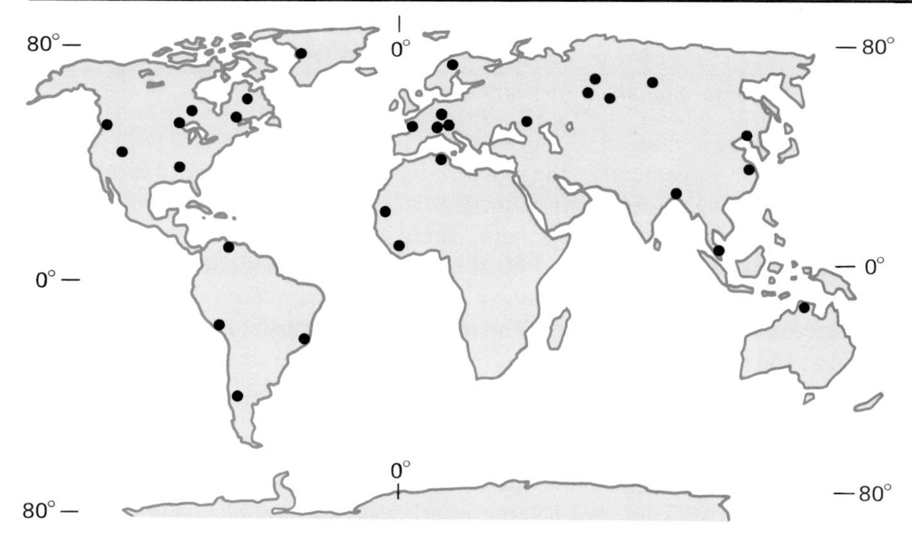


**Figure 12.2 Distribution of iron ore mining areas.**
This map shows the distribution of iron ore mining sites. Obviously, most countries in the world have no active iron mines. Although mines do not exactly correspond with the distribution of ore bodies, it does give an indication of the scattered distribution of iron ore deposits throughout the world.

## Mineral resources

Nature does not respect political boundaries. The geological and biological processes that caused the formation of mineral resources in specific places occurred many millions of years ago. Subsequently, land masses have been divided into various political entities. Some have a greater wealth of mineral resources than others. (See fig. 12.2.) Because no country has within its territorial limits all of the mineral resources it needs, an international exchange

**Table 12.1**
Distribution of some mineral resources

| Mineral | % imported that U.S. consumed 1982 | Rank of major supplier countries (1978–1981) |
|---|---|---|
| Columbium | 100 | Brazil, Canada, Thailand |
| Mica (sheet) | 100 | India, Brazil, Madagascar |
| Strontium | 100 | Mexico, Spain |
| Manganese | 99 | South Africa, Gabon, Brazil, France |
| Tantalum | 90 | Thailand, Canada, Malaysia, Brazil |
| Bauxite | 97 | Jamaica, Australia, Guinea, Surinam |
| Cobalt | 91 | Zaire, Belgium-Luxemburg, Zambia, Finland, Canada |
| Chromium | 88 | South Africa, USSR, Zimbabwe, Turkey, Philippines |
| Platinum group | 85 | South Africa, USSR, United Kingdom |
| Asbestos | 74 | Canada, South Africa |
| Tin | 72 | Malaysia, Thailand, Indonesia, Bolivia |
| Nickel | 75 | Canada, Norway, New Caledonia, Dominican Republic |
| Potassium | 71 | Canada, Israel |
| Cadmium | 69 | Canada, Australia, Mexico |
| Zinc | 53 | Canada, Mexico, Spain, Honduras |
| Mercury | 43 | Algeria, Spain, Italy, Canada, Yugoslavia |

Source: Data from *Statistical Abstract of the United States,* 1984.

has developed. In particular, the industrially developed countries import many of the things they need from countries with mineral resources but without economic resources to develop them. Table 12.1 shows several minerals and the countries that supply them. Most of the supplier countries are less developed. Oil is probably the best known example of the uneven distribution of a resource. The Middle Eastern countries have very large oil reserves, compared to what the rest of the world has. (See chapter 9 for a detailed discussion of fossil fuel reserves.)

But there are other distributions that are just as striking. Table 12.1 also indicates the world distribution of some mineral resources. Notice the large number of materials that the United States lacks and must import from other countries. It is striking that a large number of the supplier countries are in less-developed areas of the world.

Not only are mineral resources unevenly distributed, but those that are easiest to use and are the least costly to extract have been exploited. Therefore, we should expect, if we continue to use mineral resources, that they will be harder to find and more costly to develop. As with energy, the United States is one of the primary consumers of the world's mineral resources. Reasonable estimates are that the United States uses about 30 percent of the minerals produced in the world each year.

If the United States consumes 30 percent of the world's resource production annually and has only 5 percent of the world's population, it is clear that the United States receives a disproportionate share of the world's bounty.

## *Costs associated with mineral exploitation*

There are always costs associated with the exploitation of any natural resource. These costs fall into three different categories. First, the **economic costs** are those monetary costs that are necessary to exploit the resource. Money is needed to lease or buy land, build equipment, pay for labor, and buy the energy necessary to run the equipment.

A second way to assess the costs of exploiting a resource is to establish its **energy cost.** Obviously, some of the monetary cost is attributable to the cost of the energy that is used. As the cost of energy rises, some currently profitable processes may become uneconomic. For example, today it may make sense economically to talk about extracting oil from shale. But if more calories of energy must be invested than are actually obtained, it is not a practical endeavor. As fossil fuels become rarer, economic costs will rise, and it may be necessary to reevaluate current thinking.

A third way to look at costs is in terms of environmental effects. Air pollution, water pollution, animal extinction, and loss of scenic quality are all **environmental costs.** These environmental costs are being converted to economic costs as more strict controls on the pollution of the environment are enacted and enforced. It takes money to clean up polluted water and air. It also takes money to reclaim land that has been removed from biological production by mining. All three of these categories of costs (economic, energy, and environmental) are associated with the several steps that lead from the mineral resource in its undisturbed state to the manufacture of a finished product. These steps are exploration, mining, refining, transportation, and manufacturing.

## *Steps in mineral exploitation*

The costs associated with locating new sources of minerals (exploration) are primarily economic because it takes time and new technology. There are also some energy costs and some very small environmental costs. As the better sources of mineral resources are used up, it will be necessary to look for them in areas that are more difficult to explore, such as under the oceans. Therefore, both the economic and energy costs will increase. In addition, some areas, such as national parks and preserves, have been off-limits to mineral exploration. As current reserves of mineral resources are used up, pressures will build to explore in these natural areas. When this happens, environmental costs will increase. (See chap. 1.)

Once a mineral resource has been located and the decision has been made to exploit the resource, it must be taken from the earth. Mining involves large

**Figure 12.3 A strip-mining operation.**
It is easy to see the important impact a mine of this type has on the local environment. Unfortunately, many mining operations are associated with areas that are also known for their scenic beauty.

expenditures of money to pay for labor and the construction of machines and equipment. Energy must be purchased for operation. In addition to these two kinds of costs, there are significant environmental costs. Mining affects the environment in several ways. All mining operations involve the separation of the valuable mineral from the surrounding rock. The surrounding rock must then be disposed of in some way. These mine tailings are usually piled on the surface of the earth, where they are unsightly. It is also difficult to get vegetation to grow on these deposits. Some mine tailings contain materials (such as asbestos, arsenic, lead, and radioactive materials) that can be harmful to humans and other living things.

Many types of mining operations require vast quantities of water for the extraction process. The quality of this water is degraded, so it is unsuitable for drinking, irrigation, or recreation. Since mining disturbs the natural vegetation in an area, water may carry soil particles into streams and cause erosion and siltation. Some mining operations, such as strip mining, rearrange the top layers of the soil, which lessens its productivity or reduces it to zero for a long period of time. (See fig. 12.3.) Strip mining has disturbed approximately 75,000 square kilometers of land in the United States. This is equivalent to an area the size of Maine.

After ore is mined, it must be processed to remove the desired materials from the rocks in which they are embedded. Table 12.2 shows several ores and the percentage of desired material that is present in each ore. Processing ore requires economic costs: physical facilities and equipment must be built, people must be employed, and fuel must be purchased for the operation. The energy costs are also significant because refining is basically the process of concentrating the desired material. Natural physical processes tend to disperse materials; therefore, energy must be expended to concentrate them. In many cases,

**Table 12.2**
Metals in ores

| Metal | % metal needed in the ore for profitable extraction | Price range per kilogram |
|---|---|---|
| Iron | 30 | Less than a dollar |
| Chromium | 30 | |
| Aluminum | 20 | |
| Nickel | 1.5 | Several dollars |
| Tin | 1 | |
| Copper | 0.5 | |
| Uranium | 0.1 | Tens of dollars |

**Table 12.3**
Energy cost of producing 1 metric ton of raw steel

| Energy source | Amount used | | Conversion factor | | Kilocalories of energy used |
|---|---|---|---|---|---|
| Coal | 591 kilograms | x | 7200 kilocalories / kilogram | = | 4,255,200 |
| Oil | 0.33 barrels | x | 1,600,000 kcal. / barrel | = | 528,000 |
| Natural gas | 5,130 cubic ft. | x | 4110 kilocalories / cubic foot | = | 21,084,300 |
| Electricity | 381 kilowatt hours | x | 860 kilocalories / kilowatt hour | = | 327,660 |
| | | | TOTAL | | 26,195,160 |

Source: Data from *Report to the President on Prices and Costs in the United States Steel Industry.* (1977)

this energy expenditure is the major cost associated with obtaining the mineral. As energy costs increase and ores become more dilute, the economics of producing some materials may change so drastically that it becomes necessary to look for substitute materials that require less energy expenditure. Table 12.3 shows the energy that must be expended to get a metric ton of raw steel. Since iron atoms are not concentrated, considerable energy must be expended to collect them into an essentially pure state. The amount of energy needed to produce a metric ton of raw steel is about the same as that needed to keep a person alive at the subsistence level for thirty years.

Extracting materials from ores also involves environmental costs in the form of air and water pollution. These environmental costs are being converted to economic costs as regulations on industry require less environmental damage than had been tolerated previously.

Transportation is another component of the overall cost of extracting minerals from the earth. Transportation is involved in the actual mining process, in getting the ore to the refinery, and in moving the concentrated mineral to

Figure 12.4 **Steel manufacture.** The manufacture of steel requires several steps. Each step involves the expenditure of energy to concentrate the ore, the production of pollutants, and labor and capital costs.

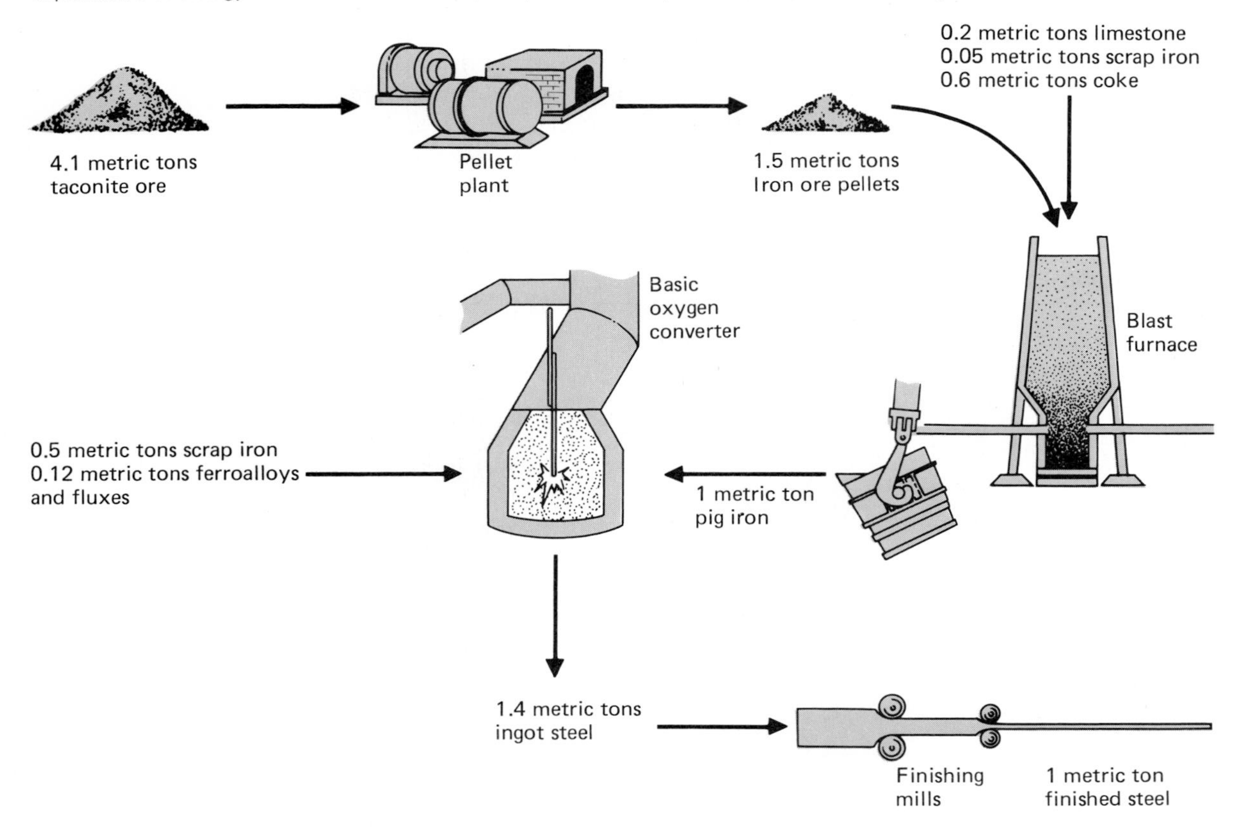


a site where it will be made into a finished product. Transportation costs are primarily in the form of money and energy.

Finally, there is the cost of making the finished product from the concentrated ore. Figure 12.4 reviews this entire process by looking at what happens from the time iron atoms are mined until finished steel is produced.

## *Recycling as an alternative*

To fully appreciate the mineral resources situation, it is important to consider the concept of **recycling.** Many minerals extracted from the earth are not actually consumed or used up; they are just held temporarily within a structure or a process.

Empty aluminum beverage cans are no longer useful. However, the aluminum atoms can be reprocessed into new cans or other aluminum products. In designing a recycling system, it is important to recapture the obsolete items before they have dispersed into the environment. For example, waste oil is relatively easy to capture at the gasoline station where engine oil is replaced, but it is impossible to do so if the oil is dumped on the ground or into a sewer.

Certainly there are costs from all three categories (economic, energy, and environmental) involved in reprocessing, but the costs are often less because mining or refining is unnecessary. This is somewhat misleading, because in many industries the economic cost is still lowest if new materials are mined and refined rather than if used materials are reprocessed. For example, it is more costly to disassemble a building and reuse its parts than to use new construction materials. However, when the energy costs and environmental costs are taken into account, it is generally less expensive to recycle materials than to use and discard them.

Government subsidies also influence the profitability of recycling. Government subsidies occur in a variety of forms. Some are direct subsidies, such as special lower prices for the transportation of new materials and price penalties for the transportation of used materials. Most metal ores are shipped at special preferred low rates. The government purchases huge quantities of goods and sets very exact specifications for these commodities. These specifications often discourage the use of recycled materials. This is an indirect subsidy but has as great an effect as many direct subsidies.

Another reason we do not recycle more materials is that historically monetary costs for energy have been extremely low. As the cost of energy continues to rise, more attention will be given to recycling minerals rather than mining new ones. Also, we are all forced to pay for the environmental costs of altering our surroundings. This is reflected in the price tag of anything we buy. It will become economically advantageous to manufacture products in an environmentally harmonious way.

## Forest resources

In general, forest resources are more evenly distributed than many other resources. Table 12.4 shows that North America, Latin America, and the USSR contain about three-fourths of the growing wood in the world. However, these three regions were responsible for only about 40 percent of the wood produced. (See table 12.5.) The discrepancy between the distribution of wood and the production of wood products exists because Latin America produced relatively little wood in comparison to its capacity. Asia produced over 30 percent of the world's wood with only 12 percent of the world's growing timber. The USSR produced 12 percent of the world's wood with 22 percent of the world's growing timber.

The areas of the world that could increase timber production are the tropical forests, and the northern forests of the USSR and Canada. Both of these areas present some unknowns. The far north of the USSR and Canada is largely forested, but these remote areas require extensive transportation expansion. In addition, these are slow-growing coniferous forests. There are questions about how intensively these forests can be exploited before a point is reached

**Table 12.4**
Distribution of forest resources

| Region of the world | Billion cubic meters of growing trees | % of the world total | |
|---|---|---|---|
| Latin America | 122.9 | 34.4 | 73.1 |
| USSR | 79.5 | 22.2 | |
| North America | 59.0 | 16.5 | |
| Asia (except Japan) | 40.9 | 11.4 | |
| Africa | 34.9 | 9.8 | |
| Europe | 13.4 | 3.8 | |
| Pacific area | 5.0 | 1.4 | |
| Japan | 1.9 | 0.5 | |
| World total | 357.5 | 100.0 | |

Source: Data from FAO of the UN.

**Table 12.5**
Wood production

| Region of the world | Production in million cubic meters of round wood 1980 | | % of the world total | |
|---|---|---|---|---|
| South America | 315.2 | | 10.4 | |
| USSR | 356.0 | | 11.8 | |
| North America | 530.8 | | 17.6 | |
| United States | | 322.3 | | 10.7 |
| Canada | | 161.4 | | 5.3 |
| Other | | 47.1 | | 1.6 |
| Asia | 1,017.1 | | 33.7 | |
| Africa | 433.9 | | 14.4 | |
| Europe | 334.3 | | 11.1 | |
| Pacific area | 33.0 | | 1.0 | |
| World total | 3,020.3 | | 100 | |

Source: *United Nations Statistical Yearbook*, 1981

when they are not growing as fast as they are being harvested. Good forest management seeks to have a sustained yield from the forest. This requires that the timber be harvested at the same rate that it is regrowing.

The tropical forests of Asia, South America, and Africa represent a second source of under-utilized timber. (See box 5.1 for a discussion of laterite soils.) Tropical forests, however, are much more diverse in tree species than temperate forests. This requires a different technology to allow for efficient logging. Also, tropical forests are not as easy to reestablish following the traditional logging practices used in temperate forests. If tropical forests do not regenerate following logging, they will become a nonrenewable resource. Once cut, they will not be available again. Tropical ecosystems are more complex; therefore, the traditional forestry practices developed in Europe and North America do not seem to be usable in the tropics. A new set of forestry principles may need to be developed to allow for the establishment of a renewable tropical forest industry. Another complicating factor is that in these tropical countries, population pressures are the greatest. More people need more food, which means more forest land will be taken for agriculture, thus reducing the stocks of timber available and also reducing the land available for forest production.

In 1982, the United States consumed approximately 375 million cubic meters of wood. This was approximately 12 percent of the total world production and considerably above the world average per capita usage. The United States imports wood products. It produces about 85 percent of what is needed and imports the remaining 15 percent. Eighty percent of the wood products imported comes from Canada and consists of lumber and paper products. Most of the remaining imports are specialty hardwoods and veneers or finished products processed in Korea, Japan, and Taiwan. The demand for wood products is continuing to increase faster than the rate of production. Eventually, the United States may become as dependent on foreign forest products as it is on foreign oil.

## *Costs associated with forest exploitation*

The same three cost categories (economic, energy, and environmental) should be considered when discussing forest exploitation. They do, however, take somewhat different forms from those discussed in the section on mineral exploitation. First of all, natural processes have collected dispersed elements (carbon, hydrogen, and oxygen) and assembled them into what we call wood.

Forests are known quantities; therefore, we have a pretty good idea of how much our reserves of wood products are. (See table 12.5.) There are still costs involved in cutting and transporting the trees from the site where they are grown to the place where they will be processed into a final product. The monetary costs are associated with the equipment and labor needed to harvest and transport the trees. The energy costs are also rather obvious, since the machines must have fuel. It is the environmental costs that are most difficult to assign a value to. Logging removes the trees and, therefore, removes the habitat for many kinds of animals that require mature stands of timber. Human activity will also drive out some animals that require a "wilderness" habitat. Also, many people dislike seeing areas that have recently been logged. Obviously, wilderness and logging cannot coexist. Therefore, it becomes necessary to designate forests as either wilderness areas to be untouched or as commercial forests that will eventually be logged. It is possible to manage forest lands so a variety of needs are met. The forest industry would prefer even-aged stands of a single species of tree to enable efficient harvesting. Wildlife managers would like to see forests with a variety of tree species at different stages of maturity to encourage species of wildlife. The integration of both uses of forests (as wildlife habitats and as a source for wood products) can occur if each of these managers is willing to make some adjustments to satisfy the needs of the other.

**Figure 12.5 Clear-cutting.**
Clear-cutting is a forest harvesting method that is economical and reasonably sound environmentally on small sites that have little slope and where regrowth is rapid.

There are several other environmental costs. For example, removal of trees from an area exposes the soil to increased erosion. Because the soil's water-holding ability is related to the amount of organic matter and roots in soil, denuded land allows water to run off rather than sink into it. Soil particles can wash into streams and cause siltation. The loss of soil particles reduces the fertility of the soil, which is certainly undesirable. The particles that enter the stream may cover spawning sites and eliminate fish populations. The stream may also become warmed if the trees that shaded it are removed. Many species of stream animals are affected by slight changes in temperature. Runoff from the land is also more rapid and often results in flooding that would not occur otherwise.

In addition, roads must be built to allow trucks to carry the trees to the milling site. Constant travel over these roads removes any vegetation and exposes the bare soil to more rapid erosion. When the roads are not properly located and constructed, they eventually become gulleys because of the water that runs down them.

Most of these environmental costs can be minimized by using proper harvest methods and carefully engineered roads. This increases the monetary costs or, more correctly, converts an environmental cost into a monetary cost. One of the most controversial forest logging practices is **clear-cutting.** As the name implies, all of the trees in a large area are removed. (See fig. 12.5.) Because less travel is involved in harvesting the trees, clear-cutting has obvious economic advantages. For some sites, this is a reasonable technique, but for others it can be disastrous. On sites with gentle slopes and where regrowth will occur rather rapidly, little erosion occurs, and streams running through the area are only slightly changed. This is especially true if some mature trees are left along stream courses, where their roots help to retain stream banks and the trees shade the stream. Clear-cutting can be very destructive on sites with a steep slope or sites where there is slow regrowth of vegetation. Other areas may allow for **patchwork clear-cutting.** In this method of harvesting timber, smaller areas are clear-cut among patches of untouched forest. This reduces many of the problems associated with clear-cutting and also improves conditions for many species of game animals that flourish in successional forests but not in mature forests. Deer, grouse, and rabbits are heavily hunted in the eastern United States. All three of these species do best in regrowing areas. Therefore, it is possible to integrate forest harvest practices with wildlife habitat management.

**Selective harvest** of tree species is also possible, but from the point of view of those in the forest industry it is not very efficient. It does allow for mature, individual, high value trees to be removed without causing major disruptions to watersheds or significantly altering the forest ecosystem.

Both foresters and wildlife managers are concerned about the control of fire and forest pests. Fire in a mature forest destroys the timber and changes the stage of succession. Fire may have beneficial effects when it is used repeatedly to reduce the buildup of leaves and branches that could support major fires at a later time.

Pests that kill trees, cause defoliation, or weaken trees can significantly reduce the timber yield but may not have a drastic effect on the wildlife species in the area. In fact, many wildlife species use weakened trees for shelter or sources of food.

## *Reforestation and processing costs*

When a forest is cut, reforestation is another cost to consider. **Reforestation** is the process of establishing new trees on a site. Depending on the site and the tree species involved, it may be necessary to actually plant tree seedlings. In other cases, seeding an area from the air or simply allowing natural regrowth is sufficient. Some species, such as aspen and coastal redwood, will quickly send up new growth from the stumps and roots, and reforestation basically takes care of itself. Pine forests, on the other hand, usually require seeding or planting to speed the rate of reforestation. (See fig. 12.6.)

Following the logging and transportation of the logs to the sawmill, there are costs associated with converting the wood into a usable product. Lumber is wood in its original condition that has been sliced into shapes and sizes that are standard for the industry. (A two-by-four originally measured 2 inches by 4 inches, but today it actually measures 1½ inches by 3½ inches. This represents a size reduction of 34 percent and allows for the utilization of smaller trees. This shows what happens when our demand for lumber is greater than our supply.) This process requires money and labor but currently has very little environmental cost associated with it.

Paper products, however, require a structural rearrangement of the wood fibers, which requires considerable quantities of water for processing. Since

**Figure 12.6 Reforestation.**
To maintain some kinds of forests, it is necessary to plant trees to replace the ones that are removed. This is more expensive than simply allowing for natural regrowth.

the wood must be treated with chemicals (such as sodium sulfides and sodium hydroxide) to separate the wood fibers from one another, the chemicals and water become mixed and the water is polluted. In addition, many papers are bleached or dyed, which can further pollute the water being used.

In the past, most paper mills simply dumped this polluted water into streams or lakes. Today, because of the enactment and enforcement of water pollution control laws, most paper mills have significantly reduced water pollution. This has resulted in a higher price for paper products, because the paper mills pass along their additional costs to the consumer.

Recycling forest products can reduce our pressure on the forests. Because paper currently represents almost 30 percent of our annual consumption of wood products, recycling these products has obvious advantages. Many kinds of paper products can be recycled into insulation, newsprint, and toilet paper. In addition to lumber and paper products, forest resources are also used to manufacture rayon, turpentine, some plastics, and specialty products, such as medicines and flavorings.

## Wildlife resources

Today it is difficult to imagine the wealth of natural resources that was available to colonial Americans: the sight of forests stretching as far as the eye could see, flocks of migratory birds that darkened the skies, spawning runs of fish that seemed almost endless, and millions of hectares of untouched land. From such a vantage point, it is only natural that the settlers assumed that there was no end to the bounties available to them.

There was another factor that encouraged the unrestricted use of natural resources. Most of our ancestors came from countries where the forests and game were the property of the landowners or royalty. In fact, the title to wild game in this country was originally held in trust for the people by the crown colonies under English common law. Later, when it was transferred to the states, wildlife became public domain.

For over one hundred years, animals such as elk, deer, grouse, and waterfowl were hunted for commercial markets. In addition to the market hunters, there were three other factors responsible for the decline in the amount of game: the individual hunter, federal and other governmental programs, and changing land-use patterns.

By the end of the 1800s the unrestricted killing of wildlife had taken its toll. Less than a half million deer remained in the entire country. Many states had no deer left. The population of elk had declined from ten million to approximately fifty thousand. The total number of big herbivores had declined by about 80 percent.

## *The restoration of wildlife in the United States*

Historically, the first step in the restoration of wildlife was to provide protection. This was partially accomplished by the passage of laws regulating or restricting hunting.

Along with the protection provided by hunting laws, there was a policy of active predator control. The idea underlying these practices was that wildlife would thrive if given a suitable habitat and protection from enemies.

The 400,000-hectare Kaibab Plateau Refuge in Arizona was established in 1906 to protect a population of 3,000 mule deer. Hunting was not allowed, and steps were taken to eliminate all types of deer predators. In a twenty-five year period, the timber wolf was eradicated from the refuge, and government hunters killed 781 mountain lions and over 5,000 coyotes.

Under such conditions the deer prospered. By 1924, their number in the refuge had increased to nearly 100,000. With no form of population control, the herd began to exceed its food supply. Tens of thousands of deer died of starvation; by 1930, 70,000 had died. Eventually the population stabilized at 15,000. Although the herd of deer has increased from 3,000 to 15,000 in the refuge, the individual animals do not seem to be as large or as healthy as those in the original herd. Thus protection was not the total solution to wildlife management.

## *Wildlife management techniques*

Currently, a number of techniques are used in wildlife management. These include (1) game and habitat analysis, (2) population census techniques, (3) stocking, (4) predator control, (5) game refuges, and (6) habitat restoration.

**Figure 12.7 Habitat requirements.** In order for these young birds to survive, all the necessary habitat requirements must be provided.

Animals have specific requirements. Therefore, in managing a particular species, it is essential that the habitat needs of the organisms be determined. (See fig. 12.7.)

An animal's habitat must provide five necessities:

1. Food and water
2. Cover for escaping from enemies
3. Cover for protection against the elements
4. Cover for loafing and sleeping
5. Cover for mating and the rearing of young

These various kinds of **cover** provide protection from enemies and the elements.

Animals are highly specific in their habitat requirements; not every area satisfies all their demands. For example, bob-white quail must have a winter supply of food that protrudes above the snow. In addition, quail require a supply of grit (small rock particles), which the birds use in their gizzards to help in the grinding of food. During the majority of the year, they need a field of tall grass and weeds as cover against natural predators. However, such protection is not available during the winter. During this season, a thicket of brush is necessary. The thicket that provides escape from the predators also shelters the birds against the winter cold.

Most areas provide suitable cover for loafing or resting, but for sleeping, quail prefer an open, elevated location. This allows the bird to take flight if attacked at night. Nesting quail prefer a grassy area that provides some bare ground for the chicks to use to dry themselves when they become wet. All of

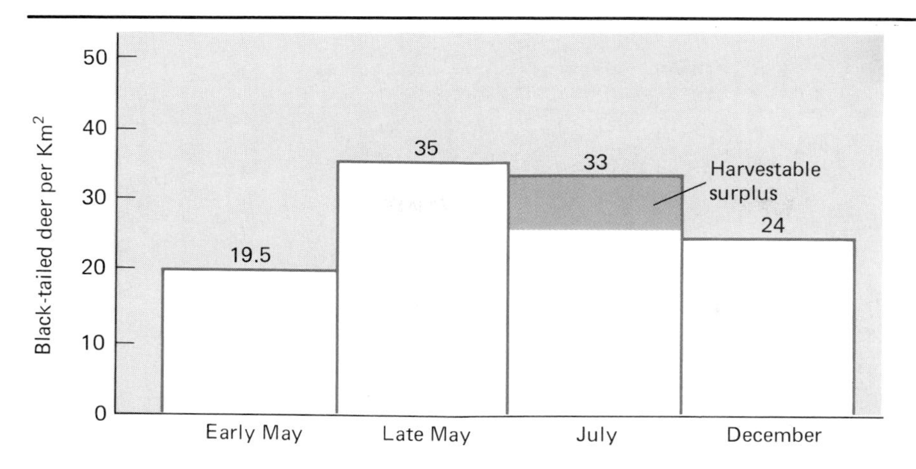


**Figure 12.8 Seasonal population changes.** The seasonal changes in this population of black-tailed deer is typical of many game species. The hunting season is usually timed to occur in the fall so surplus animals will be harvested prior to winter when the carrying capacity is lower. (Data from Taber, R. D., and Dasmann, R. F., "The Dynamics of Three Natural Populations of the Deer *Odocoileus hemionus columbianus*," *Ecology* 38, no. 2 (1957): 233–46.

these requirements must be available within a radius of four hundred meters, which is the daily range of quail. Quail, like all animals, have highly specific habitat requirements. For a population to survive, all of the habitat requirements must be readily available.

Once the habitat requirements are known, it is necessary to ascertain the desirable population size. The population must be checked several times to determine if the number is within the desired limits. Population censuses provide this data. If the population is below the desired number, organisms may be artifically introduced. This is referred to as stocking. If the population is being overkilled by hunting or natural predators, corrective measures may be taken.

The purpose of regulating hunting and establishing specific open seasons on game species is to have the largest possible healthy population at the time of the designated hunting season. Hunting seasons usually occur in the fall before the onset of winter, because many animals normally die during the winter. Winter taxes the animal's ability to stay warm and is also a time of low food supplies in most temperate regions. A well-managed wildlife resource allows for a large number of animals to be harvested in the fall and still leaves a healthy population to live through the winter and reproduce the following spring. (See fig. 12.8.)

Many species may require the establishment of refuges in which the environment is manipulated in order to provide optimal conditions for survival and reproduction of the species. (See fig. 12.9.)

The Kirtland warbler is an endangered species that nests in the central part of Michigan. (There is now some evidence of some individuals nesting in Wisconsin and Minnesota.) The nesting cover required for this species is young even-aged stands of jackpine. In order to maintain these even-aged stands, it is necessary to burn portions of their nesting area periodically to assure future nesting sites, because the cones of jackpine release seeds only after fire. Obviously, when these areas are burned, many species of animals are displaced from the area.

**Figure 12.9 Habitat restoration.**
As a result of protection from grazing, the area on the left side of the fence has been protected from sheep grazing and provides habitat for many animals native to the area.

The burning of old stands of jackpine to allow for regrowth is an example of **habitat management.** Once the food habits, predators, and cover requirements of a species are well understood, it is possible to alter the habitat to optimize it for the specific desired species. Habitat management may take the form of encouraging some species of plants that are preferred food items for the game species desired. When managing an area for deer, it is essential that there be lots of young trees, saplings, or brush and shrubs, which the deer use as food and cover. This condition may be encouraged by cutting the timber in an area and allowing natural regrowth to supply the food the deer need. In this particular instance, we can actually integrate two different uses—timber production and deer herd management. Some other populations of wildlife, such as squirrels, will be excluded or severely reduced if one manages for deer, because these species may require mature forests that produce nuts or acorns, which are necessary food items.

Given suitable habitats and protection, most wild animals can maintain a sizeable population. In general, organisms produce more offspring than can survive. Figure 12.10 illustrates the reproductive potential of both quail and whitetail deer. High reproductive potential, protection from hunting, and restoration of habitats reversed the drastic population declines of wildlife in the United States. In Pennsylvania, where deer were once extinct, the number is now in excess of six hundred thousand. Even in predominately agricultural states, the deer have made a comeback. In Indiana, a herd of nearly one hundred thousand deer has grown from an initial stocking of thirty-five individuals.

Other animals have shown comparable changes in population size. The wild turkey population in the United States has increased from twenty thousand birds in 1890 to two million birds today. In Alaska, the bald eagle was considered to be a predator of salmon; at the urging of the canning industry, a bounty was placed on them. From 1917 to 1952, one hundred and twenty-

| Quail | Adults | Young | Total |
|---|---|---|---|
| 1st year | 2 | 14 | 16 |
| 2nd year | 16 | 112 | 128 |
| 3rd year | 128 | 896 | 1024 |

| Deer | Adults | Yearlings | Young | Total |
|---|---|---|---|---|
| 1st year | 2 | 0 | 2 | 4 |
| 2nd year | 2 | 2 | 2 | 6 |
| 3rd year | 4 | 2 | 4 | 10 |
| 4th year | 6 | 4 | 6 | 16 |
| 5th year | 10 | 6 | 10 | 26 |

**Figure 12.10 Reproductive potential.** If provided with the proper habitat, animals have the reproductive potential to maintain the population in that area.

eight thousand eagles were killed for the bounty money in Alaska. Even though the practice of paying a bounty was discontinued in 1952, the population of eagles continued to decline. However, with the banning of DDT in 1972, a harmful factor was eliminated from the eagles' environment. This example of habitat restoration, coupled with protection from hunting, has resulted in a present population of thirty thousand bald eagles in Alaska.

## *Management of waterfowl*

Waterfowl (ducks, geese, swans, rails, etc.) present some special management problems because they are migratory. **Migratory birds** can fly thousands of kilometers and, therefore, can travel north in the spring to reproduce during the summer months and return to the south when cold weather freezes the ponds, lakes, and streams that serve as their summer home. (See fig. 12.11.) Because many of these waterfowl nest in Canada and winter in parts of the United States and Central America, it is necessary to have an international agreement among the involved countries to manage (prevent destruction of) this wildlife resource. Habitat management has taken several different forms. In Canada, where much of the breeding occurs, efforts have been made by government and private organizations to prevent the draining of small ponds and lakes that provide nesting areas for the birds. In addition, new empoundments have been created where it was practical.

Because birds migrate southward during the fall hunting season, a series of wildlife refuges provide havens from hunting pressure. In addition, these wildlife refuge areas may be used to raise local populations of birds.

During the winter, the birds congregate in the southern United States. It is important that these regions have sufficient areas set aside to allow the birds to find food and cover during the winter months.

**Figure 12.11 Flyways.** Migratory waterfowl follow traditional routes when they migrate. These have become known as the Atlantic, Mississippi, Central, and Pacific flyways. This map shows the general location of these flyways.

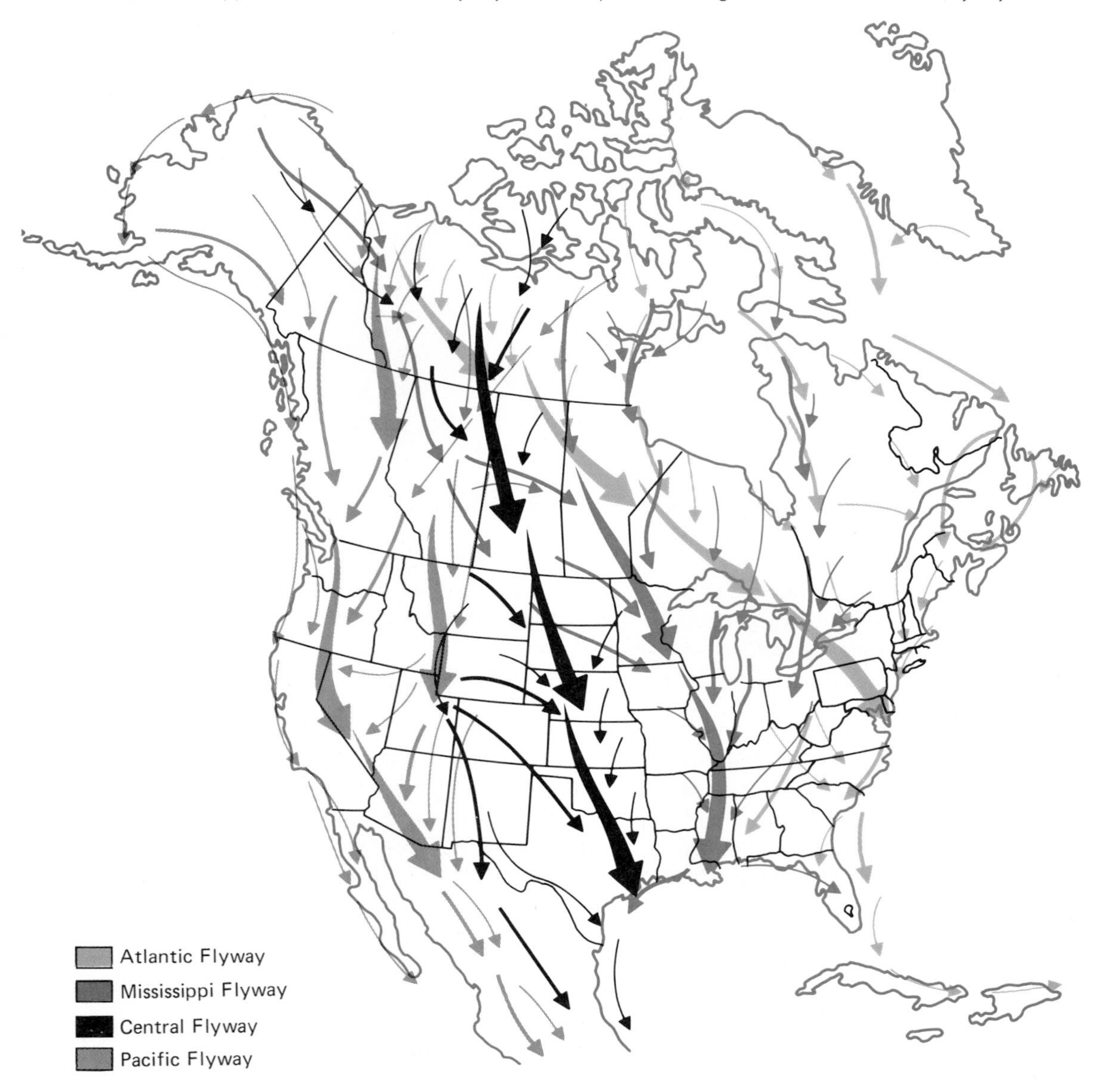


## *Fish management*

The management of fish populations is similar to that of other wildlife populations, because fish have the same basic needs as do other wildlife. They require cover, such as logs, stumps, rocks, and weed beds, so they can escape predators. They also need special areas for spawning and raising young. (This

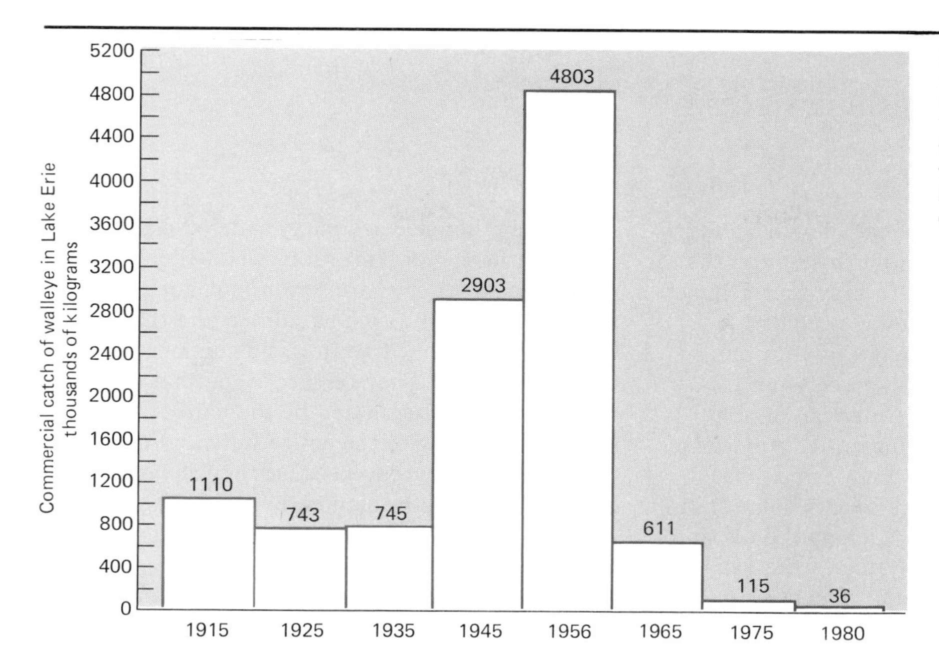


**Figure 12.12 Commercial catch of walleye in Lake Erie.** This graph shows the tremendous drop in the walleye population in Lake Erie as the water quality deteriorated.

may be a gravel bed in a stream, a sandy area in a lake, or an estuary along the coast.) Aquatic situations have the same food chain relationships that are typical of terrestrial situations. Therefore, the fisheries biologist can provide management for specific species of fish by assuring that the appropriate food, cover, and spawning areas are available.

In addition to these basic concerns, a fisheries biologist will give special attention to water quality. Whenever people use water or disturb the land near water, water quality is affected. For example, toxic substances kill fish directly, and organic matter in the water may reduce the oxygen present. Use of water by industry or the removal of trees lining a stream warms the water and makes it unsuitable for certain species. Poor watershed management results in siltation, which covers spawning areas, clogs the gills of young fish, and changes the bottom so food organisms cannot live there. As a result, the fisheries biologist is probably as concerned about what happens outside the lake or stream as with what happens in it.

Much of fish management has been concerned with encouraging the growth of populations of desirable sport species and discouraging less desirable species. The Great Lakes fishery will illustrate this point. At one time, the Great Lakes had large populations of such species as lake trout, walleye, sturgeon, and whitefish. As a result of deteriorated water quality, overfishing, and the inadvertent introduction of nonnative species (such as the alewife, smelt, and parasitic lamprey), the quality of fishing and the quantity of fish harvested in the Great Lakes was reduced drastically. (See fig. 12.12.) The lamprey in the Great Lakes was a problem because it is parasitic on lake trout and other

## Box 12.1
## Native American fishing rights

Throughout many parts of the United States, particularly in the Pacific Northwest and the Great Lakes states, a controversy over fishing is raging. This conflict, however, is unique because the real issue involves treaties that were made 100–150 years ago between the United States and native American nations.

The controversy involving native American fishing rights has become a major political, economic, social, and legal issue in some states (such as Washington and Michigan) and has involved the entire court system, from local courts to the Supreme Court. The precise question revolves around the interpretation of treaty language. This controversy has on several occasions turned to violence and has divided the population of many communities.

On the one side are the native Americans who claim to have the legal authority to engage in commercial fishing enterprises openly, even though such fishing may be restricted or banned altogether to the general public. According to the wording of many treaties entered into in the 1800s, the rights of the native Americans to engage in fishing would not be infringed upon by the states.

On the other side are many state officials and sport fishing organizations who believe that the native Americans are seriously endangering populations of fish (such as salmon and trout) by their uncontrolled netting and commercial fishing practices. They further argue that many species of fish being netted by the native Americans belong to the entire state and not only to a certain group because the fish were originally stocked or planted by the states. Another argument deals with the fishing techniques that are used by the native Americans. In the 1800s when the treaties were signed, commercial fishing technology was limited. Today, however, the native Americans use nylon nets, power boats, and other current technology that enable them to catch larger quantities of fish than in the past.

In 1978, a United States district court judge in Grand Rapids, Michigan, upheld fishing rights of two Chippewa tribes. These rights were first granted under 1836 and 1855 treaties with the United States government. When sport fishers complained that stocks of fish in Lake Michigan were being depleted because of the native Americans gill net fishing, Michigan

species. (See fig. 12.13.) The lamprey problem was brought under control by the use of a very specific larvacide that killed the lamprey larvae (immature lamprey) in the streams. It was possible to use this method because the lamprey migrate up streams to spawn and the larvae spend a year or two in the stream before migrating downstream to the lake.

The alewife, a small fish with little commercial or sport value, was brought under control by the introduction of various species of salmon that use small fish like alewives as food. The salmon were chosen because they are a highly valuable sport fish. The salmon, however, migrate up streams to spawn and thus disrupt the breeding of other species of fish. In addition, many species of salmon die shortly after spawning, which causes local odor problems. Furthermore, it is strongly suggested that the salmon have hindered the recovery state officials tried to regulate that fishing. The federal judge ruled that the state had no authority in the matter. Because the issue in question involved a federal treaty, Congress held the deciding power.

In 1979, the United States Supreme Court ruled on a similar case involving the state of Washington. In a six-to-three decision, the Supreme Court voted to uphold federal treaties that technically entitled native Americans to half of the salmon caught in the Puget Sound area of the state of Washington. A United States district judge in Tacoma, Washington, interpreted the treaties of 1854 and 1855 and ruled that the native American fishermen could catch up to 50 percent of the salmon that passed through their tribal lands on their way to other parts of the state. In the face of protests by the nonnative commercial fishing industry, the federal judge had taken over the regulation of salmon fishing in the state. Commercial fishermen, who feared for their livelihood, outnumbered native American fishermen sixty-six hundred to eight hundred.

The United States Supreme Court backed the federal district judge's interpretation but further held that the 50 percent figure had to include the salmon that the native Americans caught on their lands for home consumption and religious ceremonies as well as the fish they caught for commercial use. The high court also voted to uphold the right of the district court to continue supervising the fishing industry in light of the state's resistance.

Clearly the issues surrounding native American fishing rights are complex and broad in scope. There are purely biological questions involving a resource, there are economic issues involving families' livelihoods, there are cultural and religious issues pertaining to the native Americans, there are legal questions involving federal-state conflicts, and there is the moral question of trying to make amends for past injustices. In such conflicts, there is seldom one right answer. While the courts have ruled on the Michigan and Washington controversy, the problem still exists and is likely to for years to come.

of the lake trout population in the Great Lakes, because they compete for the same food source. Recovery of the lake trout is especially desirable in the Great Lakes because it is one species that is fished commercially.

Fish management in much of the world includes management for both recreational purposes and commercial food production purposes. Therefore, the fisheries resource manager must try to satisfy two different interests. Both the person who fishes for fun and the one who harvests for commercial purposes must adhere to regulations, but the regulations are often different. Usually, commercial fishing regulations allow fishing during different seasons and also allow different methods of capture.

The costs of maintaining fish and wildlife populations are primarily economic costs rather than energy or environmental costs. There are some minor

**Figure 12.13 The lamprey.**
The lamprey entered the Great Lakes in 1932. Because it is an external parasite on lake trout, it had a drastic effect on the population of lake trout in the Great Lakes. As a result of programs to prevent the lamprey from reproducing, the number of lamprey now seems to be under control, and the lake trout population is recovering somewhat.

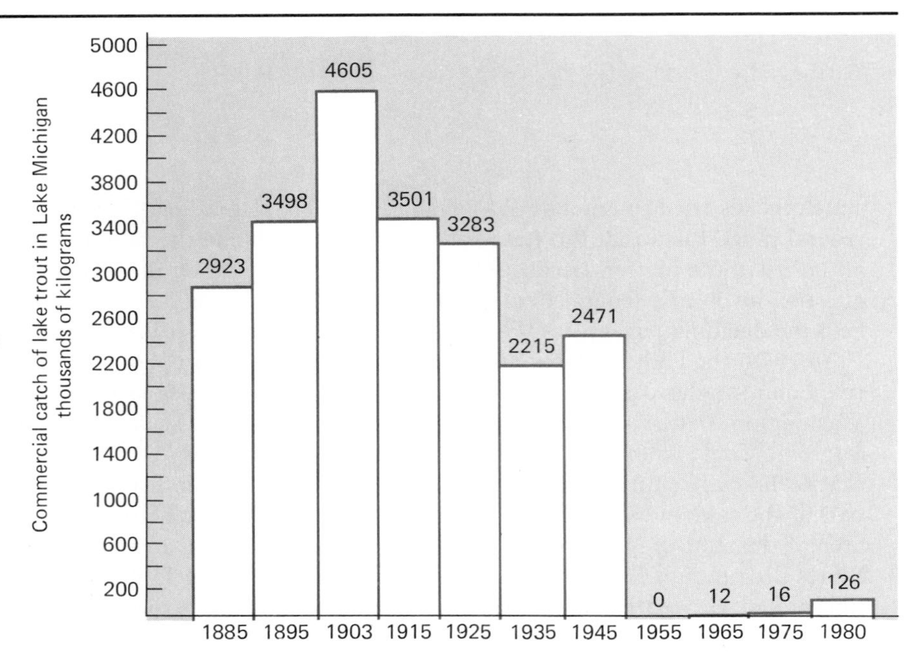


energy costs in the modification of habitat and regulatory activities, but environmental costs are almost zero, because natural populations of animals are maintained by preserving natural ecosystems. Much of the economic cost is borne by the people who use the resource. Hunting and fishing licenses are required, and their sale generates approximately $381 million annually. In 1937, The Pittman-Robertson Wildlife Restoration Act was passed in the United States. It placed an 11 percent excise tax on the sale of arms and ammunition. This generates approximately $100 million each year.

The environmental costs that do arise are usually the result of competing uses for the same piece of land. When land is managed for one particular species, other species are often adversely affected.

## Summary


Natural resources are those structures and processes that can be used by people but cannot be created by them. Renewable resources can be regenerated or repaired, while nonrenewable resources are consumed.

Mineral resources must be extracted from ores, a process that requires expenditures of energy and money. This process also changes the environment. Mineral resources are not evenly distributed, so most countries must purchase some of their minerals from foreign sources. The major steps of mineral exploitation are exploration, mining, processing the ore, transportation, and manufacturing of the finished product. Ultimately, all costs of mineral exploitation are reduced to monetary costs. Recycling reduces the demand for new sources of mineral deposits but does not necessarily save money.

## CONSIDER THIS CASE STUDY
### Knox County, Illinois

Knox County in Illinois has excellent agricultural land. It also has layers of coal a small distance below the surface of the soil. Only three other states, Pennsylvania, West Virginia, and Kentucky, mine more coal than Illinois. Much of the coal mined in central and southern Illinois is mined by strip mining. Huge shovels or draglines remove the topsoil to get at the coal beneath. Many farmers no longer own the mineral rights to their farms. These rights were sold to coal companies back in the early 1900s when times were hard and farmers needed cash just to keep going. Today, the coal company is claiming its right to mine the coal. Although strip-mining companies have made tremendous strides in reclaiming the land they disturb when they mine, the crop yields on land they have reclaimed doesn't approach the crop-producing ability of undisturbed farmland.

This conflict has many parts. Legally, the coal company has the right to mine the coal. After all, the company did pay for the rights and the landowners or their parents didn't refuse the money when it was paid to them. Environmentally, a conflict exists between the preservation of a renewable resource, agricultural land, and the exploitation of a nonrenewable resource, the coal. What is more important—feeding the world or satisfying our energy needs? Finally, there is the conflict over the method of mining. Strip-mining companies have a long history of environmental indifference. Today, regulations require the reclaiming of land disturbed by coal mining, but it cannot be totally reclaimed, at least in the short term. The fact remains that there is no other way to mine this coal.

Should this coal be mined?
What is more important—feeding people or satisfying our energy demands?
Put yourself in the position of the coal company president. How would you feel?
How would you feel if you were a farmer whose father sold the mineral rights in 1901 because he needed the money?

Forest resources are distributed fairly evenly throughout the world. Two major areas of under-utilized forests are the northern parts of the USSR and Canada, and the tropical forests. Tropical forests will require new forestry practices to allow for their use as a source of timber. The major steps in the lumbering process are cutting the trees, transportation, milling, and reforestation. Clear-cutting may be permitted if slope, soil, and forest type will allow for rapid replacement of the vegetation. Recycling of paper can significantly reduce the need for cutting trees for pulp.

Wildlife is all of the undomesticated animals that live in an area. They have ecological value, aesthetic value, and commercial value. Animals require food, water, and cover. Management for certain wildlife species often necessitates modifications in the habitat to supply these requirements. Wildlife management techniques seek to optimize the harvestable surplus of game animals. Most modifications favor one species but might harm another.

Fish management includes management for both sport and commercial fishing. Waterfowl management requires international agreements because waterfowl migrate across international boundaries.

## Review Questions

1. Why is recycling usually more energy efficient than mining new raw materials?
2. List three renewable and three nonrenewable resources.
3. What are three kinds of costs associated with resource exploitation?
4. In what parts of the world are forests under-utilized?
5. What is clear-cutting? Under what conditions should it not be used?
6. Describe six wildlife management techniques.
7. Compare the United States consumption of wood products and mineral resources with the world average per capita consumption.
8. Describe the different kinds of cover an animal needs.
9. Why must energy be expended to extract mineral resources?
10. Can the United States be self-sufficient? Why or why not?
11. What is habitat management? Why is it used?
12. How do wildlife managers generate a harvestable surplus?

## Suggested Readings


Dasmann, Raymond F. *Wildlife Biology*. New York: John Wiley & Sons, 1964. Deals with principles of wildlife populations and management of wildlife populations.

Giles, Robert H. *Wildlife Management Techniques*. Washington, D.C.: Wildlife Society, 1969. A "how to" book that describes techniques typically used in wildlife management studies.

Leopold, Aldo. *Game Management*. New York: Charles Scribner's Sons, 1933. The first book written in the field of wildlife management, as timely today as when it was written.

———. *A Sand County Almanac with Essays on Conservation from Round River*. Oxford University Press, 1966. Must reading by a master storyteller and ecological thinker.

McDivitt, James F., and Manners, Gerald. *Minerals and Men*. Baltimore: The Johns Hopkins University Press, 1974. Presents a concise treatment of the minerals picture.

Sharpe, Grant W. *Introduction to Forestry*. 4th ed. New York: McGraw-Hill Book Company, 1976. Deals with all aspects of forest practices from a forester's point of view.
Teague, Richard D., and Decker, Eugene, ed. *Wildlife Conservation: Principles and Practices*. Washington, D.C.: Wildlife Society, 1979. A collection of articles dealing with wildlife management.
The Council on Wage and Price Stability. *Report to the President on Prices and Costs in the United States Steel Industry,* 1977. Compares the United States and Japanese steel industries from an economic point of view.

Since the data presented in this chapter will change, you should be aware of three readily available sources of recent information.

Bureau of the Census. *Statistical Abstract of the United States*.
United Nations. *Statistical Yearbook*.
United States Bureau of the Mines. *Minerals Yearbook*.

## Chapter Outline

I. Soil
  A. Soil formation
  B. Soil properties
II. Erosion
  Box 13.1 Land capability classes
III. Soil conservation practices
  A. Contour farming
  B. Strip farming
  C. Terracing
  D. Waterways
  E. Windbreaks
IV. Other land uses
  Box 13.2 Conservation tillage
V. Consider this case study: Soil erosion in Virginia

## Objectives

List the physical, chemical, and biological factors that are responsible for soil formation.

Describe the various layers in a soil profile.

Differentiate between soil texture and soil structure.

Explain how texture and structure influence soil atmosphere and soil water.

Explain the role of living things in soil formation and fertility.

Explain the importance of humus to soil fertility.

Describe the processes of soil erosion by water and wind.

Explain how contour farming, strip farming, terracing, waterways, and windbreaks reduce soil erosion.

Understand that the misuse of soil reduces soil fertility, pollutes streams, and requires expensive remedial measures.

Explain how land not suited for cultivation may still be productively used for other purposes.

## Key Terms

contour farming

erosion

friability

horizons

humus

land

leaching

loam

parent material

soil

soil profile

soil structure

soil texture

strip farming

terracing

waterways

windbreak

# Chapter 13
# Soil structure and use

## Soil

Soil and land are often thought of as being the same. However, **land** is the part of the world that is not covered by the oceans, and **soil** is an organized mixture of minerals, organic material, living organisms, air, and water, all of which support plant life. Soil is a covering over the land. Farmers are primarily concerned with soil because each type of soil has an impact on what crops may be grown. Soil is also of vital concern to the urban population. If the soil is abused and can no longer grow crops or is allowed to erode and pollute our streams, urban dwellers (as well as farmers) suffer as a result. Because soil is important, everyone should be concerned with using the soil wisely. To use soil properly, it is necessary to understand the process of soil formation.

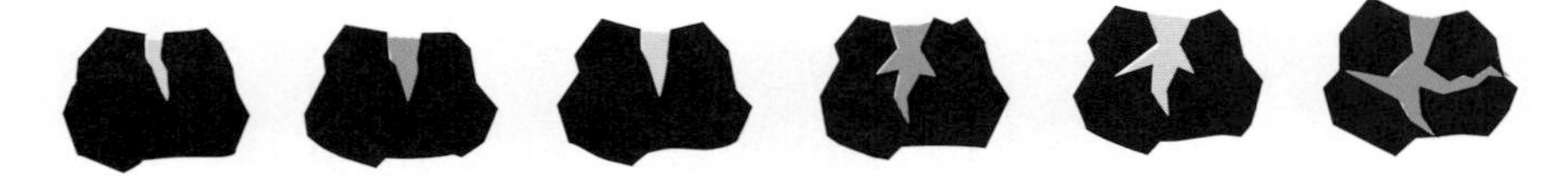

**Figure 13.1 Physical decomposition of parent rock.**
The crack in the rock fills with water. As the water freezes and becomes ice, the pressure of the ice enlarges the crack. The ice melts and water again fills the crack. The water freezes again and widens the crack. Alternating freezing and thawing splits the rock into smaller fragments.

## *Soil formation*

A combination of physical and biological events is involved in the formation of soil. Soil building begins with the physical breakdown of the **parent material,** which consists of ancient layers of rock or more recent geologic deposits from lava flows or glacial deposits. The kind of parent material determines the kind of soil that is formed. Physical factors that can bring about fragmentation are temperature changes and abrasion. Moving ice, water, and wind can transport particles that grind away the parent material.

Repeated freezing and thawing of water is another process that results in fragmentation. A similar process causes the formation of potholes in city streets. Water seeps into small cracks in rock or pavement. When the temperature drops, the water freezes, the ice expands, and pressure enlarges the crack. (See fig. 13.1.) As the temperature rises, the ice melts, and water fills the enlarged crack. When freezing and thawing is repeated numerous times, the road or rock eventually breaks down into smaller pieces.

In addition to ice, other physical conditions (such as heat) can cause the splitting of rocks. Heating a large rock can cause it to fracture because rock does not expand evenly. Heating causes the concrete of a roadway to expand until the road buckles and breaks. Many expressway drivers are familiar with this problem.

Forces that cause rock particles to move and rub against each other also cause the physical breakdown of rock. A glacier will cause rock particles to grind against one another and further reduce their size. The rock fragments carried by the glacier are deposited as the ice melts. In addition to glaciers, moving water will also transport rocks and bring about a reduction in their size through abrasion. Have you ever noticed that rocks and pebbles in a stream or along a shoreline are usually smooth? The moving water has caused the rocks to grind against each other, which has removed their sharp edges.

Although wind seems to be rather unimportant in the breakdown of rock, it can have a role in reducing particle size. As wind-carried particles collide with one another and other objects, their sharp edges are removed in the same way that glaciers and moving water polish surfaces. This has the same effect as sandblasting.

Glaciers, moving water, and wind also are responsible for the transportation and deposition of small particles. For example, the landscape of the Painted Desert was created by a combination of wind and moving water that removed some particles but left others. (See fig. 13.2.) In addition to the forces of wind, water, glaciers, and changing temperature, certain chemical activities alter the size and composition of the parent material.

**Figure 13.2 The Painted Desert.**
This landscape was created by the action of wind and moving water. The particles removed by these forces were deposited elsewhere.

Small rock fragments exposed to the atmosphere are oxidized; that is, they combine with the oxygen from the air and chemically change into different compounds. Oxidation is what causes the bright, shiny finish on a car to become dull; the paint is oxidized. In addition to reacting with oxygen, the rock may undergo hydrolysis, which is the chemical alteration of rock by the addition of water. Some of the compounds produced are removed by going into solution. As a result, the parent material is changed in size and chemical composition. The first organisms to gain a foothold in this modified parent material also contribute to the process of fragmentation. Lichens often form a pioneer community that traps dust and chemically alters the underlying rock.

Eventually the small rock fragments resulting from these physical and chemical processes will be incorporated with organic matter from biological processes to form soil. Dead organic material, **humus,** is mixed with the top layers of rock particles. Humus supplies some of the nutrients needed by plants, and it also increases acidity in the soil so inorganic nutrients that are more soluble under acid conditions become more available. For example, wheat and corn grow best in a soil with a pH of 5.5 to 7.0. A soil with a pH above 7 would become more productive if humus were added. Adding humus would lower the pH of the soil and make more nutrients available to the plants. But changes in pH may also cause certain nutrients to go into solution and be washed away. Finally, humus in the soil modifies the texture so it can hold water better and allow movement of air through the soil.

Burrowing animals, soil bacteria, fungi, and the roots of plants are also part of the biological process of soil formation. The most important burrowing animal is the earthworm. One hectare of land may support a population of a half million earthworms. These animals literally eat their way through the soil. This results in further mixing of organic and inorganic material and increases the amount of nutrients available for plant use. Soil aeration and drainage are also improved by the burrowing of earthworms. The earthworms in a hectare of soil can process as much as nine metric tons of soil a year. The presence of a large population of these animals improves the soil fertility. Other animals, fungi, and bacteria also break down organic matter and mix it with other soil particles. Fungi and bacteria are decomposers that serve as important links in the mineral cycles. (See chap. 3.)

Over a period of time, this complex array of physical, chemical, and biological processes has formed the soils we have today. Under ideal conditions, soft parent material may develop into a centimeter of soil within fifteen years. Under poor conditions, a hard parent material may require hundreds of years to build an equivalent amount of soil. Obviously, soil formation is a slow process.

## *Soil properties*

Soil properties include the texture, structure, chemical composition, soil atmosphere, soil moisture, and biotic content. **Soil texture** is determined by the size of the rock particles within the soil. The largest soil particles are gravel, which consists of fragments larger than 2.0 millimeters in diameter. Particles between 0.05 and 2.0 millimeters are classified as sand. Silt particles range from 0.002 to 0.05 millimeters in diameter. Clay particles are the smallest and are less than 0.002 millimeters in diameter.

Large particles, such as sand and gravel, allow water and air to penetrate the soil because there are many tiny spaces between particles. Both air and water tend to flow through the spaces. Water drains from this kind of soil very rapidly.

Clay particles tend to be flat and are easily packed together to form waterproof layers. Soils with a lot of clay do not drain well and are poorly aerated. Because water does not flow through clay very well, minerals are retained much better in clay than in sand.

Rarely does a soil consist of a single-sized particle. Various particles are mixed, and a variety of soil textures can be classified. (See fig. 13.3.) An ideal soil is **loam,** which combines the good aeration and drainage properties of large particles with the nutrient retention of clay particles.

Soil structure is different from texture, but it is dependent upon the soil texture. **Soil structure** refers to the way various soil particles clump together. Sand particles do not clump, and sandy soil lacks structure. A good soil forms small clumps and crumbles easily. The **friability** of soil, its ability to crumble, is determined by the soil structure and moisture content. Sandy soils are very friable, and clay has very little friability. If clay soil is worked when it is too wet, it can stick together in massive chunks that will not break up for years. A good soil will crumble and has spaces to allow air and water to mix with the soil. In fact, the air and water content of a soil depends upon the presence of these spaces. (See fig. 13.4.)

**Figure 13.3 Soil texture.** Texture depends upon the percentage of clay, silt, and sand particles in the soil. Soils with the best texture for most crops are loams. (Source: Soil Conservation Service.)

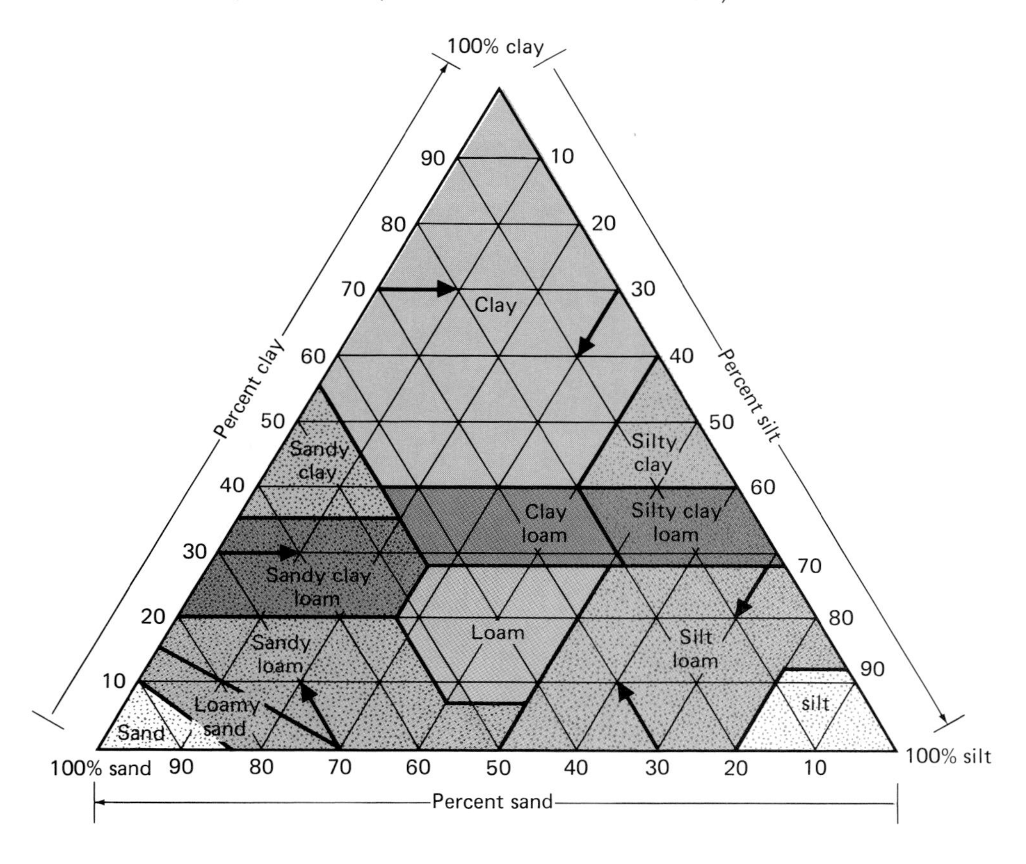


**Figure 13.4 Pore spaces and particle size.** The soil on the left, which is composed of particles of various sizes, has spaces for both water and air. The soil on the right, which is composed of uniformly small particles, has less space for air. Since roots require both air and water, the soil on the left will be better able to support crops than that on the right.

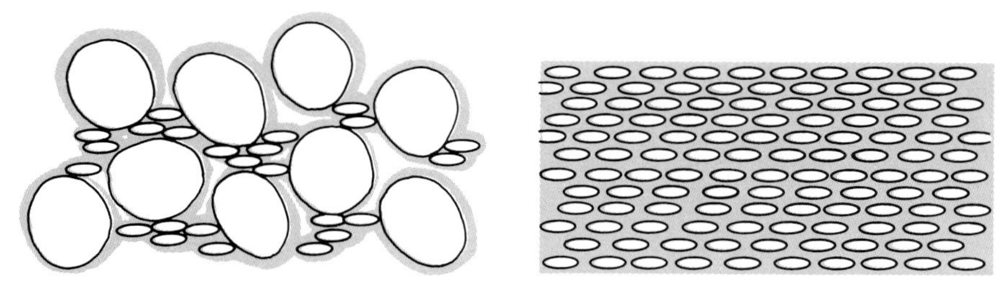

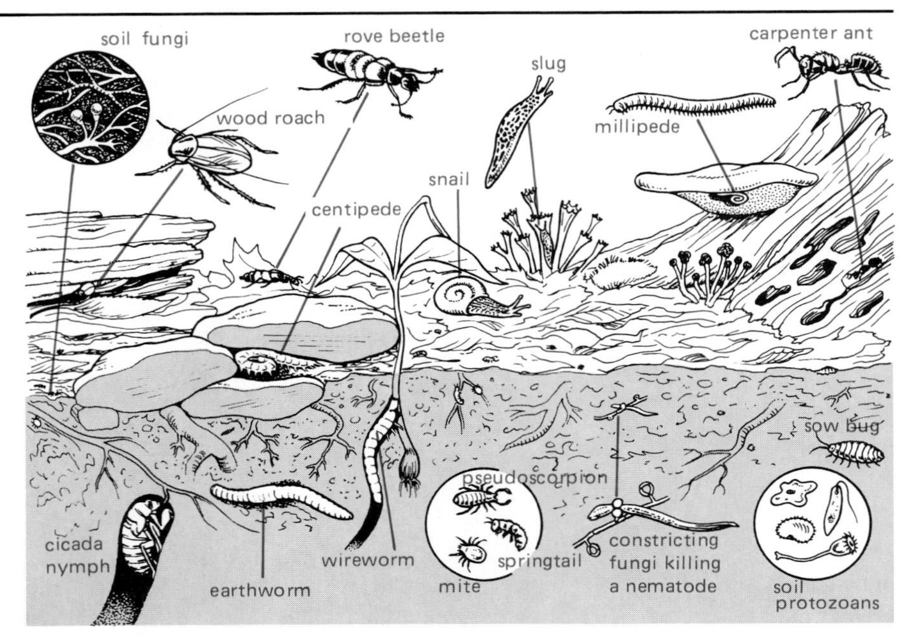


**Figure 13.5 Soil organisms.**
All of these organisms occupy the soil and contribute to it by rearranging soil particles, participating in chemical transformation, and recycling dead organic matter.

In good soil, about two-thirds of the spaces will contain air after the excess water has drained. The air in these spaces provides a source of oxygen for plant root cells. Water for the roots occupies the remaining soil space. The relationship between the amount of air and water is not fixed. After a heavy rain, all of the spaces may be filled with water. If some of the excess water does not drain from the soil, the plant roots may die from a lack of oxygen. They are literally drowned. Also, if there is not enough soil moisture, the plants wilt from a lack of water. Soil moisture and air are also important in determining the numbers and kinds of soil organisms.

Protozoa, nematodes, earthworms, insects, bacteria, and fungi are typical inhabitants of soil. (See fig. 13.5.) The role of protozoa in the soil is not firmly established, but they seem to act as parasites on other forms of soil organisms and, therefore, help to regulate their population size. The nematodes (roundworms) may aid in the recycling of dead organic matter. Some soil nematodes are parasitic on the roots of crops. Even though insects contribute to the soil by forming burrows and recycling organic materials, they are also major crop pests. The bacteria and fungi are particularly important in the decay and recycling of materials. Nutrients are released from the organic matter as a result of the action of these microorganisms. Their chemical activities change complex organic materials into products that can be used by plants. One of the nutrients made available by these microorganisms is nitrogen. This is accomplished by converting the nitrogen contained in the protein component of organic matter into nitrate, a nitrogen compound that is usable by plants. The amount of nitrate produced will vary with the type of organic matter, type of microorganisms, drainage, and temperature.

**Figure 13.6 Soil profile.** A soil has layers that differ physically, chemically, and biologically. The top layer is known as the *A* horizon and contains most of the organic matter. The *B* horizon accumulates minerals and particles as water moves from the *A* to *B* horizon. The *C* horizon consists of weathered parent material.

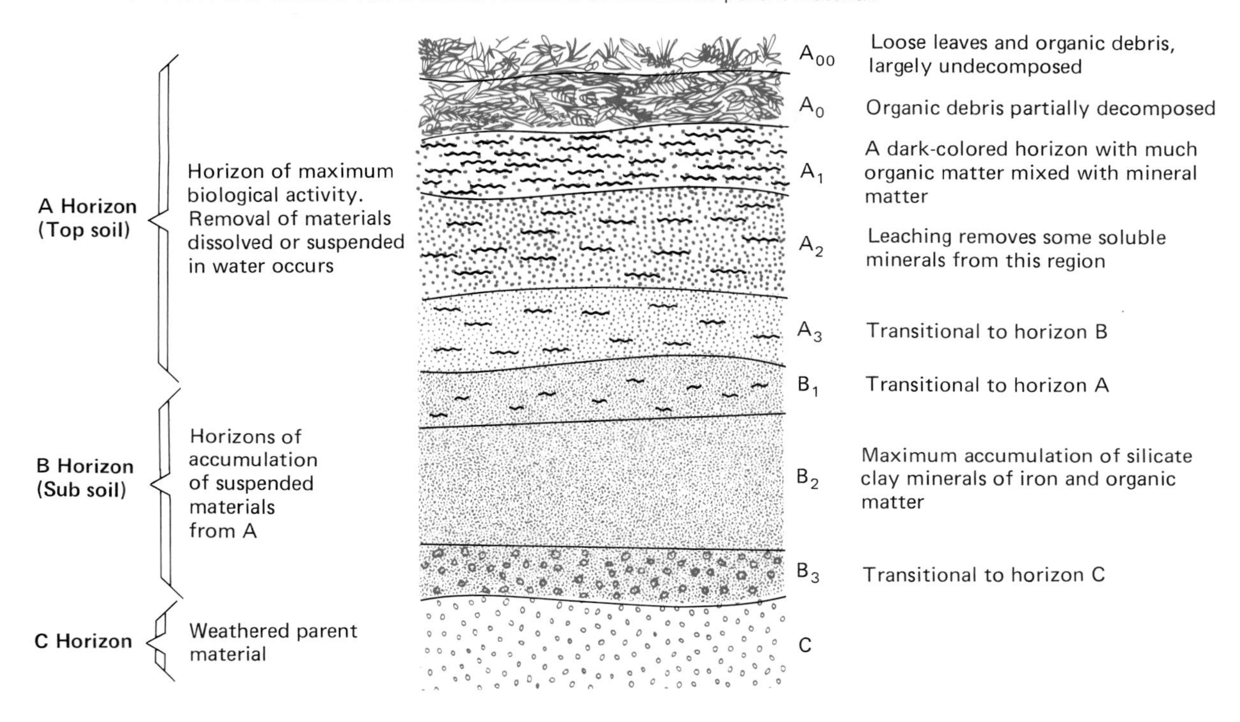


The uppermost layer of soil contains more nutrients and organic matter than do the lower soil layers. The **soil profile** is composed of the various soil layers called **horizons.** (See fig. 13.6.) The uppermost horizon is the *A* horizon, which is commonly referred to as topsoil. The thickness of the *A* horizon may vary from less than a centimeter on the steep slopes of the Rockies to over a meter in the prairie of Nebraska. The upper portion of this horizon is rich in organic materials, nutrients, and living organisms. The lower part of the *A* horizon may contain few plant nutrients because of leaching. **Leaching** occurs when soluble materials are removed from the *A* horizon as the water drains downward. The material leached from the *A* horizon accumulates in the *B* horizon, which is often referred to as the subsoil. In many types of soils, the accumulation of minerals and organic materials leached from the topsoil makes the subsoil a valuable source of nutrients. Such subsoils will support a well-developed root system.

The area below the subsoil, the *C* horizon, consists of the weathered parent material. This parent material contains no organic materials, but it does contribute to some of the properties of the soil. The chemical composition of the *C* horizon determines the pH of the soil. The *C* horizon may also influence the soil's rate of water absorption and retention. In the United States, 97 percent

**Figure 13.7 Soil types.** Soils formed in grasslands (chernozem soils) have a deep *A* horizon. The shallow *B* horizon does not have sufficient nutrients to support root growth. In forest soils (podzol soils), both the *A* and *B* horizons possess enough nutrients to allow for root growth.

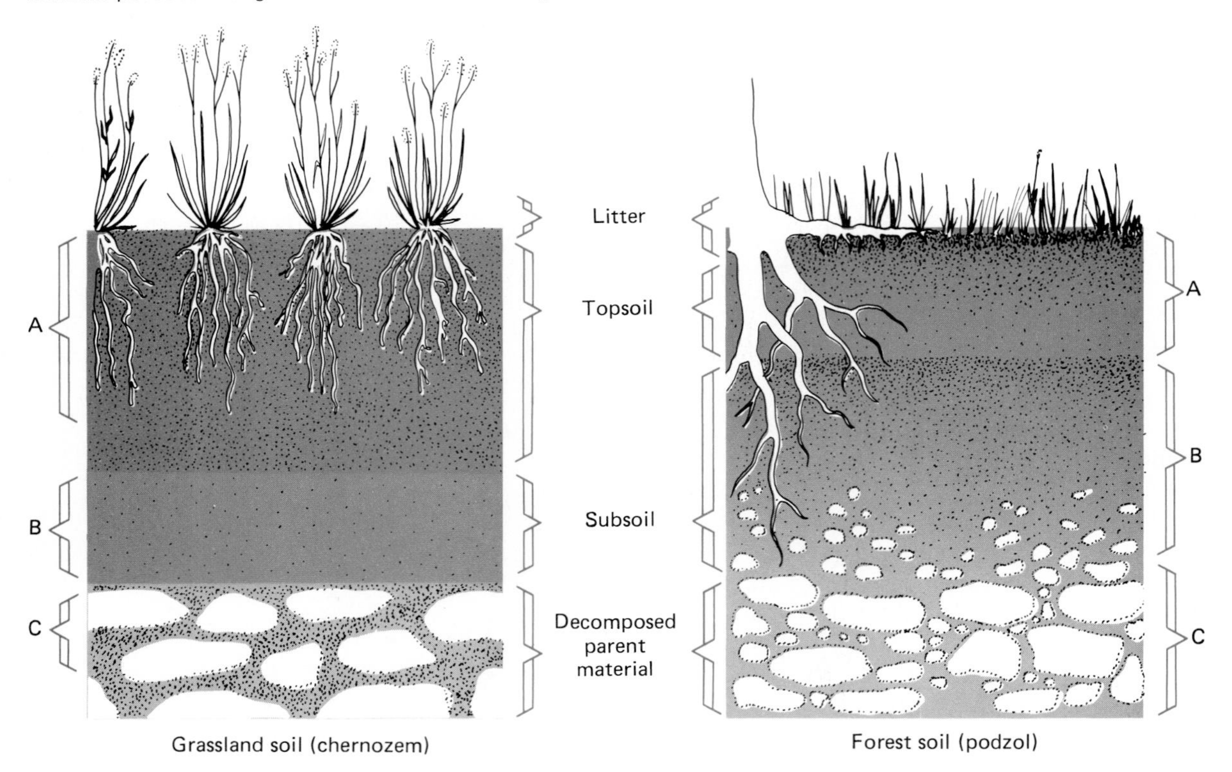


of the *C* horizon consists of weathered rock that was transported to its present location by wind, water, or glacial action.

The soil profiles and the factors that contribute to soil development are extremely varied. Over fourteen thousand separate soil types have been classified in the United States. However, most of the cultivated land in this country is either grassland soil or forest soil. (See fig. 13.7.)

The grassland soils usually have a deep *A* horizon. The low amount of rainfall in the grassland areas limits the amount of leaching from the topsoil. As a result, the *A* layer is deep and supports most of the root growth. This lack of leaching also results in a thin layer of subsoil. The subsoil is low in mineral and organic content and supports little root growth.

Forest soils develop in areas of more abundant rainfall. This results in topsoil that is usually not as thick as grassland soil, but the material leached from the topsoil forms a subsoil that supports a larger root growth. One of the materials that accumulates in the *B* horizon is clay. In some soils, particularly forest soils, this clay will accumulate and form an impermeable "hardpan" layer.

**Figure 13.8 Natural soil erosion.**
The topsoil formed on a large area of the hillside is continuously transported and collected in the region at the bottom of the hills. The resulting "bottomland" will be highly productive because it has a deep, fertile layer of topsoil.

In addition to the differences caused by the kind of vegetation and rainfall, topography will influence the soil type. (See fig. 13.8.) On a relatively flat area, the topsoil formed by soil-building processes will collect in place and will gradually increase in depth. The topsoil formed on rolling or steep slopes is often transported away as rapidly as it is formed. On such slopes, the accumulation of topsoil may not be sufficient to support a cultivated crop. The topsoil removed from these slopes is eventually deposited in the flat floodplains. As a result, the river bottom or delta regions have a greater depth of topsoil. These regions collect the topsoil that was formed over a large area. This type of soil is usually highly productive agricultural land.

## Erosion

**Erosion** is the wearing away and transportation of soil by water or wind. The Grand Canyon of the Colorado River, the little gullies on hillsides, and the rich river bottomlands were all created by the action of water moving soil. Anyone who has seen muddy water after a thunderstorm has observed soil being moved by water. (See fig. 13.9.) The force of the moving water enables it to carry large amounts of soil. Each year, the Mississippi River transports over 325 million metric tons of soil from the central region of the United States to the Gulf of Mexico. This is equal to the removal of a layer of topsoil approximately one millimeter thick from this region. This movement of soil by water is repeated by every stream and river in the country. Dry Creek, a small stream in California, has only 500 kilometers of mainstream and tributaries; however, each year it removes 180,000 metric tons of soil from a 340 square kilometer area. When soil is badly eroded, all of the topsoil and some of the subsoil is removed. Badly eroded land is no longer productive farmland. Most current agricultural practices result in the loss of soil faster than it is being replaced. Farming practices that reduce erosion (such as contour farming or terracing) should be encouraged.

**Figure 13.9 Water erosion.**
The force of moving water is removing the soil from this land. Whenever you see a muddy stream, you are viewing water that is transporting soil.

In addition to water, wind is an important mover of soil. Most of you have seen dust being blown by the wind. This is soil movement. Under certain conditions, wind can move large amounts of soil. (See fig. 13.10.) Wind erosion may not be as evident as water erosion, because it does not leave gullies on the surface of the soil. Nevertheless, it is a serious problem. Wind erosion is most common in the Great Plains. (See fig. 13.11.) There have been four serious Dust Bowls since that area was settled in the 1800s. If this area receives less than thirty centimeters of rain per year, there is not enough moisture to support crops. Several years of low rainfall is called a drought. The crops die, the soil is plowed, and the soil is left exposed. The loose, dry soil is then unprotected from the wind and is easily carried away. During the Dust Bowl era of the 1930s, wind destroyed 3.5 million hectares of farmland and seriously damaged an additional 30 million hectares of farmland.

**Figure 13.10 Wind erosion.**
The dry, unprotected topsoil from this field is being blown off of this farmer's land. The force of the wind is capable of removing all of the topsoil and transporting it for a distance of several thousand kilometers.

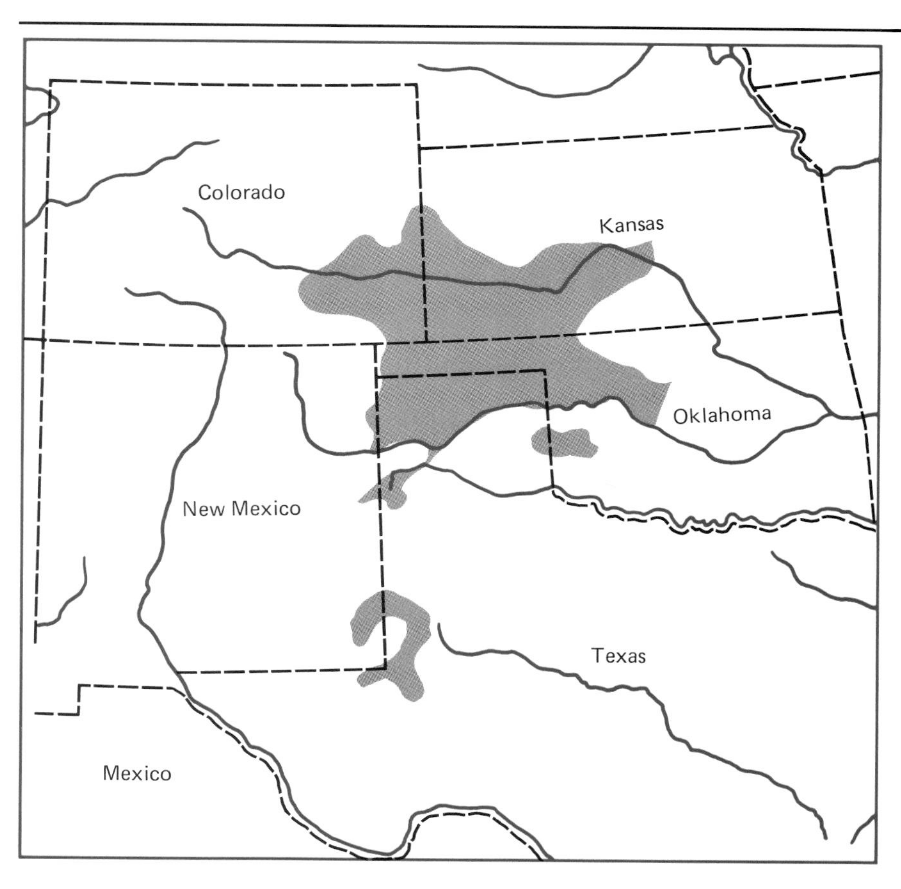


**Figure 13.11 Areas damaged by wind erosion.**
Although wind erosion is a factor in most of the country, the greatest problem area is in the Great Plains. Over thirty million hectares of land have been damaged by wind erosion. These areas are indicated on the map.

**Figure 13.12 Proper land use vs. land abuse.** The land in the top photo is no longer productive. The farm in the bottom photo will continue to supply food and provide a living for the farmer.

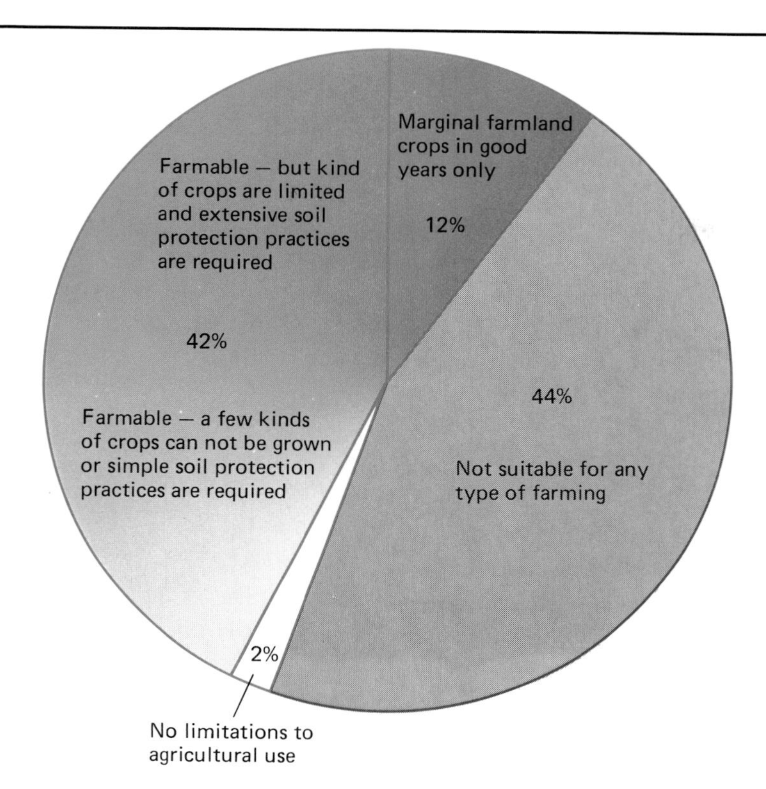


**Figure 13.13**
**Agricultural potential of land.**
Less than half of the United States land is suitable for regular cultivation. For the most part, the rest of the land requires careful soil protection practices. Some land can raise occasional crops and 44 percent is not suitable for farming. Only 2 percent can be farmed without some restrictions on its use.

## Soil conservation practices

Whenever soil is lost by water or wind erosion, the topsoil is the first layer to be removed. This topsoil is the most productive layer. Proper conservation measures should be employed to minimize the removal of topsoil. Figure 13.12 contrasts proper land use with land abuse. Within a few years, the soil from one of the farms was completely depleted. The land that remained was no longer productive. The other farmer used soil conservation practices that were appropriate to the soil characteristics. What can be done with a given piece of land depends upon soil structure, drainage, fertility, rockiness, slope, rainfall, and other climatic conditions.

As seen in figure 13.13, only 2 percent of our nation's land can be cultivated without using some type of conservation practice. Much of the remaining agricultural land, which represents 42 percent of our land, is suitable for farming but requires special care. Therefore, most of the land that we cultivate should be farmed in conjunction with good soil conservation methods. This means using farming methods to reduce the amount of erosion.

# Box 13.1
## Land capability classes

Not all land is suitable for farming, grazing, or urban building. Such factors as degree of slope, soil fertility, rockiness, erodibility, and other characteristics determine the best use for a parcel of land. In an attempt to encourage people to use land wisely, the Soil Conservation Service has established eight land classes. This classification lists the capabilities and precautions that should be observed when using the various types of land.

Unfortunately, many of our homes and industries are located on types I and II land. This does not make the best use of the land. Zoning laws and land-use management plans should consider the land-use capabilities and institute precautions to assure that the land will be used to its best potential.

| | Land Class | Characteristics | Capability | Conservation Measures |
|---|---|---|---|---|
| **Land suitable for cultivation** | I | Excellent, flat, well-drained land | Agriculture | None |
| | II | Good land, has minor limitations such as slight slope, sandy soil, or poor drainage | Agriculture<br>Pasture | Strip cropping<br>Contour farming |
| | III | Moderately good land with important limitations of soil, slope, or drainage | Agriculture<br>Pasture<br>Watershed | Contour farming<br>Strip cropping<br>Waterways<br>Terraces |
| | IV | Fair land, severe limitations of soil, slope, or drainage | Pasture<br>Orchards<br>Limited agriculture<br>Urban<br>Industry | Farm on a limited basis<br>Contour farming<br>Strip cropping<br>Waterways<br>Terraces |
| **Land not suitable for cultivation** | V | Use for grazing and forestry slightly limited by rockiness, shallow soil, or wetness | Grazing<br>Forestry<br>Watershed<br>Urban<br>Industry | No special precautions if properly grazed or logged. Must not be plowed. |
| | VI | Moderate limitations for grazing and forestry, moderately steep slopes | Grazing<br>Forestry<br>Watershed<br>Urban<br>Industry | Grazing or logging may be limited at times |
| | VII | Severe limitations for grazing and forestry, very steep slopes vulnerable to erosion | Grazing<br>Forestry<br>Watershed<br>Recreation<br>Wildlife<br>Urban<br>Industry | Careful management required when used for grazing or logging |
| | VIII | Unsuitable for grazing and forestry because of steep slope, shallow soil, lack of water, or too much water | Watershed<br>Recreation<br>Wildlife<br>Urban<br>Industry | Not to be used for grazing or logging. Steep slope and lack of soil presents problems. |

**Figure 13.14 Contour farming.**
Farming across the slope holds the water on the land and reduces the amount of soil erosion.

## *Contour farming*

Tilling across the slope, which is called **contour farming,** is the simplest method of preventing soil erosion. This practice is useful on gentle slopes. Contour farming produces a series of small ridges at right angles to the slope. (See fig. 13.14.) Each ridge acts as a dam to hold water on the slope. This slows the runoff of water and allows it to soak into the ground. Contour farming reduces soil erosion by as much as 50 percent and, in drier regions, increases crop yields by conserving water.

## *Strip farming*

When the slope is too steep or too long, contour farming alone may not be an effective method of preventing soil erosion. On such land, a combination of contour and **strip farming** is often employed. (See fig. 13.15.) In strip farming, strips of closely sown crops (hay, wheat, or other small grains) are alternated with strips of row crops (corn, soybeans, cotton, or sugar beets). The closely sown crops retard the flow of water, which reduces soil erosion and allows more water to be absorbed into the ground. The types of soil, steepness of slope, and length of slope determine whether strip contour farming is practical and dictate the width of the strips.

## *Terracing*

On very steep land, the only practical method of preventing soil erosion is **terracing.** Terraces are built across the slope on a level grade. (See fig. 13.16.) In this practice, each terrace is constructed at a particular elevation. This allows each terrace to retain water and greatly reduces the amount of erosion.

**Figure 13.15 Strip-farming.**
In rolling land, a combination of contour and strip-farming is used to prevent excessive soil erosion. The strips are planted at right angles to the slope. The bands of *close-sown crops,* such as wheat or hay, alternate with bands of *row crops,* such as corn or soybeans.

a

b

**Figure 13.16 Terraces.**
Expensive terraces seen in photograph *a* are important for food production in some countries. In the United States, a modified type of terrace is built for use with modern farming methods. Six such terraces are visible in photograph *b*.

Terracing has been used for centuries in nations with a shortage of level farmland. The type of terraces seen in figure 13.16a is not suitable for the mechanized farming used in the United States. In the United States, a modified terracing operation is used on some land. (See fig. 13.16b.) Terracing is an expensive method of controlling erosion. Many factors, such as length and steepness of slope, type of soil, and amount of precipitation, determine whether terracing is feasible for use. Even when these practices are used, it is sometimes necessary to provide channels for the movement of water.

**Figure 13.17 Control of water erosion.**
(*a*) The movement of water across the unprotected land has converted this waterway into a gulley. (*b*) Plowing does not extend into the waterway; a covering of grass protects it from damage.

a

b

## *Waterways*

**Waterways** are depressions on sloping land that allow the water to flow off the land. When not properly maintained, these areas are highly susceptible to erosion. (See fig. 13.17.) If a waterway is maintained with a permanent sod covering, the speed of the water is reduced and soil erosion is decreased. Contour farming, strip farming, terracing, and waterways reduce erosion by slowing water runoff.

a

b

**Figure 13.18 Windbreaks.**
(*a*) In sections of the Great Plains, trees may provide protection from wind erosion. The trees along the road are planted in a north-south direction to protect the land from the prevailing westerly winds. (*b*) In parts of the South, the use of temporary strips of ryegrass serve as windbreaks in cotton fields.

### *Windbreaks*

Wind erosion can be reduced if the force of the wind can be reduced. If land that is susceptible to wind erosion is to be cultivated, some protective measure must be employed. This usually involves the use of some type of **windbreak.** (See fig. 13.18.) Windbreaks reduce the velocity of the wind and thereby lessen the amount of erosion from the wind. However, in much of the potential Dust Bowl area, this method is not sufficient to reduce the amount of erosion. The only solution for this land is to keep it permanently covered with grass rather than to cultivate it.

Soil conservation is not only related to farming; it should be of concern to all of us. The loss of soil reduces the amount of cropland, and in the long run,

**Figure 13.19 Misuse of land.**
A productive farm provides an income for the farmer and food for the urban population. If the farmer abuses the land, it will no longer provide income or a supply of food. When farmland is misused, the consequences are detrimental to everyone. (Courtesy of the *Des Moines Register.*)

will reduce our ability to produce food. (See fig. 13.19.) The movement of excessive amounts of soil into streams has several undesirable effects. First, a dirty river is less aesthetically pleasing than a clear stream. Second, a stream laden with sediment affects the fish population. Fishing may be poor because unwise farming practices are used hundreds of miles upstream. Third, the soil carried by a river is later deposited somewhere. In some cases, this soil must be removed by dredging to clear shipping channels. We pay for dredging with our tax dollars, and it is a very expensive operation. It would be more economical to use farming methods that prevent soil erosion from occurring.

## Other land uses

Only 44 percent of our land is suitable for continuous cultivation. Most of this land requires some form of conservation practice when it is cultivated. About the same amount should not be plowed and farmed at all, but it is suitable for other uses. By employing proper soil conservation techniques, much of this land can be productively used for grazing, wood production, recreation, and wildlife. Figure 13.20 illustrates a section of land that is not suitable for cultivation but is capable of producing grass for cattle or sheep. Such land, if properly used, will produce food for this generation and coming generations.

**Figure 13.20 Noncultivated land use.** As long as it is properly protected, this land can produce food through grazing, but it should not be farmed.

**Figure 13.21 Forest and recreational use.** Although this land is not capable of producing crops or supporting cattle, it furnishes lumber, habitat for wildlife, and recreational opportunities.

The land shown in figure 13.21 is not suitable for crops or grazing. It is still a productive piece of land, however, because it furnishes lumber, recreation, and habitat for wildlife.

All land is not equal. Each parcel of land has its capabilities and limitations. Proper land use can eliminate a number of our current environmental problems and assure an adequate amount of suitable land to provide for the needs of future generations.

## Box 13.2
## Conservation tillage

During a typical storm in March, approximately six thousand metric tons of soil enter the western basin of Lake Erie. Along with the soil, two metric tons of phosphorus also enter the lake. This causes pollution and siltation problems in the lake. Similar soil and nutrient losses can be cited for many areas of the world. Conventional tillage methods contribute to this erosion problem. For five thousand years, plowing and cultivation have been used as primary methods of weed control. In conventional tillage, the land is plowed, disked several times, and smoothed to make a planting surface. Recently, more farmers have done their plowing in the fall. Fall plowing allows earlier planting in the spring. However, the soil is left unprotected during the winter and early spring. This subjects the unprotected land to an increased amount of soil erosion. It is estimated that over 50 percent of the cultivated cropland in the United States suffers from excessive soil erosion.

Reduced tillage uses fewer mechanical methods of controlling weeds and uses more herbicides. The type of tillage used determines the cost in labor, energy, and money. There are three major reduced tillage methods.

1. Plowing with reduced secondary tillage usually involves plowing followed immediately by planting.
2. Strip tillage involves tilling only in the row that is to receive the seeds.
3. No-tilling involves special planters that plant seeds in slits formed in the residue-covered soil.

No-till farming requires less fuel and less time but requires more use of herbicides. Yields are comparable with many kinds of crops that are produced by conventional tillage.

In addition to reducing erosion there are other effects of reduced tillage. There is an increase in the amount of winter foods and cover available for wildlife. This could result in increased wildlife populations. A reduction in the amount of siltation in waterways has several benefits, such as decreased need for expensive dredging to maintain shipping channels and cleaner water for recreation. Reduced tillage encourages planting more row crops on hilly land and enables the farmer to convert low-yielding pasture land to profitable row crops. Because fewer trips are made over the field, petroleum is saved even if the

## Consider This Case Study
## Soil erosion in Virginia

In a court case in Virginia, a city attorney filed suit against a land developer for damages. The city contended that the developer was responsible for property damages as a result of erosion from his construction site. During the trial, it was shown that the developer had constructed his subdivision without considering the possibility of soil erosion. As a result, the mud from his project covered the established yards, sidewalks, and drives in an area of established homes. As much as ten meters of soil had eroded from the developer's area and was carried downhill to cover portions of the community.

Speaking about the developer, the city attorney stated: "After gambling and principally winning over a period of fifteen months, it is apparent that you are attempting

Comparison of various tilling methods

| Tillage method | Hours required per 100 hectares | Liters of fuel per 100 hectares | Cost of herbicide per 100 hectares (1973 $) |
|---|---|---|---|
| Conventional | 200 | 2,915 | $2,717 |
| Reduced secondary tillage | 125 | 1,390 | 2,717 |
| Strip tillage | 95 | 1,020 | 2,717 |
| No tillage | 62 | 375 | 4,000 |

Source: J. E. Beuerlein and S. W. Bone. "Selecting a Tillage System" Extension Publication, Ohio State University. And D. H. Soster, "Economics of Alternative Tillage Systems" Bull. Ent. Soc. Amer. 22 (1976):297.

feedstock necessary to produce herbicides is taken into account. Reduced tillage may allow for double cropping (growing two crops in a single year). In parts of the midwest, immediately after harvesting wheat, some farmers have used reduced tillage methods to plant soybeans in the wheat stubble.

In 1972, 12 million hectares were conservation tilled, of which 1.3 million hectares were no-tilled. Just ten years later, in 1982, United States farmers practiced conservation tillage on 40 million hectares, of which 4.2 million hectares were no-tilled. It is estimated that by the year 2010, 95 percent of the United States cropland will be under reduced tillage methods. The switch to reduced tillage will require farmers to pay more attention to the types of soils and pests to be dealt with. The farmer must have a thorough knowledge of the soil type. Well-drained soils are the best type to use with reduced tillage methods.

The mulch from previous vegetation can delay the warming of the soil, which may in turn delay planting of some crops for several days. The mulch reduces evaporation and slows the upward movement of water and soil nutrients needed by the plants. However, the mulch increases the potential for an outbreak of insects and various types of plant diseases, which may necessitate the use of more pesticides.

Reduced tillage is not the total answer, but it is one part of the solution to reducing soil erosion and reducing the energy cost of raising food.

to go with your winning streak; and flood, storms, and weather hold no terror for you." The city claimed that the developer had acted in a negligent manner and that he was therefore responsible for the damages caused by the deposited mud. Therefore, he should be ordered to pay for the cleanup. The developer claimed no responsibility. He cited the grading of large tracts of land as a common construction procedure and that the weather was "an act of God" for which he could not be held accountable.

If you were a member of the jury hearing the case, would you think the developer acted in a negligent manner by leaving the soil exposed for a long period of time?

Do you believe that the property owners can hold a person responsible for actions taken on an area away from their property?

Do you believe that the erosion and subsequent deposit of mud was in fact "an act of God"?

Would you find the developer guilty or not guilty? Why?

## SUMMARY

Soil is an organized mixture of minerals, organic material, living organisms, air, and water. Soil building begins with the physical breakdown of the parent material, which is accomplished by temperature changes and the grinding effects of glaciers, moving water, and wind. Hydrolysis and oxidation also chemically alter the parent material. Organisms affect soil by burrowing, decomposition, mixing, and releasing soil nutrients.

Topsoil contains a mixture of humus and inorganic material, both of which supply soil nutrients. Soil fertility is determined by the inorganic matter, organic material, water, and air spaces in the soil. Soil consists of mixtures of sand, silt, and clay particles, along with organic matter.

A soil profile consists of the *A* horizon, which is rich in organic matter; the *B* horizon, which accumulates materials leached from the *A* horizon; and the *C* horizon, which consists of slightly altered parent material.

Soil erosion is the wearing away and transportation of soil by water or wind. By properly using such conservation methods as contour farming, strip farming, terracing, waterways, and windbreaks, soil erosion can be reduced. Misusing soil reduces soil fertility and causes water pollution. It also requires costly remedial measures. Land unsuitable for farming can be used for grazing, lumber, wildlife habitats, or recreation.

## REVIEW QUESTIONS

1. Define erosion.
2. How are soil and land different?
3. Describe the process of soil formation.
4. Name five physical and chemical processes that break parent material into smaller pieces.
5. Name the four major soil components.
6. In addition to fertility, what other characteristics determine the usefulness of soils?
7. How does soil particle size affect texture and drainage?
8. Besides cropland, what are other possible uses of the soil?
9. Describe three soil conservation practices that help to reduce soil erosion.
10. Describe a soil profile.
11. What kinds of information are important in determining land use?

## Suggested Readings


Berry, Albert R., and Cline, William R. *Agrarian Structure and Productivity in Developing Countries*. Baltimore: The Johns Hopkins University Press, 1979. Discusses the level of productivity of a farm in relation to the farm's size. It explores the productivity and rate of employment as influenced by farm size.

Brown, Lester R. "Vanishing Croplands." *Environment* 20: 6–15, 33–35. Explores on a worldwide basis the fact that nonagricultural use of farmland is increasing and that the remaining farmlands are being over-farmed.

Coats, Robert N. "The Road to Erosion." *Environment* 20: 16–20, 37–39. Due to timbering, the river basins in California have an erosion rate that is ten to one hundred times higher than the average rate in the United States.

Fitzpatrick, E. A. *An Introduction to Soil Science*. Edinburgh: Oliver & Boyd, 1974. Explains soil formation, properties of soil, and a geography of world soils.

Giere, John P.; Johnson, Keith; and Perkins, John M. "A Closer Look at No-till Farming." *Environment* 22: 14–20, 37–41. One possible solution to the farmer's erosion and weed control problems is minimum tillage (no-till) farming. But because no-till farming as practiced in the United States is highly dependent upon the use of pesticides, the risks could outweigh the benefits.

Jones, Ulysses S. *Fertilizer and Soil Fertility*. Reston, Va.: Reston Publishing Company, 1979. Explains the role of fertilizer and plant growth, and all aspects of fertilizer production and application.

Paton, T. R. *The Formation of Soil Materials*. London: George Allen and Unwin, 1978. Basic text concerning the formation of soil. It deals with soil formation in all parts of the world.

Soil Conservation Society in America. *Resource-Constrained Economics: The North American Dilemma*. Ankeny, Ia.: Soil Conservation Society of America, 1980. The mixed economies of North America developed in a climate of ample, even surplus, resources. Limits have suddenly become a reality. Presents a series of twenty-eight papers from the 1979 annual meeting of the Soil Conservation Society. Explores the issues of land, water, food, and energy-resource-constrained economies.

U.S. Department of Agriculture. *The World Food Situation and Prospects to 1985*. Foreign Agricultural Economic Report 98. Washington, D.C. An in-depth analysis on the world food situation.

## Chapter Outline

I. Early farming practices
II. Mechanized farming
   A. Crop rotation
   B. Monoculture
III. Energy vs. labor
   A. Fertilizer
   B. Pesticides
   C. Insecticides
   D. Chlorinated hydrocarbons
   E. Organophosphates
   F. Problems with insecticides
      Box 14.1 Registration of pesticides
IV. Herbicides
   A. Organic farming
V. Fungicides and rodenticides
      Box 14.2 Agent Orange
VI. Integrated pest management
      Box 14.3 Food additives
VII. Consider this case study: Pesticides and pelicans

## Objectives

Understand that the native American form of agriculture did not permanently alter the ecosystem.

Explain how improper farming techniques destroyed the productivity of large portions of United States farmland.

List the social, economic, and governmental attitudes that encouraged the misuse of land.

Explain how the invention of new farm machinery encouraged the development of a monoculture type of farming.

List the advantages and disadvantages of monoculture farming.

Explain why chemical fertilizers are used.

Understand that while fertilizer does replace plant nutrients, it does not furnish organic materials for the soil, and the soil characteristics are altered.

Explain why modern agriculture requires the use of chemicals.

Explain how poor management of soil can have an impact on aquatic ecosystems.

Differentiate between hard pesticides and soft pesticides.

Define biological amplification.

Explain why integrated pest management depends upon a complete knowledge of the pest's life.

Explain how chemicals can be used to delay or accelerate the harvesting of a crop.

List the uses of food additives.

## Key Terms

algal bloom

auxins

biocide

biological amplification

chlorinated hydrocarbons

crop rotation

fungicide

hard pesticide

herbicide

insecticide

integrated pest management

monoculture

organophosphates

persistent pesticide

pest

pesticide

rodenticide

soft pesticide

target organism

# Chapter 14
# Crops and chemistry

## Early farming practices

Primitive people acquired their food by hunting for animals and gathering edible plants. Life began to change about 10,000 B.C. when humans first began to cultivate land in the Tigris and Euphrates river valleys. By 3000 B.C., cultivating crops and raising domestic animals were practiced throughout most of Europe and Asia. A system of agriculture developed that allowed farming in the same fields year after year. Early settlers brought these farming practices with them to the New World.

Although it was often thought of as only a hunting and gathering society, native American culture included many kinds of agriculture.

**Figure 14.1 Native American farmers.**
As seen in this illustration, native Americans depended upon hunting and farming as a source of food.

Farming evolved in both North and South America around the year 5000 B.C. Various tribes in the eastern portion of North America depended upon cultivating crops for as much as half of their food supply. Agriculture was such an integral part of the Huron tribal culture that the need for suitable land was a major factor in determining the location of their villages. Like many native Americans, the Hurons cultivated corn, beans, squash, sunflowers, and pumpkins. The common agricultural practice was to cut down the trees about one meter above the ground and burn the stumps, logs, and branches. The women would sow the seeds in the open area.

Preparing the land for farming was a difficult, time-consuming process. Once cleared, the cropland was kept free of weeds and maintained for a number of years. The tools were primitive, and the fields were small. (See fig. 14.1.) The land remained in cultivation until it became less productive. Then the village was moved to a new location.

Since only some of the one million native Americans engaged in agriculture, their gardens did little to disrupt the land. Once they abandoned a field,

**Figure 14.2 Early colonial farm.**
The trees were cut to provide an area for crops. The farmer planted crops among the stumps. This primitive method of farming was done entirely by hand.

natural succession restored it to its original state. The arrival of the early settlers and their use of a European form of agriculture eventually resulted in the disruption of the natural ecosystems of North America.

The virgin forests of the eastern United States occupied an area with rich, fertile soil. Accumulation of leaves and branches from the trees produced a thick layer of humus and rich topsoil. Settlers had to cut down the trees in order to plant their crops. They used some of these trees to build homes and to furnish fuel (see fig. 14.2), but most were simply burned. At first, the settlers adopted native American methods of planting corn and other crops in hills between stumps. Plows and horses were not used in this type of farming; the hoe was the most widely used farm implement.

After much labor, the stumps were eventually removed. This allowed use of the horse and plow in the same way they had been used in Europe. It was only natural for these settlers to duplicate the farming methods they had learned in the Old World. However, differences in the physical environment between Europe and the United States caused problems. In western Europe soils were mainly heavy clays, while the American soils were rich, loose-textured loams. In addition, while the rains in Europe were fairly regular and heavy rainstorms were rare, somewhat irregular rain and thunderstorms were common in the New World. Although colonial farmers harvested large yields for a few years, the exposed topsoil was quickly eroded by the violent thunderstorms. Within fifty years, much of the land in New England was no longer suitable for farming. The solution to this problem was simple. When the land

**Figure 14.3 The westward movement.** When soils were exhausted in the East, farmers could simply move out West and start new farms. Advertisements such as this one from the McCormick Reaper reflect this attitude.

"played out," the farmer merely moved west to a new tract of virgin soil and started over again. (See fig. 14.3.) However, farming practices were not changed. As a result, this land was also ruined, and the farmers moved west once more.

Everyone thought there was an endless amount of land in the New World. Few people gave any thought to the idea of conservation. Even the government's policies encouraged waste. In 1862, Congress passed the Homestead Act, which made public land available for many purposes, including farming. Anyone could file for 160 acres of public land. If they occupied it for five years, the land became theirs. Between 1860 and 1900, almost all of the public land east of the Mississippi had become privately owned.

## Mechanized farming

Beginning in the 1830s, homesteaders began to use machinery that was unavailable to the colonial farmer. In 1831, the reaper was invented. (See fig. 14.4.) By 1837, the self-cleaning steel plow was patented. These pieces of equipment enabled the farmer to become more labor efficient in tilling and harvesting grain. As these and other labor-saving devices were introduced on the farm, it became possible for a smaller percentage of the population to engage in growing the necessary food.

Figure 14.4 **Progress.** Before the invention of the reaper, the harvesting of grain was done completely by hand. The horse-drawn reaper enabled the farmer to be more productive.

## *Crop rotation*

Almost 100 percent of the early settlers were farmers, but by 1900 only 40 percent of the nation's population were farmers. Mechanization and modern farming practices allowed fewer farmers to feed more people. However, most of the farms were still family enterprises. The family farm was often divided into four fields—one field was used to grow a row crop such as corn, a second was used for a grain crop such as wheat, a third for first-year hay, and the fourth for second-year hay. **Crop rotation** was an accepted practice. (See fig. 14.5.)

A typical rotation practice was to plant a cornfield in the spring. The corn was harvested in the fall, and wheat was then planted in that field. In the second year, the wheat was harvested, and the field was given over to hay. This hay crop was usually a mixture of grasses and legumes. The hay was harvested for two years and supplied needed food for horses and cattle. After the second year, the hayfield was plowed, which returned organic material and nutrients to the soil. Corn would be planted in this field the next spring. The corn and wheat furnished the farm with a cash crop (sold for money). Often the grain was fed to animals, which also were a source of income.

**Figure 14.5 Crop rotation.**
In the early 1900s in the United States, a section of land (250 hectares) often supported four farm families. The crops in the fields were changed each year. A row crop—corn, sugarbeets, or soybeans—was followed the next year by a grain crop—wheat or oats. This was followed by a year or two of hay or other grasses. Such farming practices in small fields reduced the pest population and maintained the fertility and structure of the soil.

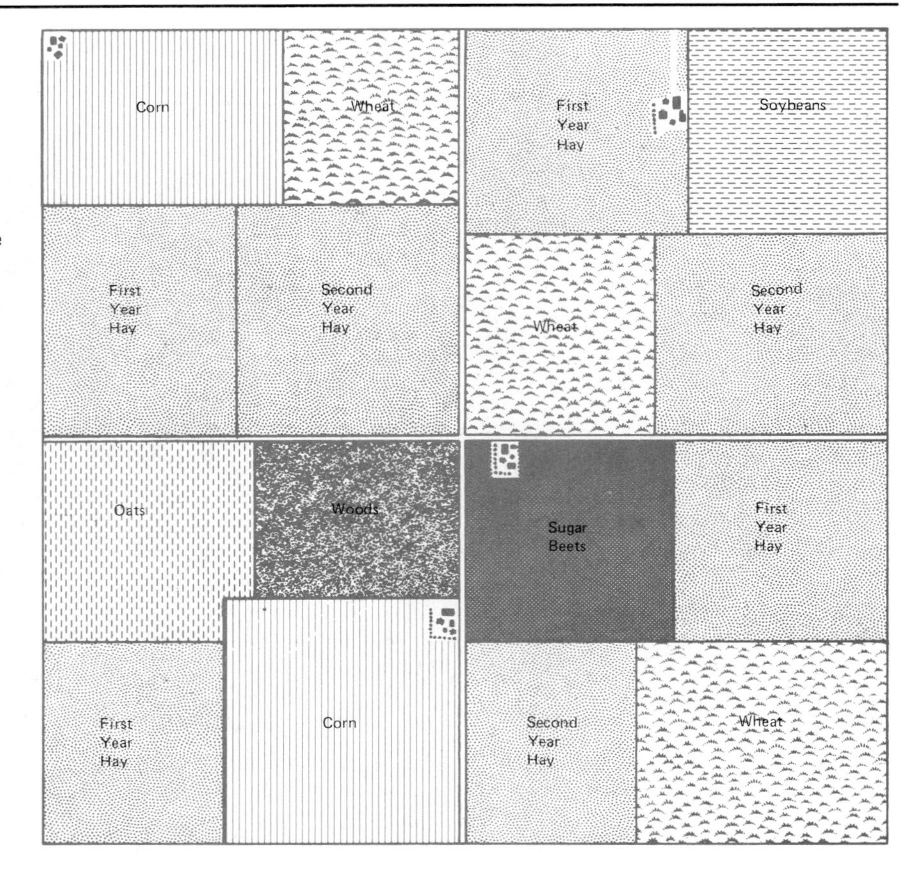


This system of crop rotation replaced organic material and prevented the depletion of essential nutrients from the soil. However, in the 1920s, when the tractor replaced the horse, there was less need for animal feed. The amount of land devoted to hay declined, and the conversion of hayfields to cash crops resulted in less crop rotation.

## *Monoculture*

The tractor also enabled the farmer to farm larger fields and therefore be more productive. These large fields were often planted in a single crop, a practice called **monoculture.** The "wheat belt" or the "corn belt" represent areas of monoculture agriculture. (See fig. 14.6.)

Even though monoculture is an efficient method of producing food, it is not without serious faults. With monoculture, vast amounts of land are plowed, planted, and harvested at the same time. This is made possible by the use of large farm machines. When large tracts of land are not covered by vegetation, soil erosion increases. The Dust Bowl of the 1930s occurred because of the large amount of unprotected plowed land in the wheat belt.

**Figure 14.6 Monoculture.** This wheat field is an example of monoculture. This kind of agriculture is highly mechanized and requires large fields. Monoculture allows for large yields with a small amount of human labor.

The monoculture method of farming results in the removal of much of the organic material each year. Today there are few special crops grown to replace this organic matter. This alters the physical, chemical, and biological properties of the soil. In addition, when large amounts of land are planted with the same crop year after year, the likelihood of disease and insect damage increases. With monoculture, pests have a vast source of food. This allows pest populations to increase rapidly unless extensive use of chemical control agents or other costly measures are employed.

To insure that a crop can be planted, tended, and harvested efficiently, farmers rely on hybrid varieties with very little genetic diversity. When all of the farmers in an area plant the same varieties, pest control becomes a serious problem. If diseases or pests begin to spread, the magnitude of the problem becomes devastating, because all of the plants are susceptible to the same diseases. If genetically diverse crops were planted, this problem would not be as great. Crop rotation further reduces pest problems and returns organic matter to the soil, but our society depends upon monoculture farming to provide for our food needs.

**Table 14.1**
Fertilizer and food production

| Country | Percentage of work force in agriculture | Fertilizer (Kg/ha) | Yields 100 Kg/ha | |
|---|---|---|---|---|
| | | | Wheat | Rice |
| United States | 3 | 83 | 21 | 51 |
| Brazil | 39 | 20 | 10 | 16 |
| India | 64 | 11 | 12 | 17 |

Source: Population Reference Bureau (1981)

## Energy vs. labor

The family farms of the early 1900s were labor intensive. The entire family was involved in the manual labor of raising crops. In 1913, it required 135 hours of labor to produce 2,500 kilograms of corn. Today, it requires 15 hours of labor to produce 2,500 kilograms of corn. The energy supplied by fossil fuels replaced the equivalent of 120 hours of labor.

It has been estimated that if we were to return to a form of farming that was solely dependent upon human and animal energy, 70 percent of the United States work force would be required to produce the needed food. Mechanized farming is an agricultural system that is heavily subsidized by fossil fuels. Any change in the availability of these fuels for agricultural requirements could change our way of life. Energy is needed for tilling, planting, harvesting, transportation, pumping water for irrigation, and the production and application of fertilizers and pesticides.

### *Fertilizer*

The elimination of crop rotation practices in favor of monoculture requires more fertilizer. It requires about five metric tons of fossil fuel to produce about one metric ton of fertilizer.

Table 14.1 shows that we use eighty-three kilograms of fertilizer per hectare of arable land. By comparison, Brazil uses 39 percent of its labor force in farming and uses only twenty kilograms of fertilizer per hectare of arable land. Without the use of chemical fertilizers, modern agriculture couldn't produce the amount of food we require. Approximately one quarter of the world's crop yield is directly attributed to the use of chemical fertilizers. Thus, if the world stopped using chemical fertilizers, food production would decline by 25 percent. At present, the world demand for fertilizer is doubling every ten years. The increased demand for fertilizer is part of our current energy problem. A lack of energy for the production of fertilizer would result in severe famine in parts of the world.

Chemical fertilizers replace soil nutrients removed by plants. The nutrients become a part of the plant body. If the plant body is removed from the field, these soil nutrients are also removed. Certain other micronutrients may also become a limiting factor if the same crop is planted year after year. Some

**Figure 14.7 Harvest time.**
Whenever crops are harvested, some of the elements are removed from the field. Today, the most practical method of replacing these elements is through the use of chemical fertilizers.

plant nutrients, such as carbon, hydrogen, and oxygen, are easily replaced by carbon dioxide and water. Others, such as nitrogen, phosphorus, potassium, and some trace elements, are less readily replaced and are often in short supply. It is these elements that are replaced by chemical fertilizers. When purchasing fertilizer, a farmer is buying primarily nitrogen, phosphorus, and potassium. Trace elements, which are required in very small amounts, can also be applied to the soil if needed. For example, harvesting a metric ton of potatoes removes over ten kilograms of nitrogen from the soil but only thirteen grams of the trace element boron. (See fig. 14.7.)

Although chemical fertilizers replace inorganic nutrients, they do not replace organic material. Organic material (humus) modifies the structure of the soil. It prevents soil compaction and maintains pore spaces. Organic matter is also important to maintain proper soil chemistry. Soil bacteria and other organisms use the organic material as a source of energy. Thus, total dependency upon chemical fertilizers changes the physical, chemical, and biotic properties of the soil.

As water moves through soil, it gathers soil nutrients and carries these nutrients to streams and lakes. Widespread use of fertilizers and heavier and more frequent applications have resulted in large amounts of nutrients entering lakes and streams. Large amounts of phosphorus and nitrogen compounds cause increased growth of algae. A rapid accumulation of algae is known as an **algal bloom.** When these cells die, they are decomposed by decay bacteria, in a process that uses oxygen. The reduction in the dissolved oxygen content of the water can severely stress or kill other aquatic organisms. In addition, unpleasant odors are often associated with this decay, and water quality is diminished.

### *Pesticides*

In addition to chemical fertilizers, mechanized monoculture requires the use of large amounts of other agricultural chemicals, such as pesticides. A **pesticide** is a chemical that kills a pest. A **pest** is considered to be any unwanted organism. A safe, efficient pesticide is one that controls only a specific **target organism** and does not harm any nontarget forms of life.

Pesticides are named according to the type of target organism they are supposed to control. **Insecticides** are supposed to kill insects. The growth of fungus is controlled by **fungicides.** Mice and rats are killed by **rodenticides. Herbicides**

## Box 14.1
## Registration of pesticides

*The Environmental Protection Agency (EPA) is charged by Congress to protect the nation's land, air, and water systems. Under a mandate of national environmental laws focused on air and water quality, solid waste management and the control of toxic substances, pesticides, and noise and radiation, the Agency strives to formulate and implement actions which lead to a compatible balance between human activities and the ability of natural systems to support and nurture life.*

In order to fulfill this mandate, the EPA is charged with the enforcement of various federal laws and acts. One such act governs the registration of pesticides.

Section 3 (c)(1) of the Federal Insecticide, Fungicide, and Rodenticide Act (FIFRA) states:

*Procedure for Registration—*

1. *Statement required. Each applicant for registration of a pesticide shall file with the Administrator a statement which includes—*
   (A) *the name and address of the applicant and of any other person whose name will appear on the labeling;*
   (B) *the name of the pesticide;*
   (C) *a complete copy of the labeling of the pesticide, a statement of all claims to be made for it, and any directions for its use;*
   (D) *except as otherwise provided in subsection (c) (2) (D) of this section, if requested by the Administrator, a full description of the tests made and the results thereof upon which the claims are based, or alternately a citation to data that appears in the public literature or that previously had been submitted to the Administrator and that the Administrator may consider in accordance with the following provisions;*
   (E) *the complete formula of the pesticide; and*
   (F) *a request that the pesticide be classified for general use, or restricted use, or for both.*

If a corporation wants to manufacture and market a pesticide within the United States, the pesticide must not adversely affect the environment. There is no specific format to test a product's effects on the environment, but eliminate undesirable vegetation. Unfortunately, most pesticides are not very specific and will affect a wide variety of nontarget organisms. Therefore, they could be called broad spectrum **biocides.**

### *Insecticides*

Insects have always consumed a portion of the human food supply. As long as there was an ample supply of food for both humans and insects, there was no problem. When this competition became intense, humans sought methods of reducing insect populations.

some of the tests conducted include the following:

1. *Degradation*—A determination of the physical and chemical methods of decay and an identification of the decay products.
2. *Metabolism*—A determination of the organisms that act on the pesticide and the metabolic products released.
3. *Mobility*—A determination of where the molecules are likely to go and what route they will follow in the ecosystem.
4. *Accumulation*—Do the pesticide molecules accumulate in living tissue? If so, in what form and in what quantities?
5. *Hazard*—Evaluation of the possible hazard to plants, microorganisms, wildlife, and humans, whether target or nontarget organisms.

After receiving the information submitted in the procedure for registration, the Administrator of EPA may register the insecticide for use under the provisions of Section 3(c) (5) of the FIFRA:

*Approval of Registration—*
*The Administrator shall register a pesticide if he determines that, when considered with any restrictions imposed under subsection (d)—*

(A) *its composition is such as to warrant the proposed claims for it;*
(B) *its labeling and other material required to be submitted comply with the requirements of this Act;*
(C) *it will perform its intended function without unreasonable adverse effects on the environment; and*
(D) *when used in accordance with widespread and commonly recognized practice it will not generally cause unreasonable adverse effects on the environment.*

The administrator also publishes all applications and supporting data so that various governmental agencies and the public have an opportunity to comment.

The FIFRA is designed to protect the environment from being damaged by the use of pesticides. The law places the burden upon the manufacturer to prove that a pesticide is safe. One source estimates that eight to twelve years and $10 million to $15 million are needed to bring a major new pesticide from discovery to first registration. These figures *do not* include the capital cost of constructing a facility to manufacture the pesticide.

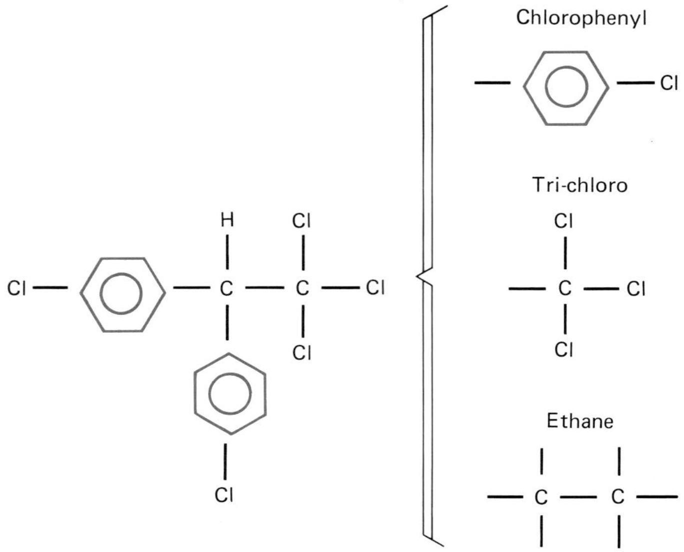

**Figure 14.8 Structural formula of DDT.**
This diagram shows the arrangement of the atoms in a molecule of DDT.

Chlorophenyl

Tri-chloro

Ethane

In addition to crop damage, insects harm humans because they are often carriers of diseases. Sleeping sickness, bubonic plague, and malaria are some examples of human diseases transmitted by insects. Mosquitos carry over thirty diseases. It is estimated that within the first ten years of use (1942 to 1952) DDT saved five million human lives. In spite of this success, mosquitos are still a problem. In 1965, mosquitos were the cause of carrying disease to 100 million people. The desire to increase the human food supply and decrease human illness encouraged the development of synthetic pesticides, which were originally hailed as a major advancement. Nearly three thousand years ago, the Greek poet Homer mentioned the use of sulfur to control insects.

The first synthetic inorganic insecticide, Paris green, was formulated in 1867. This mixture of acetate and arsenide of copper was used to control Colorado potato beetles. Prior to the development of synthetic organic pesticides in the 1940s, the organic pesticides used were the natural products of plants—nicotine from tobacco, rotenone from tropical legumes, and pyrethrum from chrysanthemums. However, because they were difficult to apply and their effect was short-lived, research was conducted to find better methods of insect control. Today there are over sixty thousand compounds synthesized for use as insecticides. These can be divided into two basic categories—chlorinated hydrocarbons and organophosphates.

## *Chlorinated hydrocarbons*

**Chlorinated hydrocarbons** are a group of pesticides that contain carbon, hydrogen, and chlorine and that have a complex, stable structure. Therefore, they are difficult for organisms to break down, and they tend to accumulate in ecosystems. The best known example of a chlorinated hydrocarbon is dichlorodiphenyltrichloroethane (DDT). (See fig. 14.8.) Other forms are

**Table 14.2**
A list of pesticides that are used to control organisms considered to be harmful or undesirable

| | Example | Application | Action on target organism |
|---|---|---|---|
| **Herbicide** | | | |
| Artificial auxins | 2, 4-D MCPA | Broad-leaf weeds | Starvation |
| | 2, 4, 5-T | Woody plants | Certain plant enzymes convert this into a toxic material |
| Chlorophyll destruction | Monolinuron, linuron | Broad-leaf weeds and grasses | Interferes with photosynthesis |
| | Fenuron, monuron, diuron, nebrun | Any vegetation | Interferes with photosynthesis |
| **Fungicide** | | | |
| Copper mixture | Bordeaux mixture | Potato blight | Poisons germinating fungal spores |
| Phenolic compounds | Dinocap, binapacryl, thiram | Powdery mildew | Kills overwintering fungi |
| **Insecticide** | | | |
| Organochlorides | Chlordane, DDT, dieldrin, heptachlor | Large variety of insects | Interferes with nervous reaction |
| Carbamate | Dimetelan | Houseflies | Interferes with cholinesterase enzyme used for nervous reactions |
| | Aminocarb | Snails, slugs | Interferes with cholinesterase enzyme used for nervous reactions |
| | Propoxur | Mosquitos, ants, flies, cockroaches | |
| Organophosphates | Parathion, malathion, schradan, demeton | Wide range of insects, aphids, mites | Interferes with cholinesterase enzyme used for nervous reactions |
| **Molluscides** | | | |
| | Metaldehyde | Snails | Dehydrates |
| **Rodenticides** | | | |
| | Warfarin | Rodents | Prevents blood clotting |

chlordane, aldrin, heptachlor, dieldrin, and endrin. How chlorinated hydrocarbons work is not fully understood. It is believed that the animal's nervous system is affected by these compounds.

Another major problem is that chlorinated hydrocarbons are **persistent pesticides** and remain active in the environment for a long time. DDT has a half-life of ten to fifteen years. (The half-life is the amount of time required for half of the chemical to decompose.) The half-life varies depending on soil type, temperature, the kinds of soil organisms present, and other factors. If a metric ton of DDT were sprayed over an area, ten to fifteen years later five hundred kilograms would still be present in that area. Thirty years from the date of application, two hundred fifty kilograms would still be present. As a result, these **hard pesticides** tend to accumulate in the soil and food chain. This has led to the search for **soft pesticides,** which decompose to harmless products in a short time. (See table 14.2.) However, like hard pesticides, the soft pesticides also kill nontarget organisms. They are more toxic to humans and other vertebrates than are hard pesticides. The main disadvantage of soft pesticides from the farmer's point of view is that they are only effective for a short time. Therefore, more frequent applications are needed. This requires more labor, fuel, and expense.

## *Organophosphates*

**Organophosphates** are soft pesticides, which function by disrupting the nervous system of an organism. A nerve impulse is conducted from one nerve to another by means of a chemical known as acetylcholine. When this chemical is produced at the end of one nerve cell, it causes an impulse to be passed to the next cell, thereby transferring the nerve message. As soon as this transfer is completed, an enzyme known as cholinesterase destroys the acetylcholine, and the nerve cell is ready to receive another message.

Organophosphates destroy the enzyme cholinesterase. Therefore, the acetylcholine is not destroyed, and the nerve cell is continuously stimulated. As a result, the animal goes into a state of nervous uncoordination that results in death. Although organophosphates are not as persistent as chlorinated hydrocarbons, they vary from moderately to highly toxic. Because they have a short half-life, the organophosphates are not a long-term problem in the environment. In addition to organophosphates, a group of nerve poisons known as carbamates are used to control some pests. They are not persistent, having a half-life of only a few days or weeks.

Parathion and malathion are both organophosphates. Malathion is widely used for such projects as mosquito control, but parathion is a restricted organophosphate because of its high toxicity to humans.

## *Problems with insecticides*

Although there has been a trend away from using hard pesticides, several are still used in the United States. Many other countries have not prevented the use of the worst of the persistent pesticides. Because of their stability, these chemicals have become a permanent problem. They may be transported to other parts of the world from the place where they were originally applied. For example, persistent insecticides can become attached to small soil particles. Wind or water erosion may move these soil particles and the attached insecticides to any environment in the world. Thus, chemicals originally sprayed on a Midwest cornfield may be carried into the ocean and eventually become a problem in the Arctic or Antarctic ecosystems. Another problem associated with this persistence is that the insecticide may accumulate in the food chain and kill organisms at higher trophic levels, even though organisms at lower trophic levels are not injured. This is called **biological amplification.**

When DDT is applied to an area, it is usually dissolved in an oil or fatty compound. It is then sprayed over an area and falls on plants that the insect population uses for food, or it may fall directly on the insect. The insect takes the DDT into its body, where it interferes with the normal activity of the organism. If small quantities are taken in, the insect will digest and break down the DDT as it would any other organic chemical compound. Since the DDT is soluble in fat or oil, the insect stores the DDT or its broken-down products in the fatty tissues of its body. If an area has been lightly sprayed with the pesticide, some insects will die, but others will be able to tolerate the DDT. As much as one part per billion DDT or its breakdown products may be stored in insect tissue. This is not very much DDT!

If an aquatic habitat were sprayed with a small concentration of DDT, simple water organisms may accumulate a concentration of DDT that is up to two hundred and fifty times greater than the concentration of DDT in the surrounding water. These organisms are eaten by shrimp, clams, and small fish, which are in turn eaten by larger fish. If you measure the DDT in the large fish, it may be two thousand times more concentrated than the original concentration in the water. Birds that feed on fish may accumulate as much as eighty thousand times the original concentration. (See fig. 14.9.) What was a very small initial concentration has now become so concentrated that it could be fatal to animals at higher trophic levels.

Birds generally cannot tolerate high levels of DDT. DDT interferes with the production of the eggshells, and the resulting eggs are much more fragile and easily broken. This problem is more common in carnivorous birds because they are at the top of the food chain.

DDT was originally used to kill insects responsible for spreading malaria and yellow fever. Instead of just killing the pests, it selects for survival those insects that can tolerate it. Instead of harming only the insect pests, it kills useful insects, accumulates in other forms of life, and interferes with reproduction in birds. It presents serious health concerns to humans, who may have accumulated up to three to five parts per million DDT or its breakdown products in fat cells. All in all, DDT has not been the gift that it was originally thought to be. In the early 1970s, the sale of DDT was prohibited in the United States. Heptachlor, aldrin, and dieldrin, which have similar properties to DDT, have also been banned.

In addition to DDT, other stable compounds can become amplified as they pass through food chains. Mercury, aldrin, chlordane, and PCBs are common examples.

Another problem associated with insecticides is the ability of insect populations to become resistant. All organisms within a given species are not identical. Each individual has a slightly different genetic composition from other members of the same species. If an insecticide is used for the first time on a particular insect, it kills all the insects that are genetically susceptible. However, the individuals with a different genetic composition may not be killed by the pesticide.

Suppose that in a population of insects, 5 percent of the individuals possess genes that make them resistant to an insecticide. The first application of the insecticide would, therefore, kill 95 percent of the population. However, the tolerant individuals would then constitute the majority of the breeding population. This would mean that many insects in the second generation would be tolerant. The second use of the insecticide on this population would not be nearly as effective as the first.

**Figure 14.9 Biological amplification.** Note how the concentration of DDT increases as it passes through the food chain. (Source: Brookhaven National Laboratory)

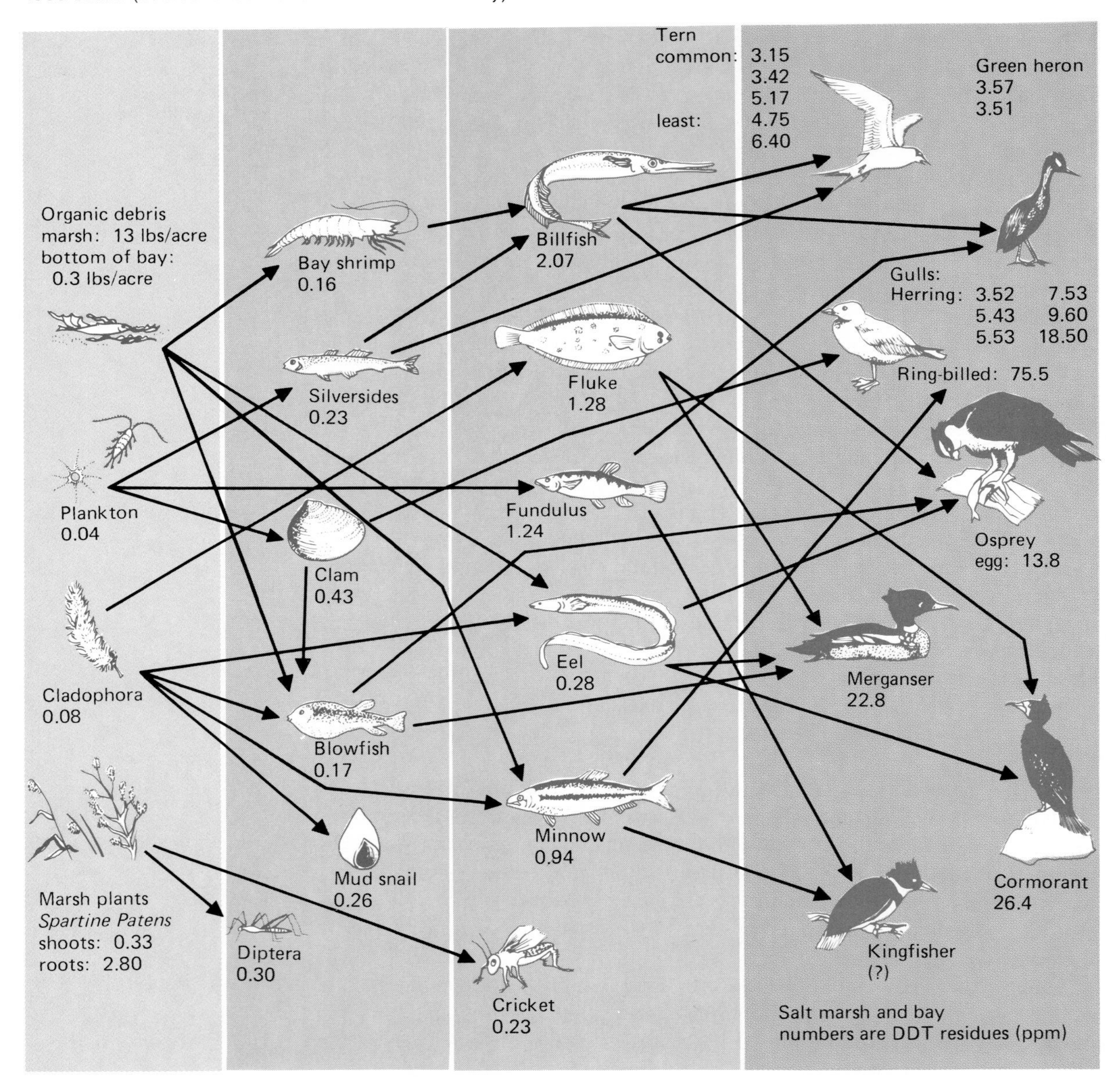

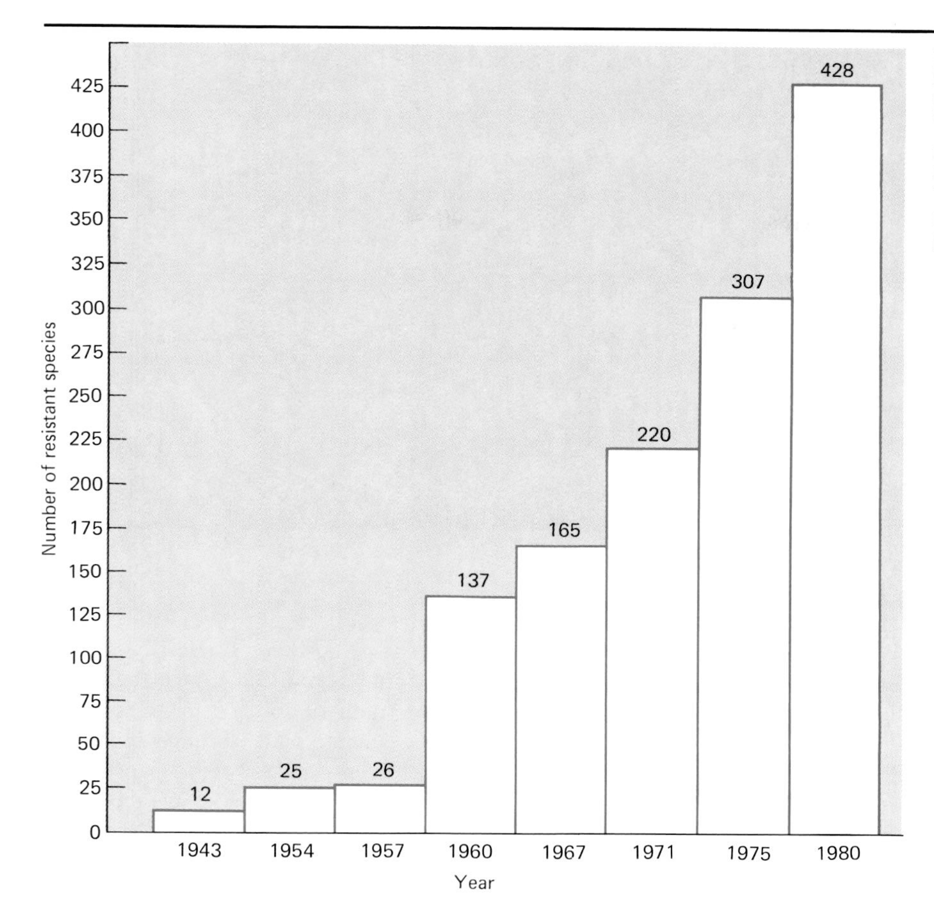


**Figure 14.10 Resistance to insecticides.**
The continued use of pesticides has increased the number of resistant species. This reduces the effectiveness of the pesticides.

Certain species of insects will produce a generation each month. If the insecticide continues to be used, each generation will become more tolerant. In organisms with a short generation time, 99 percent of the individuals can become resistant to an insecticide within five years. As a result, that particular insecticide is no longer effective on that species of insect. Figure 14.10 indicates that over four hundred species of insects have populations that are resistant to insecticides.

Because most insecticides are not specific to only pest insects, many beneficial organisms are killed with the pest species. This often removes natural controls on pest species. With the loss of natural controls, such as predators and parasites, the only control on the pest is insecticides.

**Figure 14.11 Herbicide damage.**
These weeds have been treated with a mixture of herbicides. One of the herbicides stops photosynthesis and the other interferes with germination.

## Herbicides

Another major class of chemical control agents is herbicides. (See fig. 14.11.) Herbicides are widely used to control unwanted vegetation. Since the beginning of agriculture, farmers have sought methods of controlling weed populations. A weed is a plant growing in an undesirable place. Bluegrass in your lawn is desirable, but bluegrass in a cornfield is a pest, a weed.

During the growing season, weeds compete with crops for space, nutrients, and water. At harvesttime, the weeds and crops must be separated. Originally, this process was accomplished through human labor and the use of a hoe or similar implement. Eventually, horses and tractors supplied the energy to physically remove the weeds. These methods are either labor or energy intensive. Therefore, research was conducted to discover methods to control the growth of weeds. Today, chemicals are widely used to control weeds and manipulate plant processes.

**Auxins** are plant growth regulators. Two common synthetic plant growth auxins are 2, 4-dichlorophenoxyacetic acid (2, 4-D) and 2, 4, 5-trichlorophenoxyacetic acid (2, 4, 5-T). When applied to broadleaf plants, these chemicals cause the metabolism of the plant to exceed its food-producing potential. As a result, the plant grows so rapidly that it exhausts its food reserves.

In addition to being used as a herbicide, synthetic auxins are used to facilitate the mechanical harvesting of cotton. A synthetic auxin that causes leaves to fall off is sprayed on the plants. This reduces clogging in the mechanical picker.

Many other chemicals are also used in farming. In addition to using fertilizers to increase yields and pesticides to control plant and animal pests, farmers use various chemicals in harvesting crops. By using NAA (naphthaleneacetic acid), a fruit grower can prevent immature apples from dropping

**Figure 14.12**
**Mechanical harvesting.**
By using chemicals and machinery, this farmer can rapidly harvest the cherry crop. This practice reduces the amount of labor required to pick the cherries but requires the application of chemicals to loosen the fruit. It also requires fossil fuel energy to run the machinery.

from the trees and being damaged because of bruising. This practice keeps the apples on the trees for up to ten extra days. This allows for a longer harvest period. Under other circumstances, it may be advantageous to promote an earlier harvest.

By applying the chemical Ethephon to cherries, the grower promotes loosening of the fruit and facilitates mechanical harvesting. (See fig. 14.12.) The use of Ethephon and mechanical harvesting lowers the labor cost. Chemicals are also used to prolong the storage life of fruits and vegetables after they are harvested.

Herbicides are also widely used in nonagriculture areas. Utility companies, railroads, and highway departments use herbicides to control plant growth on their rights-of-way. (See fig. 14.13.)

A variety of chemicals disrupts the photosynthetic activities of a plant. Depending on the concentration, some of these are toxic to all plants, and some are very selective about the plant species they affect. One such herbicide is diuron. At the proper concentration, this herbicide is used to control annual grasses and broadleaf weeds in over twenty crops. At a high concentration, it is used to kill all vegetation in an area.

Fenuron is a herbicide that kills woody plants. In low concentration, it is used to control woody weed plants in cropland. In high concentrations, it is used on noncroplands such as power line rights-of-way. In addition, the farmers can mix two or more herbicides in a combination that is effective against specific weeds.

**Figure 14.13**
**Herbicides.**
The use of chemicals for plant control is not limited to croplands.

**Table 14.3**
Yield differences between organic and conventional farming (metric tons/hectare)

| Crop | Organic | Conventional |
|---|---|---|
| Corn | 6.45 | 7.00 |
| Soybeans | 2.44 | 2.57 |
| Wheat | 1.88 | 3.28 |

## *Organic farming*

Before the invention of synthetic fertilizers, pesticides, herbicides, and the wide variety of other chemicals, all farming was organic. However, the initial advantages of various chemicals changed farm practices. The use of chemical fertilizers replaced manure as a source of soil nutrients. The reduction in the number of animals used on the farm required less land to produce hay. This allowed farmers to grow more "cash crops" (such as corn, wheat, and cotton), which usually resulted in a greater pest problem. Chemical pesticides helped to solve this problem.

Just as there were a number of factors that caused the change from organic farming to modern farming, there are factors that may be causing some farmers to return to the organic farming concept. These factors include the increased price of fuel, high costs of agricultural chemicals, and concern about the problems caused by the release of chemicals into the environment.

A recent study by Lockeretz, Shearer, and Kohl lists some comparisons between conventional and organic farming. This study found that organic farms used conventional cultivation instead of chemicals for weed control, crop rotation as the main method of pest control, and manure instead of chemicals to restore soil fertility. From an economic viewpoint, the conventional farmer received a higher return, netting $384 per hectare as compared to $333 per hectare for the organic farmer. The crop yield for the two types of farms is given in table 14.3. One of the greatest differences is in energy consumption. The organic farm used only 40 percent of the energy used by conventional farming methods.

There are problems associated with a return to organic farming. The use of legumes in crop rotation reduces the amount of land in cash crops, but this is offset by an increase in the number of cattle raised on the hayfields. Critics of organic farming say that such methods cannot produce the amount of food required for today's population. Proponents disagree and stress the fact that organic farming reduces the release of chemicals into the environment and reduces soil erosion. Regardless, it is evident that as fuel and chemical costs increase, organic farming may be viewed as one alternative to reducing cost.

## Box 14.2
## Agent Orange

During its involvement in the conflict in Southeast Asia, the United States used more than 105 million metric tons of herbicides to defoliate large areas so that the North Vietnamese and Viet Cong soldiers could not use these areas as hiding places. Herbicides were also used to kill crops that supplied food for these soldiers.

The most common chemical, Agent Orange, was an equal mixture of 2, 4-D (2, 4-dichlorophenoxyacetic acid) and 2, 4, 5-T (2, 4, 5-trichlorophenoxyacetic acid). (The herbicide got its name from the bright orange stripes on the steel drums that contained it.) The United States first sprayed Agent Orange in 1965 and eventually sprayed over 2,400,000 hectares of land.

Despite the vast amounts of money and effort that were expended, the defoliation program was a failure in most respects. It did have a detrimental effect on the native people, however, because their crops were destroyed. Some of the 2, 4, 5-T was contaminated with an extremely toxic compound known as dioxin (2, 3, 7, 8-tetrachlorodibenzo-p-dioxin) or TCDD. Exposure to Agent Orange and its contaminants has been linked with skin disorders, psychological problems, liver damage, cancer, and birth defects. By the time the United States stopped using Agent Orange in 1970, between 50,000 and 60,000 members of the U.S. armed forces were exposed to the chemical, as were thousands of Vietnamese citizens.

Despite dioxin's adverse effect on laboratory animals, it has not been conclusively proven that dioxin-contaminated Agent Orange has the same effect on humans. In humans, it has not been found to be directly responsible for ailments more serious than chloracne, a disfiguring skin problem. But after a 1976 explosion at an Italian chemical plant, which spread dioxin over a village and resulted in chloracne and the widespread death of animals, many veterans became convinced that Agent Orange was responsible for most of their ailments.

After many years of litigation, the issue reached the federal court in 1984. Before the case was to be heard, the seven chemical companies who produced Agent Orange agreed to place $180 million into a fund that will be used to compensate victims and their families. The fund is expected to last twenty-five years. The chemical companies denied any liability for the veterans' illnesses; their position was that Agent Orange had not caused the health problems and that they had merely manufactured the defoliant according to the military's instructions. The companies also maintain that if herbicides are correctly formulated, no contamination by dioxin will occur. Furthermore, they say that if the herbicides are used according to the specifications published by the manufacturers, they are safe and valuable agricultural chemicals.

## Fungicides and rodenticides

Fungi are natural decomposers that break down complex organic compounds into simpler materials. Other fungi are parasites on plants. Fungicides are used to control the spoilage of fruits and vegetables and to eliminate the parasites. One common fungicide—methylmercury—is widely used for improving seed survival. When these treated seeds are planted, the mercury is incorporated into the environment.

Like fungi, rodents are harmful because they destroy food supplies. Several different kinds of poisons have been developed to control rodents. However, the best control is to prevent their access to the food supply.

## Integrated pest management

The perfect pesticide would have the following four characteristics:

1. It would be inexpensive;
2. It would affect only the target organism;
3. It would have a short half-life; and
4. It would break down into harmless materials.

The perfect pesticide has not yet been created, but many other techniques have been used that seem to be more harmonious with natural ecosystems. These techniques are known as **integrated pest management.**

That the agricultural community has become heavily dependent upon chemical pesticides is not surprising. In the United States, 33 percent of the crops are consumed by pests. (The loss worldwide is 35 percent.) This represents an annual loss of $18.2 billion in the United States alone. The farmers, grain storage operators, and the food industry continually seek methods to reduce this loss.

The chemical industry emerged with a method of preventing this loss: chemical control. Thousands of salespeople touted the benefits of their products. They promised a "spray and save" program. But as we have seen, this method has its shortcomings. Agriculture is seeking a better method of control.

As the name implies, integrated pest management depends upon the use of a number of methods of control, rather than a single method. It depends upon a complete understanding of all ecological aspects of the crop and the particular pest to which it is susceptible. Information about the time of hatching, other hosts used, overwintering survival rates, natural enemies, and many other factors must be considered in using integrated control. Chemical pesticides also have a role in this type of control.

An outstanding example of the use of integrated pest management is the Canete Valley in Peru. This valley consists of twenty-two thousand hectares of cultivated land surrounded by an arid region. Since 1920, about 60 percent of the land has been planted in cotton. As a result of this monoculture approach, cotton pests increased rapidly. Until 1949, heavy metals and natural

**Figure 14.14 Integrated pest management.**
As a result of integrated pest management, the Peruvian farmers in this valley had a 30 percent increase in cotton production.

organic insecticides were used to control the insects. Beginning in 1949, synthetic organic pesticides were used to control the insects; the results were disastrous.

The wide use of synthetic chemicals reduced the number of natural insect predators and created resistant strains of pests. At first the chemicals were applied every fifteen days; this was shortened to every eight days, and then to every three days. This dependency on chemicals became very expensive. In addition, insects that had not previously been considered pests now began to threaten cotton plants. In spite of the chemicals, in 1956 the cotton yield dropped to its lowest level in the decade.

The farmers of Canete Valley were organized to implement an integrated pest management system, which included strict planting and harvesting dates that deprived the pests of a food supply during part of the year. In addition, marginal farmland was taken out of production so only the most healthy plants were grown. Irrigation practices were altered to hinder pest population growth. Synthetic pesticides were banned or reduced, and natural predators were encouraged. After six years of integrated pest management, cotton production increased almost 30 percent. (See fig. 14.14.)

One aspect of integrated pest management is to employ specific biological techniques. These include sex attractants, male sterilization, release of natural predators or parasites, and the development of resistant strains of crop organisms. In some species of insects, a chemical is released by the female to attract the males. Males can detect the presence of a female for a distance of as much as three kilometers. Gyplure is a synthetic sex attractant that has proven very successful in controlling the gypsy moth. Male moths are attracted to the scent and are lured into traps. With the males inside the traps and the females outside, the reproductive rate drops, and the insect population is brought under control.

Another technique that reduces reproduction is male sterilization. In southern United States, the screwworm fly weakens or kills large grazing animals, such as cattle, goats and deer. The female screwworm fly lays eggs in open wounds, and the larvae feed on the animal. Control of the screwworm fly was accomplished by releasing huge numbers of sterile males. Since a female screwworm fly mates only once, a female mated with a sterile male will not produce offspring. In Curacao, an island sixty-five kilometers west of Venezuela, a program of introducing sterile male screwworm flys eliminated this disease from the twenty-five thousand goats on the island. In parts of southwestern United States, the sterile male technique has been very effective; however, because of the reintroduction of screwworm flys from Mexico, periodic releases of sterile males are required. Male sterilization uses a natural characteristic of the species to control a pest population. This control is made possible by our knowledge of the biology of the screwworm fly (females mate only once).

Manipulation of predator-prey relationships can be used to control pest populations. For instance, the ladybird beetle is a natural predator of aphids and scale insects. (See fig. 14.15.) Artificially increasing the population of ladybird beetles is one method of controlling aphid and scale populations. In California during the early 1900s, scale insects on orange trees greatly affected the health of the trees and reduced crop yields. The introduction of a species of ladybird beetles from Australia quickly brought the pests under control. Years later, when chemical pesticides were first used in the area, so many ladybird beetles were accidentally killed that the scale insects once again became a serious problem.

In 1961, the grape growers in California's San Joaquin Valley were troubled by a grape leaf hopper. To combat this pest, the growers applied DDT. However, the leaf hopper quickly developed a resistance to the DDT, and other insecticides had to be employed. Thus, in 1961 the growers spent over $8 million to control leaf hoppers and only established a resistant population.

Continued research into the leaf hopper problem revealed that a particular species of parasitic wasp was the natural predator of the leaf hopper. The female wasp deposits an egg on the egg of a leaf hopper. The larval wasp used the leaf hopper egg as a source of food. In one growing season, the wasp produces nine or ten generations compared to three for the leaf hopper. This keeps the leaf hopper under control.

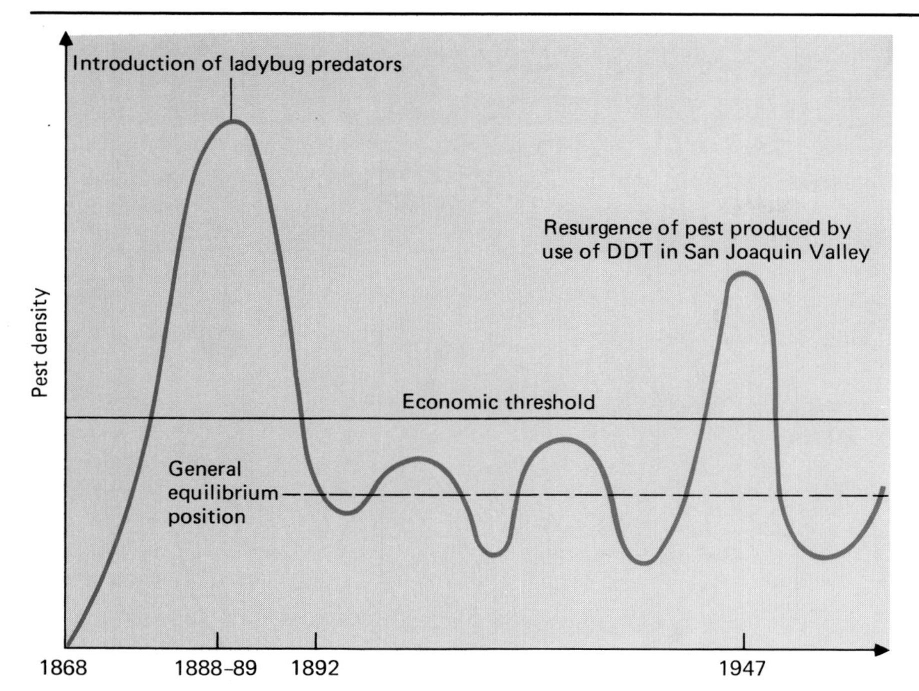


**Figure 14.15 Insect control with natural predators.**
In 1889, the introduction of ladybugs brought the cottony cushion scale under control. In the 1940s, DDT reduced the population of ladybugs and the cottony cushion scale population increased. Stopping the use of DDT allowed the ladybug population to increase and this reduced the pest population.

It was also discovered that the grape leaf hopper spent the entire year in the vineyards. The wasp only lived in the vineyards during the summer. In the winter, the wasp required an alternate host. The alternate host was a non-economic leaf hopper species normally found on blackberry bushes. In the natural ecosystem or when grapes were grown on a limited area, the parasitic wasp could readily move from the blackberries to the grapes. With the establishment of a monoculture of grapes in much of the valley, most blackberry bushes were destroyed. Because of this, the distance between the vineyards and the remaining blackberry patches became too great. Thus, the wasp was no longer a predator on the grape leaf hopper in most of the valley. The use of pesticides and a monoculture type of farming caused an increase in the grape leaf hopper population. (See fig. 14.16.)

Biological methods other than the use of predators have been used to control insect populations. For example, marigolds are planted to reduce the number of soil nematodes, and garlic plants are employed to check the spread of Japanese beetles.

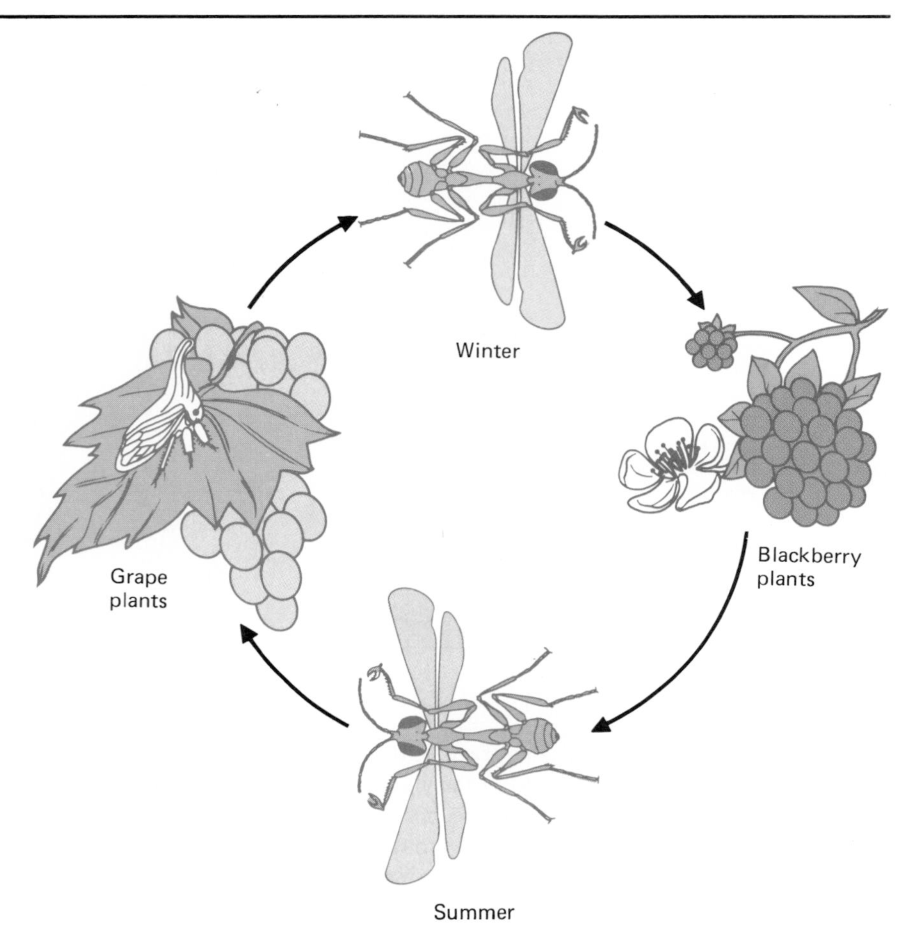


**Figure 14.16 Life cycle of a parasitic wasp.** During the summer, the population of leafhoppers that feeds upon the grape leaves is controlled by a parasitic wasp. These wasps migrate to blackberry bushes to overwinter but return to the grapes during the summer. If the distance is too great, the wasps are unable to migrate between grapes and the blackberry bushes. As a result, the grape leafhoppers are unchecked, and they damage the grape plants. Therefore, the removal of blackberry patches to clear land for the growing of grapes increases the amount of damage that the leafhoppers inflict upon the grapes.

Sometimes, through experimentation and selective breeding, a population of crop plants can be developed that is resistant to a specific pest. For a long period of time, wheat rust (a plant parasite that grows on wheat) lowered the wheat yield. Today, new strains of wheat are planted that are resistant to the rust. Also, certain new strains of wheat have a short life span and ripen before the rust can become established.

These examples of biological control give an idea of the potential that natural methods have for controlling pests. These techniques are usually very specific and have fewer side effects than chemical control agents. However, extensive research into the biology of pest species is needed so these kinds of controls can be developed. This research is expensive and must be completed for each pest species. Also, all organisms change through time, which can cause both pesticides and natural controls to become obsolete.

## Box 14.3
### Food additives

Food additives are chemicals that are added to food before its sale. The purpose of these additives is to prolong the storage life of food, to make it more attractive by adding color or flavor, or to increase its nutritional value. One example of a nutritional additive is iodine in table salt. Iodine is a trace element required for proper thyroid function. Individuals suffering from a lack of iodine often develop goiter, an enlargement of the thyroid gland. The addition of iodine to table salt has eliminated goiter in the United States.

Calcium propionate is added to baked goods to increase shelf life. Baked goods can become contaminated with airborne spores of molds and bacteria, which begin to grow and spoil the product before it is used. Certain chemicals will retard this germination process.

Another chemical, Diethylstilbestrol (DES), was used in the poultry industry to produce fatter birds. Because there are indications that DES is carcinogenic, the Food and Drug Administration banned the use of DES in chickens and declared that it was potentially hazardous to humans. Further studies were conducted to determine whether DES is safe to use in connection with raising beef. (With DES in the diet, an animal gains much more weight in less time than an animal not fed DES.) As a result of these studies, DES has been banned from all animal feed used in the United States and Europe.

Sometimes additives are used just to make food look better. For years, Red Dye II was used to color a variety of foods. Its use was discontinued when it was found to be carcinogenic. Other food coloring is still widely used to increase the appeal to the consumer. Other chemicals, such as monosodium glutamate, are used to alter or improve the flavor of foods.

All of these methods are used to make the food production industry more efficient and profitable. However, they should be used with forethought and an understanding of the whole range of impacts they can have.

## Summary

The agricultural methods practiced by the native Americans did little to permanently disrupt the ecosystems in North America. The early settlers practiced European methods of farming that were not compatible with the types of soil and weather in the United States. This resulted in ruined land. These practices were encouraged because settlers seemed to believe that the supply of virgin land was endless. The Homestead Act resulted in a move westward.

Mechanized farming replaced horse-drawn implements, and crop rotation was abandoned. Although these practices increased production, they were not without problems. Energy is needed for tilling, planting, and harvesting and for the production and application of fertilizers and pesticides.

Monoculture involves planting large areas in the same crop year after year. This causes problems with plant diseases and soil erosion. Although fertilizer replaces soil nutrients that are removed when the crop is harvested, it does not replace soil organic material.

# Consider This Case Study
## Pesticides and pelicans

Since the late 1960s, a number of carnivorous birds at the end of long food chains have been producing eggs with thin shells. Some are so thin that they collapse when the birds incubate them. Analysis of fat in such eggs has revealed the presence of high levels of DDE (dichloro-diphenyl-dichloroethane), a pesticide closely related to DDT. This chemical is a metabolite of DDT. It is now known that DDE interferes with the action of a liver enzyme, carbonic anhydrase, which causes calcium to mobilize from the birds' bones and to become incorporated into the eggshells. The greater the quantity of DDE, the less effective the enzyme, and the thinner the shells.

In May 1968, two researchers from the Smithsonian Institution visited Anacapa Island, located just offshore from Ventura, California. May is the most active nesting time for brown pelicans. In an area where there should have been at least fifteen hundred active nests, with up to three young birds in each, the researchers found no young at all. Within a short time, other scientists became aware of the pelican nesting failure. Many visitors went to Anacapa Island in 1969, and the final count for pelicans that season was six hundred nests and only two young.

Analysis of the fat in the pelican eggs on Anacapa Island revealed levels of DDE as high as 2,500 ppm. Information was also collected from other islands at various distances from Los Angeles. The table below relates distance from the Los Angeles area, ppm DDE, and shell thickness (normal shell is about 0.57 mm).

Modern agriculture is heavily dependent upon the control of pests by chemicals and biological methods. Hard pesticides are stable and persist in the environment. Soft pesticides degrade readily but are extremely toxic to humans. DDT and other molecules can accumulate in food chains and eventually threaten high-level carnivores.

With a proper understanding of an organism's ecology, populations may be controlled by natural processes such as predators, mating females with sterile males, sex attractants, and breeding resistant strains of crops. These are called biological controls.

Chemicals are also added to food to reduce spoiling, enhance attractiveness, alter flavor, stimulate growth, and increase nutritional value.

| Location of island | ppm DDE | Shell thickness |
|---|---|---|
| Anacapa Island | 1223 | 0.32 mm |
| West Coronado Island | 1158 | 0.34 mm |
| San Martin Island | 429 | 0.45 mm |
| San Benitos Island | 128 | 0.51 mm |
| Gulf of California | 13 | 0.57 mm |

It is apparent that pelicans who breed farther from Los Angeles produce eggs with less DDE and thicker shells.

As more and more data accumulated, it seemed that the waters around Los Angeles may have contained abnormally high levels of DDE. In searching for the source, it was determined that in this highly urbanized area, agricultural runoff was not a significant factor. When municipal sewage outfalls were checked between San Francisco and San Diego, only one showed high levels of DDE. Los Angeles' outfall was discharging about fifty kilograms per day. All of the other cities combined discharged less than twenty-three kilograms per day.

By monitoring the sewer lines that flow to the ocean outfall, the DDE was traced to its source, a chemical company a few kilometers inland that was manufacturing DDT. Following a court order issued in April 1970, the chemical company stopped releasing its wastes into the sewer system and began trucking them to a nearby landfill.

The list below documents the decline of pelican breeding success and the subsequent recovery after the correction of the problem.

Pelican nesting success on Anacapa Island, California*

| | | |
|---|---|---|
| 1968 | no active nesting | no young |
| 1969 | 600 nests | 2 young |
| 1970 | 552 nests | 1 young |
| April 1970, dumping of DDE into sewer system banned | | |
| 1971 | 600 nests | 7 young |
| 1972 | nests not counted | 56 young |
| 1973 | 597 nests | 134 young |
| 1974 | 1,286 nests | 1,185 young |

While numbers of nests and young in 1974 had not yet reached levels known before the discovery of the DDE problem, it can be seen that significant recovery had occurred.

Has the pelican problem been solved?
Does trucking wastes to a landfill solve the problem?
Has reproductive success of the pelican returned to normal?

*Data used with permission of Ray E. Williams, Rio Hondo College, Whittier, Ca.

## Review Questions

1. How do hard and soft pesticides differ?
2. What is crop rotation? Why was it used?
3. What are two physical environmental differences between Europe and the United States that influence farming practices?
4. What are the advantages and disadvantages of biological control?
5. List four uses of food additives.
6. Name three nonchemical methods of controlling pest populations.
7. What is biological amplification?
8. What is monoculture?
9. What elements are contained in a fertilizer? Why are they needed?
10. List three reasons why fossil fuels are essential for modern American farms.

## Suggested Readings

Ball, Gordon A., and Heady, Earl O. *Size, Structure, and Future of Farms.* Ames, Ia.: The Iowa State University Press, 1972. Twenty authors combine to present a broad overview of farming in the United States. Presents such issues as societal goals in farm size, trends in farm size, farm labor and the labor market, industrialized farming, nonfamily corporation farming, and community service in a dynamic rural economy.

Barnard, C. S., and Nix, J. S. *Farming, Planning and Control.* Cambridge, England: University Press, 1973. Although this book concerns farming in Great Britain, it explains why farming is no longer a simple way of life but a business that requires careful planning, organization, and control.

Carter, Luther J. "Organic Farming Becomes 'Legitimate.' " *Science* 209: 254–56. The Secretary of Agriculture sees the USDA's forthcoming report on organic practices as a much needed boost for a promising mode of agriculture.

Conrat, Maisie, and Conrat, Richard. *The American Farm: A Photographic History.* Boston: Houghton Mifflin Company, 1977. This book presents an excellent series of photographs depicting all aspects of farming in the United States from 1880 until the present.

Coughlin, Robert E. "Farming on the Urban Fringe." *Environment* 22: 33–40. Deals with the loss of good farmland to urban sprawl. Presents some public programs designed to reduce this loss.

Donahue, Roy L.; Follet, Roy; and Teelock, Rodney W. *Our Soils and Their Management.* Danville, Ill.: The Interstate Printers & Publishers, 1976. Deals with various careers relating to soils and includes sections on soil formation, fertilizer use, and tillage practices. A wide variety of soil uses are discussed, including homesites, pasture, field crops, orchards, and greenhouses.

Doutt, Richard L. "De-Bugging the Pesticide Law." *Environment* 21: 32–35. Examines the EPA's record of pesticide administration and the need for changes in the laws.

Fletcher, W. W. *The Pest War.* New York: John Wiley & Sons, 1974. Traces the history of pest control from earliest times to the present. In an informative, easy-to-read book that presents the use of chemicals as insecticides, herbicides, and fungicides. There is also a chapter on biological and other methods of control.

General Accounting Office. *Environmental Protection Issues in the 1980's.* Publication CED 81–83. Washington, D.C., 1982. Presents the issue of pesticides and their secondary effects.

Huffaker, Carl B., ed. *New Technology of Pest Control.* New York: John Wiley & Sons, 1980. Presents fourteen papers that are summary reports of the progress toward development of integrated pest management systems accomplished in recent years by the NSF/EPA Integrated Pest Management Project.

Johnson, Anita. "The Case Against Poisoning Our Food." *Environment* 21: 6–13. The National Academy of Science and food processors are seeking to lower some of the food safety standards. This article presents reasons for raising the standards instead of lowering them.

Johnson, Paul C. *Farm Inventions in the Making of America.* Des Moines, Ia.: Wallace-Homestead Book Company, 1976. A well-illustrated account of the development of farm machinery in the late nineteenth and twentieth centuries.

Lockeretz, William; Shearer, Georgia; and Kohl, Daniel. "Organic Farming in the Corn Belt." *Science* 211: 540–47.

Marshall, Eliot. "Health Committee Investigates Farm Drugs." *Science* 209: 481–82. Proposed legislation would enable FDA to ban the use of antibiotics in animal feeds.

Ordish, George. *The Constant Pest: A Short History of Pests and Their Control.* New York: Charles Schribner's Sons, 1976. An interesting account of pests and their control from Neolithic times to the present.

Pimentel, David, ed. *World Food, Pest Losses, and the Environment.* Boulder, Colo.: Westview Press, 1978. A series of nine papers presented at an AAAS symposium. Focuses on current food shortages and on the impact of pests in reducing world food supplies.

Roberts, Daniel A. *Fundamentals of Plant-Pest Control.* San Francisco: W. H. Freeman & Company Publishers, 1978. Concerns the science of plant-pest control, methods of plant-pest control, and systems approach to plant-pest control.

Wendall, Berry. *The Unsettling of America: Culture & Agriculture.* San Francisco: Sierra Club Books, 1977. The author argues that farming cannot be considered as an issue separate from our culture. He is concerned with the development of large "corporate" farms that place emphasis on the maximum production from the land.

## Chapter Outline

I. Historical growth of the American city
II. The rise of suburbia
  A. Urban sprawl
  B. Problems associated with unplanned urban growth
III. Land-use planning principles
    Box 15.1 A planned community: Reston, Virginia
IV. Planning on a larger scale
  A. Regional planning
  B. Urban transportation planning
  C. Multiple land use
V. Consider this case study: Decision making in land-use planning

## Objectives

Explain the impact water has on the location and development of cities.

List the relative amounts of land used for crops, livestock, urban development, and other uses.

Explain why farmland surrounding cities was used for housing.

Explain how taxation may influence land use.

Explain why floodplains and wetlands are often mismanaged.

List the steps involved in the development and implementation of a land-use plan.

Describe methods of enforcing compliance with land-use plans.

List the problems associated with local land-use planning.

Explain the advantage of regional planning and the problems associated with it.

List some land uses that are exclusionary and some that may be integrated, multiple land use.

## Key Terms

floodplain

floodplain zoning ordinances

land-use planning

megalopolis

multiple land use

ribbon sprawl

tract development

urban sprawl

wetlands

zoning

# CHAPTER 15
# Land-use principles: Issues and decisions

## Historical growth of the American city

In the past, land was considered a limitless resource. As the population grew and demand for food, lodging, and transportation increased, we began to realize that land is a finite nonrenewable resource.

Land development during the last two decades has destroyed many natural areas that people have long enjoyed. The Sunday drive in the countryside has become a battle to escape the ever-growing suburbs. The unique character of neighborhoods and communities has been changed by apartment complexes. Developments have sprung up along beaches, and recreational communities are common sights. Even the deserts and the marshes have not been spared; they sprout the little red flags of subdividers.

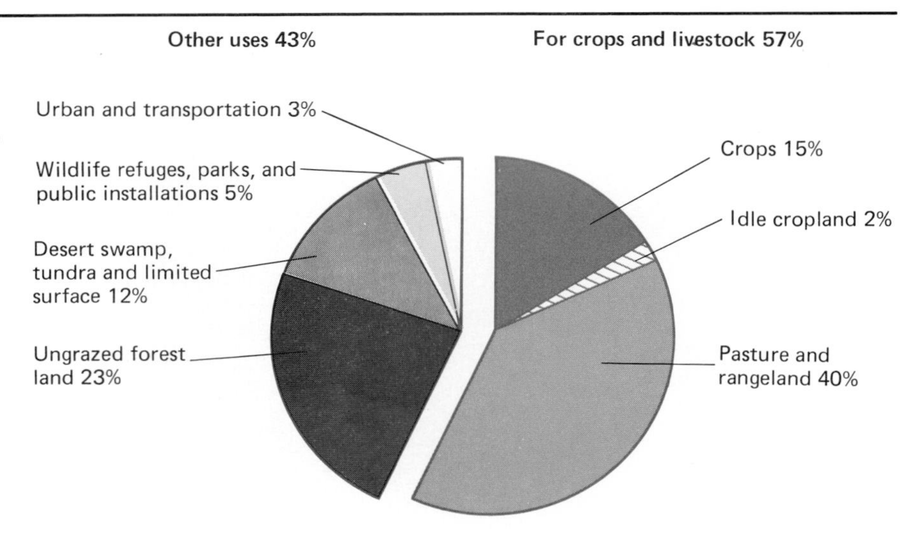


**Figure 15.1 Land use in the United States.** Fifty-seven percent of the land in the United States is used for crops and livestock, while only 3 percent is used for urban centers and transportation. (Source: *Statistical Abstracts of the United States,* 1984)

Uses of land in the United States are varied. (See fig. 15.1.) Fifty-four percent of the land is used for crops and livestock; 43 percent is forests and natural areas. The remaining 3 percent is used intensively by people in urban centers and for transportation corridors.

The United States began as a rural country. Early settlers often came from crowded European cities. The promise of open spaces and available land was a primary incentive for the immigrants who came to the United States. As the population increased, villages and cities developed.

These early American towns were usually built near water. Look at the map (fig. 15.2) and locate St. Louis, Chicago, Detroit, New York, Boston, or any other large American city. Note that they are all near water. Without access to water, these cities would not have developed. Water met many of the needs of villages and small towns. Primary needs were drinking water, transportation, power, and waste disposal. In addition, industrial use of water became important as the cities grew.

Because the country had no system of railroads or highways, the primary method of transportation was by water. Without a reliable transportation network, cities could not prosper and grow. Because access to water was so important to industry, most industrial development took place on the waterfront. Large industrial developments replaced small gristmills, sawmills, and blacksmith shops. Industry eventually took over the major portion of the waterfront. Because there was little control of their activities, the waterfront became polluted, unhealthy, and an undesirable place to live. Anyone who could afford to do so moved away from the original city center.

Land and water are intimately interrelated. Development of a water resource for industrial use or transportation will affect the land bordering the water. The type of land use similarly affects the water course. In urban centers, the "decision" to use waterfront land for industrial development was forced

Figure 15.2 **Water and urban centers.** Note that each of the large urban centers is located on water. Water is an important means of transportation and was a major determining factor in the growth of cities.

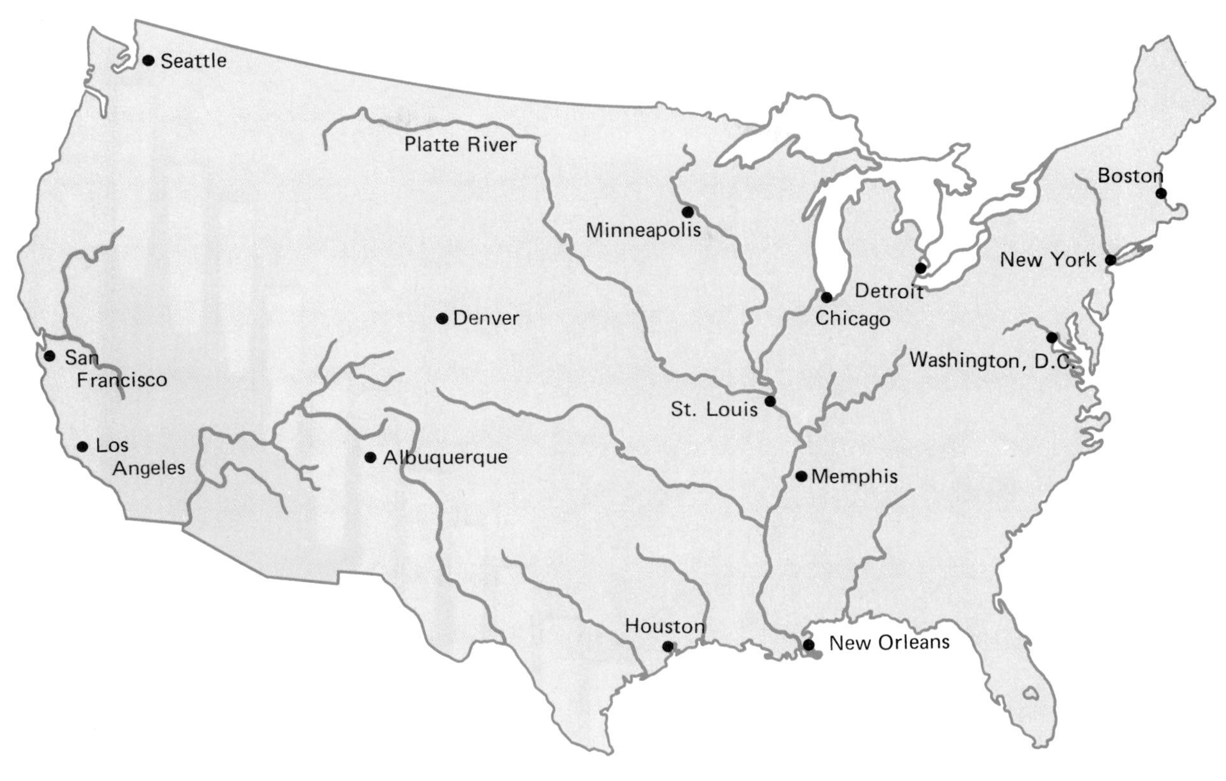


by a need for transportation and power. Once this pattern had begun, a series of events would eventually result in the development of suburban metropolitan regions.

North America remained essentially rural until industrial growth began in the last third of the 1800s. During this period, the population began to shift from rural to urban areas. (See fig. 15.3.) This large-scale migration to the cities had several causes. Improvements in agriculture required less farm labor; therefore, people moved from the farm to the city. Second, job opportunities were available in the city because industry was developing. The average American was no longer a farmer but rather a factory worker, shopkeeper, or clerk living in a tenement or tiny apartment. Most of this early migration occurred in New England and surrounding states where farms were abandoned for more secure factory work in nearby cities. As time passed, this same pattern of rural to urban migration occurred throughout the United States. From 1900 to 1910, the population of the United States increased by sixteen million. Much of this population increase was the result of new immigrants, three-quarters of whom settled in towns and cities.

**Figure 15.3 Rural to urban population shift.** In 1800, the United States was essentially a rural country. Industrialization in the late 1800s began the shift to an increasing urban population. (Modified from "U.S.A. Population Changes, 1950–60," *Population Bulletin,* vol. 19, no. 2, Population Reference Bureau, March 1963 and *World Population Data Sheet,* Population Reference Bureau, 1984.)

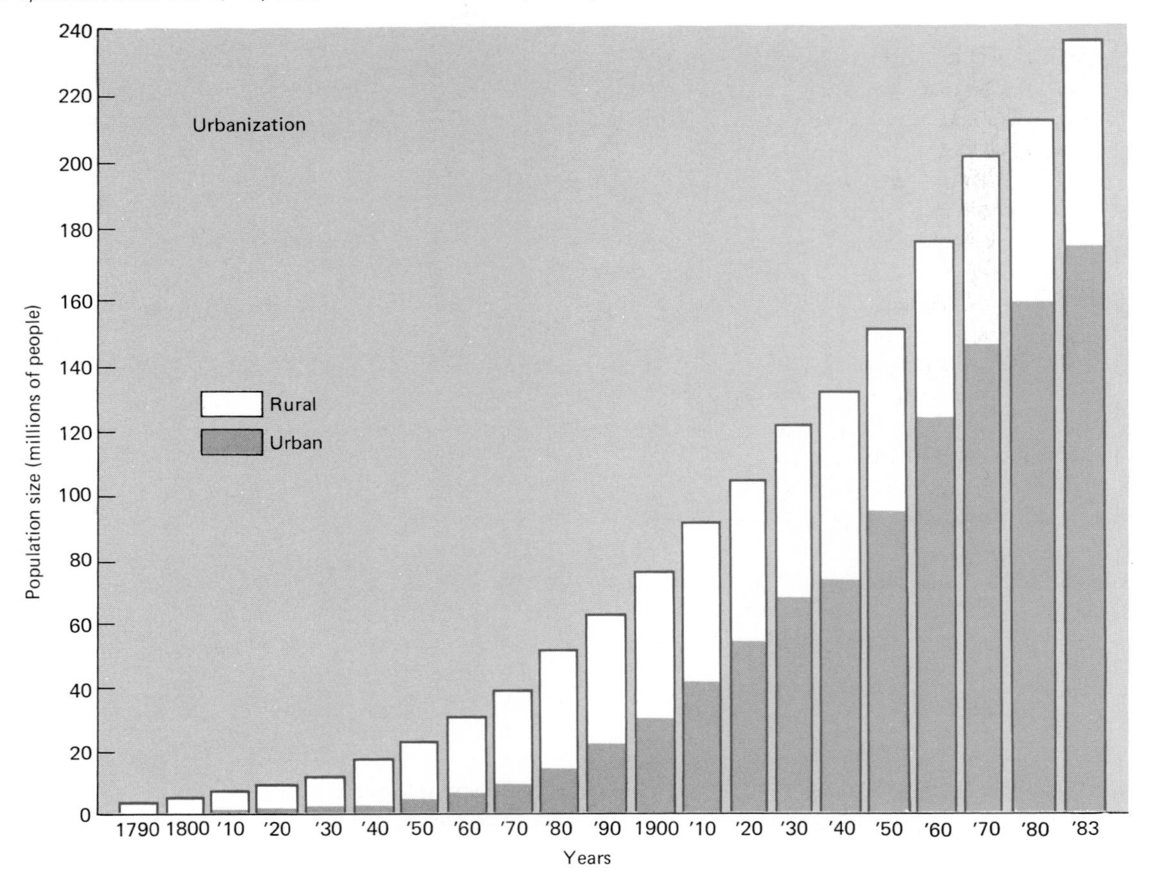


Many immigrants congregated in parts of cities. As a result, today most large cities have ethnic sections, such as Irish town, Chinatown, German town, or Greek town. These sections give American cities a unique character. Another reason for the growth of cities is that they can offer a greater variety of cultural, social, and artistic opportunities than rural communities can. This attracts people to the cities for cultural as well as economic reasons.

Because cities are usually located on water, they are often surrounded by rich farmlands. River valleys provide flat, rich land for the growth of crops. Until transportation systems became well developed, it was necessary for farms to be close to the city so the produce could be marketed. This rich farmland adjacent to the city was the most easily developed space for the expanding city. As the population of the city grew, there was an increased demand for land within the city. As city land prices rose, people and businesses began to look for cheaper land on the outskirts of the city. Real estate agents were quick

**Figure 15.4 Transportation and suburbia.**
This is a common experience for people who work in cities but live in the suburbs. The popularity of automobiles and the desire to live in the suburbs are tied closely to one another.

to respond and to assist in the acquisition and conversion of agricultural land to residential or commercial uses. Developers viewed land as a commodity to be bought and sold for a profit rather than a finite resource to be managed. As long as money could be made by converting agricultural land to other purposes, it was impossible to prevent it.

## The rise of suburbia

As cities continued to grow, certain sections within each city began to deteriorate. Industrial activity continued to be concentrated near water in the center of the city. Industrial pollution and urban crowding turned the core of many cities into undesirable living areas. In the early 1900s, people who could afford to leave began to move to the outskirts of the city. Home ownership increased after World War II. Most of the homes were in the attractive suburbs, away from the pollution and congestion of the central city. These houses were built on large lots in decentralized patterns, which increased the cost of supplying services, increased energy needs, and made it very difficult to establish efficient public transportation networks. Rising automobile ownership and improved highway systems also encouraged suburban growth. The convenience of a personal automobile encouraged decentralized housing patterns, which required better highways, which in turn led to further decentralization. (See fig. 15.4.)

In some areas throughout North America, the growth of suburbs has been slowed due to the increased cost of housing and transportation. There has been a migration back to the cities on a limited scale. This is due primarily to the lower cost of urban houses and the fact that public transportation is generally more efficient in the city than in the suburbs, thus freeing urban residents from the daily cost of commuting. This reverse migration, however, is still greatly offset by the continual growth of suburban communities.

**Figure 15.5 Urban sprawl.** Note the three different types of growth depicted in these photos. The wealthy suburbs (a) with large lots are adjacent to the city. Ribbon sprawl (b) develops as a commercial strip along highways. Tract development (c) results in neighborhoods consisting of large numbers of similar houses on small lots.

a

b

c

### *Urban sprawl*

By 1960, unplanned suburban growth had become known as **urban sprawl.** Large housing tracts surrounded cities, which made it difficult for people to find open space. A city dweller could no longer take a bus to the city limits and enjoy the open space of the countryside.

Urban sprawl occurs in three different ways. (See fig. 15.5.) One type of growth is associated with the wealthier urbanites adjacent to the city. These subdivisions of homes are on large individual lots in the more pleasing geographic areas. A second form of urban sprawl involves development along transportation routes. This is referred to as **ribbon sprawl** and usually consists of commercial and industrial building. Ribbon sprawl results in high costs for the extension of utilities and other public services. It also makes the extent of urbanization seem much larger than it actually is. A third developmental pattern is tract development. **Tract development** consists of the construction of similar residential units over large areas. Initially, these tracts are often separated from each other by farmland.

As suburbs continued to grow, cities (once separated by miles of farmland) began to merge. It is difficult to tell where one city ends and another begins. This type of growth led to the development of regional cities. Although these cities maintain their individual names, they are really just part of one large urban area. (See fig. 15.6.) Each of these large regional urban centers is called a **megalopolis.** The eastern seaboard of the United States from Boston to Washington, D.C. is an example of a continuous city. The area from Buffalo, New York, to Chicago, Illinois, is also well on the way to becoming one continuous urban area.

### *Problems associated with unplanned urban growth*

One United States Department of Agriculture study found that three-fourths of the land recently urbanized was previously used for high-value crop production. An area that is flat, well drained, accessible to transportation, and close to cities is ideal farmland. However, it is also prime development land. Land adjacent to cities that once supported crops now supports new housing

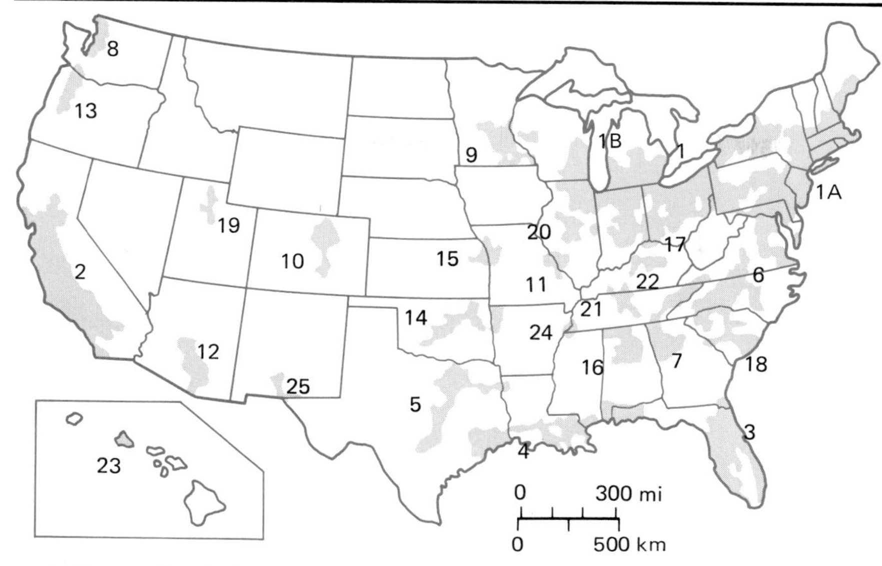


1. Metropolitan belt
   A. Atlantic Seaboard (BoWash)
   B. Lower Great Lakes (ChiPitts)
2. California Region (SanSan)
3. Florida Peninsula (JaMi)
4. Gulf Coast
5. East central Texas – Red River
6. Southern Piedmont
7. North Georgia – South Tennessee
8. Puget Sound
9. Twin Cities Region
10. Colorado Piedmont
11. Saint Louis
12. Metropolitan Arizona
13. Willamette Valley
14. Central Oklahoma–Arkansas Valley
15. Missouri–Kaw Valley
16. North Alabama
17. Blue Grass
18. Southern Coastal Plain
19. Salt Lake Valley
20. Central Illinois
21. Nashville Region
22. East Tennessee
23. Oahu Island
24. Memphis
25. El Paso–Ciudad Juarez

**Figure 15.6 Regional cities.**
By the year 2000, it is projected that twenty-five major urban regional cities will have developed. Each will have a population of at least one million people. Four super cities will each have a great deal more than one million people. (From J. P. Pickard, "U.S. Metropolitan Growth and Expansion, 1970–2000, with Population Projections" in *Population Growth and the American Future.* Washington, D.C.: U.S. Governmental Printing Office, 1972.)

developments, shopping centers, and parking lots. (See fig. 15.7.) The development of farmland is proceeding at a rapid pace in the United States. Approximately one million hectares of land is developed each year. This is equivalent to an area half the size of New Jersey.

The Fourteenth Amendment to the United States Constitution specifies that all real property shall be assessed and taxed at its highest and best use. This means that property will be taxed on what *can* be done with it, not necessarily what *is* being done with it. For example, if land can be used for both farming and residential development, it will be taxed as if it were being used for residential purposes. If a farmer sells a portion of the farm to a developer who builds five houses, local taxing authorities would consider these houses to be the "highest and best use" of the land. All of the farmer's land would probably be reassessed, and the taxes would be increased substantially. Farmers faced with this situation are often forced to sell their land because they are taxed on the commercial value of the land rather than its value as farmland. This policy encourages development and forces people out of farming.

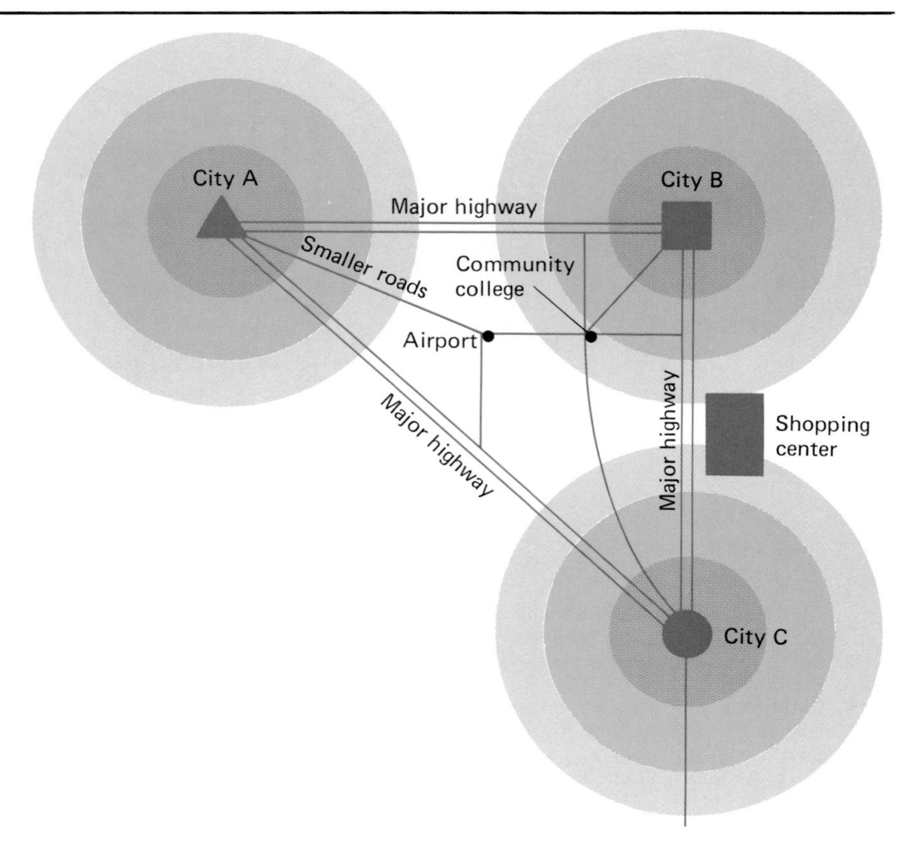


**Figure 15.7 Loss of farmland.**
As each city grows outward, they will eventually grow together to form a regional city. The land once used for farming becomes residential and commercial. Improved transportation routes and joint facilities (such as the airport, shopping center, and community college) hasten this loss of farmland.

A number of states have developed different taxation practices to relieve this inequity. These new policies assess taxes based on *current* use of the land and not on the land's highest *potential* use.

Farmland is valuable to the urban dweller for its open space. Because nearly three-fourths of our population live in urban areas, open space breaks up the sights and sounds of the city and provides a place for many kinds of recreation. Because there has been inadequate land-use planning, the open spaces in or near urban areas are rapidly being converted to other uses. Until recently, the creation of a new park within a city was considered to be an uneconomical use of the land. Attitudes are slow to change, and people are only now beginning to realize the need for parks and open spaces.

Some cities have recognized the need for open space and allocated land for parks. New York City set aside approximately two hundred hectares for Central Park in the late 1800s. (See fig. 15.8.) Boston developed a park system that provides a variety of urban open space. However, other cities have not dealt with the need for open space.

Many cities are located near rivers in areas called floodplains. A **floodplain,** the lowland on either side of a river, is periodically covered with water. Some

**Figure 15.8 Urban open space.**
This is an aerial view of Central Park in New York City. If the land had not been set aside in the late 1800s, it would have been developed like the rest of Manhattan.

areas are flooded every year, and other higher areas may flood less regularly. The best use of these areas may be for open space or recreation.

Floods are natural phenomena. Contrary to popular impressions, there is no evidence to support the premise that floods are worse today than they were one hundred or two hundred years ago, except perhaps on small, localized watersheds. What has increased, of course, is the economic loss from floods. This loss reflects the fact that cities have been built on floodplains, and also that industries, railroads, and highways have been constructed on floodplains. Sometimes there were no alternatives, but often risks have been ignored. Because flooding causes loss of property and life, floodplains should be used only for agricultural and recreational purposes. In addition, our tax monies are used to repair the damage that results from the unwise use of these areas.

Many communities have enacted **floodplain zoning ordinances** to restrict future building in floodplains. Although such ordinances may prevent further economic losses from flooding, what happens to those individuals who already live in floodplains? Floodplain building ordinances usually allow current residents to remain. However, these residents often find it extremely difficult to obtain property insurance. Relocation, usually at a financial loss, is the only alternative. While such situations are unfortunate, it can only be hoped that in the future they can be prevented by proper planning.

**Wetlands** are other areas that are frequently misused. Swamps, tidal marshes, and coastal wetlands provide habitats for many diverse species of wildlife. The coastal areas adjacent to large bodies of water are extremely important as fish spawning areas. Commercial and sport fisheries are dependent upon the species that use wetlands for spawning and feeding. Most wetlands receive constant inputs of nutrients from the water that drains from the

**Figure 15.9 Wetlands.** These natural areas provide habitat for wetland wildlife and often provide recreational sites near an urban center. Because cities have developed near water, all too often wetlands are drained to allow for further development, to get rid of mosquitoes, or to allow for more direct transportation routes.

surrounding land. This land is highly productive and an excellent place for the rapid growth of aquatic species. Human impact on the coastal environment has severely degraded or eliminated these spawning and nursery habitats.

Besides providing a necessary habitat for fish, wetlands also provide natural filters for sediments and runoff. This natural filtration process protects water quality. (See fig. 15.9.) In addition, wetlands often protect shorelines from erosion. Often, when wetlands are destroyed, costly artificial measures (such as breakwaters) are built to protect the shoreline.

Wetlands also breed mosquitos, may smell unpleasant, and provide barriers to movements; therefore, they are often drained or filled. The pressure to destroy these natural areas is greatest where there is a large population. Because these lands have long been thought to be useless, it is easy to see why they are filled or drained to be put to other uses. In the past, many cities have used these sites for landfills.

Until recently, most communities looked upon new construction and additional residents as positive additions to the local tax base. People and industries pay taxes, so people in growing communities anticipate lower taxes. However, every addition to the community requires services that are paid for with tax dollars. Rarely do taxes decrease. Poorly planned growth will bring more tax dollars to a community, but it may actually result in a net loss because the cost of services may exceed the tax monies generated. Communities are beginning to recognize that economic analysis is important to urban land-use planning, especially at the urban fringe. A land-use plan minimizes public costs and prevents chaotic development, which wastes money, land, and energy. When tract developments are spaced far apart and far from the city, the costs of providing services becomes even greater. These costs are borne by the community and the taxpayer. When growth is properly planned, these costs can be kept to a minimum.

## Land-use planning principles

**Land-use planning** is the construction of an orderly list of priorities for the use of available land. The development of any plan involves a data-gathering process in which current use, geological data, biological data, and sociological information are surveyed. From these data, projections are made about what human needs will be. All of the data collected are integrated with human needs projections, and each parcel of land is evaluated and assigned a best use under the circumstances. Finally, there must be mechanisms for implementing the plan. These mechanisms can be divided into two categories—purchase of the land or regulation of the land.

Probably the simplest way to insure protection of certain desirable lands is to purchase them. This method has often been used by the government or by conservation-minded organizations or individuals. The major problem with purchasing land is the cost. Many communities are not in a financial position to purchase lands; therefore, they attempt to regulate land use by zoning laws.

**Zoning** is a common type of land-use regulation. When land is zoned, it is designated for specific potential uses. Several common designations are agricultural, commercial, residential, recreational, and industrial. (See fig. 15.10.) Often local governments lack the funds to hire professional planners. As a result, zoning regulations are frequently made by people who see only the short-term gain and not the possible long-term loss. The land is simply zoned for its current use. Even when well-thought out land-use plans exist, they are usually modified to encourage local short-term growth rather than to provide for the long-range needs of the community. The community needs to be alert to variances from established land-use plans because once the plan is compromised, it becomes easier to accept future deviations that may not be in the best interests of the community. Many times individuals who make zoning decisions are realtors, developers, or local business people. These individuals wield significant local political power and are not always unbiased in their decisions. Concerned citizens must try to combat special interests by attending zoning commission meetings and by participating in the planning process.

**Figure 15.10 Zoning.** Most communities have a zoning authority that designates areas for particular use. These signs are indications that decisions have been made about the "best" use for these lands.

## Planning on a larger scale

### *Regional planning*

Regional planning is more effective than local land-use planning, because political boundaries seldom reflect the geological and biological data base used in planning. Larger planning units can afford to hire professional planners, which the local units cannot afford. The concept behind regional planning is coordination, because problems do not respect political boundaries. For example, the location of airports should be based upon a regional plan that incorporates the several local jurisdictions in the region. Three cities only thirty kilometers apart should not build three separate airports when one regional airport could serve their needs better at a lower cost to the taxpayers.

**Figure 15.11 Hawaii land-use plan.**
Hawaii was the first state to develop a comprehensive land-use plan. This development in an agricultural area was stopped when the plan was implemented in the 1960s.

Although regional planning is increasing in the United States, the majority of regional governmental bodies are presently voluntary and lack any power to implement programs. Their only role is to advise the member governments on policy. Unfortunately, members of regional organizations still seem to put their own units above the goals of the region and view policy from a narrow perspective. An elected, multipurpose, regional government, on the other hand, would be ideal for implementing land-use policy. Such governments exist in only a handful of places and show few signs of spreading.

One way to encourage regional planning is to develop planning policies at the state level. The first state to develop a comprehensive state land-use program was Hawaii. During the early 1960s, Hawaii began to see much of its natural beauty being destroyed to build houses and apartments for the increasing population. The same land that attracted tourists was being destroyed to provide hotels and supermarkets for them. (See fig. 15.11.) Local governments had failed to establish and enforce land-use controls. Consequently, the Hawaii State Land-Use Commission was founded in 1961. This commission designated all land as urban, agricultural, or conservation. Each parcel of land could be used only for its designated purpose. Other uses are allowed only by special permit. To date, the record of Hawaii's action shows that it has been successful in controlling urban growth and preserving the natural beauty of the islands.

Several states are attempting to follow Hawaii's lead in state land-use regulation. Some states have passed legislation dealing with special types of land use. Examples are wetland preservation, floodplain protection, and scenic and historic site preservation. Although direct state involvement in land-use regulation is relatively new, it is expected to grow. Only large, well-financed levels of government can afford to pay for the growing cost of adequate land-use planning. State and regional governments are also more likely to have the power to counter the political and economic influences of land developers, lobbyists, and other special interest groups when conflicts over specific land-use policies arise.

## Box 15.1
## A planned community: Reston, Virginia

Historically, towns and cities grew as a natural by-product of people choosing to live in certain areas for agricultural, business, or recreational reasons. Beginning in the 1920s, private and governmental planners began to think about how an ideal town would be planned. These communities would be completely built before houses were offered for sale. This concept of pre-planning, designing, and building an ideal town was not fully developed until the 1960s. By 1967, there were about forty-three towns that could be classified as planned "new towns."

One example of a new town is Reston, Virginia, which is located about forty kilometers west of Washington, D.C. Reston began to accept residents in 1964 and has a projected population of eighty thousand. Because developers tried to preserve the great natural beauty of the area and the high quality of architectural design of its buildings, Reston has attracted much attention. Reston also has innovative programs in education, government, transportation, and recreation. For example, the stores in Reston are within easy walking distance of the residential parts of the community, and there are many open spaces for family activities. Because Reston is not dependent upon the automobile, noise and air pollution have been greatly reduced. Recent research indicates that the residents of Reston have rated their community much higher than residents of less well-planned suburbs.

Although the concept of properly planned new towns is appealing, such communities are still rare. There are several reasons for this. To attract citizens, new towns must be located near large urban centers where land availability is limited. The cost of creating new towns is extremely high, and many people still prefer the charm of a traditionally developed community.

### *Urban transportation planning*

A growing concern of urban governments is to develop comprehensive urban transportation plans. While the specifics of such a plan might vary from region to region, urban transportation planning usually involves four major goals.

1. Conserve energy and land resources
2. Provide efficient and inexpensive transportation within the city for the elderly, young, poor, and handicapped
3. Provide suburban people with the opportunity to commute efficiently
4. Help to reduce urban pollution

Any successful urban transportation plan should integrate all of these goals, but funding and intergovernmental cooperation are needed to achieve this. To date, both of these factors have been limited in the United States. The problems associated with current urban transportation will certainly not disappear overnight, but comprehensive planning is the first step to help solve problems.

**Figure 15.12 Mass transportation.**
The graph indicates that mass transportation has continued to decline since the 1940s.

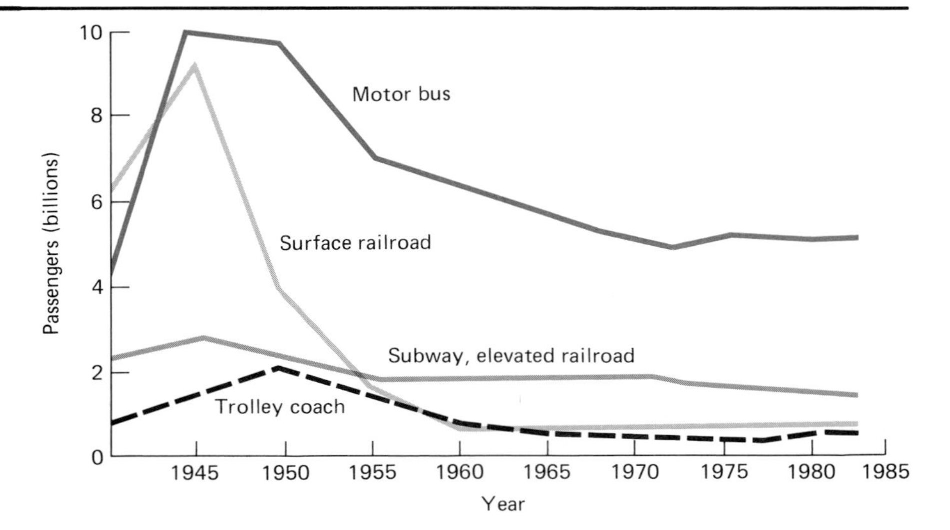


A primary method of transportation in most urban centers is the automobile. The automobile has advantages such as convenience and freedom of movement. However, many urban areas are beginning to recognize that its disadvantages may outweigh its advantages. The disadvantages of urban automobile transportation include increased air pollution, substantial land use for highways and parking areas, and encouragement of urban sprawl. Recognizing these disadvantages, some cities have discouraged automobile use. Toronto, London, San Francisco, and New York, for example, have attempted to dissuade automobile use by developing mass transit systems.

The major types of urban mass transit systems are railroads, subways, trolleys, and buses. Such systems, however, are not without their problems. The major problems associated with mass transit are that it is

1. Economically feasible only along heavily populated routes
2. Less convenient than the automobile
3. Extremely expensive to build and operate
4. Often crowded and uncomfortable.

Although mass transit provides for a substantial portion of the urban transportation in some parts of the world, such as western Europe and the Soviet Union, mass transit use in the United States has continued to decline since the 1940s. (See fig. 15.12.) A variety of forces has caused this decline. As people became more affluent, they could afford to own automobiles, which are convenient individualized methods of transportation. The government encouraged automobile use by financing highways and expressways and by withdrawing support for most forms of mass transportation. The government still encourages automobile transportation with hidden subsidies but maintains that rail and bus transportation should pay for themselves.

Rising gas prices and increasing competition for parking will prompt more Americans to seek alternatives to private automobile use. Car and van pools and dial-a-ride systems have been accepted by many individuals. Other types of transportation, such as the bicycle, should not be overlooked in urban transportation planning.

### Multiple land use

In many cases, a particular piece of land may be suitable for a wide variety of uses. There are examples of single exclusionary uses for land. Once a land-use decision is made, it may be irreversible. For example, once a highway is constructed, it is virtually impossible to use that land for any other purpose. If land is designated as wilderness, no human development can be allowed. Other exclusionary uses of land are agriculture, recreation, waste storage, housing, or industrial development.

Some land uses do not have to be exclusionary. When two or more uses of the land occur at the same time, it is called **multiple land use.** Perhaps the best example of multiple use of land takes place in our national forests. In 1960, the Multiple Use Sustained Yield Act was passed. Under this act, use of national forests was divided into four categories: wildlife habitat preservation, recreation, lumbering, and watershed protection. This act encouraged both economic and recreational use of the forests.

Another example of multiple land use involves the development of parks on floodplains. These parks provide open space, recreational land, and storage reservoirs for storm sewer runoff. Storage reservoirs can even be blended into the park landscape as ponds or lakes.

Once land is recognized as a nonrenewable resource, it may become more fashionable to engage in meaningful land-use planning. Such plans must consider the long-term needs of the region.

## Consider This Case Study
## Decision making in land-use planning

The following situation has happened thousands of times during the last twenty years.

A developer has just announced plans to build a large shopping mall on the outskirts of your city in what is now prime farmland. Many jobs will be created by the construction and operation of the proposed shopping mall. In addition to the stores, three new theaters will be built. Presently, your city has some unemployment, only one theater in the downtown area, and little variety in its retail businesses. On the surface, the proposed mall seems to offer only good news. Is this the case?

Before you answer, look at the entire situation. What will happen to the downtown area when the mall is opened? If you owned a downtown business, would you favor building the mall? What will happen to the taxes on the farms near the new mall? If you were a farmer, would you favor the project? What effect will paving prime farmland have? How will storm water runoff be affected? How will future housing development be influenced? Is the proposed project all good news after all? What other problems can be associated with building the complex? If you were the mayor or city manager, would you favor construction of the mall? Why?

## Summary

Fifty-four percent of our land is used for crops and livestock. Forty-three percent is forests and natural areas. The remaining 3 percent is used for urban centers and transportation corridors. Historically, our large urban centers began as small towns located near water. The water served the needs of the town in many ways, especially transportation. As towns became larger, the farmland surrounding the towns became suburbs surrounding industrial centers.

Many problems have resulted from unplanned urban growth. Current taxation policies encourage residential development of farmland, which results in a loss of valuable farmland. Floodplains and wetlands are two areas that are often mismanaged. Loss of property and life results when people build on floodplains. Wetlands protect our shorelines and provide a natural habitat for fish and wildlife.

Land-use planning involves gathering of data, projecting needs, and developing mechanisms for implementing the plan. Purchasing and zoning the land are two ways to enforce land-use planning. However, local planning is often not on a large enough scale to be effective, because problems may not be confined to the local political boundaries. Regional planning units can afford professional planners and are better able to withstand political and economic pressures.

Some examples of exclusionary land use are housing and highways. Multiple land use is possible where various uses can be integrated.

## Review Questions

1. Why did urban centers develop near waterways?
2. Describe the typical changes that have occurred in cities from the time they were first founded until now.
3. Why do some farmers near urban areas sell their land for residential or commercial development?
4. What is a megalopolis?
5. What land uses are suitable on floodplains?
6. Describe the steps necessary to develop a land-use plan.
7. What are the advantages of regional or state planning?
8. What is multiple land use?
9. List three benefits of land-use planning.
10. Why do people want to move to the suburbs?

## Suggested Readings

Alonso, W. *Location and Land Use.* Cambridge, Mass.: Harvard University Press, 1964. Interdisciplinary work on land use and water use.

Breckenfield, G. "Downtown Has Fled to the Suburbs." *Fortune,* October 1972. Interesting article on the problems of ever-increasing urbanization.

Bryant, William R., and Conklin, Howard E. "New Farm Land Preservation Program in New York." *Journal of the American Institute of Planners* 41 (1975): 390–96. Description of a unique contemporary approach to help preserve valuable farmland from development.

Gundars, Rudzitis, and Schwartz, Jeffrey. "The Plight of the Parklands." *Environment* 24: 6–11. Presents changes in the national parks as a result of increased demand and development.

Harris, C. D., and Ullman, E. L. "The Nature of Cities." *Annals of the American Academy of Political and Social Sciences,* 1945. A classical study on the growth and problems associated with cities.

McHarg, Ian. *Design with Nature.* Garden City, N.Y.: Natural History Press, 1971. The theme of this book is that ecological principles should be applied to land-use planning. A well-written and illustrated book.

Reilly, William K., ed. *The Use of Land.* New York: Thomas Y. Crowell Co., 1973. A series of well-written essays discussing a wide range of land-use problems and issues.

Urban Land Institute. *Management and Control of Growth.* Washington, D.C.: Urban Land Institute, 1975. Analysis of various professionals in the planning field and their methods of controlling undesirable growth.

## Chapter Outline

I. The water issue
II. The hydrologic cycle
III. Kinds of water use
    A. Domestic use of water
    B. Agricultural use of water
    C. Instream use of water
    D. Industrial use of water
    Box 16.1 Waterless toilets
IV. Water-use planning issues
    A. Preserving scenic water areas and wildlife habitats
    B. Groundwater mining
    C. Salinization
    D. Water Diversion
    E. Managing urban water use
V. Consider this case study: The California Water Plan

## Objectives

Explain how water is cycled through the hydrologic cycle.

Explain the significance of groundwater, aquifers, and runoff.

Explain how land use affects infiltration and surface runoff.

List the various kinds of water use and the problems associated with each.

List the problems associated with water impoundment.

Describe the problem of waste dilution in water and the potential dangers involved.

Explain how federal laws control water use and prevent misuse.

List the problems associated with water-use planning.

Explain the rationale behind the federal laws that attempt to preserve specific water areas and habitats.

List the problems associated with groundwater mining.

Explain the problem of salinization that is associated with large-scale irrigation in arid areas.

List the kinds of water-related services provided by local governments.

## Key Terms

aquifers

domestic water

groundwater

groundwater mining

hydrologic cycle

industrial uses

infiltration

instream uses

irrigation

potable waters

runoff

transpiration

water diversion

water table

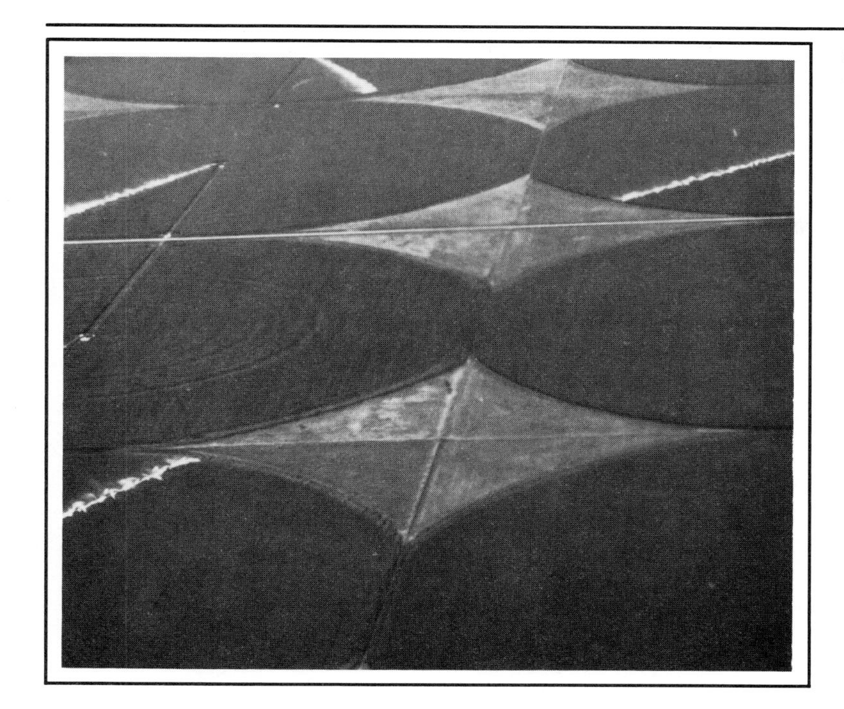

# Chapter 16
# Water-use principles: Issues and decisions

## The water issue

The importance of water to living organisms cannot be overemphasized. All living organisms are composed of cells that contain at least 70 percent water. Organisms can only exist where there is access to adequate supplies of water. Water, however, is also a unique and necessary resource because it has remarkable physical properties. Water's ability to act as a solvent and its capacity to store heat are perhaps its most useful properties. As a solvent, water can dissolve and carry substances ranging from nutrients to industrial and domestic wastes. A glance at any urban sewer will quickly point out the value of water in dissolving and transporting wastes.

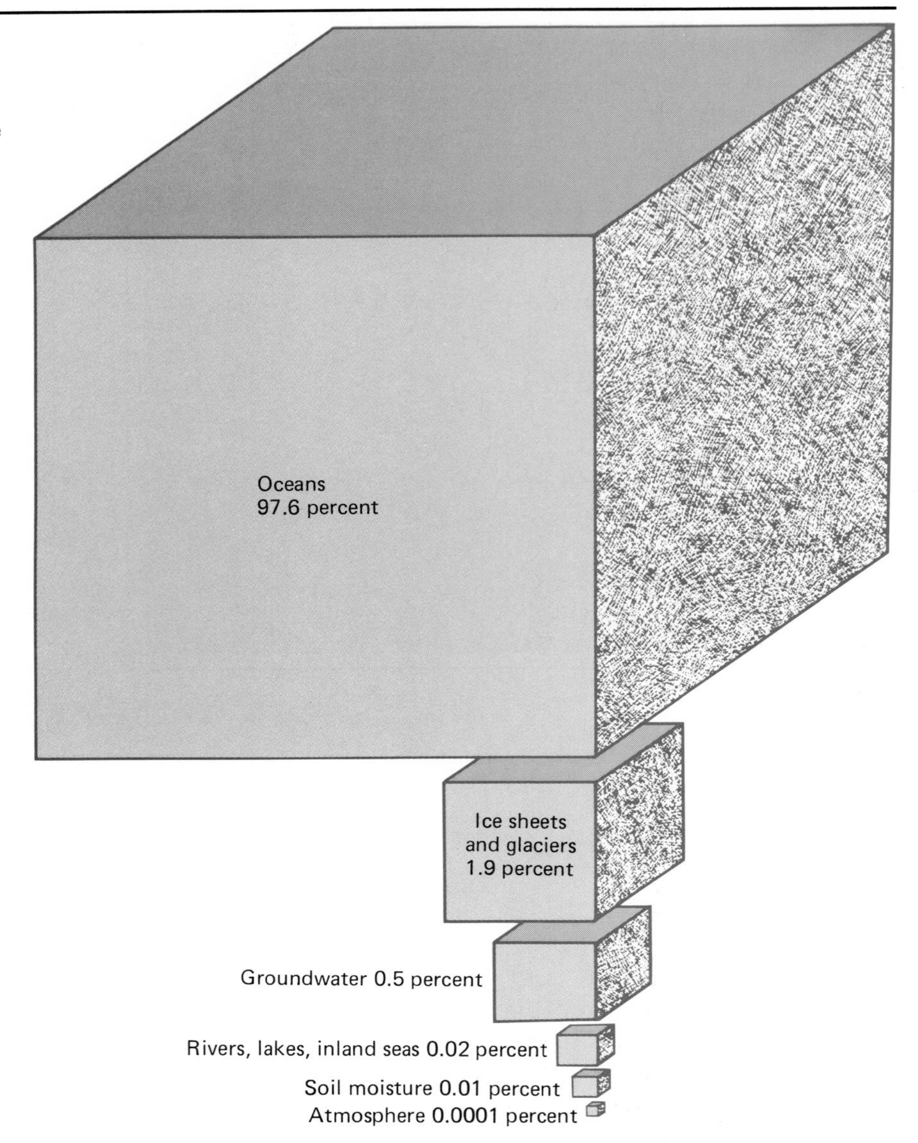


**Figure 16.1 Free water storage on the earth.** The majority of the world's supply of water is in the oceans. The readily available fresh water is found as groundwater in porous rock beds. Although ice sheets and glaciers hold a great deal of fresh water, their turnover is too slow to be usable.

Its ability to store or contain heat is another unusual physical property of water. Because water heats and cools more slowly than most other substances, it is used in large quantities for cooling purposes in electrical power plants and industrial processes. Areas near large bodies of water have their local climate conditions modified by the heat retention ability of the water. They do not have the wide temperature changes characteristic of other areas.

For most human uses, as well as some commercial ones, the quality of the water is as important as its quantity. Water must be substantially free of salinity, plant and animal wastes, and bacterial contamination to be suitable for human consumption. Unpolluted fresh water supplies are known as **potable waters.** Early human migration routes and settlement sites were certainly influenced by the availability of drinking water. Today, despite advances in drilling, irrigation, and purification, the location, quality, quantity, ownership, and control of potable waters remains a most important human concern.

Over 70 percent of the earth's surface is covered by water. The vast majority of this water, however, is ocean salt water, which has limited use. (See fig. 16.1.) Salt water cannot be consumed by humans or used for many kinds of industrial processes. But clean fresh water supplies have always been considered inexhaustible. Only now are we beginning to understand that we will probably exhaust our usable water supplies in some areas because of both human and natural factors. Human factors that have affected usable water supplies include a steadily increasing demand for fresh water for industrial, agricultural, and personal needs.

Shortages of potable water throughout the world can also be directly attributed to human abuse in the form of pollution. Water pollution has negatively affected water supplies in almost all of the world's densely populated industrialized nations, including Japan, western Europe, the Soviet Union, and the United States. The specifics of water pollution will be discussed in chapter 20.

Unfortunately, the outlook for the world's fresh water supply is not very promising. According to studies by the United Nations and the International Joint Commission, many sections of the world will experience shortages of potable water by the year 2000. Areas of particular concern include Mexico, India, Europe, and the United States.

Another serious problem is the fact that the world's supply of water is unevenly distributed. Fresh water sources such as lakes, rivers, and underground water are continually being replenished by rainfall. Rainfall, however, varies significantly throughout the world. (See fig. 16.2.)

Parts of the world, particularly parts of Africa, continue to suffer massive droughts because of long periods of extremely limited rainfall. In other parts of the world, floods result from too much rainfall.

## The hydrologic cycle

All water is locked into a constant recycling process called the **hydrologic cycle.** (See fig. 16.3.) Solar energy evaporates water from the ocean surface. Additional quantities of water are absorbed from the soil and evaporated from the soil and from the surfaces of plants. This water loss from plants is called **transpiration.** The air containing water vapor moves across the surface of the earth. As this warm, moist air cools, water droplets form and fall to the land as precipitation. Although some precipitation may simply stay on the surface until it evaporates, most will either sink into the soil or flow downhill until it eventually returns to the ocean. The water that infiltrates the soil may be stored for long periods in underground reservoirs. This water is called **groundwater,** as opposed to the surface water that enters a river system as **runoff.**

**Figure 16.2 Global average annual precipitation.** The world's precipitation is not uniformly distributed. Although some areas of the world such as Southeast Asia and Central America have extremely high levels of precipitation, other areas such as northern Africa and the Middle East have very little annual precipitation.

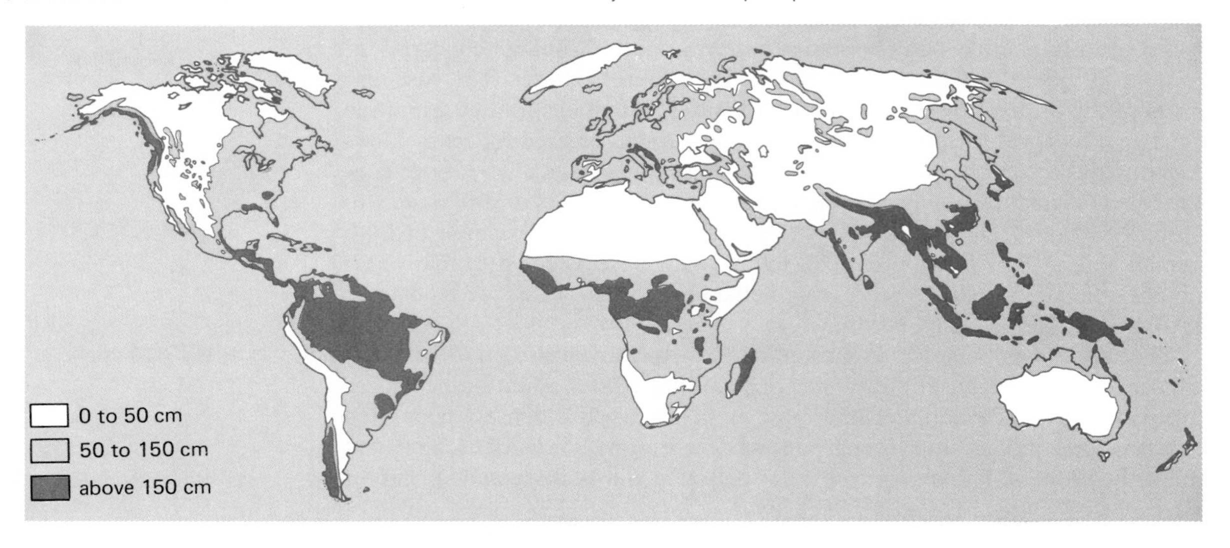


**Figure 16.3 The hydrologic cycle.** This cycle is powered by solar energy. Water evaporates from the oceans, collects in the atmosphere, and then is moved by the wind. When the air cools, the water molecules condense into droplets that fall as precipitation. Some of it falls on the land and runs off the surface. This water eventually returns to the sea.

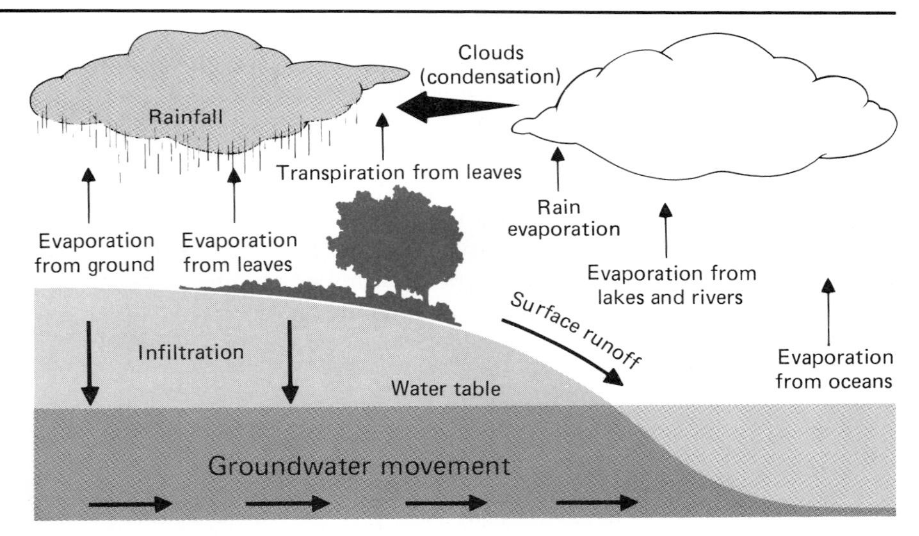


The way in which land is used has a significant impact on evaporation, runoff, and the rate of **infiltration.** When water is used for cooling purposes in power plants or for the irrigation of crops, the rate of evaporation is intensified. Water impounded in reservoirs also evaporates rapidly. This rapid evaporation can affect local atmospheric conditions. Water infiltration is also greatly affected by human activity. Urban complexes with paved surfaces and storm

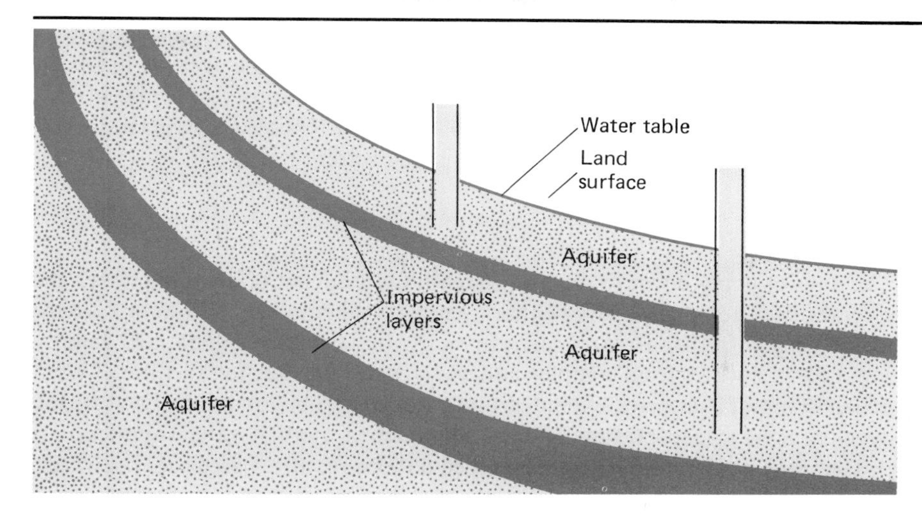


**Figure 16.4 Aquifer.** An aquifer exists when water seeps into the ground and is stored in porous rock layers. Its bottom is an impervious layer of rock so the water cannot seep down further. The top is the water table.

sewers increase runoff and reduce infiltration. Because the demand for underground water in urban areas is usually high, urban developments simply increase the gap between supply and demand.

Water that enters the soil and is not picked up by roots of plants will move slowly downward until it reaches an impervious layer. Because the water cannot pass through this impervious layer, it accumulates in porous strata called **aquifers.** The top of this saturated layer is called the **water table.** Water within the aquifer can still move slowly through the rock. If the underlying rocks are stressed and folded, one end of an aquifer might be higher than the other. (See fig. 16.4.) Water moves by gravity to the lower levels, often over great distances. Aquifers are extremely important in supplying water for industrial, agricultural, and municipal use. Most of the larger urban areas in the western part of the United States depend upon underground water for their water supply. This groundwater can continue to supply water as long as it is not used faster than it can be replaced. Determining how much water can be used and what the uses should be is often a problem.

## Kinds of water use

The use of water can be classified into four categories.

1. Domestic use
2. Agricultural use
3. Instream use
4. Industrial use

Water uses can be measured by either the amount withdrawn or the amount consumed. Water withdrawn for use is diverted from its natural course. It may be withdrawn and then later returned to its source so it can be used again in the future. For example, when a factory removes water from a river for cooling purposes, it returns the water to the river so it can be used again. Water that is incorporated into a product or lost to the atmosphere through evaporation and transpiration cannot be reused in the same geographic area. It has been consumed. Much of the water used for irrigation evaporates. Some of this water is removed with the crop when it is harvested. Therefore, irrigation must be considered to both withdraw and consume water.

### *Domestic use of water*

The days of going down to the local stream for a bucket of drinking water are essentially gone in the United States. Many rural residents still obtain a safe water supply from untreated private wells, but urban residents are usually supplied with water from complex and costly water purification facilities. When urban communities have extended and merged, problems have arisen in the development, transportation, and maintenance of quality water supplies.

Domestic activities in highly developed nations require a great deal of water. This domestic use includes drinking, air conditioning, bathing, washing clothes, washing dishes, flushing toilets, and watering lawns and gardens. On an average, each person in an American home uses 300–400 liters of water each day. To satisfy this demand, municipalities must provide for both the domestic supplies and the treatment of the waste water following its use. Both processes are expensive and require trained personnel to operate them.

The major problem associated with domestic use of water is maintaining an adequate suitable supply for growing metropolitan areas. The demand for water in urban areas sometimes exceeds the immediate supply. This is particularly true where local surface water serves as the domestic supply. During the summer, water demand is high and precipitation is often low.

Most **domestic water** is used for washing and carrying away wastes, with only a small amount being used for drinking. (See fig. 16.5.) Yet all water that enters the house has been purified and treated to make it safe for consumption. Until recently the cost of water in almost every community has been so low that there was very little incentive to conserve. Increasing purification costs have raised the price of domestic water. We waste more domestic water than we consume. As the demand for clean fresh water increases, the cost of water will rise. Perhaps then, people will begin to conserve water and may reduce wasteful use.

The loss of water from domestic supplies is amazingly large. Nearly 20 percent of the water withdrawn from public supplies is lost before it can be used (mainly through leaking water pipes and water mains). Another major cause of water loss has been that of public attitudes. As long as we consider water to be a limitless, inexpensive resource, there will be little effort to conserve. As our attitude towards water changes, so will our usage and our efforts to conserve.

Most domestic water is used as a solvent to carry away wastes. Concentration of people in urban areas compounds the problems of waste water. Natural processes cannot cope with wastes when they are concentrated into a small area. The result is unsightly, smelly, and a potential health problem.

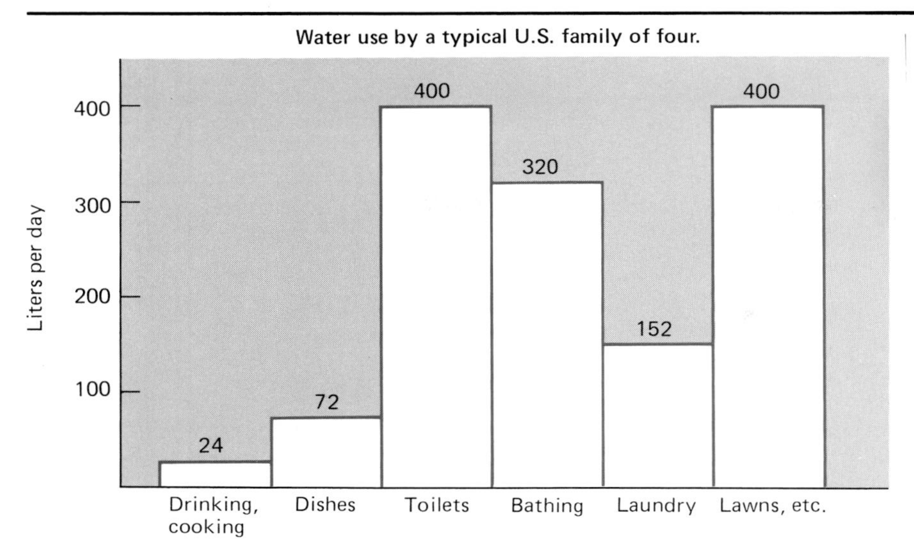


**Figure 16.5 Urban domestic water uses.** Over one hundred billion liters of water are used each day for urban domestic purposes in the United States.

The use of watercourses for waste disposal can degrade the quality of the water and may reduce its usefulness for other purposes. If a watercourse becomes an open sewer, its value for other purposes is significantly reduced. Although we cannot expect to find absolutely pure, clean water flowing in streams in major urban centers, we should not tolerate abuse of this resource.

## *Agricultural use of water*

The major consumptive use of water in the United States is for agricultural purposes. The primary use of water in agriculture is for **irrigation.** In 1984, for example, irrigation accounted for nearly 80 percent of all the water consumed in the nation. The amount of water used for irrigation and livestock continues to increase. Future agricultural demand for water will depend on the cost of water for irrigation; the demand for agricultural products, food, and fiber; federal policies; and the development of new technology.

In some areas, irrigation is a problem because there is not a supply of water nearby. This is particularly true in the western United States, where about fourteen million hectares of land are irrigated. (See fig. 16.6.) In some sections of the West, water must be piped hundreds of kilometers in order to irrigate.

Because most of the world's consumptive use of water results from irrigation, it is becoming increasingly important to modify irrigation practices to use less water. There are many ways to reduce water loss from irrigation. Increasing the cost of water will stimulate conservation of water by farmers. Another way is to reduce the amount of water-demanding crops grown in dry areas. Switching to trickle irrigation will also reduce water loss. This method delivers the water directly to the roots of the plants rather than flooding entire fields.

**Figure 16.6 Irrigation.** Large areas of the United States require irrigation in order to be farmed economically. The circular fields in this photograph have a long pipe, which slowly rotates about a central pivot. In this way, the field is automatically irrigated. Each unit in this photograph irrigates sixty-five hectares.

Irrigation requires a great deal of energy. It has been estimated that 40 percent of the energy devoted to agriculture in Nebraska is used for irrigation. Increasing energy costs may force some farmers to reduce or discontinue irrigation. In addition, much of western Nebraska relies on groundwater for irrigation, and the water table is dropping rapidly. If a shortage develops, land values will decline. Land use and water use are interrelated and cannot be viewed independently.

## *Instream use of water*

When the flow of water in streams is interrupted or altered, the value of the stream is changed. Major **instream uses** of water are for hydroelectric power, recreation, and navigation. Electricity from hydroelectric power plants is an important energy resource. Presently, hydroelectric power plants produce about 13 percent of the total electricity generated in the United States. Hydroelectric power plants do not consume water and do not add waste products to it. However, dams have definite disadvantages, including the high cost of construction and the resultant destruction of the natural habitat in streams and surrounding lands. While dams reduce the amount of flooding, they do not eliminate it. (See fig. 16.7.) In fact, building a dam often encourages people to feel secure and to develop the flood plain. As a result, when flooding occurs, there is a greater potential for loss of property and of lives.

The sudden discharge of the impounded water can seriously alter the downstream environment. If the discharge is from the top of the reservoir, the stream temperature rapidly increases. Discharging the colder water at the bottom of the reservoir causes a sudden decrease in the water temperature of the stream. Such changes are harmful to aquatic life in the stream.

The impoundment of water also reduces the natural scouring action of a flowing stream. If water is allowed to flow freely, the silt accumulated in the river is carried downstream during times of high water. This maintains the

**Figure 16.7 Dams interrupt the flow of water.**
The flow of water in most of our large rivers is controlled by dams. Most of these dams provide electricity. In addition, they prevent flooding and provide recreational areas. However, dams destroy the natural river system.

river channel and carries nutrient materials to the mouth of the river. But if a dam is constructed, the silt deposits behind the dam. Eventually this fills the reservoir with silt.

In addition, impounded water has a greater surface area, which increases the amount of evaporation. In areas where water is scarce, the amount of water lost through such evaporation may be serious. This is particularly evident in hot climates. Furthermore, the flow is often intermittent below the dam, which alters the water's oxygen content and interrupts the migration of fish. The populations of algae and other small organisms are also altered. (See chap. 9.) Therefore, careful planning must be done when considering dam construction.

Water tends to be a focal point for recreation activities, both in rural and urban areas. (See fig. 16.8.) Sailing, waterskiing, swimming, fishing, and camping all require water of reasonably good quality. Water is used for recreation in its natural setting and often is not physically affected. Even so, it is necessary to plan for recreational use, because overuse or inconsiderate use can degrade water quality. For example, waves generated by powerboats can accelerate shoreline erosion and cause siltation.

Dam construction creates new recreational opportunities, because reservoirs provide sites for boating, camping, and related recreation. However, this is at the expense of a previously free-flowing river. Some recreational opportunities, such as river fishing, have been lost.

Most of our major rivers are used for navigation. The United States now has more than forty thousand kilometers of commercially navigable waterways. Waterways used for navigation must have sufficient water to insure passage of transport vessels. Canals, locks, and dams are employed to guarantee this. Often dredging is necessary to maintain the proper channel depth. Although water quality may not be impaired, the flow within the hydrologic system is changed, which in turn affects the water's values for other uses.

**Figure 16.8 Recreational use of water.**
All of these people are enjoying the recreational value of water. Not all recreational uses of water are compatible with one another.

Most large urban areas rely on water to transport needed resources. During recent years, the inland waterway system has carried about 10 percent of the nation's goods, such as grain, coal, ore, and oil. The traffic on inland waterways has continued to increase because waterways have been improved for navigation. Federal expenditures for the improvement of the inland waterway system have totalled billions of dollars.

In the past, almost any navigation project was quickly approved and funded, regardless of the impact on other uses. Today, however, such decisions are not made until the impact on various other uses are carefully analyzed.

## *Industrial use of water*

Water for **industrial use** accounts for more than half of the total water withdrawals. Ninety percent of the water used by industry is for cooling. Most industrial processes involve heat exchanges. Water is a very effective liquid for carrying heat away from these processes. Electric-generating plants use water to cool steam so that it changes back into water. If the heated water that is used in industrial processes is dumped directly into a watercourse, it significantly changes the water temperature in the watercourse. This affects the aquatic ecosystem by increasing the metabolism of the organisms and reducing the water's ability to hold dissolved oxygen. Therefore, industries are being required to construct cooling towers or ponds in order to reduce thermal pollution. (See fig. 16.9.) (See chap. 20 for a discussion of thermal pollution.)

**Figure 16.9 Thermal pollution.** Industrial use of water for cooling can significantly increase the temperature of the water. Photograph *a* shows hot water being discharged from a power plant. Industries are now required to cool the water before returning it to the watercourse. The cooling towers in photograph *b* allow the heat to dissipate into the atmosphere.

a

b

Industry also uses water to dissipate and transport waste materials. In fact, many streams are now overused for this purpose, especially watercourses in urban centers. The use of watercourses for waste dispersal degrades the quality of the water and may also reduce its usefulness for other purposes. This is especially true if the industrial wastes are toxic.

During the past thirty years, many laws have been passed that severely restrict industrial discharge of wastes into watercourses. The federal role in maintaining water quality began in 1948 with the passage of the Federal Water Pollution Control Act. This act provided federal funds and technical assistance to strengthen local, state, and interstate water-quality programs. Through amendments to the act in 1956, 1965, and 1972, the federal role in water pollution control was increased to include establishing water-quality standards, financing areawide waste-treatment management plans, and establishing the framework for a national program of water-quality regulation. All of these activities require an integrated water-use plan.

## Water-use planning issues

In the past, wastes were discharged into waterways with little regard to the costs imposed on other users by the resulting decrease in water quality. With today's increasing demands for high-quality water, unrestrained waste disposal will lead to serious conflicts about water uses and will cause social, economic, and environmental losses.

Increased demand for water will force increased reuse of existing water supplies. In many areas where water is used for irrigation, both the water and the soil become salty because of evaporation. When this water returns to a stream, it lowers the quality of the water.

## Box 16.1
## Waterless toilets

As fresh water becomes a more rare resource in parts of the world, we will be forced to look for alternatives to many of our current uses of water. A waterless toilet would save considerable amounts of water, reduce the need for sewage treatment, and provide compost.

Rikard Lindstrom was the developer of the Clivus Multrum organic waste-treatment system over thirty years ago in Sweden. The Clivus Multrum system employs the principles of gravity, natural draft, and unaided microbial decomposition. The container is nine feet long, four feet wide, and five feet high—large enough to hold all the organic wastes of a family for several years while the wastes decompose. It can be located in a basement (directly below the toilet and garbage inlet), or it can go in the ground outside the house.

The inclination of the floor of the container and the air flow through the mass make turning of the wastes unnecessary. After decomposition by bacteria, the compost is safe to use in gardens. In addition, there are no odors in the bathroom or kitchen.

This method of treating organic wastes has important economic implications. It could virtually eliminate central waste-water treatment facilities, resulting in a substantial savings for the taxpayers. The system, however, is not without some problems, most notable of which are cost and restrictive local health codes.

Currently the Clivus Multrum system is expensive ($1,500 to $2,000 per unit). This cost, however, would be offset by cheaper water and sewage rates. The cost will, no doubt, decrease when mass production of the units begins. Unfortunately, in many sections of the United States, outdated health codes prohibit the use of such systems. But it is likely that these codes will also be modified when the price of water rises and when the "waterless toilets" prove to be a safe, clean, and viable alternative to our present system of waste removal.

In other areas, wells provide water for all categories of use. If the groundwater is pumped out faster than it is replaced, the water table is lowered. In coastal areas, seawater may intrude into the aquifers and ruin the water supply. The demand for water-based recreation is increasing dramatically and requires high-quality water, especially for water recreation involving total body contact such as swimming.

### *Preserving scenic water areas and wildlife habitats*

As was mentioned earlier, the use of land influences water quality. This is particularly true along shorelines and river banks, where the land and water meet. Some bodies of water have unique scenic value. To protect these resources, the way in which the land adjacent to the water is used must be consistent with preserving these scenic areas.

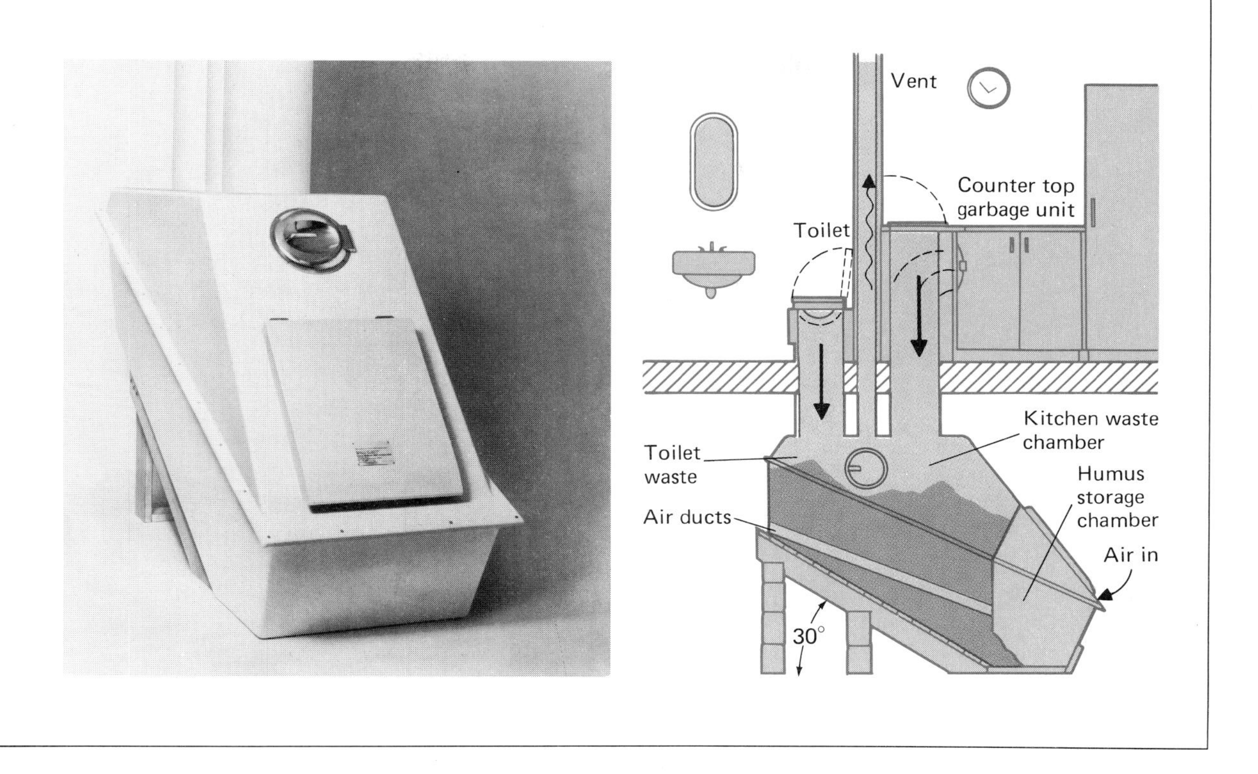


The Federal Wild and Scenic Rivers Act of 1968 is aimed at preserving some of these resources. This act established a system to protect wild and scenic rivers from development. All federal agencies must consider the wild, scenic, or recreational values of rivers in planning for the use and development of rivers and adjacent land resources. The process of designating a river or part of a river as wild or scenic is complicated. It often encounters local opposition from businesses dependent on growth. Following reviews by state and federal agencies, rivers may be designated as wild and scenic by action of either Congress or the secretary of the interior. Sections of fifty-five streams have been designated as wild or scenic.

Many unique and scenic shorelands have also been protected from future development. Until recently, the nation's estuaries and shorelands have been subjected to significant physical modifications. Modifications, such as dredging and filling, may improve conditions for navigation and construction, but this

**Figure 16.10**
**Shoreland protection.**
This map of the Cape Cod Bay area indicates the shoreland areas of particular scenic, historic, or wildlife value that are protected by governmental action from development.

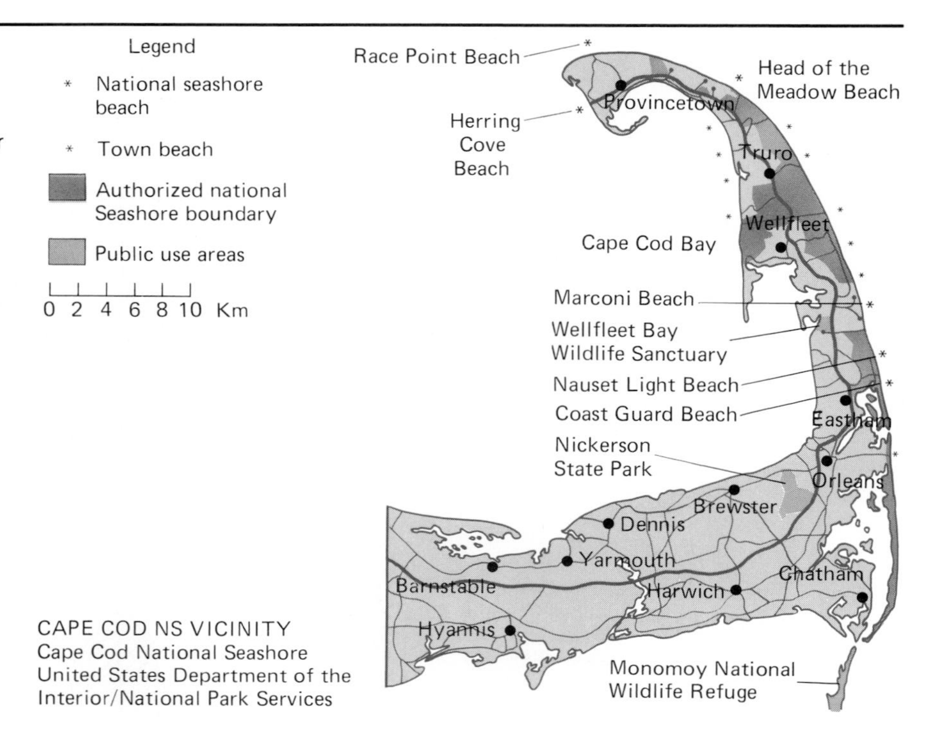


alteration destroys fish and wildlife habitats. Recent action at both the state and federal levels has attempted to restrict the development of shorelands. Development has been restricted in some particularly scenic areas. Cape Cod National Seashore in Massachusetts is one such example. (See fig. 16.10.)

Historically, poorly drained areas were considered worthless. Subsequently, many of these wetlands were filled or drained. The natural and economic importance of wetlands has been recognized only recently. In addition to providing necessary spawning and breeding habitat for many species of wildlife, wetlands act as natural filtration systems. These areas trap nutrients and pollutants and prevent them from entering adjoining lakes or streams. Wetlands slow down the force of flood waters and permit nutrient-rich particles to settle out. In addition, wetlands can act as reservoirs and release water slowly into lakes, streams, or aquifers, thus preventing floods. Coastal estuarine zones and adjoining sand dunes also provide significant natural flood control. Sand dunes act as barriers and absorb damaging waves caused by severe storms.

## *Groundwater mining*

**Groundwater mining** removes water from an aquifer faster than it is replaced. This problem becomes serious when this practice continues over a long period of time because the water table eventually declines. Groundwater mining is currently used in areas of the West due to the growth of cities and the use of increased amounts of water for irrigation. In aquifers with little or no recharge, virtually any withdrawal constitutes mining, and sustained withdrawals will eventually exhaust the supply. This problem is particularly serious in communities (such as Garden City, Kansas) that depend heavily upon groundwater for their domestic needs.

For twenty million years, much of the precipitation that fell on the Great Plains infiltrated the sand and gravel aquifer surrounding Garden City, Kansas. For nearly one hundred years, wheat (a low-water demanding crop) was the predominant crop. But in 1960 the farmers began to tap the groundwater in the Ogallala Aquifer, and because of this availability of water, corn (which requires more water than wheat) quickly replaced wheat as the main crop. For $4.50, a farmer could distribute one meter of water over one hectare of land.

An economic boom resulted when the farmers began to irrigate and grow corn. Today, nearly twenty-five thousand wells are irrigating 1.4 million hectares of land, and the corn is being fed to feedlot cattle. Five large packing plants in the area process enough cattle in a single day to feed a million people.

There is much controversy regarding the amount of irrigation in the area. Although irrigation has been in operation for only twenty-five years, the cost of pumping one meter of water to one hectare of land could increase to $250. The increase in cost is a result of higher fuel costs, inflation, and the decreased amount of water in the aquifer. In twenty-five years of pumping, the water level in the aquifer has dropped four meters. At the present rate of pumping, it is estimated that the aquifer will be dry by the year 2000.

Soil conservationists predict that when the water is no longer available from the aquifer, farmers will not be able to produce a crop and that the conditions will be similar to those of the Dust Bowls of the 1930s. Farmers, cattle growers, and slaughterhouse owners are resisting any attempts to limit the amount of water allowed for irrigation. They are certain that new sources of water will be found.

Groundwater mining can also lead to problems of settling or subsidence of the ground surface. This happens when withdrawal of groundwater exceeds its replenishment rate by rainfall. Removal of the water allows the ground to become compacted and large depressions may result.

For example, in the San Joaquin Valley of California, groundwater has been withdrawn for irrigation and cultivation since the 1850s. In the last forty years, groundwater levels have fallen over one hundred meters. More than one thousand hectares of ground have subsided, some as much as six meters. Currently, the ground surface in that area is sinking thirty centimeters per year. In 1981, in Winter Park, Florida, a large area of subsidence or "sinkhole" occurred because of excessive water withdrawal. (See fig. 16.11.)

Groundwater mining is a serious problem in western Texas. This area depends on irrigation for agriculture. The population of this area has also increased dramatically over the last twenty-five years. The ever-increasing demand for groundwater has led to its rapid depletion. Precipitous declines in agricultural production are forecast within the next ten years. As the groundwater is depleted, the land values will decline. Land use then directly affects

**Figure 16.11 The development of a "sinkhole."**
If the water table is lowered as a result of groundwater mining or a severe drought, the land surface can subside. The space formerly occupied by water can collapse and create sinkholes. This photo shows only one of many sinkholes that occurred in Florida in 1981 as the result of the combined effects of excessive groundwater use and a long drought.

water use, and water use and availability directly influence land use. London, Mexico City, Venice, Houston, and Las Vegas are some other cities that have experienced subsidence as a result of groundwater withdrawal.

Groundwater mining poses a special problem in coastal areas. As the fresh groundwater is pumped from wells along the coast, the saline groundwater moves inland, replacing fresh groundwater with unusable salt water. (See fig. 16.12.) Saltwater intrusion is a serious problem in heavily populated coastal areas, such as New York, New Jersey, southern California, and Florida.

## *Salinization*

Another water-use problem results from the increasing salt concentrations in soil. When plants extract the water they need, the salts present in all natural waters become concentrated. Irrigation of arid farmland makes this problem more acute because so much water is lost due to evaporation. Irrigation is most common in hot, dry areas, which normally have high rates of evaporation. This results in a concentration of salts in the soil and in the water that runs off the land. (See fig. 16.13.) Every river increases in salinity as it flows to the ocean. Salinity of the Colorado River water increases twenty times as it passes through irrigated cropland between Grand Lake in north central Colorado and the Imperial Dam in southwest Arizona. The problem of salinity will continue to increase as irrigation increases.

**Figure 16.12 Salt water intrusion.** This figure illustrates how salt water can intrude upon fresh groundwater. When this occurs, the groundwater becomes unusable for human consumption and for many industrial processes.

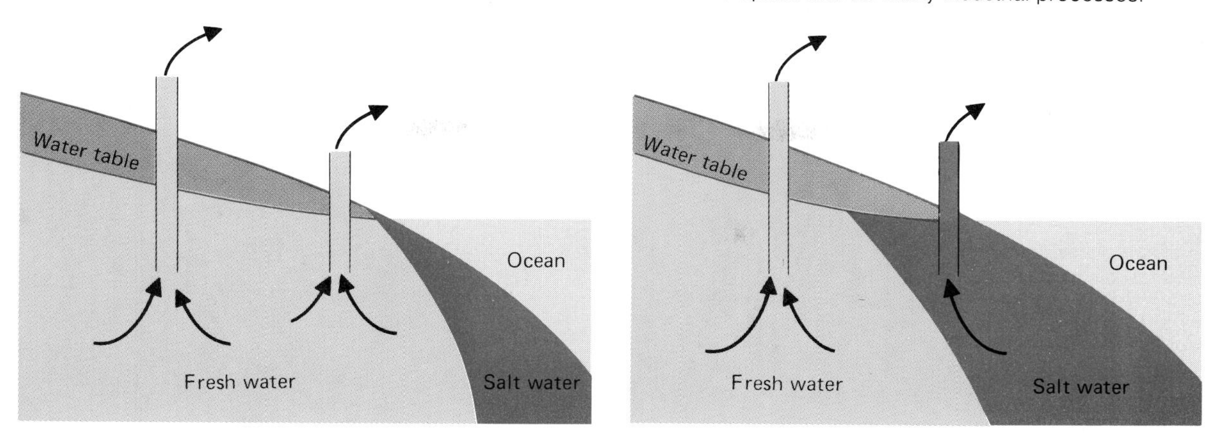

**Figure 16.13 Salinization.** As water evaporates from the surface of the soil, the salts it was carrying are left behind. In some areas of the world, this has permanently damaged cropland. Other areas must flush the soil to rid it of salt in order to maintain its fertility.

## *Water diversion*

**Water diversion** is the physical process of transferring water from one area to another area. Early examples of water diversion were the aqueducts of ancient Rome. Thousands of diversion projects have been constructed since then. New York City, for example, diverts water from Pennsylvania (250 kilometers away), and Los Angeles obtains part of its water supply from the Colorado River.

While diversion is necessary in many areas of the world, it can be misused or detrimental. An example of this is the proposed Garrison Diversion Unit, which is designed to divert water from the Missouri River system for use in irrigation. (See fig. 16.14.)

**Figure 16.14 The Garrison Diversion Unit.** This unit would divert water that flows into the Missouri River through the McClusky Canal to the Sheyenne River and eventually into Hudson Bay.

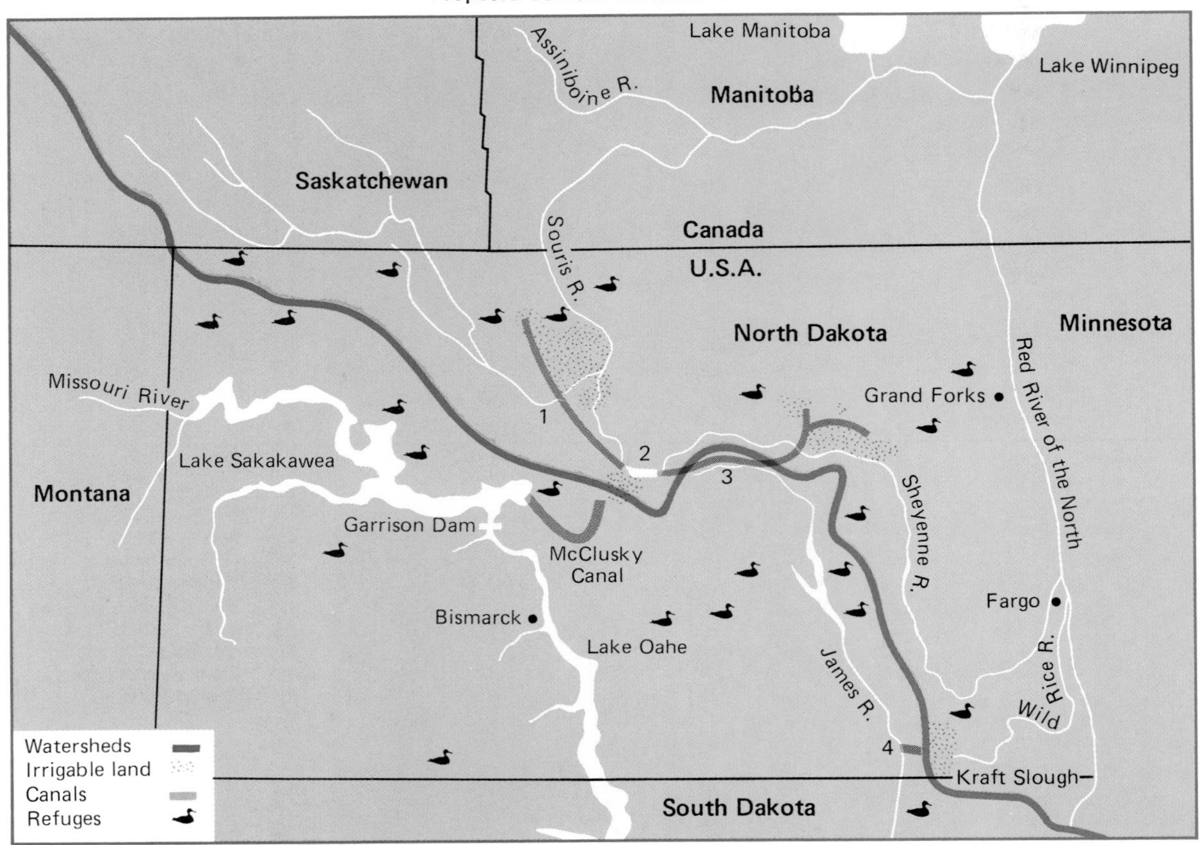


This action would have enormous environmental consequences. Wildlife refuges, native grasslands, forests, and waterfowl breeding marshes would be damaged or destroyed. In addition, hundreds of kilometers of streams could be seriously damaged. There is considerable opposition to this plan both by private groups and governmental agencies. In fact, the Canadian government has fought against the project.

In contrast, the people who need the water for irrigation see the Garrison Diversion Unit plan as beneficial. If this plan was implemented, they could raise corn (a more profitable crop), rather than the dryland crop of wheat. The corn could then be used for cattle feed.

The initial plan to irrigate this portion of the Great Plains began during the Dust Bowl era of the 1930s. The Federal Flood Control Act of 1944 authorized the construction of the Garrison Diversion Unit, but controversy concerning it has continued since its proposal.

This controversy has resulted in various changes in the political climate, which has altered the scope of the project. Cecil Andrus, Secretary of Interior for President Carter, called it a "dog." James Watt, President Reagan's Secretary of Interior, viewed the undertaking more favorably. In 1981, he proposed that the irrigation of thirty-five thousand hectares of the project be completed within a ten-year period. However, in 1982, the House of Representatives killed any funds for this purpose, although the Senate still favored its construction.

In 1984, William Clark (who succeeded Watt) appointed a twelve-member commission to resolve the conflict concerning the Garrison Diversion Project. The commission recommended a $906 million expenditure for irrigation of forty thousand hectares and cancellation of the building of two major reservoirs.

When Donald Hodel succeeded Clark, he stated that the Department of Interior supported the project. However, he did not clarify whether this support was for the original proposal or for the proposal that was recommended by the twelve-member commission. While the political struggle continues, the citizens of North Dakota appear indifferent. A poll taken in March, 1985, indicated that 35 percent favored the project, 15 percent opposed it, and 53 percent had no opinion. Thus forty years after its proposal, the controversy over the Garrison Diversion Unit continues.

### *Managing urban water use*

Providing water services for metropolitan areas is another serious water-planning issue. Metropolitan areas must provide three basic water services.

1. Water supply for human and industrial needs
2. Waste-water collection and treatment
3. Storm-water collection and management

In order to provide water for human and industrial use, the water must be properly treated and purified. It is then pumped through a series of pipes to the consumers. After the water is used, it flows through a network of sewers to a waste-water treatment plant before it is released. Metropolitan areas must also deal with great volumes of excess water during storms. Because urban areas are paved and little rain water can be absorbed into the ground, management of storm water is a significant problem. Cities often have severe local flooding because the water is channeled along streets to storm sewers. If these sewers are overloaded or blocked with debris, the water cannot escape and flooding occurs.

Many cities have a single system to handle both sewage and storm-water runoff. During heavy runoff, the flow is often so large that the waste-water treatment plant cannot handle the volume. The waste water is then diverted directly into the receiving body of water without first being treated. Some cities have areas in which to store this excess water until it can be treated. This is expensive and, therefore, is not usually done unless governmental grants are available.

## Consider This Case Study
### The California Water Plan

The management of fresh water is often a controversial subject involving social, ecological, and economic aspects. A good example is the California Water Plan.

In the early 1900s, it became clear that the growth of Los Angeles, which was then a small coastal town, would be encouraged by irrigating the surrounding land. Los Angeles looked to the Owens Valley, four hundred kilometers north, for a source of water. The Los Angeles Aqueduct connecting these two areas was completed in 1913.

Since the Owens Valley project, California has developed a statewide water program known as the California Water Plan. This water plan was necessary because most of the state's population and irrigated land are found in the central and southern regions, but most of the water is in the north. The plan details the construction of aqueducts, canals, dams, reservoirs, and power stations to transport water from the north to the south. (See the map on the next page.) In addition to supplying water for southern regions, the aqueducts provide irrigation for the San Joaquin Valley. Eventually, the new land made available for agriculture will amount to about four hundred thousand hectares.

The California Water Plan, which has spanned some thirty years, has been one of the most controversial programs ever undertaken in California. Adoption of the plan instigated a sectional feud between the moist "north" and the dry "south." Southern California was accused of trying to steal northern water, but because the population concentration in the southern part of the state carried the vote, the plan was adopted. Environmentalists still claim that the project has scarred the countryside irreparably and upset natural balances of streams, estuaries, vegetation, and wildlife. They argue that providing water to southern California would promote population growth, which would lead to further urbanization and land development.

Many questions and controversies about the water plan center on whether the water is really needed. Ninety percent of the water used in southern California is for irrigation; there is evidently abundant water for domestic and industrial use. Although the cost is borne by all the rate payers, most of whom are urban, California is now one of the nation's most productive agricultural areas.

Allocation of water resources is a matter of economics as well as a matter of technology. The California water project has been criticized for using public funds to increase the value of privately held farmland. Furthermore, technological advances in desalination plants may give a new dimension (unforeseen when the water plan was devised) to the problem of water resources. Although more aqueducts, canals, and pumping plants are planned for the 1980s, it is uncertain whether they will be completed.

What are the major advantages of such a plan?
What problems develop when water is transported to arid regions?

In a recent drought, San Francisco was short of water and southern California was not.

What implications does this have for expansion of the existing plan?
Should water from northern California be sent to southern California?

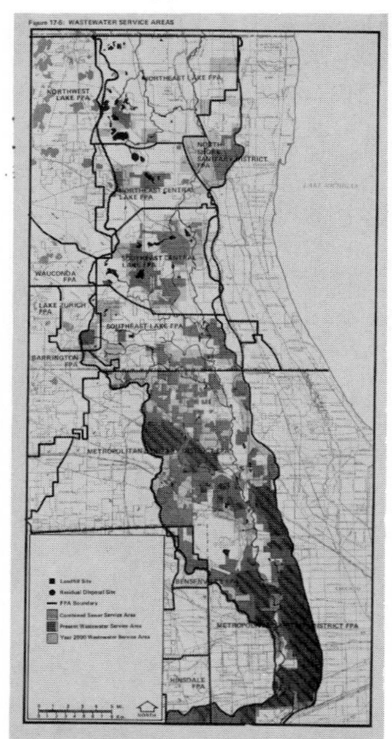


**Figure 16.15 Metropolitan water-use planning.**
The Chicago metropolitan area has done a good job of managing its water services. This area still has 349 separate water supply systems and 135 waste-water disposal systems to serve its people. This map shows some of the overlapping jurisdictions.

All water services provided by metropolitan areas are expensive. These services must be provided with an understanding that there are limited water supplies. Also, there is a limited ability of water to dilute and degrade pollutants. Proper land-use planning is essential if these objectives are to be met.

In pursuing these objectives, city planners encounter many obstacles. Large metropolitan areas often have hundreds of local jurisdictions (governmental and bureaucratic layers) that divide responsibility for management of basic water services. The Chicago metropolitan area is a good example. This area is composed of 6 counties and approximately 2,000 local units of government. It has 349 separate water supply systems and 135 separate waste-water disposal systems. Imagine the complications and frustrations in trying to implement a water-management plan when so many layers of government are involved. (See fig. 16.15.)

## Summary

Water is a renewable resource that circulates continually between the atmosphere and the surface of the earth. The energy for the hydrologic cycle is provided by the sun. Water loss from plants is through transpiration. Water that infiltrates the soil and is stored in underground reservoirs is called groundwater, as opposed to surface water that enters a river system as runoff. The way in which land is used has a significant impact on the evaporation, the runoff, and the rate of infiltration.

The four human uses of water are domestic, agricultural, instream, and industrial. Water uses are measured by either the amount withdrawn or the amount consumed. Domestic water is in short supply in many metropolitan areas. Most domestic water is used for waste disposal and washing, with only a small amount being used for drinking. The largest consumptive use of water is for agricultural irrigation. Major instream uses of water are for hydroelectric power, recreation, and navigation. Most industrial uses of water are for cooling and for dissipating and transporting waste materials.

Reduced water quality can seriously threaten land use and in-place water use. Federal legislation helps to preserve certain scenic water areas and wildlife habitats. Shorelands and wetlands provide valuable services as buffers, filters, reservoirs, and wildlife areas. Water management concerns that are of growing importance are groundwater mining, increasing salinity, and planning urban use of water. Urban areas face several problems, such as providing suitable drinking water, waste-water treatment, and handling storm-water runoff in an environmentally sound manner. Water planning involves many governmental layers, which makes effective planning difficult.

## Review Questions

1. What are the three major water services provided by metropolitan areas?
2. How does irrigation increase salinity?
3. List several uses of water that are nonconsuming and nonwithdrawing.
4. What is the major industrial use of water?
5. Describe the hydrologic cycle.
6. What is the Federal Wild and Scenic Rivers Act? Why is it important?
7. Define groundwater mining.
8. What are the similarities between domestic and industrial water use? How are they different from instream use?
9. How is land use related to water quality and quantity?
10. Why is storm-water management more of a problem in an urban area than in a rural area?

## Suggested Readings

National Water Commission. *Water Policies for the Future.* Final Report to the President by the National Water Commission. Water Information Center, Inc., 1973. A good account of all the problems associated with water.

Sheridan, David. "The Underwatered West: Overdrawn at the Well." *Environment* 23: 7–13. Concerned with irrigation and other water problems in the West.

———. "The Desert Blooms—At a Price." *Environment* 23: 5–20. Excellent presentation of many of the problems facing the use of water in the West.

U.S. Bureau of Reclamation. *Colorado River Water-Quality Program.* Washington, D.C.: U.S. Government Printing Office, 1973. An excellent case study of water problems.

U.S. Water Resources Council. *The Nation's Water Resources.* Washington, D.C.: U.S. Government Printing Office, 1976. Comprehensive overview of federal water policies.

Wollman, Nathaniel, and Bonem, Gilbert. *The Outlook for Water: Quality, Quantity, and National Growth.* Published for Resources for the Future, Inc. Baltimore, Md.: by the Johns Hopkins Press, 1978. An in-depth study with supporting data.

# PART 5

# Pollution

The quality of the environment is something that most people have strong feelings about. However, these feelings vary greatly, people do not agree on what is good or bad for environmental quality. Economics and politics have become interwoven in the environmental quality equation. Great differences of opinion exist about what environmental quality should be or what the current quality of the environment is. This section deals with pollution, which is invariably defined in terms of negative effects.

It is important to understand that the degree of pollution can often be quantified and that pollution may have different degrees of seriousness depending on a variety of local conditions. Although pollution by definition is harmful, different amounts of pollution can be tolerated, depending on the specific situation.

Chapter 17 discusses the interrelated forces that led to a technological society. The development of major pollution problems as a natural consequence of public attitude is a central theme. When pollution became a public health problem, the public's attitude toward it began to change. This led to the initiation of major legislative changes.

Chapters 18, 19, and 20 deal with specific kinds of pollution problems and the approaches used in dealing with them. Throughout this section, economic and political realities are discussed as a part of the pollution equation.

## Chapter Outline

I. Historical basis of pollution
  A. Unlimited resource availability
     Box 17.1 Early industrial community
  B. Large capacity of resources to absorb abuse
  C. Growing populations overwhelm recovery rate of resources
  D. High technology = high pollution
II. Changing attitudes of the public
  A. Pressure groups representing "quality of life"
  B. Political activism
  C. Governmental regulations
  D. People pay
III. Consider this case study: Oregon bottle bill

## Key Terms

biodegradable

environmental quality

pollution

public resources

technological advances

## Objectives

Explain how technological progress and increased population density have caused pollution.

Define pollution.

Explain how the seemingly endless natural resources of North America fostered an attitude of waste.

Describe the role of water and other natural resources in the early industrial development of the United States.

Recognize that even though natural resources can tolerate a great deal of abuse, they can be irreversibly damaged.

Explain how concerns about living and working conditions in the 1800s resulted in improved environmental quality.

Recognize that increased technology not only results in more pollution but also generates nonbiodegradable pollutants.

Explain how intense lobbying by pressure groups and court action have brought about a number of improvements in the environment.

Explain how the media are used by environmental pressure groups.

Explain how all levels of government have become involved in regulating environmental quality.

Explain how to pay for improved environmental quality.

# CHAPTER 17
# People, progress, and pollution

## Historical basis of pollution

Humans are always concerned about potential threats to their welfare. Ancient people attempted to describe these forces in their folklore and mythology. "The Four Horsemen of the Apocalypse" represents such an attempt: Four riders—famine, pestilence, disease, and war—were pictured as riding across the earth destroying human life. Three of the riders—famine, pestilence, and disease—are natural phenomena. The fourth, war, is a product of human activity.

Throughout history there have been numerous attempts to eliminate the misery caused by hunger and disease. In general, we rely on science and technology to improve the quality of life. However, technological progress often offers a short-term solution to a specific problem but in the process can create an additional problem—pollution. While scientific advancement and technology attempts to alleviate human misery, they might actually degrade the quality of life for future generations. For example, improving sanitary conditions in Thailand does decrease the spread of waterborne diseases but, in turn, results in an increase in the population, which must then be fed, housed, and clothed. This attempt to improve life resulted, ultimately, in increased crowding and hunger.

Not everyone defines pollution in the same way, so there will be differences of opinion about what a pollutant is. **Pollution** is sometimes defined as something that people produce in *large enough quantities* that it interferes with our health or well-being. In many cases, it is difficult to prove whether a substance is causing harm or whether the amount produced is large enough to cause damage.

In 1981, a small city in the midwest discovered traces of arsenic in its groundwater supply. It still has not been determined whether trace amounts can be considered hazardous to the health of the local population or what the specific source of the arsenic is. In this case it is unclear whether arsenic should be considered a pollutant.

If we accept the definition of pollution as materials that can be hazardous to health, certainly, toxic heavy metals such as arsenic should be considered toxic, but we just do not know how much arsenic contamination is allowable before harm occurs. Over a period of time, are the small quantities of arsenic in groundwater great enough that they will accumulate and do damage to the population? Is it likely that one part of the population might be more susceptible than another part? Will children with smaller body masses be more affected than adults? How did the arsenic get into the groundwater? Was it the result of intense spraying of apple trees in past generations, the result of disposal of a waste from an industry no longer in operation, or the result of the dissolution of the parent rock material? The answers to these questions are necessary before we can label the arsenic as a pollutant. Once we determine that a material is a pollutant, we still need to identify potential problems connected to the pollutant and how to deal with those problems. A first step in this determination is to decide whether the contaminant is merely an annoyance or if it is truly a health hazard. (See figure 17.1.)

Early Roman history cites examples of air and water pollution; however, widespread pollution began more recently. Increases in the human population and the advance of technology are the main factors that contribute to modern pollution. When the human population was small and people lived in a primitive manner, the wastes produced were biological and so dilute that they often did not constitute a pollution problem. People used what was naturally available and did not manufacture many products. Humans, like any other animal, fit into their natural ecosystems. Their waste products were **biodegradable** materials. Biodegradable materials are a source of food for decomposers, which organisms break the material down into simpler chemicals, such as water and carbon dioxide.

**Figure 17.1 Forms of pollution.**

Health Hazard

Annoyance

Fish kill—an indication of water contamination

Nuclear power plant cooling towers—possible thermal pollution and radiation hazards

Smoke from stack—contains particulate material, which could cause lung problems

Smog—an indication that thermal inversion has kept the air contamination in the valley

Traffic—fumes (HC, $NO_x$, PAN, etc.) from internal combustion engines cause eye irritation

Feed lot—odor pollution as well as a source of water contamination from surface runoff

Strip—visual pollution, which is an annoyance but not a health hazard

Litter—an indication that we need to become more aware of how we dispose of materials

**Figure 17.2 Factory town.**
Factory towns developed during the Industrial Revolution. People lived in houses close to the factories. The workers and their families were exposed to constant air pollution.

Pollution began when human populations became so concentrated that their waste materials could not be broken down as fast as they were produced. As the human population increased, people began to congregate and establish cities. The release of large amounts of smoke and other forms of waste into the air caused an unhealthy condition because the pollutants were released faster than they could be absorbed by the atmosphere.

Another example of early pollution was the dumping of large amounts of human waste into waterways. It was thought that a river, in about twenty kilometers, would degrade wastes sufficiently so the water could be reused. If large concentrations of people use the same waterway for waste disposal, the river may not be able to clean itself. Cities must be far apart or waste purification techniques must be used.

During the Industrial Revolution, the population became further concentrated around factories. (See fig. 17.2.) These early factories and the resulting towns were located along rivers. The river provided power for machinery, a means of transporting material, and a site for waste disposal.

## *Unlimited resource availability*

The United States attracted many immigrants because they heard of its abundant resources. This country was pictured as having unending forests extending from the Atlantic to the Pacific. (Early speculators did not mention the plains.) The early settlers must have truly believed that the nation did indeed possess "unlimited" resources. The streams yielded a rich harvest of fish, provided water power, and a means of transportation. The endless stands of uncut timber promised to last forever. The land was so plentiful that the government gave it away. In fact, people showed no concern for conserving the resources at hand. The philosophy of the times was one of "consume the resources and prosper."

Some resources, such as land, water, and air, have been considered **public resources.** Because everyone "owned" these resources, everyone felt that he or she should be able to use them freely. Because the population was small and the resources were vast, few questioned how they were to be used.

In the classic essay, *The Tragedy of the Commons,* Garrett Hardin points out that when there is common ownership of a resource, no individual assumes responsibility for husbanding it. Hardin used the example of cattle grazing on public land. No single owner of cattle wanted to restrict the use of the area because this would give an economic edge to neighbors. As a result, overgrazing became a problem for all the cattle owners.

In much of the country, lumber interests purchased public land and clear-cut all the timber. Having cut the timber, they had no further use for the land, so they let it revert to government ownership. In other sections of the country, mining companies removed ore from the ground without regard to the value of the land. Air and water were also used by industry as if they were free and infinite.

During the 1800s, "factory towns" developed. The factory was constructed first. Everything else was built around it. The best riverfront sites were occupied by the new factories. The business district, homes, and schools occupied the remaining land. If suitable building sites were in short supply, the buildings in the town were crowded close together. Today a drive through much of New England shows the prominence of the factory in the development of these communities.

These early industries used water freely. In 1856, Charles Cowley wrote *A Hand Book of Business in Lowell, with a History of the City* (see box 17.1). This article reflects the attitude of the times. He states that "the wild Merrimac, which once rolled unchecked to the Atlantic, has been tamed to the purposes of man, domesticated to labor, and charmed into bondage to the wizards of mechanism." Nowhere does he mention pollution. The fact that a double row of factories and two hundred thousand people were disposing their refuse into the water was not a concern. The policy of one community using the water in a stream and polluting it was an accepted practice. The next town downstream used this slightly polluted water and polluted it even more. (See fig. 17.3.)

Industry still requires large amounts of water for the processing of materials. (See table 17.1.) Availability of water for processing determines the location of many types of industry. As the number and size of factories increased, the amount of pollution increased until the rivers were no longer able to cleanse themselves.

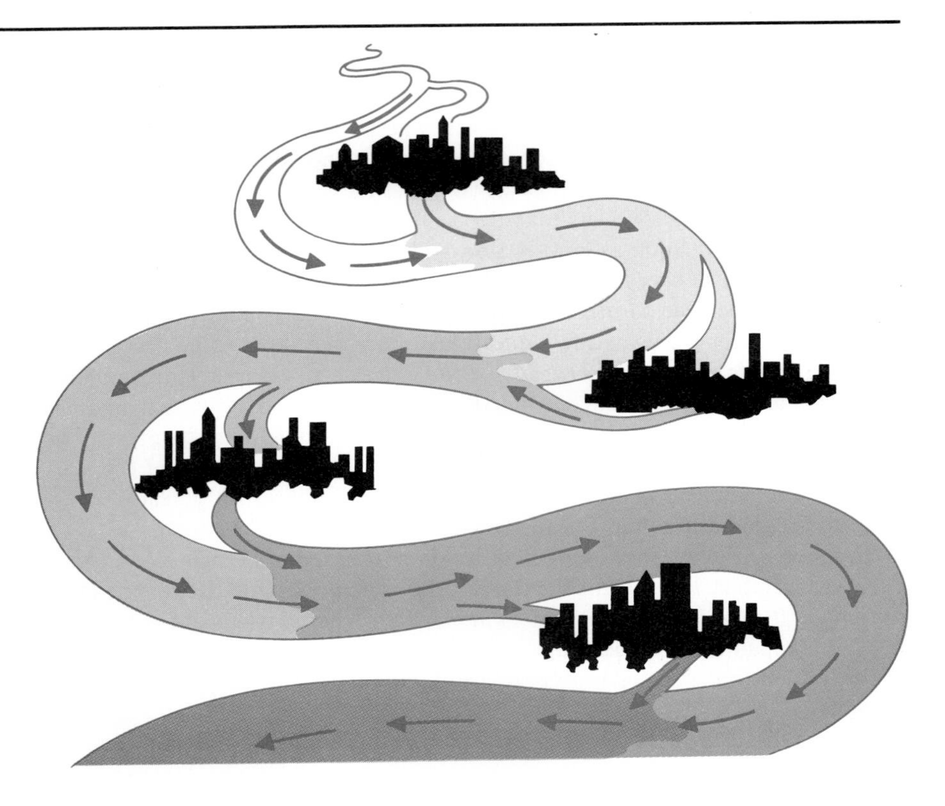

**Figure 17.3 Abuse of a public resource.**
Each community used the water but had little concern for the quality of the water that flowed to the towns downstream. Such a policy was typical of the lack of regard for water resources well into the 1900s.

**Table 17.1**
Industrial uses of water

| One metric ton of processed | Liters of water required |
| --- | --- |
| Steel | 400,000 |
| Cardboard | 350,000 |
| Sugar | 100,000 |
| Cotton | 60,000 |
| Soap | 16,000 |

## *Large capacity of resources to absorb abuse*

The uncaring attitude toward public resources was due in part to the ability of these resources to absorb abuse. The land that once supported vast stands of timber was damaged by the logging industry. However, the exposed soil left by the loggers soon was covered by young trees. People saw no permanent damage as a result of logging operations, even though an entire ecosystem had been destroyed.

Other natural resources also were able to absorb large amounts of abuse. In areas of heavy air pollution, the concentrations of smoke and other particles in the air were constantly blown away from the polluted area and diluted in the atmosphere. The wind and rain continually cleansed the atmosphere.

## Box 17.1
## Early industrial community

In 1856, Charles Cowley wrote the following account in *A Hand Book of Business in Lowell, with a History of the City:*

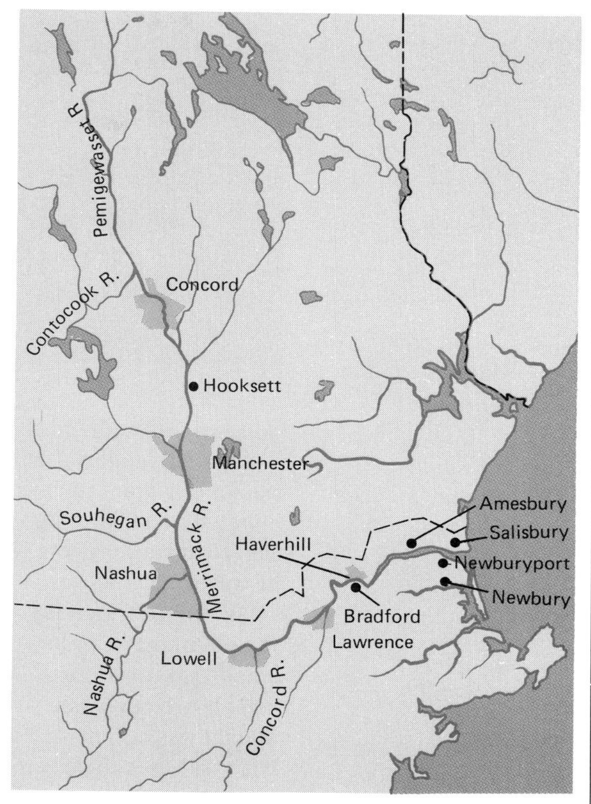


The head of this river is on the northerly border of Merrimac County, in New Hampshire. Here the Winnipissawkee, the outlet of the lake of that name, unites with the Pemigewasset, which rises in the White Mountains, that "milk the clouds." The union of these two streams forms the great river of the valley. The tributary waters of the Contoocook, Souhegan, Nashua, Concord, and a hundred lesser streams are received by the Merrimac at various points on its course. The general direction of the river, during the first 80 miles of its career, is southerly but after entering Massachusetts, it bends to the northeast. Having described a journey of a 110 miles, it discharges into the Atlantic, a brief distance below Newburyport.

Its course is interrupted by numerous waterfalls, which furnish incipient agents of mechanism, that will endure till manufacturers perish, or these waters cease to flow. By means of dams and canals, these natural resources of water power have been augmented and improved. The wild Merrimac, which once rolled unchecked to the Atlantic, has been tamed to the purpose of man, domesticated to labor, and charmed into bondage to "the wizard of mechanism." Populous cities, great beehives of industry, have sprung up all along its banks, like the enchanted palaces of the Arabian tales. The whole valley, from lake to sea, teems with the sights and sounds of the spindle and the loom.

At the head of this valley stands Concord, the capital of New Hampshire, "The Switzerland of America." Below Concord is Hooksett, just rising to the eminence of her sister cities. A little lower is Manchester, a fine miniature of her English namesake. Where a single sawmill stood twenty years ago, now stand the Stark Mills, which consume more cotton and weave more cloth than any other similar establishment of which the world can boast. Fifteen miles lower is Nashua, and still lower looms up the Queen-City of the valley—Lowell—the subject of this sketch. Ten miles lower on the river is—Lawrence—fitly named after the great Merchant Prince, the Medici of America. Ten years ago, a few sandy farms were all that it contained. Now, it displays a double row of factories, among which is the largest mill in the world; and the sites of many others yet to be. Below Lawrence are Haverhill, Bradford, Newbury, Salisbury, and the two Amesburys, with Newburyport, the key of them all.

With a population of two hundred thousand souls—all engaged in the industry of the factory or of the farm—what an aggregate of productive force is here!

**Figure 17.4 Ocean dumping.**
Many coastal cities still make use of ocean dumping as a means of disposing of unwanted materials.

Like the land and the air, water has a remarkable ability to cleanse itself. This is particularly true when it serves as a dump for organic waste. Large bodies of water (such as the oceans, the Great Lakes, and large rivers) contain enough bacteria to decompose vast amounts of organic waste. However, this self-cleansing process reinforces the belief that there are unlimited amounts of water.

When a large body of water disperses a pollutant over its entire volume, the pollutant is diluted. The material is still there, but the concentration is so small that it no longer interferes with our well-being. The idea that the oceans can absorb our waste is so ingrained that, as late as 1966, a Los Angeles public official was still urging the use of our oceans as a dump for sewage. New York City still uses the ocean as a dump for sewage sludge and garbage. Although many cities find that using the ocean as a dump is economically advantageous, this practice has taxed the ability of the oceans to cleanse themselves. (See fig. 17.4.) Thor Heyerdahl, who crossed the Atlantic Ocean on a reed boat in 1969, made the following comment:

> Large surface areas in the mid-ocean as well as nearer the continential shores on both sides were visibly polluted by human activity. It was unpleasant to dip our toothbrushes into the sea. Once the water was too dirty to wash our dishes in.

Technological advancements and the increase in human population on the earth has even had an effect on the vast oceans.

## *Growing populations overwhelm recovery rate of resources*

A population of one million native Americans living across the expanse of the United States created no pollution problem. Their fires for cooking and warmth were scattered over more than eight million square kilometers of land. This

**Figure 17.5 Donora, Pennsylvania.**
This valley was the scene of a serious air pollution incident. The pollutants from the industry were trapped by an inversion, and almost half of the population became ill.

did not even pose a serious local air pollution problem. The atmosphere absorbed the pollutants almost as rapidly as they were produced. Also, this small population was so widely scattered that human waste was no threat to their well-being. These people were not concentrated in a single area, so waste disposal never became a problem. Most of these people moved periodically as fields failed or game became scarce.

However, if all one million native Americans had lived on Manhattan Island, the area around New York City would have been polluted long before the arrival of the first European settlers. Pollution becomes a problem when a large population becomes concentrated in a limited area. When wastes are produced so rapidly that air and water resources cannot recover, these resources are damaged.

Whenever humans dump more into the environment than can be recycled, there will be changes in the ecosystem and pollution occurs. A well-documented case of pollution that was harmful to human health occurred in Donora, Pennsylvania. (See fig. 17.5.) The city of Donora is located in a valley. In October, 1948, the pollutants from a zinc plant and steel mills became trapped in the valley, and a dense smog formed. Within five days, 17 people died, and 5,910 persons became ill. The polluted atmosphere affected nearly 50 percent of the city's 12,300 inhabitants. The technology that provided jobs for the people was also killing them.

## *High technology = high pollution*

**Technological advances** increase the demand for energy. As fossil fuels are burned, pollutants are released into the atmosphere. The development of the internal combustion engine is another cause of pollution. The combustion of gasoline releases various waste products into the atmosphere. Various highly toxic products are formed from a series of photochemical reactions. These chemicals produce smog. Technology had invented a new form of pollution.

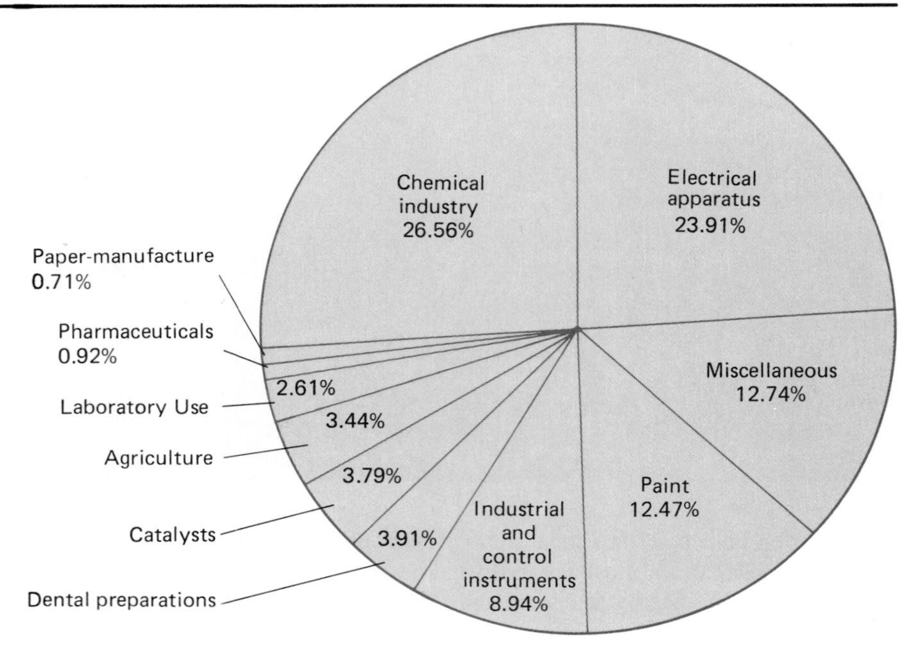


**Figure 17.6 Sources of mercury pollution.** Mercury has many uses in industrial and chemical processes. In addition, it is used as a poison.

Early forms of pollution were largely organic materials. The waste from humans, their domestic animals, slaughterhouses, tanneries, sawmills, and paper mills were capable of being recycled by natural events. Synthetic organic compounds create a special problem because many cannot be broken down by decay organisms. Plastics, nylon, sandwich wrap, pesticides, polychlorinated biphenyls, and many other synthetic organic compounds accumulate after use because they do not easily biodegrade.

The development and use of large amounts of inorganic compounds is also a problem. For example, mercury compounds are widely used in industry and agriculture. The production of chemicals, paper, electrical equipment, medicine, paint, plastics, and many other products requires the use of mercury. In agriculture, mercury compounds are used as fungicides on seeds. (See fig. 17.6.) Mercury released into the environment can be converted by certain bacteria into methylmercury, a very toxic compound. Methylmercury enters the food chain of aquatic animals. The concentration of mercury increases as the compound passes up the food chain. For years, mercury was released into the water with no knowledge of its consequences. In 1967 it was learned that free mercury in the water could be converted into methylmercury, which could enter the food chain. Mercury can interfere with the transmission of nerve impulses in humans. (See Consider this case study in chap. 3.) In 1970, the FDA confiscated 89 percent of the swordfish available for sale in the United States because the mercury level in the fish exceeded the 0.5 parts per million standard set by the FDA. The World Health Organization uses a more rigorous standard of 0.05 parts per million. Mercury represents only one of many materials dumped into our waterways that have eventually proved to be harmful.

**Table 17.2**
Stream pollution

| River | State | Type of pollution |
|---|---|---|
| Apalachicola | Ga., Fla. | Textile, poultry packing |
| Big Horn | Wyo., Mont. | Oil, mining, sugar beets |
| Canadian | Okla., Tex., N. Mex. | Food processing, agricultural |
| Des Moines | Minn., Iowa, Mo. | Meat packing, dairy |
| Hocking | Ohio, Pa. | Paper, gravel processing |
| Hoosac | Mass., Vt., N.Y. | Chemicals, paper, dairy, tanning |
| Lost | Calif., Oreg. | Pesticides from agricultural runoff |
| Montreal | Mich., Wis. | Iron mining |
| Palouse | Idaho, Wash. | Sawmills, agricultural |
| Republican | Nebr., Kans. | Municipal and agricultural |
| Tamgipohoa | Miss., La. | Chemicals, sand, gravel |
| Yellowstone | Mont. | Oil refineries |

Almost every section of our nation has streams that have been polluted in this way. Table 17.2 lists some rivers and the materials that have been dumped into them. It shows the variety of types of pollution that are discharged into our waters.

## Changing attitudes of the public

In the past, people tolerated pollution as a necessary price to pay for progress. When coal replaced waterwheels in the factories, smoke polluted the air. The use of fossil fuels for energy caused more pollution, but it also resulted in more jobs. Dirty skies meant that jobs were plentiful. However, several developments in England during the nineteenth century began to change attitudes about pollution. One of the most significant developments was the Chadwick Report. In 1833, Edwin Chadwick, a lawyer and statistician, was named Chief Commissioner of the Poor Law Commission established by the British Parliament. (See fig. 17.7.) One of the recommendations of the commission was to provide a death certificate listing *the cause of death*. Crowded living conditions and poor working conditions resulted in many persons being infected with communicable diseases. In 1844, all of the boys and ninety-one out of a total of ninety-four girls in one factory were infected with tuberculosis. The statistics collected and other activities of the commission resulted in the Public Health Act of 1848, which established the National Board of Health in England.

**Figure 17.7 Edwin Chadwick.**
Edwin Chadwick was instrumental in establishing the National Board of Health in England. His statistical analysis of the causes of death were widely read and led to changes in working and living conditions throughout the world.

Chadwick is considered to be the founder of public health and modern sanitation methods. The Chadwick Report was widely read in the United States. In 1845, the New York Association for Improving Conditions of the Poor was founded. Two years later, the American Medical Association appointed a standing committee on public health. As a nation, the United States was becoming more concerned with the health of its population. Groups of individuals began to work toward improving working conditions.

**Figure 17.8 Christmas seals.** In 1907, Miss Emily Bissels (an active member of the Red Cross) sought a means of raising three hundred dollars to keep a tuberculosis facility open. To meet this goal, she printed fifty thousand of the world's first TB Christmas seals and distributed them with the letter as shown here. (Courtesy of the American Lung Association.)

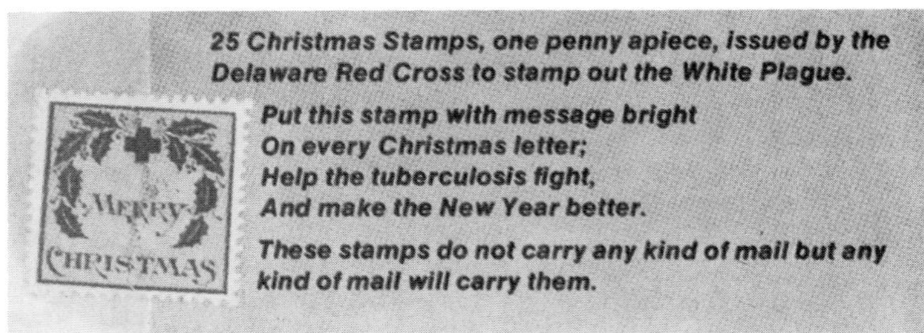


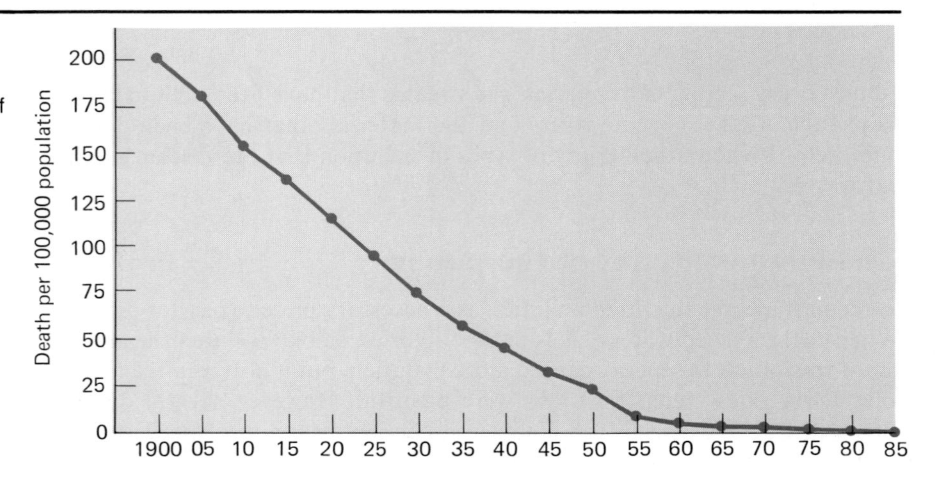


**Figure 17.9 Tuberculosis deathrate.** This graph shows that a concern for the prevention of tuberculosis and better treatment of tuberculosis victims has greatly reduced the number of deaths from this disease. (Source: U.S. National Center for Health Statistics, *Vital Statistics of the United States.*)

## *Pressure groups representing "quality of life"*

With the knowledge of how communicable diseases are transmitted, various organizations were formed to control the spread of disease. In 1885, a sanatorium movement for the treatment of tuberculosis was founded in New York, and in 1904, the National Tuberculosis Association was organized. (See fig. 17.8.) The success of these efforts is seen in figure 17.9.

In addition to health associations, various other groups were organized to address **environmental quality.** Some groups concentrated on a specific problem, and others focused on a wider range of concerns. During the drought period of the 1930s, there was a serious decline in the waterfowl population of North America. A group of citizens formed Ducks Unlimited. Their goal was to provide more favorable breeding areas for waterfowl. They have established over seventeen hundred breeding areas, totaling over 1,200,000 hectares. This organization has raised over $100 million over the years to implement its policies.

In 1969, Friends of the Earth (FOE) was organized. This group is concerned about all environmental issues. They work in all levels of government,

# Environmentalists oppose water-quality plan

**BY STEVEN ANDERSON**
***The Idaho Statesman***

Proposed weakening of state water-quality standards drew fire from conservationists and the federal government in a public hearing Thursday, but were backed by irrigators, Idaho Power Co., and a fish farm.

The changes were proposed by the state Department of Health and Welfare's Division of Environment in response to comments at public hearings around the state in September and October.

If adopted by the Board of Health and Welfare, the changes could end a longtime dispute between the state and the operations of the American Falls hydroelectric dam and reservoir—Idaho Power and the American Falls Reservoir District.

There were indications Thursday that the federal Environmental Protection Agency might veto weakening of the standards.

In a public hearing at the Idaho Supreme Court building Thursday afternoon, more than 30 persons testified on either side of the issue. The board is expected to act on the changes when it meets Jan. 9.

Most controversial among the approximately 50 proposed changes in water-quality regulations is a proposal to lower the standard for dissolved oxygen in the Snake River between the American Falls Dam and Lake Walcott, from 6 parts per million to 5 ppm.

The Idaho Legislature last session passed a resolution calling for reducing the dissolved oxygen standard for water below hydroelectric dams from 6ppm to 5 ppm statewide. Gov. John Evans refused to sign the bill, saying it was invalid for technical and legal reasons.

The new American Falls Dam, built during 1975–78, came under fire when state officials said the proportion of oxygen in water below the dam had dropped because the dam directed water through generators, rather than over spillways where it formerly was aerated.

**Figure 17.10 Political activism.**
Environmental issues are usually many sided. Political activism involves all parties presenting their views. Such action will lead to a better solution to the problem.

lobbying for better environmental protection. FOE has also taken its cause to court by initiating lawsuits to protect the environment. The League of Conservation Voters is a committee of FOE that actively seeks to have environmentally concerned persons elected to office.

In 1970, Zero Population Growth was established to address the population problem facing the world. The Sierra Club, Resources for the Future, and the Audubon Society, as well as many other local and national organizations, actively engage in securing a better environment. See Appendix 1 for a more complete list of active environmental organizations. There are also politically active groups who attempt to use the courts as an additional way of protecting the environment.

## *Political activism*

The main thrust of most environmental groups is to enlighten the public about environmental problems. Certain groups are most interested in single issues, such as water quality as it relates to sport fishing or waterfowl hunting. These groups devote their time and energy to the identification of pollutants and the sources of pollutants in streams and waterways. They attempt to get legislation passed that regulates the use of water and preserves it for their specific interests. Other groups are more interested in maintaining scenic landscapes. Still others work for purity of the air we breathe. To be effective, each of these groups must become active both environmentally and politically. Well trained legal experts are often employed by these groups. (See fig. 17.10.)

To secure decisions favorable to the environment, concerned citizens and organizations put pressure on appropriate governmental offices. This pressure may include persuading a legislator, city council member, or county commissioner to vote in a particular way. At times, members of the executive branch may be the target for persuasion, because they make appointments, allot funds, and perform other functions that can influence the environment.

The manner in which the news media reports or comments on judicial decisions and the activities of environmental groups can greatly affect public opinion. Often these situations are newsworthy because they take the form of a *David and Goliath* confrontation. All forms of the media report environmental concerns. Because news coverage is almost instantaneous, detailed descriptions of events and incidents are not likely to be hidden from public view. Leading magazines regularly report on environmental issues. Television produces documentary specials concerning the environment. Cartoons and photographs can be used to comment on the issues. Because of media coverage, the public is better informed on environmental problems. However, media commentary can sometimes be biased.

## *Governmental regulations*

In a democracy, people cause the government to act. Often this action is in the form of laws and regulations. Early regulations were aimed at improving living conditions. In 1867, the state of New York passed the Tenement House Law. It provided regulations for the welfare of the tenants, such as provisions for fire escapes, room ventilation, and indoor plumbing.

Nearly one hundred years later, in 1964, New York City enacted an ordinance aimed at reducing the amount of air pollution by lowering the sulfur content of fuel oil. Today, the fuel oil being used in New York City contains less than 1 percent sulfur. This has significantly decreased the sulfur dioxide content in the atmosphere of New York City. (See fig. 17.11.)

Today most governmental agencies have adopted regulations related to the health and well-being of the public. School boards require vaccinations to protect students against disease. Nearly all cities have requirements about water purification treatment and waste-water disposal. Some states have placed limits on the amount of noise to which a person may be subjected. The federal government limits the amount of exposure to radiation. Automobile exhaust emissions must also meet federal standards. From these examples it is apparent that governments have assumed a responsibility for alleviating problems of health and welfare.

## *People pay*

Whenever there is a regulation, there is a corresponding monetary cost. An example is the taxes paid for waste-water treatment and solid-waste disposal. Catalytic converters increase the price of automobiles. The costs incurred by industry to meet state and federal environmental regulations have been passed on to the consumer in the price of the product or service. Health and safety

**Figure 17.11**
**Government regulations.** The photograph of New York City shows a concentration of smog on November 24, 1966. Today, the smog problem in the city is not as severe. City regulations require a lower amount of sulfur in fuel oil, and as a result, the sulfur is taken out of the oil before it is burned. Therefore, it becomes a valuable by-product at the refinery and does not pollute the air.

regulations have reduced productivity of coal mines and increased the cost of coal. Most people are willing to accept these increased costs. However, some industries, particularly the United States automobile industry and power companies, have repeatedly expressed their inability to meet air and water pollution standards and have requested variances from the laws.

Not all of the costs involved in pollution control are monetary costs. Many products that were very effective, such as DDT, cyclamate (an artificial sweetener), and high-phosphate detergents, have been removed from the market. For the most part, the substitutes have not been as effective. Due to air-quality regulations, many cities have banned open burning, and the public must collect and dispose of leaves and other refuse. The number of fireplaces in some areas has been restricted for the same reason. Because there is a potential water pollution problem in Lake Tahoe, construction is closely regulated and certain types of development are prohibited.

Pollution was "invented" by humans. As population increases, there is a danger of increased pollution. Misuse and overuse of technology also increases this danger. Although pollution can never be eliminated, it can be controlled by keeping pollutants at tolerable levels. This requires public concern and a willingness to share the cost of a cleaner environment.

## Summary

Pollution is the result of technological advancements and increased population density. Pollution is anything produced by humans in a quantity that interferes with the health or well-being of an organism.

## Consider This Case Study
### Oregon bottle bill

In October, 1972, Oregon became the first state to enact a "Bottle Bill." This statute required a deposit of two to five cents on all beverage containers that could be reused. It banned the sale of one-time-use bottles and cans for beverages. The purpose of the legislation was to reduce the amount of litter. It was estimated that beverage containers made up about 62 percent of the state's litter. The bill succeeded in this respect; within two years after the bottle bill went into effect, beverage-container litter decreased by about 49 percent.

Those opposed to the bottle bill cited a loss of jobs that resulted from the enactment of the bill. One hundred and forty-two persons in two small can manufacturing plants did lose their jobs, but hundreds of new employees were needed to handle the returnable bottles. Other arguments against the bottle bill included that it would be a "major" inconvenience to have the consumers return bottles and cans to the store and that the retailers would have storage problems. Neither of these problems has proven to be serious.

The passage of bottle bills in other states, including Vermont, Connecticut, Iowa, Maine, Michigan and South Dakota, has prompted people to consider a national bottle bill. A national bottle bill would reduce litter, save energy, save money, and create more jobs. It would also save natural resources. A national container bill could save approximately seven million metric tons of glass, two million metric tons of steel, and a half million metric tons of aluminum each year.

At the present time, there is much concern about the amount of energy consumed in the United States. Can the United States afford to use the equivalent of 115,000 barrels of oil per day merely because of the convenience of throwaway containers?

Are you willing to tolerate the litter generated by discarded beverage containers? Count the number of discarded beverage containers you see in one day. What are some undesirable effects of discarding such containers?

How would a national container bill make the United States less dependent upon foreign supplies of oil and mineral ores? How would such a bill affect the economy of the nation?

Do you think that such a law singles out one industry—the beverage industry—and places an undue hardship on them while not treating other food processing industries equally? Should food containers also have a deposit required?

Wasting natural resources has been a philosophy of the United States throughout our history. Early settlers thought that the nation possessed an unlimited supply of natural resources, which was a prime factor in our early industrial development.

Early forms of pollution were largely organic materials, which can be broken down by decay organisms. Synthetic organic compounds create a special problem because many cannot be broken down by decay organisms and they tend to accumulate.

About one hundred years ago, people began to take action to improve the quality of life. Enactments of laws and court actions that were initiated by environmental pressure groups aided by the media have led to many environmental improvements. Improved environmental quality results in both inconvenience and additional costs.

## Review Questions

1. Why did an increased population result in increased pollution?
2. Why were Edwin Chadwick's ideas important in the history of pollution control?
3. How is political activism important for securing a better environment?
4. Define pollution.
5. What role does the media play in the development of environmental regulations?
6. Who pays for pollution control?
7. What are the basic functions of environmental organizations?
8. Name three public resources and how they have been abused.
9. Why is technological development responsible for increased pollution?
10. What were some early industrial uses of water?

## Suggested Readings


Bennett, Charles F. *Man and Earth's Ecosystem*. New York: John Wiley & Sons, 1976. Discusses the changes in the earth's natural environment that have been caused by human activity.

Hardin, Garrett. *Exploring New Ethics for Survival*. Penguin Books, 1973. Explores social attitudes and their impact on the environment.

Morris, David, and Hess, Karl. *Neighborhood Power: The New Localism*. Boston: Beacon Press, 1975. Explains how to provide political activism at a local level.

Petulla, Joseph M. *American Environmentalism: Values, Tactics, Priorities*. College Station, Texas: Texas A&M University Press, 1980. Details our historical, sociological, and philosophical patterns concerning various environmental issues.

Savage, Donald, et al. *The Economics of Environmental Improvement*. Boston: Houghton Mifflin Company, 1974. Discusses the economic costs of correcting environmental problems.

Smith, W. E., and Smith, A. M. *Minamata*. New York: Holt, Rinehart & Winston, 1975. Details how the release of mercury into the water caused problems for a rapidly developing industrial nation, Japan. Illustrates the increased danger of environmental problems that accompanies increased technology.

Spencer, J. E., and Thomas, W. L. *Introducing Cultural Geography*. 2d ed. New York: John Wiley & Sons, 1977. An account of the use of natural resources in the advancement of human civilization.

Traverer, W. B., and Luney, P. R. "Drilling, Tankers, and Oil Spills on the Atlantic Outer Continental Shelf." *Science* 104: 791–96. Illustrates how increased technology of off-shore oil production causes increased pollution.

## Chapter Outline

I. Kinds of pollution
  A. Persistent vs. nonpersistent pollutants
    Box 18.1 Public health advisory
  B. Noise pollution
  C. Aesthetic pollution
II. Ultimate disposition of pollutants
  A. Natural sinks
  B. Solid waste
    Box 18.2 Lead and mercury poisoning
III. Hazardous wastes
  A. Methods of treating hazardous wastes
    Box 18.3 The control of hazardous and toxic substances
    Box 18.4 A contaminated community
  B. Radioactive waste
IV. Consider this case study: Love Canal

## Key Terms

compost system

hazardous waste

natural sinks

nonpersistent pollutant

persistent pollutant

radioactive waste

recycling

sanitary landfill

solid waste

synergism

threshold levels

## Objectives

Define a persistent pollutant.

Define a nonpersistent pollutant.

Recognize that many widely used chemicals are persistent pollutants.

List the dangers caused by the concentration of heavy metals in the environment.

Define and describe aesthetic pollution.

Explain why it is difficult to control aesthetic pollutants.

Define a natural sink.

Explain why the disposal of solid waste is a serious problem in the United States.

Explain how the "packaging revolution" affects the amount of solid waste generated and consumer buying habits.

Explain why incineration or secured landfills are the best ways to deal with toxic wastes.

List the problems of radioactive waste disposal.

# CHAPTER 18
# Dealing with pollution

## Kinds of pollution

Pollution is defined as anything produced by humans that is present in such quantities that it interferes with the health or well-being of people. Some forms of pollutants may poison or injure organisms. Other physical alterations, such as heat or sound, can cause either physical or psychological injury. Still others are offensive to our senses and affect us emotionally rather than physically.

When dealing with the issue of pollution, a question that frequently arises is, How much is too much? Unfortunately this is an extremely difficult question to answer given the many unknown effects of pollution on living organisms. We need to understand that the

**Figure 18.1 Use of synthetic materials.** All of these items have at least part of their structure composed of slowly degrading or nonbiodegradable synthetic compounds. Because of this, they constitute a waste disposal problem.

effects of certain pollutants do not have to be immediate to be harmful. For example, lead has been used in paints for many years, but it was only recently discovered that it has harmful effects, particularly when paint chips are eaten by children. The **threshold levels** of pollutants must also be taken into consideration. Small quantities of a pollutant, such as fecal bacteria (1 to 2 per 100 milliliters), in drinking water may present no adverse health effects, but increased levels may result in illness among sensitive individuals. Threshold levels vary significantly among species as well as among members of the same species. In addition, certain combinations of pollutants cause more serious problems than do individual pollutants. This is referred to as **synergism.** For example, uranium miners who smoke tobacco have unusually high incidences of lung cancer. Apparently the radioactive gases found in uranium mines interact synergistically with the carcinogens found in tobacco smoke.

## *Persistent vs. nonpersistent pollutants*

The methods used to deal with forms of pollution depend upon the pollutant's character and persistence. **Persistent pollutants** are those that remain in the environment for many years in an unchanged condition. Most of the persistent pollutants are human-made materials. It has been estimated that there are thirty thousand synthetic chemicals used in the United States. They are mixed in an endless variety of combinations to produce all types of products. These products are used in every aspect of daily life. They are part of our food, transportation, clothing, building materials, home appliances, medicine, recreational equipment, and many other items. Our way of life is heavily dependent upon synthetic materials. (See fig. 18.1.)

An example of a persistent pollutant is DDT. It was used as an effective pesticide worldwide and is still used in some countries because it is so inexpensive. However, once released into the environment, it accumulates in the food chain and causes death when its concentration is high enough. (See chap. 14 for a discussion of DDT as a pesticide.)

Another widely used group of synthetic compounds of environmental concern are polychlorinated biphenyls (PCBs). PCBs are highly stable compounds that resist changes from heat, acids, bases, and oxidation. These same characteristics make PCBs desirable for industrial use but also make them persistent pollutants when they are released into the environment. About half of the PCBs are used in transformers and electrical capacitors. Other uses include inks, plastics, tapes, paints, glues, waxes, and polishes.

PCBs are harmful to fish and other aquatic forms of life because they interfere with reproduction. In humans, PCBs produce liver ailments and skin lesions. In high concentration, they can damage the nervous system, and they are suspected carcinogens. In 1970, the production of PCBs was limited to those cases where satisfactory substitutes were not available. Figure 18.2 shows the fate of PCBs produced since 1929.

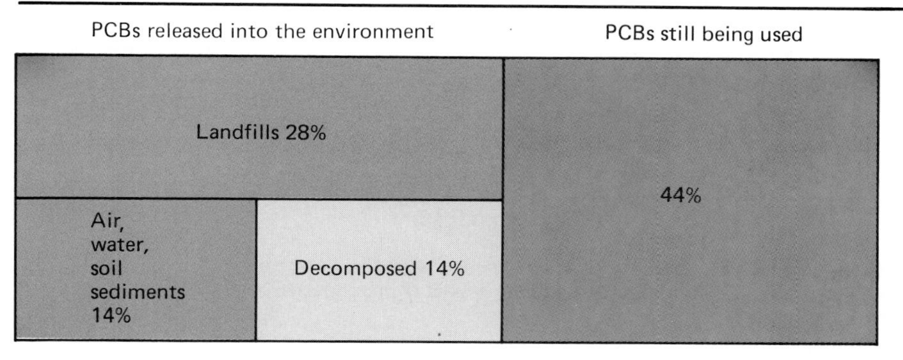


**Figure 18.2 The fate of PCB in the United States.** Of the 330 million kilograms of PCB produced in the United States since 1929, 56 percent has been released into the environment. Only one quarter of this has decomposed; because of the persistent nature of the chemical, the majority of PCBs released into the environment are still present.

In addition to synthetic compounds, heavy metals are used for many purposes. Mercury, beryllium, arsenic, lead, and cadmium are examples of heavy metals that are toxic. When released into the environment, they enter the food chain and become concentrated. In humans, these metals can produce kidney and liver disorders, weaken the bone structure, damage the central nervous system, cause blindness, and lead to death. Because these materials are persistent, they can accumulate in the environment even though only small amounts might be released each year. When industries use these materials in a concentrated form, it presents a hazard that is not found naturally.

A **nonpersistent pollutant** does not remain in the environment for a long time. Most nonpersistent pollutants are biodegradable. Others decompose as a result of inorganic chemical reactions. A biodegradable material is chemically changed by living organisms and often serves as a source of food and energy for decomposer organisms, such as bacteria and fungi. Waste from food processing plants, garbage, human sewage, animal wastes, and other remains of organisms are examples of nonpersistent pollutants that are biodegradable.

Besides the naturally occurring nonpersistent pollutants, there are also synthetic nonpersistent pollutants. Many of these break down chemically by oxidation or hydrolysis. These include the "soft biocides." For example, organophosphates are a type of pesticide that usually decomposes within several weeks. As a result, organophosphates do not accumulate in food chains, because they are pollutants for only a short period of time.

## *Noise pollution*

Noise is referred to as unwanted sound. However, noise can be more than just unpleasant sound. Research has shown that exposure to noise can cause physical, as well as mental, harm to the body. The loudness of noise is measured by decibels (db). Decibel scales are logarithmic rather than linear. Because they are logarithmic scales, a tenfold increase in sound loudness or power occurs with a 10 db rise. Thus, the change from 40 db (a library) to 80 db (a dishwasher or garbage disposal) represents a ten thousandfold increase in sound loudness.

## Box 18.1
## Public health advisory

**PUBLIC HEALTH ADVISORY**

Some sport fish contain chemical contaminants. Although levels of some contaminants have markedly declined, uncertainties about the impact of prolonged exposure dictates the following advice:

**Do not eat any fish**—Deer Lake (Marquette Co., Ishpeming Twp.), Tittabawassee River, Saginaw River, Pine River (downstream from St. Louis), Chippewa River, (downstream from Chippewa Road), Raisin River (downstream from Monroe Dam), Kalamazoo River (downstream from Kalamazoo). Portage Creek (downstream from Milham Park) and Shiawassee River (M-59 to Owosso).

**Do not eat certain fish**—Cass River (downstream from Bridgeport) avoid catfish, Grand River (Clinton Co.) avoid carp, Lake Macatawa avoid carp, Hersey River (Reed City area) avoid bullheads and trout, St. Joseph River (downstream from Berrien Springs Dam) avoid carp.

**Certain Great Lakes fish should not be eaten**—by children, women who are pregnant, nursing or expect to bear children. **Limit consumption** by all others to no more than 1 meal per week. Lake Michigan—carp, catfish, salmon[3], trout[3] and whitefish;[1] Lake Superior—lake trout; Lake Huron—carp,[2] catfish,[2] muskellunge,[1] salmon,[1,3] trout,[1,3] Lake St. Clair and the Detroit and St. Clair rivers—muskellunge; and Lake Erie (western edge)—carp, catfish and muskellunge.

[1] Southern half of lake only.
[2] Saginaw Bay area only.
[3] Advisory also applies to tributaries into which these species migrate.

NOTE: Fatty fish continue to show higher contaminant levels than lean fish. Cleaning fish by skinning, filleting, and trimming off fatty portions, reduces contaminant levels. Baking on a rack, barbecueing, poaching or frying in vegetable oil also reduce contaminant levels.



This public health advisory was printed in the 1984 Michigan Fishing Guide. It was distributed with fishing licenses. It illustrates that some kinds of chemical wastes accumulate in water. Furthermore, such chemicals can be concentrated within the food chain of certain species of fish and can render these fish harmful to humans. The various letters refer to specific chemical contaminants.

PBB is a chemical that was released into the environment from a chemical plant. Although DDT has been banned for ten years, it still persists in the environment. PCB has been restricted in use but is still used for some electrical purposes. Mercury entered the water when it was discharged by various industries. TCDD is a highly toxic chemical that has entered some of the state's streams.

This type of health advisory is not unique to Michigan. Other states may warn those who fish about the chemical dangers of eating their catch. If humans persistently ignore the fact that waste will accumulate somewhere, large parts of our environment will be seriously damaged. If the wastes are persistent, it may require several generations before the damage can be corrected.

**Table 18.1**
Intensity of noise

| Source of sound | Intensity in decibels |
|---|---|
| Jet aircraft at takeoff | 145 |
| Pain occurs | 140 |
| Hydraulic press | 130 |
| Jet airplane (160 meters overhead) | 120 |
| Discotheque | 120 |
| Unmuffled motorcycle | 110 |
| Subway train | 100 |
| Farm tractor | 98 |
| Gasoline lawnmower | 96 |
| Food blender | 93 |
| Heavy truck (15 meters away) | 90 |
| Heavy city traffic | 90 |
| Vacuum cleaner | 85 |
| Hearing loss occurs with long exposure | 85 |
| Garbage disposal unit | 80 |
| Dishwasher | 65 |
| Window air conditioner | 60 |
| Normal speech | 60 |

The frequency or pitch of a sound is also a factor in determining the degree of harm that it may cause. We know that high-pitched sounds are the most annoying. The most common sound pressure scale for high-pitched sounds is the A scale, whose units are written "dbA." Hearing loss begins with prolonged exposure (eight hours or more) to 80 to 90 dbA levels of sound pressure. Sound pressure becomes painful at around 140 dbA and can kill at 180 dbA. (See table 18.1.)

In addition to hearing loss, noise pollution is linked to a variety of other ailments, ranging from nervous tension headaches to neurosis. Research has also shown that noise may cause blood vessels to constrict, which reduces the blood flow to key body parts, disturbs unborn children, and sometimes causes seizures in epileptics. The EPA has estimated that noise causes about forty million Americans to suffer hearing damage or other mental or physical effects. It has also been estimated that up to sixty-four million people live in homes that are affected by aircraft, traffic, or construction noise.

The Noise Control Act of 1972 was the first major attempt made in the United States to protect the public health and welfare from detrimental noise. This act also attempted to coordinate federal research and activities in noise control, to set federal noise emission standards for commercial products, and to provide information to the public. Subsequent to the passage of the Noise Control Act, many local communities enacted their own noise ordinances. While such efforts are a step in the right direction, the United States is still controlling noise less than are many European and Scandinavian countries. Several European countries have developed quiet construction equipment in conjunction with strongly enforced noise ordinances. For example, in the Soviet Union, factory noise levels above 85 dbA are prohibited as are noise levels

**Figure 18.3 Visual pollution.**
Although many would consider this view to be visually obnoxious, the signs do provide information for travelers and are an inexpensive form of advertising.

above 30 dbA in residential areas. The West Germans and Swiss have also established maximum day and night noise levels for certain areas. Regarding noise pollution abatement, North America has much to learn from European countries.

## *Aesthetic pollution*

Visual pollution is a sight that offends us. This is highly subjective and is, therefore, very difficult to define or control. To most people, an open garbage dump is a form of visual pollution. (See fig. 18.3.) A dilapidated home or building may also be offensive. If that same building was in an area of high-priced homes, it would be an eyesore. A badly littered highway or street is aesthetically offensive, and litter along a wilderness trail is even more unacceptable.

To many, roadside billboards are offensive, but they are helpful to the advertiser and to travelers looking for information. This conflict is typical of aesthetic forms of pollution. People do not agree on what is offensive. Therefore, it becomes very difficult to control or regulate many forms of visual pollution.

Some of the chemicals discharged into our waterways have a definite taste. If the chemical does no biological injury but has an unpleasant taste, it is a taste pollutant. If minute quantities of certain chemicals enter drinking water or food, we may taste them. In fish, a concentration of one hundred parts per million of phenol can be tasted. Although this small amount of chemical is not harmful biologically, it does make the fish very unappetizing.

Odor pollution may originate from several sources. Various airborne chemicals have a specific odor that may be offensive. People who live near stockyards, paper mills, chemical plants, steel mills, and many other types of industry may be offended by the odors originating from these sources. Slaughterhouses, food processing plants, chemical plants, paper mills, and other industries may discharge waste materials into the water that decompose or evaporate and cause odor pollution.

People who are constantly exposed to an odor are not as offended by it as are people who are newly exposed to it. When an odor is constantly received, the brain ceases to respond to the stimulus. In other words, the person is not aware of the odor.

Aesthetic pollutants, such as noise, sights, tastes, and odors, are extremely difficult to define. Each of us has our own idea of what is offensive. This makes it very difficult to establish aesthetic pollution standards. Controlling this type of pollution can be accomplished by educating the public so generally unacceptable levels of aesthetic pollution are eliminated.

## Ultimate disposition of pollutants

All pollutants follow a path that leads to their final disposition. Some pollutants, such as heat and sound, are forms of energy that dissipate and, therefore, are not persistent pollutants. Other forms of pollution either decompose or are deposited in water, on land, or in the atmosphere.

### *Natural sinks*

Water, land, and atmosphere are **natural sinks** that serve as storage reservoirs for unwanted material. (See fig.18.4.) As more of a nondecomposing substance is released, its concentration within the sink increases. If the pollutant is persistent, it remains in a sink for a long time. If the pollutant is a nonpersistent type, it only remains in the sink until it is recycled or degraded to a different compound.

Most gaseous pollutants and small solid particulates are released into the atmosphere. Rain may eventually wash these materials from the air. For example, sulfur dioxide is released into the atmosphere. When sulfur dioxide combines with water to form acids, it returns to the surface of the earth as acid rain. It may be incorporated into the soil or run off into lakes, seas, or oceans. (See box 19.2.)

Water is the sink for many kinds of materials. Various salts are water soluble. As more of these are discharged into the water, the concentration becomes higher. Most of the salts remain in solution and are transported to the oceans. Nonsoluble materials, such as glass, metals, and plastics, are also dumped into the water. The physical action of the water breaks down some of these materials and carries them in suspension. Larger particles of nonsoluble materials settle to the bottom.

**Figure 18.4 Nat[illegible] sinks.**
Although the air, water, and land receive pollutants, the pollutants may flow from one compartment to another before final containment. Pollutants that are released into the air may eventually enter water and ultimately be deposited on land.

**Table 18.2**
United States solid-waste production (1975)

| | Metric tons (millions) |
|---|---|
| Mining waste | 1,700 |
| Residential, commercial, and institutional waste | 250 |
| Industrial waste | 110 |
| *Total* | 2,060 |

Source: Bureau of Solid Waste, Department of Health, Education and Welfare

The land serves as a sink for materials that are not gaseous or soluble in water. The solid waste in dumps, landfills, and along the side of the road are being "stored" on land. (See fig. 18.5.) Three major types of wastes are stored on land: solid wastes, toxic chemicals, and radioactive wastes.

## *Solid waste*

**Solid wastes** may be incinerated, dumped into the ocean, or stored on land. The major types of solid waste and the amounts produced are listed in table 18.2. The slag heaps and mine tailings that remain after ores have been processed are known as mining wastes. (See fig. 18.6.) Because a large amount of useless material is produced in this way, storing it on land seems to be the only practical solution. These unsightly mounds can be used as fill or for other purposes, but this is not always practical.

Sanitary land fill

Dump

Highway litter

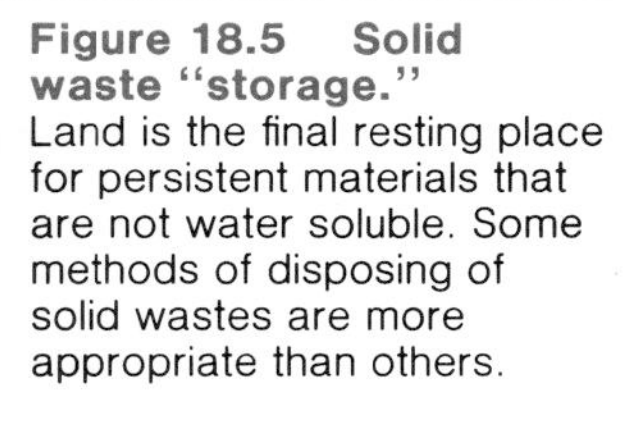

**Figure 18.5 Solid waste "storage."**
Land is the final resting place for persistent materials that are not water soluble. Some methods of disposing of solid wastes are more appropriate than others.

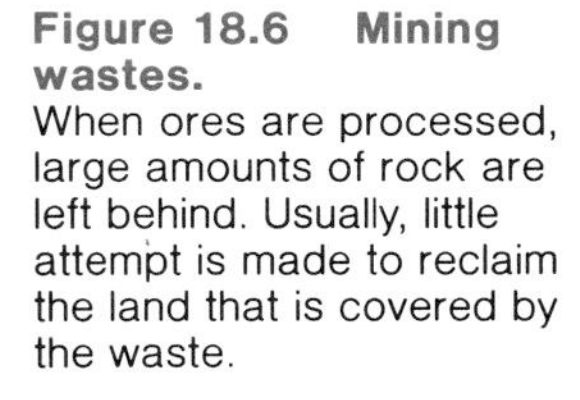

**Figure 18.6 Mining wastes.**
When ores are processed, large amounts of rock are left behind. Usually, little attempt is made to reclaim the land that is covered by the waste.

**Figure 18.7 Compost system.** In this system, certain materials are removed from the refuse for possible recycling. Most of the organic material provides nutrients and humus for agricultural use. The city of Rome operates a compost plant that handles six hundred metric tons of refuse per day.

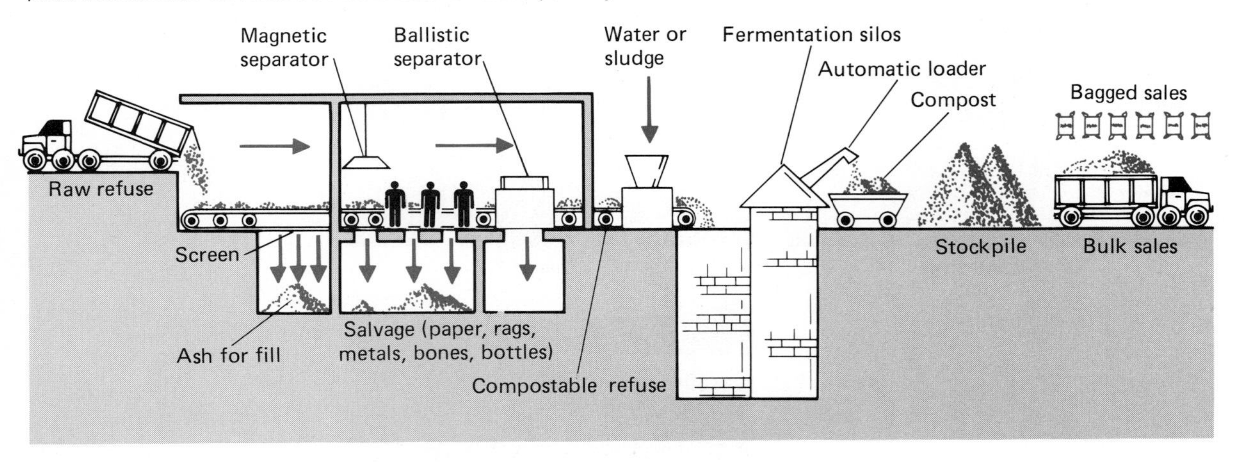


The combination of residential, commercial, and institutional solid wastes constitutes the second largest source of solid waste. (See table 18.2.) The per capita production of solid waste is 2.5 kilograms daily. Because this is a large volume of waste and it is costly to store, several proposals have been suggested to reduce the magnitude of the problem.

One approach is **recycling** part of the material and burning the part that is not salvageable. An eleven hundred-unit apartment complex in Sundyberg, Sweden, serves as a model. The apartment dwellers discard their trash into a vacuum system that carries it to a central location. Glass and metals are automatically sorted. The combustible material is burned to provide hot water for the apartments. (Similar units were constructed in the 1972 Olympic Village in Munich.) The construction costs of such a vacuum system are high, but the labor costs are much lower than the typical residential waste collection methods.

Another approach is a **compost system.** (See fig. 18.7.) In this system, the trash is collected and hauled to a central location. The metal, glass, and other salvageable objects are sorted out. The organic material is composted. This method has been used with success in many parts of the world, except in the United States. Higher labor costs and the availability of cheap land for landfill have made this an unattractive alternative in the United States.

The most frequently used method of solid waste disposal is the **sanitary landfill,** even though it has the disadvantages of high collection costs and a lack of good disposal sites. To make landfills less disruptive to the environment, care should be taken to select a site that will preclude leaching of materials into water. The waste material is covered with soil as a sanitary precaution. It also reduces visual pollution. (See fig. 18.8.) In fact, some landfills are planned to provide pleasant, useful community facilities, such as parks and playgrounds, when they are filled and secured. (See fig. 18.9.)

**Figure 18.8 Sanitary landfill.** When proper care is taken, a sanitary landfill is an acceptable method of disposing of some types of solid waste. Care must be taken to prevent runoff into the water supply. By building the site downwind and covering it with soil as it is filled, visual and odor pollution are minimized. (Courtesy of U.S. Department of Health and Human Services.)

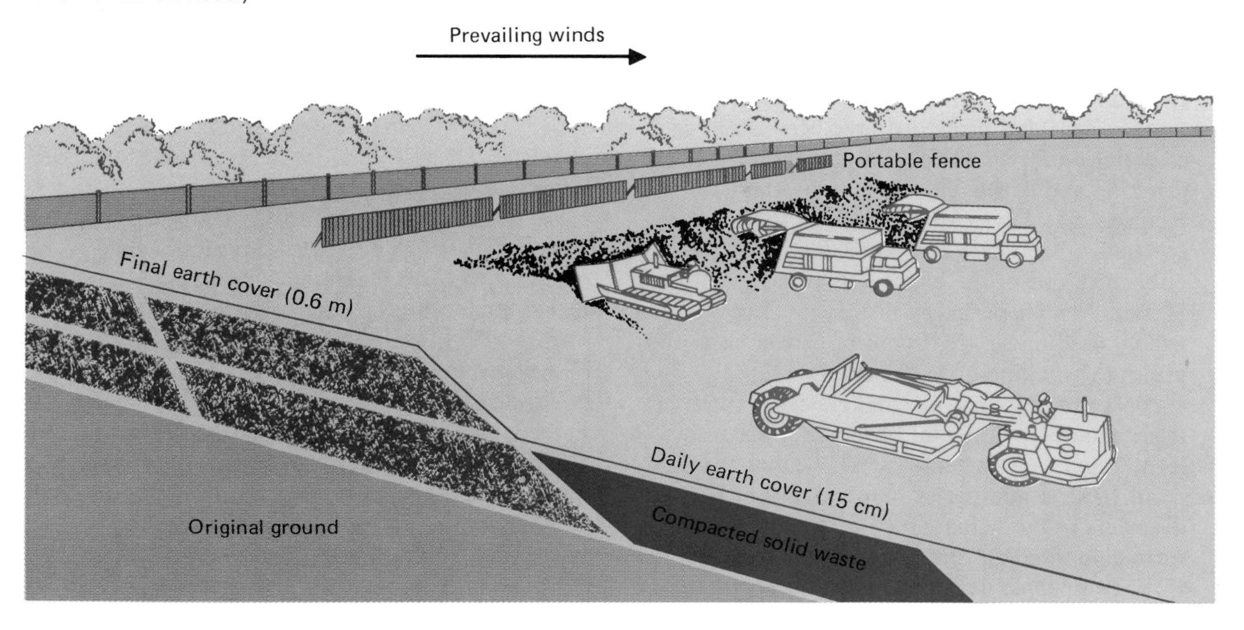

**Figure 18.9 From landfill to recreation area.**
This area was once a landfill for Jackson, Mississippi. With advanced planning, the citizens were able to provide this recreational site for all to enjoy.

## Box 18.2
## Lead and mercury poisoning

Lead and mercury are naturally present in the environment and probably have been a source of pollution for centuries. For example, the lead drinking and eating vessels used by the wealthy Romans may have caused many of their deaths. In 1865, when *Alice in Wonderland* was published, one of the characters was the Mad Hatter. At that period in history, mercury was widely used in the treatment of beaver skins for use in the making of hats. As a result of exposure to mercury, hat makers often suffered from a variety of mental problems, hence, the term "mad as a hatter."

In 1953, a number of physical and mental disorders in the Minamata Bay region of Japan were diagnosed as being caused by mercury. Fifty-two people developed symptoms of mercury poisoning; seventeen died and twenty-three became permanently disabled. In 1970, there was an outbreak of mercury poisoning in the United States. It was traced to mercury in the meat of swordfish and tuna. In both incidents the toxic material was not metallic mercury, but it was a mercurous compound, methylmercury. Bacteria in the water convert the metallic mercury into methylmercury. This compound enters the food chain and may become concentrated as the result of biological amplification. When sufficient amounts are present in humans, it can cause brain damage, kidney damage, or birth defects. Today there are regulations that reduce the release of mercury into the environment and set allowable levels in foods. The problem still persists because it is impossible to eliminate the large amounts of mercury already present in the environment, and it is difficult to prevent the release of mercury in all cases. For example, burning coal releases 3,200 metric tons of mercury into the earth's atmosphere each year, and mercury is still "lost" when it is used for various industrial purposes.

Like mercury, lead is a heavy metal and has been a pollutant for centuries. Studies of the Greenland Ice Cap indicate a 1,500 percent increase in the lead content today as compared to 800 B.C. These studies reveal that the first large increase occurred during the Industrial Revolution and the second great increase occurred after the invention of the automobile. Oil companies added lead to gasoline to improve performance. In fact, burning gasoline is a major source of lead pollution. Another source of lead pollution is older paints. Prior to 1940, indoor and outdoor paints often contained lead. In 1940, the use of lead in indoor paints was stopped, and in 1958, the use of lead in outdoor paints was reduced. The Lead Poisoning Prevention Act was passed in 1971.

There is still disagreement over what levels of lead and mercury can cause human health problems. Although ingested lead from paint can cause death or disability, such a strong correlation cannot be made for atmospheric lead. In fact, the EPA is being pressured to allow the continued use of lead in gasoline. There has also been a slight increase in the allowable amounts of mercury compounds used as fungicides.

As with most other forms of pollution, there is no agreement on how much is harmful. So it seems that we will continue to tolerate low levels of these materials in our environment and continue to disagree about the danger of these pollutants to humans.

The amount of solid waste produced is largely the result of the "packaging revolution." This is a method of using the container to promote sales. The manufacturer is trying to catch your eye and get you to purchase a certain product. For example, you require two bolts to complete a job around the house. You probably will purchase these at a local hardware store. Often the store handles no bulk materials, so you cannot buy only two bolts. But you can purchase six bolts in a plastic container. You have just been forced into buying more than you need. You also have a nonbiodegradable plastic box to add to the trash. The large quantity of material used in such packaging adds to the amount of waste generated each day. These waste materials constitute a health problem, and they can be unsightly.

## Hazardous wastes

**Hazardous wastes** are substances that, if released into the environment, could threaten life. Although some hazardous substances are found in municipal and domestic waste, the majority are residues from industrial processes. These residues can be liquid, semisolid sludge, or solid. They can be reactive (ignitable, corrosive, or explosive in certain concentrations) or toxic.

Some of the more common wastes produced by industry are acids from metallurgical processes, spent caustic from the pulp and paper industry, and the leftovers from oil refining. These wastes contain oil, phenols, arsenic, mercury, lead, and a large number of human-made chemicals. It is estimated that in the United States and Canada, hundreds of billions of kilograms of hazardous wastes are generated each year, and 90 percent of these wastes are disposed of improperly.

In addition to the wastes produced directly by industry, there are also hazardous "distressed materials." These are products like PCBs, insecticides, and herbicides that have been banned or phased out but which still require special treatment and disposal.

The U.S. Environmental Protection Agency estimates that there are as many as 50,000 dumps and 180,000 open pits and ponds contaminated with hazardous wastes. EPA officials say that at least fourteen thousand of these sites are potentially dangerous (posing fire hazards, threatening groundwater, or emitting noxious fumes). The cost to clean up this growing problem has been estimated to be in excess of $260 billion.

With the passage of air and water pollution control regulations, industry has been forced to search for disposal sites on land. However, waste disposal on land is not always feasible because the wastes may migrate with surface and groundwater. The contaminants in these wastes may then enter our drinking water or be taken into the food chain by organisms such as plants or fish. As the chemicals move up the food chain, they may become more concentrated and interfere with cell structure and reproduction. In addition to being poisonous, the chemicals may have cumulative effects and may result in mutations, cancer, and reproduction failure in organisms, including humans.

**Figure 18.10 Toxic chemical storage.**
This is a site where toxic wastes were improperly stored. The company went bankrupt. The state now assumes the problems and the cost of correcting the company's errors.

In 1980, the United States Congress set up the "Superfund" program to speed the cleanup of the most hazardous existing sites. The five-year $1.6 billion program was designed to allow the EPA to clean up the sites first and collect fines later, suing the dumpers, if necessary, to recover up to three times the cleanup costs. The law requires states to provide 10 percent of the cleanup costs for sites located on private property and 50 percent of the costs for those on public land. Only eight states have special state cleanup budgets. The other forty-two states cannot afford to pay for their share of the cleanup costs, thus they are disqualified from the Superfund project.

The demand for systems to treat and dispose of hazardous wastes is growing as our economy grows and as antipollution legislation is enforced. At the same time, there is strong community resistance to the location of proper treatment and disposal facilities. The problem is compounded by the fact that traditional landfill sites are no longer considered acceptable for these substances. The result is that untreated wastes, some highly corrosive, are being stored in drums and tanks until treatment is available. (See fig. 18.10.) Wastes are also being stored permanently in deep wells where porous rock layers below the water table absorb liquid contaminants. Some wastes are being dumped into municipal sanitary landfill sites, which may be an undesirable situation. Where permitted, certain types of wastes are incinerated, and some disappear at the hands of less scrupulous individuals, sometimes called "midnight dumpers." On the positive side, the technology for proper treatment of hazardous wastes is available. The difficult issue is to decide where the treatment facilities should be located and where the end products of the treatment process are to be deposited, presumably in chemical landfill sites. The solution to the problems of location will involve some public debate and compromise.

## Box 18.3
## The control of hazardous and toxic substances

With the appearance over the last several decades of substances such as DDT, PBB, PCB, and Kepone, fears about pollution and toxic wastes have become very real for many individuals. Toxic substances have been linked to cancer, birth defects, and many other serious health disorders. In the past, new chemicals were manufactured and released to consumers and later were discovered to have harmful side effects. This was the case with thalidomide, a sedative that was used in Europe. Before this drug was taken off the market in the early 1960s, about eight thousand pregnant women had given birth to seriously deformed infants as a result of taking thalidomide during their pregnancies.

Beginning in 1972, the federal government passed a series of laws dealing in part with the problem of hazardous and toxic wastes. The first of these was the Marine Protection Research and Sanctuaries Act of 1972, which regulated ocean dumping. This was followed by the Ports and Waterways Safety Act of 1972, which regulated oil transport. The Insecticide, Fungicide and Rodenticide Act of 1972 classified and regulated pesticides and their use. In 1974, the Hazardous Materials Transportation Act was passed to regulate the transportation of hazardous materials on land. In 1976, the Resource Conservation and Recovery Act was passed to regulate the treatment, storage, and disposal of hazardous wastes. Perhaps no law, however, has gone as far in dealing with toxic wastes as the 1976 Toxic Substances Control Act (TOSCA).

It took five years of lobbying and infighting in Congress to push through the Toxic Substances Control Act. This act is designed to control nearly one thousand new chemicals that are put into production each year. Under this new law, the Environmental Protection Agency (EPA) listed almost thirty thousand chemicals produced in the United States at that time. Any substance not on this list is considered a new chemical. (This does not apply to drugs, pesticides, cosmetics, radioactive materials, and food additives, which are covered by other regulations.) The law requires extensive premarket screening of new chemicals for potential toxicity. The EPA has the authority to block the manufacture of a new chemical for up to 180 days if it feels there are questions about the chemical's safety. If the industry objects to a continued ban beyond the 180-day period, the EPA can seek a court injunction, but it must provide proof to substantiate its concerns. After the injunction, the manufacturer must prove that the compound is safe.

While the intent of the Toxic Substances Control Act is well placed, controversy still surrounds it. Environmentalists argue that Congress has not provided sufficient funds for the EPA to enforce and monitor the law. The manufacturers of chemicals worry that the law will force public disclosure of trade secrets, which could aid competitors. The chemical industry feels that the cost of screening and testing new chemicals for toxicity (about $500,000 per chemical) could hinder the development of new and needed chemicals. As you can see, the passage of the Toxic Substances Control Act did not solve all of the problems associated with toxic and hazardous materials. The law did, however, raise many important questions that will need more consideration.

## Box 18.4
## A contaminated community

"We don't know of any other community in the nation that has ever been contaminated like this one." This statement was made by a spokesperson from the United States Center for Disease Control in Atlanta about the six hundred residents of Triana, Alabama. Triana is a community located twenty-five kilometers from Huntsville.

From 1947 to 1970, the Olin Chemical Corporation manufactured DDT on a site leased from the nearby Redstone Arsenal. During the time of production, about four thousand metric tons of DDT residue accumulated in the soil of an adjacent marsh. Today, the DDT from this soil is filtering into Indian Creek, which flows through Triana. The fish in this stream contain forty times the federal standard of DDT allowable. Because the people from Triana depend upon the catfish from the stream as a source of food, they have accumulated large amounts of DDT in their bodies.

The average person in the United States contains 8 to 10 ppm DDT in his or her body fat. People from Triana have been found to contain up to sixty times that amount. The older residents have higher levels of DDT than the younger members of the community. This may be the result of older people bioaccumulating the DDT for a longer time.

A large portion of the community's income was earned from the sale of catfish. Because these fish are not suitable for sale, this income has been lost. The wildlife from nearby Wheeler National Wildlife Refuge also suffer from the effects of DDT. Mallard ducks have been found to contain 480 ppm DDT. A wintering population of thousands of double-crested cormorants once used the refuge, but they are now absent. A heron rookery that contained three hundred nests has disappeared from the area. These and other environmental changes will occur in Triana and the surrounding area as long as the DDT contaminated soil remains in the marsh.

As yet, the only move toward a solution has been an allotment of $1.5 million by the federal government for a study to determine what to do about the problem. However, Clyde Foster, the mayor of Triana, believes that regardless of the solution, the town will never be the same because of the problems caused by the DDT.

## *Methods of treating hazardous wastes*

Current levels of technology allow for four techniques of treating hazardous wastes.

1. Physical treatment
    - Filtration to remove suspended solids
    - Evaporation of water from liquid wastes
    - Reverse osmosis
    - Separation of liquids from solids by centrifuges
    - Encapsulation of untreatable wastes in a glass or silicate shell
2. Chemical treatment
    - Neutralization by blending acidic and caustic solutions
    - Oxidation-reduction
    - Precipitation of solids
    - Ion exchange
    - Solidification
3. Thermal treatment
    - Incineration of man-made chemicals such as PCBs and DDT
4. Biological treatment
    - Degradation by bacteria

These processes reduce the volume of wastes and leave relatively inert residues, which should be deposited in special secured landfill basins that are designed to prevent contamination of ground and surface water. The basins should be lined with impermeable materials to prevent seepage and eventually must be capped to avoid overflow as a result of rainwater accumulation.

In constructing hazardous-waste treatment plants, there are a number of points to consider:

Plants should be built near areas of concentration of industrial activity so that there is enough treatable waste to insure that they operate at full capacity.

Plants should also be located near areas that generate the greatest volumes of waste so that the risk and cost of transportation can be minimized. Shipping costs represent a significant portion of the total cost of treatment.

Sites should be accessible to road and rail routes.

Geological features, such as rock formations, soil types, the groundwater table, and fault lines, must be considered.

Prevailing winds and annual precipitation rates should be taken into account.

Existing land use will be a significant factor in dictating location.

Public opinion will affect location and operation. If a site is being considered as a possible location for a plant, it is essential that the community be informed of this as soon as possible. This will give citizens the opportunity to ask questions and voice concerns about reliability, safety procedures, and liability; the real need for the treatment plant and criteria for site selection; the role of government; ownership of the plant; the effect on property values, aesthetics, and traffic levels; the benefits and the costs to the community; and the overall effect on quality of life.

The influence of public opinion in selecting sites has become the predominant issue in recent years. Lack of communication and cooperation has created what could be called a "hazardous waste standoff." In the future, all levels of government, industry, and the public must work together if we are to arrive at an acceptable compromise.

### *Radioactive waste*

Another substance that requires special treatment is radioactive waste. The increased use of nuclear-powered electrical-generating plants has produced a significant quantity of **radioactive waste.** Exposure to large amounts of radiation causes radiation sickness. The symptoms include reddening of the skin, loss of hair, and general fatigue. Within several weeks, the number of red blood cells and white blood cells will decline. There is also a decline in the number of blood platelets, which are necessary for blood clotting. Although some individuals may live for many months, death usually occurs within a few weeks.

If a person survives radiation sickness or is subjected to less than a lethal exposure, delayed reactions to the radiation often occur. It may be years after the radiation exposure before cancer develops. Other delayed effects include the development of anemia, cardiovascular disorder, cataracts, and a loss of fertility.

A one thousand megawatt nuclear plant (a standard, moderately large power plant) produces two cubic meters of waste per year. Currently this waste is being stored "temporarily" in water-filled containers on nuclear power plant sites or in one of several government storage areas. The waste is placed in containers that are stored above ground until some of the radioactivity has been dissipated. Although these sites are supposed to be carefully monitored, thousands of liters of radioactive wastes have leaked into the soil. None of these sites are considered permanent, although they have been in existence for over twenty years.

It has been proposed that placing radioactive wastes in geologically stable salt deposits could be a permanent answer to the storage problem. The waste from a nuclear plant would be placed in boro-silicate glass containers. Each glass container would be sealed in a thick-walled, stainless steel canister. Due to the high temperature of the radioactive wastes, the containers would be stored above ground for ten years. At the end of this time, the temperature of the container would have been reduced and the material could be buried in a salt deposit six hundred meters below the surface. (See fig. 18.11.)

One of the problems associated with radioactive wastes is that we cannot detect them with our unaided senses. We are, therefore, dependent on the openness and truthfulness of those people who monitor radioactivity and the nuclear power industry.

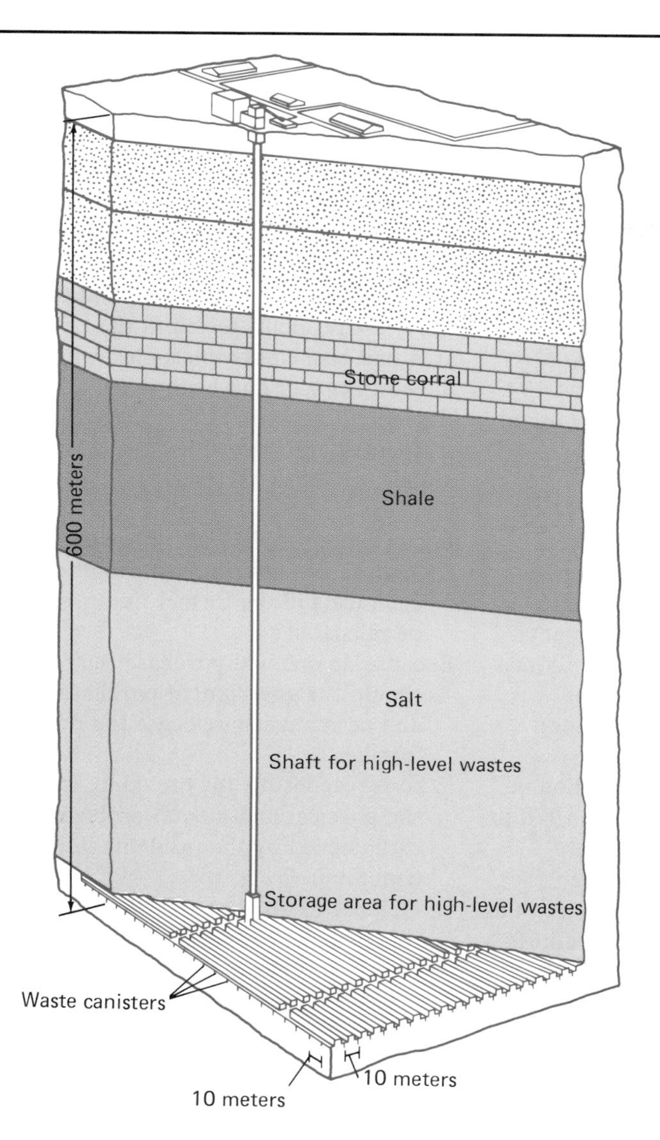


**Figure 18.11 Nuclear waste.**
After being sealed in glass containers and placed in stainless steel canisters, the waste from nuclear power plants could be deposited in salt domes six hundred meters underground. (From "The Disposal of Radioactive Wastes from Fission Reactors," by Bernard L. Cohen, June 1977. Copyright © 1977 by Scientific American, Inc. All Rights Reserved. Reprinted by permission.

## Consider This Case Study
### Love Canal

Love Canal near Niagara Falls, New York, was constructed as a waterway in the nineteenth century. It was subsequently abandoned and remained unused for many years. In the 1930s, it became an industrial dump. The Hooker Chemical Company purchased the area in 1947 and used it as a burial site for twenty thousand metric tons of chemicals.

Hooker then sold the property to the local government for one dollar. A housing development and an elementary school were constructed on the site. Soon after the houses were constructed, people began to complain about chemicals seeping into their basements. In 1978 eighty different chemicals were found in this seepage. Approximately one dozen probable carcinogens were identified among these chemicals.

As a result of these findings, $27 million in government funds were appropriated in 1978 to purchase homes and permanently relocate 237 families. The funds also provided for the construction of a series of ditches to contain the chemicals and a clay cap to prevent the fumes from entering the atmosphere. But the problems continued.

The remaining 710 families in the Love Canal area were not satisfied with the government's approach to the problem. They cited the fact that women in the area had a 50 percent higher rate of miscarriages. Of seventeen reported pregnancies in the area during 1979, two children were born normal, nine had defects, two were stillborn, and four ended in miscarriage. In addition to the abnormalities in birth, there are other biological problems in the Love Canal area.

Neurologists have determined that the speed of the nerve impulses in thirty-seven residents who were examined were slower than normal. They stated that chemical exposure could have caused this damage. In 1980, the EPA released the findings of a study that found eleven out of thirty-six residents tested in the Love Canal area had broken chromosomes, which are linked to cancer and birth defects. As a result, the federal government released $5 million to temporarily relocate Love Canal residents to motels or other quarters.

Because the people in the Love Canal area seem to be suffering as a result of the chemicals in the former dump, should they be relocated?

Because no one will purchase their homes, should the government purchase these homes and permanently relocate the remaining 710 families?

Who is responsible for providing treatment for the physical and mental problems experienced by the residents in this community?

In October of 1983 an out of court settlement was reached between 1,345 residents and Occidental Petroleum Corporation, the parent company of Hooker Chemical Company. The settlement will cost Occidental Petroleum $5–6 million and its insurance company up to $25 million to compensate the residents for damages. The EPA has instituted a $124 million lawsuit against Hooker. If the damages are awarded, how should they be applied to the residents of Love Canal?

Finally, there is a question that no one ever seems to ask. Why were permits ever awarded to construct a thousand unit housing development on top of a site known to contain twenty thousand metric tons of toxic wastes?

## Summary

Persistant pollutants are those that remain in the environment for many years in an unchanged condition. The chemically stable characteristics of many synthetic compounds make them a persistant pollutant. Many of these chemicals are toxic.

Nonpersistant pollutants are biodegradable and decompose as a result of inorganic reactions.

Aesthetic pollutants are difficult to define, so they are difficult to control. Some examples of aesthetic pollutants are noise, unpleasant sights, unusual tastes, and foul odors.

Unwanted wastes are eventually deposited in the air, in the water, or on land. Solid waste is disposed of by recycling, by composting, or by sanitary landfill. Toxic chemicals can be either incinerated or stored in a secured landfill.

Hazardous wastes are solid, liquid, or gaseous wastes that can threaten life. These wastes require special treatment methods to render them harmless or isolate them from humans. Radioactive wastes are another special category of waste that requires special handling to protect the public.

## Review Questions

1. Distinguish between persistent and nonpersistent pollutants.
2. Why are biodegradable materials easier to dispose of than those that are not biodegradable?
3. List two examples of situations where the atmosphere serves as a natural sink.
4. Describe three methods of dealing with solid wastes.
5. What are two acceptable methods for dealing with hazardous wastes?
6. How are radioactive wastes currently being disposed of?
7. What is the packaging revolution? How does it contribute to the problem of pollution?
8. List three common noises that could harm your sense of hearing.
9. Why are taste and odor pollution difficult to control?
10. How are affluence, advertising, and packaging associated with visual pollution?

## Suggested Readings

American Chemical Society. *Cleaning Our Environment: A Chemical Perspective.* 2d ed. Washington, D.C.: American Chemical Society, 1978. Presents a detailed analysis of the problems of solid waste disposal.

Breidenbach, Andrew W. *Composting of Municipal Solid Wastes in the United States.* Washington, D.C.: Government Printing Office, 1981. Presents alternative methods for disposing of municipal waste.

Burns, Williams. *Noise and Man.* Philadelphia: J. B. Lippincott Company, 1973. Details the effect of noise on humans.

Carmen, Richard. *Our Endangered Hearing*. Emmaus, Pa.: Rodale Press, 1977. Explores the causes, dangers, and solutions of noise pollution.

Council on Environmental Quality. *Annual Reports, 1980–82*. Washington, D.C.: U.S. Governmental Printing Office, 1982. Details the amounts of waste generated and the various methods of dealing with the disposal of this waste.

Hayes, Dennis. *Repairs, Reuse, Recycling; First Steps Towards a Sustainable Society*. Washington, D.C.: Worldwatch Institute, 1978. Presents detailed accounts of the advantages of resource conservation and recycling.

Kasper, William C. "Power from Trash." *Environment* 16:34–39. Explains the use of trash as a source of fuel for the production of energy.

Maugh, T. H. "Toxic Waste Disposal: A Growing Problem." *Science* 204: 819–23. Presents an overview of the problems of disposing toxic waste materials.

Metry, Amir A. *The Handbook of Hazardous Waste Management*. Westport Technomic Publishing Company, 1980. A technical report on waste management.

Taylor, Rupert. *Noise*. New York: Penguin Books, 1970. An easy-to-read book concerning the problems of noise pollution.

Wilson, Thomas W., Jr. *International Environmental Action: A Global Survey*. New York: Dunellen Publishing Co., 1971. Relates how various forms of government, from local to national, react to environmental problems.

## Chapter Outline

I. The atmosphere
   Box 19.1 Ozone
II. Primary air pollutants
III. Carbon dioxide and the greenhouse effect
IV. Secondary air pollutants
V. Control of air pollutants
   A. Acid deposition
   B. Indoor air pollutants
   Box 19.2 Acid rain: Canada vs. United States
VI. Consider this case study: International air pollution

## Objectives

Explain why air can accept and disperse significant amounts of pollutants.

List the sources and effects of the five primary air pollutants.

Describe how photochemical smog is formed and how it affects humans.

Explain how the PCV valve, APC valve, catalytic converter, scrubbers, precipitators, filters, and changing fuel types reduce air pollution.

Explain how acid rain is formed.

## Key Terms

acid deposition
carbon dioxide
carbon monoxide
carcinogenic
greenhouse effect
hydrocarbons
oxides of nitrogen
ozone
particulates
photochemical smog
primary pollutants
secondary pollutants
sulfur dioxide
thermal inversion

# Chapter 19
# Air pollution

## The atmosphere

Pollution, if you will recall, is material produced by humans that interferes with our well-being. Therefore, because we cause pollution, we may be able to do something to prevent it. In this book, we do not consider volcanic ash and gases as pollution because we do not cause the problem and cannot control it. We do consider automobile emissions and odors and factory smoke to be air pollution. We will focus on those things we do influence and will look at how we have changed the environment.

If it were possible to live as single-family units on farms where each family would grow its own food, air pollution would not be a

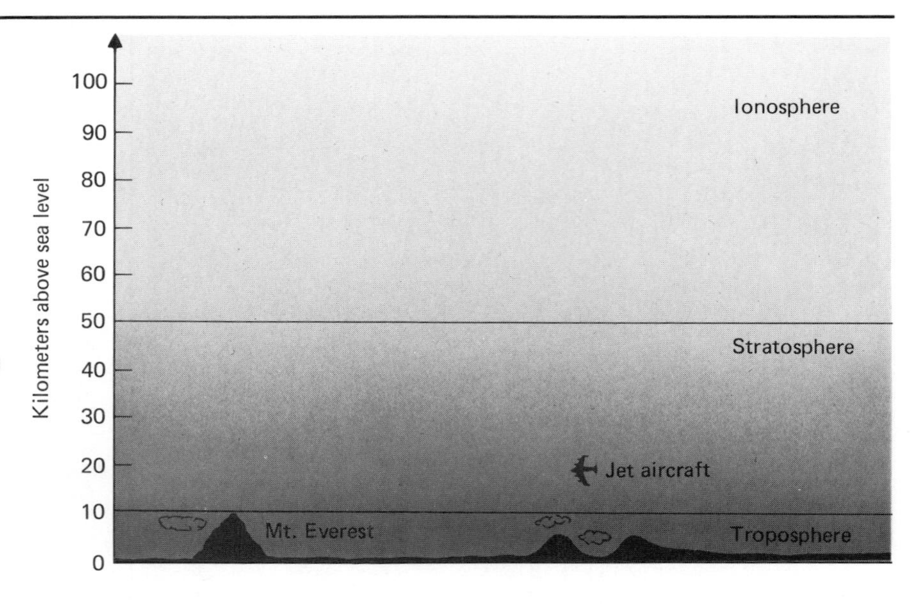


**Figure 19.1 The atmosphere.**
The atmosphere is divided into several layers. The troposphere is the relatively dense layer of gases close to the earth. Weather is confined to this layer. The stratosphere has essentially the same atmosphere as the troposphere, but it has a lower density. There is not sufficient oxygen in this layer to support most kinds of life. The ionosphere is composed of ionized gases.

problem. The amount of waste put into the atmosphere would be diluted, so no one's air quality would be seriously altered. However, we are an urbanized, industrialized civilization with a growing population and a history of increasing use of fossil fuels and technological aids. When we manufacture the products needed by this concentrated population, we also produce byproducts and release them into our environment.

The atmosphere is normally composed of 79 percent nitrogen, 20 percent oxygen, and a 1 percent mixture of carbon dioxide, water vapor, and small quantities of several other gases. Most of the atmosphere is held close to the earth by the pull of gravity. The atmosphere gets thinner with increasing distance from the earth. (See fig. 19.1.)

Even though gravity keeps the air near the earth, the air is not static. As it absorbs heat from the earth, it expands and rises. When its heat content is radiated into space, the air cools and becomes more dense and flows toward the earth. Besides this circulation of air due to heating and cooling, the air moves horizontally over the surface of the earth because the earth rotates on its axis. The combination of all air movements creates the specific wind patterns that are characteristic of different regions of the world. (See fig. 19.2.) Therefore, gases or small particles released into the atmosphere are likely to be mixed, diluted, and circulated, but they are likely to stay near the earth due to gravity. When we put a material into the air, we do not get rid of it; we just dilute it and move it out of the immediate area. When people lived in small groups, the smoke from fires was diluted; the smoke was in such low concentrations that it did not interfere with neighboring groups downwind. In industrialized urban areas, the pollutants cannot always be diluted before the air reaches another city. The polluted air from Chicago is further polluted by Gary, Indiana, supplemented by the wastes of Detroit, Michigan, and Cleveland, Ohio, and finally moves over New England to the ocean. (See fig. 19.3.)

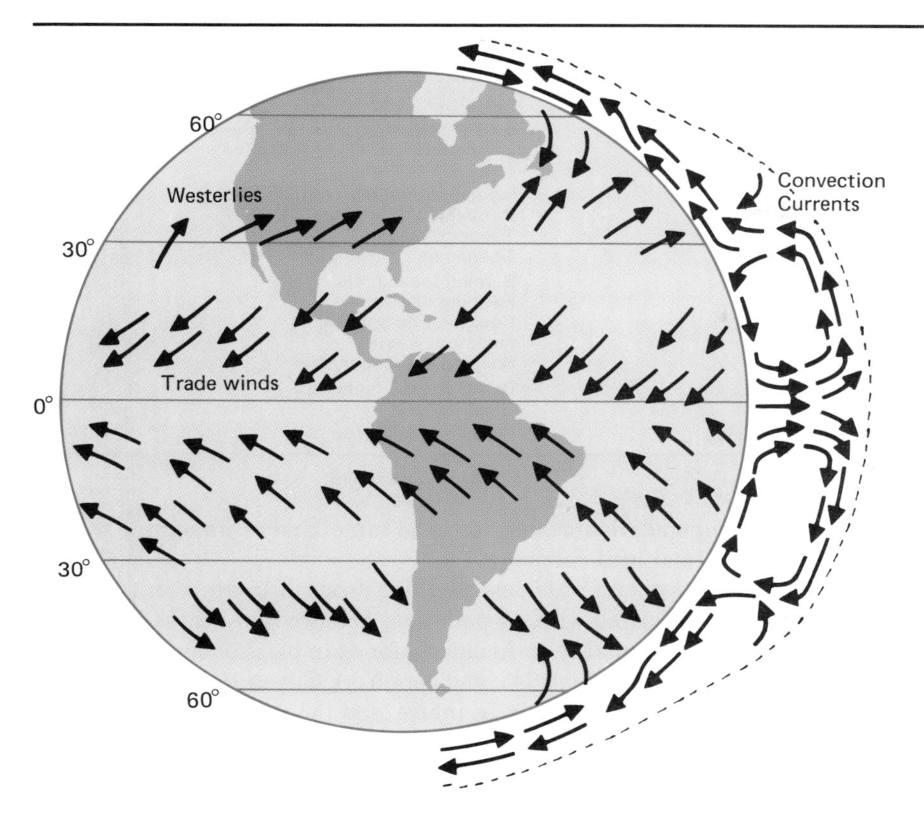


**Figure 19.2 Global wind patterns.**
Wind is the movement of air caused by temperature differences and the rotation of the earth. Both of these contribute to the patterns of air movement in the world. In the North American continent, most of our winds are westerlies (from the west to the east).

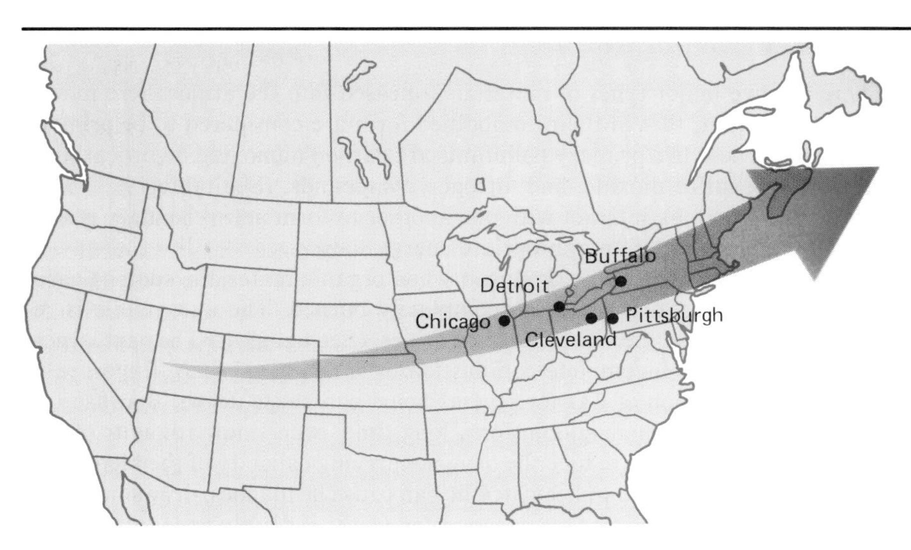


**Figure 19.3 Accumulation of pollutants.**
As an air mass moves across the continent from west to east, each population center adds its pollutants to the air.

**Table 19.1**
Sources of primary air pollutants

| Pollutant | Sources |
|---|---|
| Carbon monoxide | Incomplete burning of fossil fuels<br>Tobacco smoke |
| Hydrocarbons | Incomplete burning of fossil fuels<br>Decaying vegetation<br>Tobacco burning<br>Chemicals |
| Particulates | Burning fossil fuels<br>Farming operations<br>Construction operations<br>Industrial waste<br>Building demolition |
| Sulfur dioxide | Burning fossil fuels<br>Smelting ore |
| Nitrogen compounds | Burning fossil fuels |

While not every population center adds the same kind or amount of waste, each adds to the total load carried.

In addition to aesthetic problems resulting from dirty air, there are also health problems associated with air pollution. Hundreds of deaths have been directly related to poor quality air in cities—for example, London in 1952 and 1956; Donora, Pennsylvania, in 1948; and New York City in 1965. These deaths occurred primarily among the elderly, infirm, and the very young. Thousands of others have also been affected by air pollution. Bronchial inflammations, allergic reactions, and irritation of the mucous membranes of the eyes and nose all indicate that we need to reduce air pollution.

## Primary air pollutants

There are five major types of materials, released into the atmosphere in sufficient quantities, that in their unmodified form are considered to be **primary pollutants.** These five primary pollutants are carbon monoxide, hydrocarbons, particulates, sulfur dioxide, and nitrogen compounds. (See table 19.1.) Furthermore, these may interact with one another to form new **secondary pollutants** in the presence of an appropriate energy source.

**Carbon monoxide (CO)** is produced when organic materials, such as gasoline, coal, wood, and trash, are incompletely burned. The automobile is responsible for most of the CO produced in cities. (See fig. 19.4.) The next largest source is from the incomplete combustion of smoking tobacco. Exposure to 750 parts per million of CO for several hours can cause death. Because CO remains attached to hemoglobin for a long time, even small amounts of CO tend to accumulate and reduce the oxygen-carrying capacity of the blood. The amount of CO produced in heavy traffic can cause headaches, drowsiness, and blurred vision. A heavy smoker in congested traffic is doubly exposed and may experience a severely impaired reaction time as compared to nonsmoking drivers.

**Figure 19.4 Carbon monoxide source.**
The major source of carbon monoxide is the internal combustion engine, which is used to provide most of our transportation. The more concentrated the number of automobiles, the more concentrated the carbon monoxide. It is not unusual to experience CO concentrations of a hundred parts per million in rush-hour traffic in large metropolitan areas. These concentrations are high enough to cause fatigue, dizziness, and headaches.

Fortunately, CO is not a persistent pollutant. It seems that natural processes convert CO to other compounds that are not harmful. Therefore, the air can be cleared of its CO if no new CO is introduced into the atmosphere.

Automobiles also emit a variety of **hydrocarbons** (HC). Hydrocarbons are a group of organic compounds consisting of carbon and hydrogen atoms. They are either evaporated from fuel supplies or are remnants of the fuel that did not burn completely. Undoubtedly the internal combustion engine is the major culprit, although refineries and other industries do add hydrocarbons to the total atmospheric burden. Hydrocarbons in the atmosphere should be no great problem. Most of them are washed out of the air when it rains and run off into surface water. They do cause an oily film on surfaces, but hydrocarbons do not generally cause more than nuisance problems, except when they react to form secondary pollutants.

**Particulates** constitute the third largest category of air pollutants. These are small pieces of solid materials that are dispersed into the atmosphere. Smoke particles from fires, bits of asbestos from brake linings and insulation, dust particles, and ash from industrial plants contribute to the particulate load. Particulates cause problems ranging from the annoyance of soot settling on your backyard picnic table to the **carcinogenic** (cancer-causing) effects of asbestos. Particulates frequently get a lot of attention because they are so readily detected by the general public. The heavy black smoke from a factory can be seen by anyone without any expensive monitoring equipment. It generally causes an outcry, whereas the production of colorless gases like CO and $SO_2$ goes unnoticed.

**Sulfur dioxide** ($SO_2$) is a compound containing sulfur and oxygen that is produced when sulfur-containing fossil fuels are burned. Coal and oil were produced from organisms that had sulfur as one of the component parts of their living structure. When the coal or oil was formed, some of the sulfur was incorporated into the fossil fuel. As the fuel is burned, the sulfur is released as sulfur dioxide. $SO_2$ has a sharp odor and irritates respiratory tissue. It also

## Box 19.1
## Ozone

In the 1970s various sectors of the scientific community became concerned about the possible reduction in the ozone layer surrounding the earth. The presence of ozone in the outer layers of the atmosphere, thirty kilometers from the earth's surface, shields the earth from the harmful effects of ultraviolet radiation.

**Ozone** is a molecule that consists of three atoms of oxygen. This molecule in the presence of ultraviolet light is split into an oxygen atom and an oxygen molecule.

$$O_3 \xrightarrow{\text{uv light}} O_2 + O$$

Oxygen molecules are also split by ultraviolet light to form oxygen atoms.

$$O_2 \xrightarrow{\text{uv light}} 2O$$

The recombination of oxygen atoms and oxygen molecules forms ozone.

$$O_2 + O \longrightarrow O_3$$

These reactions absorb 99 percent of the ultraviolet energy from the sun, so it does not reach the earth's surface. Destruction of the ozone layer would increase the amount of ultraviolet light at the surface of the earth. This increase in ultraviolet light would result in increased mutations in plants and animals and an increase in skin cancers.

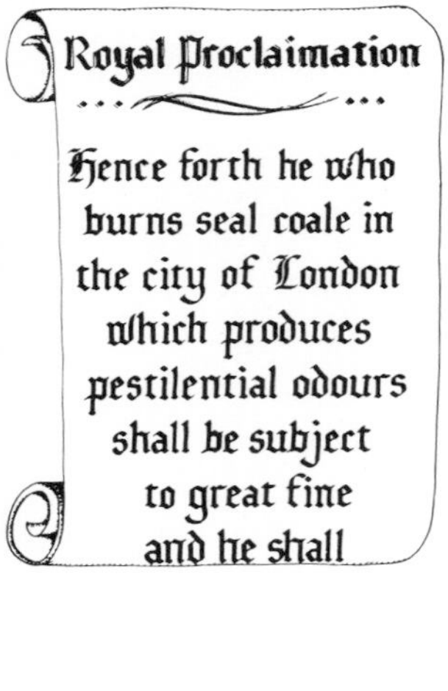

**Figure 19.5 Air pollution proclamation.** Edward I of England published a proclamation forbidding the burning of coal in 1306.

reacts with water, oxygen, and other materials in the air to form sulfur-containing acids. The acids can become attached to particles. When inhaled, such particles are very corrosive to lung tissue. Some of the earliest anti-air pollution legislation concerned the burning of high sulfur coals in London. (See fig. 19.5.)

In 1952, London, England, was covered with a dense fog for several days. During this time, the air over the city failed to mix with the layers of air in the upper atmosphere due to the temperature conditions. The factories of the city continued to release smoke and dust into this stagnant layer of air. The air became so full of the fog, smoke, and dust that people got lost in familiar surroundings. This combination of smoke and fog has become known as smog. Many individuals who lived in the city developed symptoms such as respiratory discomfort, headache, and nausea. Four thousand people died in the next few weeks, and their deaths have been associated with the high levels of sulfur compounds in the smog. Thousands of others suffered from severe bronchial irritation, sore throats, and chest pains.

The use of chloro-fluorocarbons (freon) as refrigerator gases and as aerosol propellants has resulted in the release of large amounts of these reactive molecules into the atmosphere. It was discovered that ultraviolet radiation splits the chlorine atoms from the chloro-fluorocarbon molecule. The free chlorine atoms then react with ozone molecules and break them into molecular oxygen.

$$2O_3 \xrightarrow[\text{atoms}]{\text{chlorine}} 3O_2$$

Concern developed about whether this reaction could occur in the upper layers of the atmosphere and destroy the ozone layer. This concern led to the ban on the use of freon in aerosol containers in the United States. Even though several other countries (Sweden, Denmark, and Canada) have similar bans, most nations still permit the unrestricted use of freon. Therefore, today the worldwide use of freon has been reduced by only 40 percent.

Many scientists believe that because there is such a lack of information about the ozone layer, we cannot be certain that the use of chloro-fluorocarbons has actually caused a reduction in the ozone in the upper atmosphere. Therefore, they do not feel that the use of freon presents an environmental problem.

In Donora, Pennsylvania, in 1948, a similar situation resulted in twenty deaths and widespread discomfort and ill health. These symptoms were related to the particles and sulfur dioxide in the air.

**Oxides of nitrogen ($NO$ and $NO_2$)** are also major primary air pollutants. They consist of a variety of different compounds containing nitrogen and oxygen. When combustion takes place in air, the nitrogen and oxygen molecules from the air may react with each other.

$N_2 + O_2 \rightarrow 2\ NO$ nitrogen oxide
$2\ NO + O_2 \rightarrow 2\ NO_2$ nitrogen dioxide

Oxides of nitrogen result. A mixture is usually produced, and the mixture is called $NO_x$. Together, these nitrogen compounds produce a reddish-brown color in the atmosphere and react with other compounds to produce photochemical smog. The primary source of nitrogen oxides is the automobile engine. (Catalytic converters reduce the amount of nitrogen oxides released.) Nitrogen oxides are involved in the production of secondary pollutants.

## Carbon dioxide and the greenhouse effect

In addition to $SO_2$, $NO_x$, and CO, combustion of fossil fuels releases large amounts of **carbon dioxide ($CO_2$)**. $CO_2$, a normal component of the earth's atmosphere, is released as a result of normal metabolic processes as well as by combustion. Plants trap $CO_2$ through the process of photosynthesis and convert it to organic molecules. However, there has been much speculation about the effects of artificially elevated levels of $CO_2$ in the atmosphere. One such speculation centers on the property of $CO_2$ that allows light energy to pass through the atmosphere but prevents heat from leaving. In many respects, this is similar to what happens in a greenhouse, so it has been called the **greenhouse effect.**

Light readily passes through glass, but heat passes through less readily. When light enters a greenhouse or passes through the windows of an automobile, the light is absorbed by the objects inside. The light energy is converted to heat energy, which becomes trapped inside the building or automobile. In the atmosphere, the more concentrated the $CO_2$, the more heat is retained. The amount of $CO_2$ in the atmosphere has increased over the past thirty years. At Mauna Loa observatory, the average $CO_2$ concentration increased from approximately 312 parts per million in 1957 to approximately 324 parts per million in 1971. Many people have speculated that this is part of a long-term trend and that $CO_2$ concentrations were lower in the past and will continue to increase in the future because of our increasing use of fossil fuels. Some people have tried to tie this to increasing world temperatures. However, since the 1940s the average world temperature has decreased, while the level of $CO_2$ has been increasing.

This drop in temperature has been tied to increased amounts of particulate matter in the atmosphere. It has been suggested that the particulate matter in the atmosphere acts as a screen that reduces the amount of light reaching the earth, which accounts for the lower temperature. Such speculation will continue until there are enough data to substantiate the link between $CO_2$ concentration and temperature or until it is proven that no such link exists.

## Secondary air pollutants

Secondary pollutants are compounds that result when pollutants react with each other. **Photochemical smog** is the result of the interaction of hydrocarbons, oxides of nitrogen, and sunlight, which produces a yellowish-brown haze. (See fig. 19.6.) Photochemical smog is a mixture of complex organic compounds that irritate respiratory tissue and eye membranes and stimulate the production of tears. People who live in areas where smog is common have more health problems than do those who live in less polluted areas. These areas also show altered plant growth.

**Figure 19.6**
**Photochemical smog.**
The interaction among hydrocarbons, oxides of nitrogen, and sunlight produces new compounds that are irritants to humans. The visual impact of this smog is shown in these photographs.

**Figure 19.7 Thermal inversion.**
Under normal conditions, the air at the earth's surface is heated by the sun and rises to mix with the cooler air above it. When a thermal inversion occurs, a layer of warm air is formed above the earth, and the cooler air at the surface is unable to mix with the warm air above. This problem is exaggerated in metropolitan areas that are surrounded by mountains.

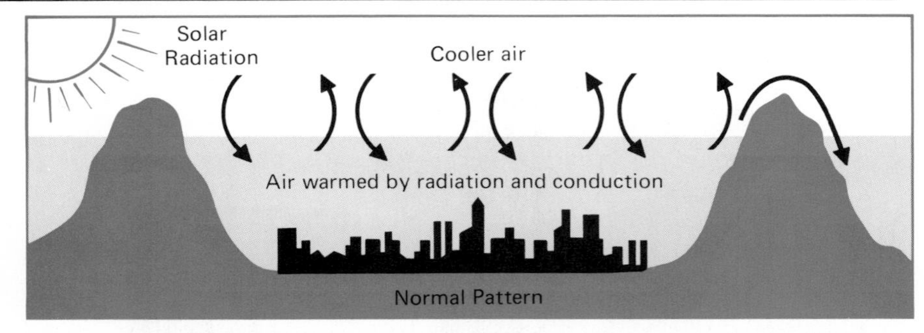


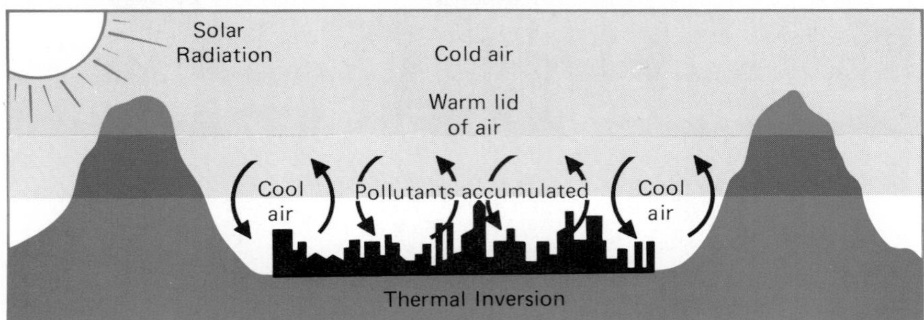


Large metropolitan areas, such as Los Angeles, Salt Lake City, and Denver, have more trouble with photochemical smog than the metropolitan areas of the East Coast. This is due to the climate and the geographic features of their locations. Each of these cities is located within a ring of mountains. The prevailing winds are from the west. As cool air flows into a valley, it pushes the warm air upward. This warm air is sandwiched between two layers of cold air and acts like a lid on the valley. This condition is known as a **thermal inversion.** The air is trapped in the valley. (See fig. 19.7.) The warm air cannot rise further because it is covered with a layer of cooler air pushing down on it. It cannot move out of the area because of the ring of mountains. Without normal circulation of air, hydrocarbons and nitrogen oxides from automobile exhausts accumulate. Sunlight acts upon these pollutants, and a mixture of ozone, aldehydes, and PAN (peroxyacetyl-nitrate) is produced. This mixture is called photochemical smog. These harmful chemicals continue to increase in concentration until a major weather change causes the warm air to move up and over the mountains. Then the underlying cool air can begin to circulate, and the polluted air is diluted. To eliminate smog problems, we would need to decrease the use of internal combustion engines (perhaps eliminate them completely) or move population centers away from the valleys that produce thermal inversions. Both of these solutions would require expenditures of billions of dollars; therefore, people will probably continue to live with the problem.

**Figure 19.8 Control of particulates.**
By installing proper pollution control devices, the amount of particulate material released into the air can be reduced significantly.

## Control of air pollution

Eliminating photochemical smog would require large-scale changes in life-style and culture, but other pollutants are much more readily controlled. Government regulations have pressured the automobile industry to engage in air pollution control research. The positive crankcase ventilation valve (PCV) and gas caps with air pollution control valves (APC) reduce hydrocarbon loss. Catalytic converters reduce oxides of nitrogen but necessitate the use of lead-free fuel.

Particulates are produced primarily by burning and can be controlled with scrubbers, precipitators, and filters. (See fig. 19.8.) These devices are effective, but they are expensive. As with auto emissions, government has required the installation of these devices. When industries install these devices, the cost of construction and operation is passed on to the consumer. This amounts to millions of dollars.

To control $SO_2$, which is produced primarily by electric power generating plants, several possibilities are available. One alternative is to change from high-sulfur fuel to low-sulfur fuel, such as natural gas, oil, or nuclear fuel. This is not a long-term solution because low-sulfur fuels are in short supply and nuclear power plants present a new set of pollution problems. A second alternative is to remove the sulfur from the fuel before it is used. This is technically possible, but it increases the cost of electricity to the rate payer. Scrubbing the gases that are emitted from a smoke stack is a third alternative. The technology is available, but, of course, these control devices are costly to install, maintain, and operate.

In the past, a common solution was to build taller smoke stacks. Tall stacks release their gases above the inversion layers and, therefore, add the $SO_2$ to the upper atmosphere where it is diluted before it comes in contact with the population downwind. This works as long as the stack is tall enough and as long as we do not add so much pollution to the air that it cannot be diluted to an acceptable level. However, as the $SO_2$ content of the upper atmosphere has increased, a new problem has developed. The $SO_2$ reacts with oxygen and dissolves in the water in the atmosphere to form sulfuric acid. This acid is washed from the air when it rains or snows. The acid damages plants and animals and increases corrosion of building materials and metal surfaces.

## *Acid deposition*

**Acid deposition** is the accumulation of potential acid-forming particles on a surface. Acids result from natural causes such as volcanos and human activities such as burning and the internal combustion engine. (See fig. 19.9.) These combustion processes produce sulfur dioxide ($SO_2$) and oxides of nitrogen ($NO_x$). Oxidizing agents, such as ozone, hydroxyl ions, or hydrogen peroxide, along with water, are necessary to convert the sulfur dioxide or nitrogen oxides to sulfuric or nitric acid. Various reactive hydrocarbons (HC) encourage the production of oxidizing agents.

The acid-forming reactants are classified as wet or dry. Wet reactions occur in the atmosphere and enter the environment as some form of precipitation, acid rain, snow, or dew. Dry deposition is the settling of the precursors for the acid on a surface. An acid does not actually form until these materials mix with water.

Acid rain is becoming a worldwide problem. Reports of high acid rain damage have come from Canada, England, Germany, France, and Scandinavia, as well as the United States. Acid rain is precipitation that is a weak solution of sulfuric and nitric acids. It is caused by emissions of sulfur and nitrogen oxides from industries, such as electric power generating plants, smelters, and factories. These oxides mix with water vapor, oxidants, and particulates to form sulfuric and nitric acid.

Rain is normally slightly acid (a pH between 5.6 and 5.7) due to the atmospheric carbon dioxide that dissolves to produce carbonic acid. But the high acid rains sometimes have a concentration of acid a thousand times higher than normal. Several extremely acid rains have been recorded. In 1969, New

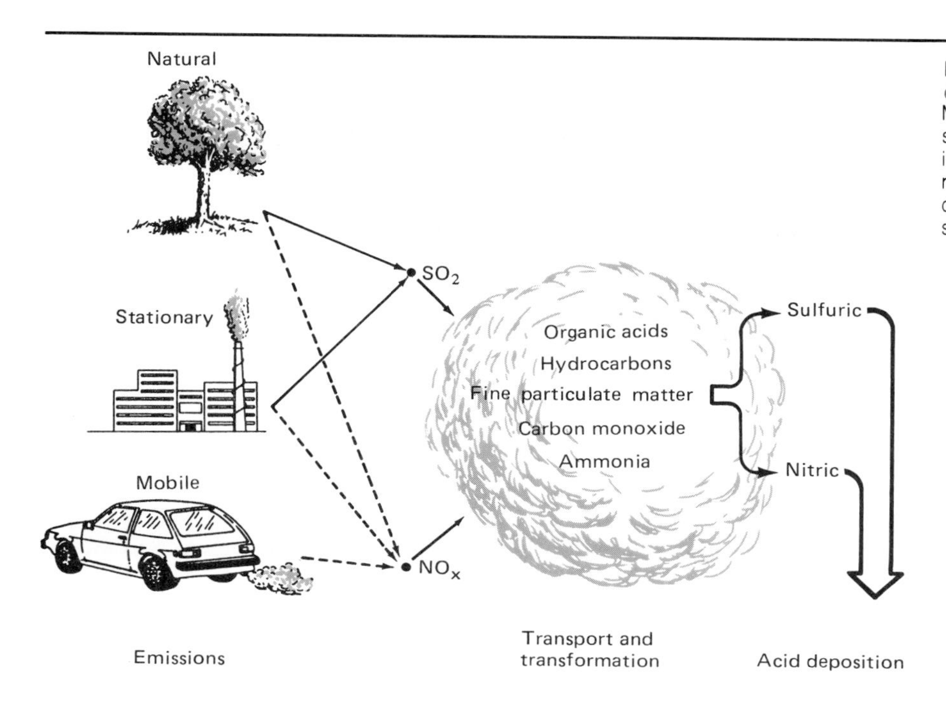


**Figure 19.9 Acid deposition.** Molecules from natural sources, burning, and the internal combustion engine react to produce the chemicals that are the source of acid deposition.

**Figure 19.10 Acid deposition damage.** Sulfuric acid ($H_2SO_4$), which is a major component of acid deposition, reacts with limestone ($CaCO_3$) to form gypsum ($CaSO_4$). Since gypsum is water soluble it washes away with rain.

1 Acid rain, or dry deposition falls

2 Crust forms

3 Crust washes off

4 Layer of stone is removed

Hampshire had a rain with a pH of 2.1 In 1974, Scotland had a rain with a pH of 2.4. The average rain in much of the northeastern part of the United States and adjoining parts of Ontario has a pH between 4.0 and 4.5.

The acid erodes stone, cement, or other types of masonry and causes chemical damage to metal surfaces. The damage to stone or masonry is caused by the acid transforming the calcite in the stone into gypsum, which is much more soluble in water. (See fig. 19.10.)

Various parts of the ecosystem are disrupted by acid deposition. For example, the acid causes lesions that reduce the growth of plants. Acids also inhibit decay bacteria, which results in an accumulation of litter and a loss of

**Figure 19.11 Effects of acid deposition.** The low pH of the water in which this fish lived caused abnormal bone development. This ultimately resulted in the death of the fish.

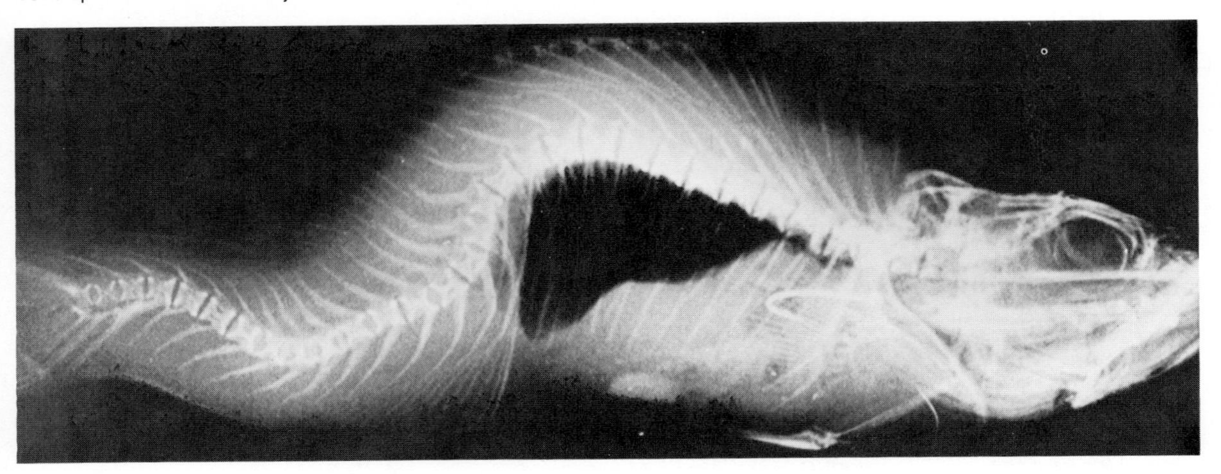

nutrients available to plants. In addition, minerals such as aluminum, magnesium, and potassium are leached from the soil. This action on the ecosystem has reduced the growth rate of spruce trees in part of the eastern United States by as much as 50 percent. In this same region, there has been a marked reduction in the number of new seedlings.

One of the most publicized effects of acid deposition in the environment has been damage in aquatic ecosystems. Acid deposition lowers the pH level of the water, which results in less calcium being available for the growth of fish bones. This causes a humpbacked fish, a condition that ultimately results in death. (See fig. 19.11.) The increased acidity also triggers the release of aluminum into the water, which impairs the function of the gills of fish. At a pH of 5.5 many female fish no longer produce eggs.

Lakes are normally slightly alkaline to neutral. When the pH drops below 5.5, many desirable species of fish are eliminated, and at a pH of 5, fish are eliminated entirely. Lakes with a pH below 4.5 are nearly sterile. In upper New York state two hundred bass and trout lakes are in danger of becoming unfit for fish, and approximately one thousand lakes in Ontario are similarly threatened. Many lakes in Scandinavia are already sterile bodies of water because of acid rain.

The extent to which acid deposition affects an ecosystem depends in a large measure on the geology of the ecosystem. (See fig. 19.12.) Parent material derived from igneous rock is not capable of buffering the effects of acid deposition, while soils derived from sedimentary rocks (such as limestone) release bases that neutralize the effects of the acids. It is estimated that in the eastern half of the United States, thousands of kilometers of streams and twenty thousand lakes are sensitive to acid deposition because they are unable to buffer the effect of acid runoff. Acid deposition has already changed the quality of

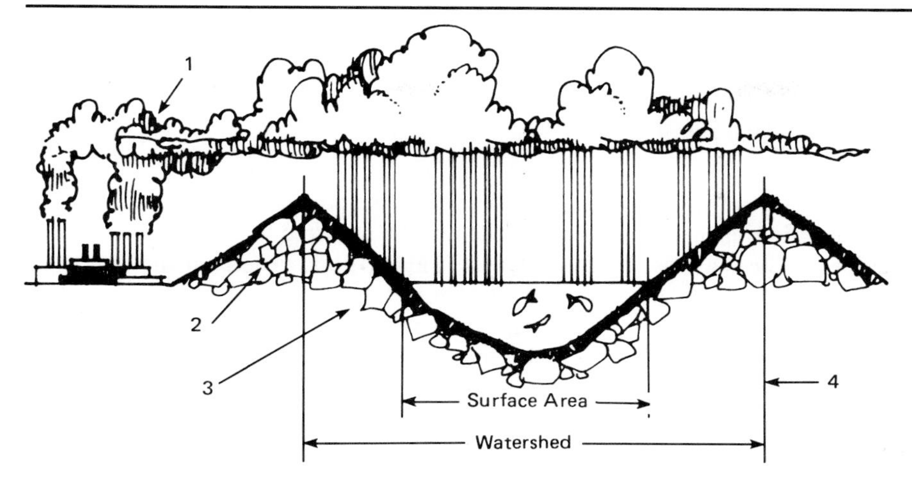


**Figure 19.12 Acid deposition damage in an aquatic ecosystem.**
In an aquatic ecosystem, there is an increased risk of damage from acid deposition if the following factors are present: (1) a lake is located downwind from a major source of air pollution; (2) the area around the lake is hard, insoluble bedrock covered with a layer of thin infertile soil; (3) the soil has a low buffering capacity; and (4) there is also a low watershed to lake surface area ratio. (From EPA's "Acid Rain.")

many of these waters. One of the primary sources of acid derived from combustion is associated with emission from smokestacks. In the past, a common solution was to build taller smoke stacks. Tall stacks release their gases above the inversion layers and, therefore, add the $SO_2$ to the upper atmosphere where it is diluted before it comes in contact with the population downwind. This works as long as the stack is tall enough and as long as we do not add so much pollution to the air that it cannot be diluted to an acceptable level. Currently, there are three options available to solve the acid deposition problem: scrubbing of the smokestack gases, removing the sulfur from coal before it is burned, or using a low-sulfur fuel.

Scrubbing removes the sulfur from the flue gases before they leave the stack. This is a successful but costly method. Large amounts of lime are used in a scrubber, and there is a large amount of waste material produced. It is estimated that in the United States, 22.5 million metric tons of sulfur are released into the atmosphere each year. Installing scrubbers on just fifty of the largest coal burning plants would reduce this amount by over one-third. By chemically or physically treating coal before it is burned, it is possible to remove nearly 40 percent of the sulfur. The sulfur can then be sold, and some of the cost of its removal can be recovered. Switching from a high-sulfur coal to a low-sulfur coal reduces the amount of sulfur released into the atmosphere by 66 percent. Changing to natural gas or oil as a fuel also reduces the amount of sulfur.

There is resistance to taking these steps. First, large utility companies (a prime source of atmospheric $SO_2$) question the effect of acid deposition on the environment and deny that they are the main source of sulfur emissions. Second, it is extremely difficult to document the adverse influence of acid on vegetation. It may be many years before sufficient data can be collected to indicate the effects of acid deposition on growth rates of forests. Nevertheless, acid deposition has increased and steps must be taken to reduce it.

## Box 19.2
## Acid rain: Canada vs. United States

"U.S. footdragging and interference in the development of scientific information has reached frustrating proportions." These words, spoken by the Honorable John Roberts, Canadian Minister of the Environment, sum up the confrontation between Canada and the United States regarding the question of acid deposition. Although the Canadians do release large amounts of sulfur dioxide and oxides of nitrogen into the atmosphere, they have long contended that much of the acid deposition in their country originated in the United States. They are very concerned because 2.5 million square kilometers of Canada are highly susceptible to acid deposition.

With the signing of the Memorandum of Intent on Transboundary Air Pollution on August 5, 1980, the United States and Canada took the first step to cooperatively reduce the amount of acid deposition. The memorandum created scientific groups to study the problems of air pollution. After two years of study, there was still no accord. The Canadians accused the Reagan administration of delaying the studies, and the United States accused the Canadians of acting too rapidly.

In fact, the dispute reached such a proportion that in 1983 the U.S. Department of Justice ruled that a Canadian-produced film on acid rain must be labeled as political propaganda before it could be shown in the United States. The U.S. Department of Justice also required that the names of U.S. groups viewing the film be reported to the Justice Department. This attitude prompted the Canadian Minister of the Environment to observe, "It sounds like something you would expect from the Soviet Union, not the United States."

In February, 1982, Canada suggested a mutual 50 percent reduction of sulfur dioxide by 1990. Citing a lack of research, the Reagan administration did not agree with the plan. This prompted Minister Roberts to state, "Always the constant refrain rings out from the administration that nothing is proven, and that an indefinite amount of further study is needed, not prompt action. Well, we can't wait. Our lakes and forests are literally dying."

In his 1984 state of the union address, President Reagan affirmed that the United States would take no direct action regarding the question of acid deposition other than to continue to research the problem. On March 6, 1984, the Canadians announced a goal of reducing acid deposition by 50 percent and trusted that the United States would join them. However, the possibility seems remote, for on May 2, 1984, a House subcommittee voted against a bill to reduce the emissions of sulfur dioxide by 10 million metric tons by 1993. This killed any United States action regarding reduced acid deposition for 1984.

**Figure 19.13 Concern for air quality.**
The fact that some states and communities have limited or banned smoking in certain areas is evidence of the growing public concern over air quality and health.

These options discussed reduce the acid deposition resulting from sulfur dioxide; they do not reduce the production of oxides of nitrogen. Oxides of nitrogen are formed when the combustion of fossil fuel occurs at high temperatures. The use of catalytic converters on automobiles is the primary method of controlling the production of oxides of nitrogen as well as carbon monoxide and hydrocarbons. The use of catalytic converters, however, requires the use of lead-free fuel, which necessitated design changes in automobile engines.

## *Indoor air pollutants*

Because there is a correlation between smoking and health, regulations often segregate smokers from nonsmokers. Providing nonsmoking sections on airplanes and in restaurants and banning smoking in public places indicate that some attention is being given to the amount of CO that we are subjected to. (See fig. 19.13.)

Even though we spend almost 90 percent of our time indoors, the movement to reduce indoor air pollution lags behind regulations governing outdoor air pollution. As a result, many people enjoy cleaner air outside than they do inside a building. Indoor air pollutants include tobacco smoke, nitrogen oxides, carbon monoxide, microorganisms, formaldehyde, aerosols, and the end products of various chemical reactions.

There are many other serious indoor air pollutants, such as asbestos, propellants and products in aerosol sprays, paint fumes, plastics, bleaches, dust, and disease-causing microorganisms. The prime reason for an increase in the amount of indoor air pollution is the weatherizing of buildings. This is done in an attempt to reduce heat loss and save on fuel costs. In most homes, there is a complete exchange of air every hour. This means that fresh air leaks in around doors, windows, cracks, and holes in the building. In a weatherized home, a complete air exchange may occur only once every five hours. Such a home is more energy efficient, but it also tends to trap air pollutants.

## Consider This Case Study
### International air pollution

Acid rain originates as a form of air pollution, but it may damage the environment in the form of water pollution. Also, air pollution may originate in one country and water pollution may be formed in another country. A particular nation may have stringent environmental controls within its own boundaries, but have environmental problems because of the actions of a neighboring country.

Recently, the phenomenon of acid rain has underscored the need for international cooperation in dealing with various environmental concerns. For example, an estimated fifty-six percent of the acid rain falling in Sweden originated outside of that country. The main sources of the pollutants are West Germany and the United Kingdom because the west coast of Sweden receives winds from these countries. When these industralized countries release more sulfur dioxide into the atmosphere, the result is more acid rain in Sweden. Since the 1930s, the lakes in western Sweden have become more acidic by a value of two pH units. Ten thousand lakes have a pH below 6.0 and five thousand lakes have a pH below 5.0. A pH below 5.5 is too acidic for many species of fish.

In the 1950s, the United Kingdom initiated a program to reduce air pollution. This program has been successful, in that the air in the United Kingdom has become cleaner. But one of their solutions was to build taller smokestacks. As a result, more of the pollutants from the United Kingdom are transported to Sweden.

Should the United Kingdom be permitted to disperse air pollutants in a manner that damages the Swedish environment?

Should there be a series of international agreements to control and regulate the movement of air-borne pollutants across international boundaries?

In addition to air pollutants, are there any other forms of pollutants that can naturally be transported across international boundaries?

## Summary

The atmosphere has a tremendous ability to accept and disperse pollutants. Carbon monoxide, hydrocarbons, particulates, sulfur dioxide, and nitrogen compounds are the major primary pollutants. They can cause a variety of health problems.

Photochemical smog is a secondary pollutant that is formed when $NO_x$ and hydrocarbons are trapped by thermal inversions and react with each other in the presence of sunlight to form PAN, aldehydes, and ozone. Eliminating photochemical smog requires changes in technology, such as scrubber precipitators and sulfur removal from fuels.

Acid rain is caused by emissions of $NO_x$ and $SO_2$ in the upper atmosphere, which form acids that are washed from the air when it rains or snows.

## Review Questions

1. List five pollutants commonly released into the atmosphere and their sources.
2. Define secondary pollutant and give an example.
3. List three health effects of air pollution.
4. Why is air pollution such a large problem in urban areas?
5. Describe three actions that can be taken to control air pollution.
6. What causes acid rain? List three possible detrimental consequences of acid rain.
7. How does energy conservation influence air quality?
8. How is $CO_2$ (a nontoxic normal component of the atmosphere) involved in air quality?
9. What is photochemical smog? What causes it?
10. What would the consequences be if the ozonosphere were destroyed?

## Suggested Readings

Bernard, Harold W., Jr. *The Greenhouse Effect.* Cambridge, Mass.: Ballinger Publishing Company, 1980. A thorough description of possible consequences of altering the heat content of the atmosphere by changing the $CO_2$ content. Looks at the problem on a region-by-region basis.

Dotta, Lydia, and Schiff, Harold. *The Ozone War.* Garden City, N.Y.: Doubleday and Company, 1978. The authoring team of a chemist and science writer provides an easy-to-read description of the controversy surrounding the quantity of ozone in the atmosphere and its importance.

Environment Canada. "Downwind: The Acid Rain Story." Information Directorate, Environment Canada, Ottawa, Ontario K1A OH3. An excellent sixteen-page booklet concerning all aspects of acid deposition.

Kirsch, Laurence S. "The Problem of Indoor Pollutants." *Environment* 25: 17–20. Presents the major sources of indoor pollution.

Rhodes, Steven L., and Middleton, Paulette. "Public Pressures, Technical Options: The Complex Challenge of Controlling Acid Rain." *Environment* 25: 6–9. Explains causes of and solutions regarding acid deposition.

Stern, Arthur, ed. *Air Pollution: Vols. I, II, and III.* 2d ed. New York: Academic Press, 1968. These three volumes are the definitive work relating to air quality. The material is written for air pollution professionals, but with some attention to detail, the advanced student can master the concepts. Excellent reference lists follow each chapter.

## Chapter Outline

I. Kinds and sources of water pollution
   A. Municipal water pollution
   B. Industrial water pollution
      Box 20.1 Marine oil pollution
   C. Thermal water pollution
II. Waste-water treatment
III. Runoff
IV. Eutrophication
    Box 20.2 International Joint Commission
V. Consider this case study: Lake Erie

## Objectives

Explain how water can accept and disperse significant amounts of pollutants.

List the major sources of water pollution.

Differentiate between point and nonpoint sources of pollution.

Define biochemical oxygen demand (BOD).

Explain how nutrients cause water pollution.

Explain how heat can be a form of pollution.

Differentiate between primary, secondary, and tertiary sewage treatment.

Describe eutrophication.

## Key Terms

activated sludge sewage treatment

agricultural runoff

biochemical oxygen demand (BOD)

eutrophication

nonpoint source

point source

primary sewage treatment

secondary sewage treatment

storm-water runoff

tertiary sewage treatment

thermal pollution

# Chapter 20 Water pollution

## Kinds and sources of water pollution

It is probably impossible in an industrialized society to maintain unpolluted water in all drains, streams, rivers, and lakes. We need to remember that water quality is related to the use we intend to make of the water. Adding material to water may cause the water to become unfit for some uses but may not affect other uses. If silt is added to a lake, the water may still be drinkable, but the lake may no longer be an acceptable place to swim. If salts are added to a lake, the water may then be less acceptable as drinking water, but the salts may not interfere with the lake's recreational value. It may

**Figure 20.1 Effect of organic wastes on dissolved oxygen.** Sewage contains a high concentration of organic materials. When these are degraded by organisms, oxygen is removed from the water. The greater the BOD, the less desirable the water is for human use. Therefore, there is an inverse relationship between sewage and oxygen in the water.

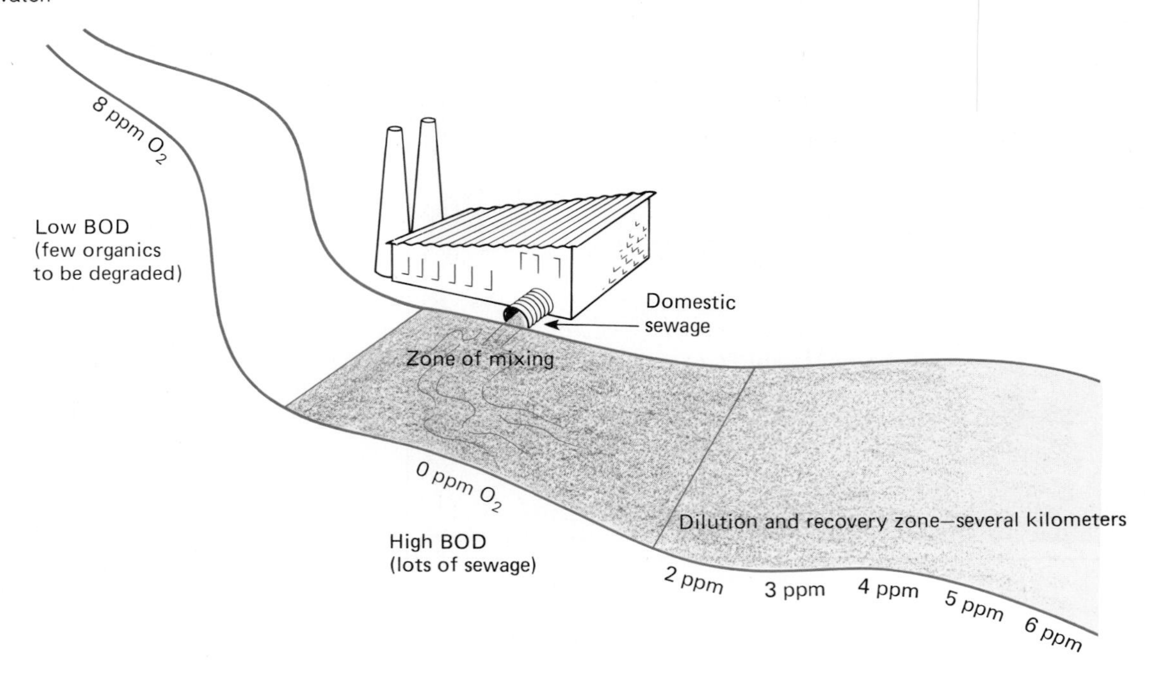


not be necessary to maintain absolutely pure water. The cost of removing the last few percentages of some materials from the water may not be justified. This is certainly true of organic matter, which is biodegradable. The value of removing all of the organic material does not equate with the cost. However, radioactive wastes and toxins that may accumulate in living tissue are a different matter. The cost of completely removing these materials is justified because of their potential to cause harm.

## *Municipal water pollution*

Municipalities are faced with the double-edged problem of providing suitable drinking water for the population and of disposing of wastes. These wastes consist of storm-water runoff, wastes from industry, and wastes from homes and commercial establishments.

Wastes from homes consist primarily of organic matter from garbage, food preparation, cleaning of clothes and dishes, and human wastes. Human wastes are mostly undigested food material and a concentrated population of bacteria, such as *Escherichia coli* and *Streptococcus faecalis.* These bacteria normally grow in the human large intestine, where they are responsible for some food digestion and for the production of vitamin K. Low numbers of these bacteria in water are not harmful to healthy people. They are used to indicate the amount of pollution from human waste, because they can be easily identified. (The number present in water is directly related to the amount of human waste entering the water.) Disease-causing bacteria from the human large intestine may also be present in amounts too small to detect by sampling, but they may cause serious illness. Therefore, these bacteria are indicators that disease-causing organisms may be present.

The nonliving organic matter in sewage presents a different kind of pollution problem because it decays in the water. Microorganisms use oxygen dissolved in the water when they degrade the organic material. As the microorganisms metabolize the organic matter, they use up the available oxygen. The amount of oxygen required to decay a certain amount of organic matter is called the **biochemical oxygen demand (BOD).** (See fig. 20.1.) Measuring the BOD of a body of water is one way to determine how polluted it is. If too much organic matter is added to the water, all of the available oxygen will be used up. Then anaerobic (nonoxygen requiring) bacteria begin to break down wastes. Anaerobic respiration produces chemicals that have a foul odor and an unpleasant taste and that generally interfere with the well-being of humans. Therefore, they are pollutants.

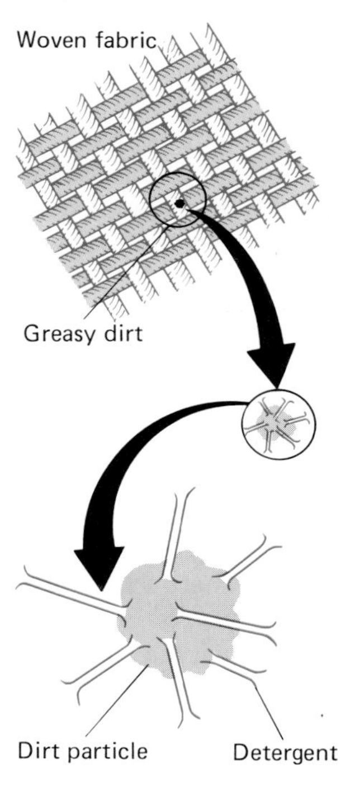


**Figure 20.2 Action of detergents.**
Dirt and grease are not soluble in water and are, therefore, difficult to remove from clothing. Detergent molecules have one end that is soluble in water and another end that dissolves in the dirt. This helps to free the dirt from clothing, and the dirt particles are washed away.

Although the water used for cleaning dishes and clothing may contain some organic material, it has a more important group of contaminants—soaps and detergents. Soaps and detergents are used because one end of the molecule will dissolve in dirt or grease and the other end will dissolve in water. When the molecules of soap or detergent are rinsed away by the water, the dirt or grease goes with them. (See fig. 20.2.) Detergents frequently contain phosphate as a part of their chemical structure. The phosphate group acts as a fertilizer when it reaches the surface water. This fertilizer promotes the growth of undesirable algae populations. An algae population explosion is called an algae bloom. Algae and other plants may interfere with the use of the water by fouling boat propellers, clogging water intake pipes, changing the taste and odor of water, and causing the buildup of organic matter on the bottom. As this organic matter decays, BOD increases, oxygen levels decrease, and fish and other aquatic species die.

Because these problems are all associated with the addition of plant nutrients such as phosphates to the water, many states have banned the sale of detergents with high-phosphate content. Detergent manufacturers point out that nutrients from **agricultural runoff** are probably more significant than detergents in adding phosphate to lakes and streams. Most domestic waste water, however, goes through a sewage treatment plant, so it is easier to measure and control the phosphate content. In those areas where agriculture is uncommon, detergents can be the major source of phosphate pollution.

## *Industrial water pollution*

Frequently a factory or industrial complex will dispose some or all of its wastes into a municipal sewage system. Depending on the type of industry involved, these wastes are likely to be a combination of organic materials, petroleum products, metals, acids, and so forth. The organics and oil add to the BOD of

## Box 20.1
## Marine oil pollution

On January 28, 1969, an offshore oil drilling platform near Santa Barbara, California, malfunctioned, resulting in the release of an estimated 940,000 liters of crude oil into the waters of the Santa Barbara harbor. This catastrophe helped to awaken the nation to the dangers of marine oil pollution. Shortly after the Santa Barbara spill, Congress responded to pressure and passed measures to control oil spills and to provide new pollution standards for platform oil drilling. Although such legislation may help to prevent new spills from happening in the United States coastal waters, the problem of world marine oil pollution is far from being resolved.

There are many sources of marine oil pollution. One source is accidents such as oil drilling blowouts or oil tanker accidents, for example, the Torrey Canyon supertanker that broke up off the coast of England in 1967, and the Amoco Cadiz supertanker that broke up off the coast of France in 1978. The Amoco Cadiz resulted in a spill of more than 254 million liters of oil. The major sources of marine oil pollution, however, are not accidental.

Nearly two-thirds of all human caused marine oil pollution comes from two sources—runoff from streets and disposal of lubricating oil from machines and automobile crankcases (primarily into sewers) and intentional oil discharges that occur during the loading and unloading of tankers. Pollution also occurs when the tanks are cleaned or the oil-contaminated ballast water is released. Oil tankers use seawater as ballast to stabilize the craft after they have discharged their oil. This oil-contaminated water is then discharged back into the ocean when the tanker is refilled. As the number of offshore oil wells and the number and size of oil tankers continue to grow, the potential for increased oil pollution also grows. Many methods for controlling marine oil pollution have been investigated and tried. Some of the more promising methods include reprocessing of used oil and grease from automobile service stations and industries, and enforcing stricter regulations on the offshore drilling, refining, and shipping of oil.

the water. The metals, acids, and other ions need special treatment depending on their nature and concentration. As a result, municipal sewage treatment plants must be designed with their industrial customers in mind. In most cases, it is preferred that industries take care of their own wastes. This allows the industry to segregate and control toxic wastes and design a waste-water treatment facility that meets the specific needs of the industry.

Most companies, when they remodel their facilities, include waste-water treatment as a necessary part of an industrial complex. However, many older facilities continue to pollute. These companies discharge acids, particulates, heated water, and noxious gases into the water. Generally, these plants are easily detected because the pollution is from a single effluent pipe or series of pipes. This pollution is said to come from a **point source.** Diffuse pollutants, such as agricultural runoff, road salt, and acid rain, are said to come from

**nonpoint sources.** Economic pressure and adverse publicity can affect those companies that continue to pollute from point sources. However, pollutants that come from nonpoint sources are very difficult to control. In addition, pollution legislation relating to them is very difficult to enforce.

The amendments to the Federal Water Pollution Control Act of 1972 (PL 92–500) have mandated changes in how industry treats water. Industries are no longer allowed to use water and return it to its source in poor condition. One of the standards regulates the temperature of the water that is returned to its source. Because many industries use water for cooling, thermal pollution has become a problem.

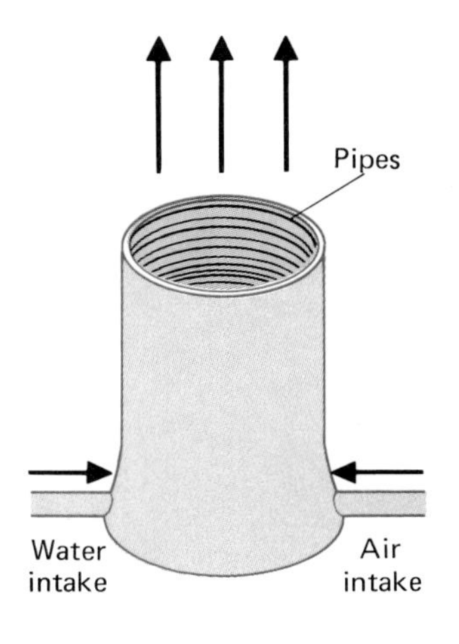


**Figure 20.3 Dry cooling tower.**
In this type of cooling tower, the water is contained within pipes. The water is cooled by air flowing across the water pipes. Although this is an expensive method of cooling water, it is important where water is in short supply because no water evaporates in the cooling process.

## *Thermal pollution*

**Thermal pollution** occurs when an industry returns heated water to its source. United States industry uses more than 225 billion liters of water annually for cooling purposes. Seventy-five percent is used by electric generating plants. Most of the remainder is used for generating steam, cooling steel, and manufacturing plastics.

Power plants heat water to convert it into steam, which drives the turbines that generate electricity. In order for steam turbines to function efficiently, the steam must be condensed into water after it leaves the turbine. This condensation is usually accomplished by using water from a lake or stream to absorb the heat. This heated water is then discharged.

Cooling water used by industry does not have to be released into aquatic ecosystems. There are three other methods of discharging the heat. One method is to construct a large shallow pond. Hot water is pumped into one end of the pond, and cooler water is removed from the other end. The heat is dissipated from the pond into the atmosphere and substrate.

A second method is to use a cooling tower. In a cooling tower, the heated water is sprayed into the air and cooled by evaporation. The disadvantage of cooling towers and shallow ponds is that large amounts of water are lost by evaporation. The release of this water into the air can also produce localized fogs.

A third method of cooling, the dry tower, does not release water into the atmosphere. In this method, the heated water is pumped through tubes and the heat is released into the air. This is the same principle that is used in an automobile radiator. The dry tower is the most expensive to construct and operate. (See fig. 20.3.)

The least expensive and easiest method is to return the heated water to the aquatic environment, but this can create problems for the inhabitants of the area. Although an increase in temperature of only a few degrees may not seem significant, some aquatic ecosystems are very sensitive to minor temperature changes. The spawning behavior of many fish is triggered by temperature changes. Lake trout will not spawn in water above 10°C; therefore, if a lake has a temperature of 8°C, the lake trout population will reproduce. But an increase of only 3°C would prevent spawning of this species and result in their eventual elimination from that lake.

Ocean estuaries are very fragile. The discharge of heated water into an estuary may alter the type of plant food available. As a result, animals with specific food habits may be eliminated because the warm water supports different kinds of food organisms. The entire food web in the estuary may be altered by only slight increases in temperature.

## Waste-water treatment

Because water must be cleaned before it is released, most companies and municipalities maintain waste-water treatment facilities. Treatment of sewage is usually classified as primary, secondary, or tertiary. **Primary sewage treatment** removes larger particles by filtering through large screens and settling in ponds or lagoons. Water is removed from the top of the settling lagoon and released. Water that has been treated in this manner has had its sand and grit removed, but it still carries a heavy load of organic matter, dissolved salts, bacteria, and other microorganisms. The organisms use the organic material for food, and as long as there is sufficient oxygen, they will continue to grow and reproduce. If the receiving body of water is large enough and the organisms have enough time, the organic matter will be degraded. In crowded areas where several municipalities take water and return it to a lake or stream within a few kilometers of each other, primary water treatment is not adequate.

**Secondary sewage treatment** usually follows primary treatment and involves holding the waste water until the organic material has been degraded by the bacteria and other microorganisms. To encourage this action, additional oxygen is required. This is accomplished by mixing the waste water with large quantities of highly oxygenated water or aerating the water directly, as in a trickling filter system. In this system, the waste water is sprayed over the surface of rock to increase the amount of dissolved oxygen. It also provides a place for the bacteria and other microbes to attach so they are exposed simultaneously to the organic material and oxygen. Most secondary treatment facilities are concerned with promoting the growth of microorganisms. These microorganisms feed on the dissolved organic matter and small suspended particles, which then become incorporated into their bodies as part of their cell structure. The bodies of the microorganisms are larger than the dissolved and suspended organic matter, so this process concentrates the organic wastes into particles that are large enough to settle out. The sludge that settles consists of living and dead microorganisms and their waste products.

In **activated sludge sewage treatment** plants, some of the sludge is returned to aeration tanks where it is mixed with incoming waste water. This kind of process uses less land than a trickling filter. (See fig. 20.4.) Both processes produce a sludge that settles out of the water. The sludge must be disposed of. It is concentrated and often dried before disposal. This is a major problem in large population centers. In the San Francisco Bay area, twenty-five hundred metric tons of sludge is produced each day. Most of this is carried to landfills and lagoons, and some is incinerated. A very small amount is composted and returned to the land as fertilizer.

a

b

**Figure 20.4 Primary and secondary waste-water treatment.**
Primary treatment is physical; it includes filtrating and settling of wastes. Photograph *a* is a settling tank in which particles settle to the bottom. Secondary treatment is mostly biological and includes the concentration of dissolved organics by microorganisms. Two major types of secondary treatment are the trickling filter and activated sludge methods. Photograph *b* is a trickling filter. The water trickles over the rocks allowing microorganisms to digest the dissolved organic material in the waste water.

Primary and secondary facilities are the most common types of sewage treatment in cities in the United States. The water discharged from sewage treatment plants must be disinfected. The least costly method of disinfection is chlorination of the waste water after it has been filtered and the organic materials have been allowed to settle. Using ultrasonic energy to mechanically break down waste may be less harmful and more effective, but it is also more expensive than chlorination. In a few sewage treatment plants, there are additional processes called tertiary sewage treatment. **Tertiary sewage treatment** involves a variety of different techniques to remove dissolved pollutants left after primary and secondary treatments. The tertiary treatment of waste water removes phosphorus and nitrogen that could increase aquatic plant growth. Tertiary treatment is very costly because it requires specific chemical treatment of the water to eliminate specific problem materials. (See table 20.1.) Certain industries are beginning to maintain their own secondary or tertiary waste-water facilities because of the specific nature of their waste products.

**Table 20.1**
Tertiary treatment

| Kind of tertiary treatment | Problem chemicals | Methods |
|---|---|---|
| Biological | Phosphorus and nitrogen compounds | 1. Large ponds are used to allow aquatic plants to assimilate the nitrogen and phosphorus compounds from the water before it is released.<br>2. Columns containing denitrifying bacteria are used to convert nitrogen compounds into atmospheric nitrogen. |
| Chemical | Phosphates and industrial pollutants | 1. Water can be filtered through calcium carbonate. The phosphate will substitute for the carbonate ion and the calcium phosphate can be removed.<br>2. Specific industrial pollutants, which are nonbiodegradable, may be removed by a variety of specific chemical processes. |
| Physical | Primarily industrial pollutants | 1. Distillation.<br>2. Water can be passed between electrically charged plates to remove ions.<br>3. High-pressure filtration through small pored filters.<br>4. Ion exchange columns. |

## Runoff

**Storm-water runoff** from streets and buildings is often added directly to the sewer system and sent to the municipal waste-water treatment facility. During heavy precipitation or spring thaws, the sewage treatment plant may be unable to handle the volume of waste water, so some of it might be discharged directly to the surface water without treatment. Modern waste-water treatment facilities have provided holding tanks to contain storm runoff and domestic wastes during heavy rains until they can be treated.

Nonpoint sources of pollution, such as agricultural runoff or mine drainage, are more difficult to detect and control than those from municipalities or industries. One of our largest water pollution problems is associated with agricultural runoff from large expanses of open fields. Precipitation dissolves materials in these areas and carries the materials away. Either the water runs over the surface and carries away exposed topsoil and nutrients (which are deposited in drains, streams, and rivers), or water seeps into the soil and carries dissolved nitrogen and phosphorus compounds into the groundwater.

There are several ways a farmer can reduce runoff. One is to leave a zone of undisturbed land near drains or stream banks. This retards surface runoff because soil covered with vegetation tends to slow the movement of water and allows the silt to be deposited on the surface of the land rather than in the streams. This is a costly process because the farmer may need to remove valuable cropland from cultivation. To retard leaching, the farmer keeps the soil covered with a crop as long as possible and carefully controls the amount and the time of application of fertilizers. This makes good economic sense because any fertilizer that runs off or leaches out of the soil has to be replaced.

Another nonpoint source of water pollution is associated with mining. When coal is mined, water that drains from these mines is often very acidic. In addition, fine coal dust particles are suspended in the water. This makes the water chemically and physically less valuable as a habitat. There are also dissolved ions of iron, sulfur, zinc, and copper in mine drainage. Control of mine drainage involves the containment of the water from the mines so it does not mix with surface water.

## Eutrophication

A body of water is like any other ecosystem. As soon as it is formed, it begins to go through a series of orderly, predictable changes called succession. Eventually, lakes will completely fill with sediment. Lakes are born, mature, and die. This is a natural process that can be speeded up by the activities of people. The enrichment of water (either natural or cultural) is known as **eutrophication.**

If you have ever watched a small pond or pool, you are aware that the entire body of water can become a green soup when the population of algae increases. In the spring or when the pool is first filled, it is clean and clear. Gradually, dirt and other debris settle into the body of water, and the water becomes cloudy as algal growth occurs. The amount of debris and microorganisms is defined as the turbidity of the water and is frequently used as an indicator of the level of pollution. Changes in turbidity occur in all bodies of water; but conditions of climate, size of lake, rate of input of nutrients (especially nitrates and phosphates), and depth of the lake influence the rate of these changes. In some very large bodies of water, etrophication may take hundreds of years to occur.

One simplified method of describing lakes is related to their productivity. A very deep lake with cold water at the bottom is not as productive as a shallow, warm lake. The deep, unproductive lake is classified as oligotrophic. At first, the water in an oligotrophic lake is clean. Such a lake has little organic material and supports a small amount of life or biomass. Over the years, as nutrients enter the lake by way of rain and runoff, the plant population increases. Plants are at the base of the food chain, and, therefore, the animal populations increase as well. The amount of solid material (such as sand, silt, and other

**Figure 20.5 Eutrophic and oligotrophic lakes.** Lakes with high amounts of nutrients are very productive and are called eutrophic lakes. This type of lake may not produce the kinds of fish or the aesthetic characteristics we would like to see. Oligotrophic lakes, which are low in nutrients, are not as productive and support different populations of fish.

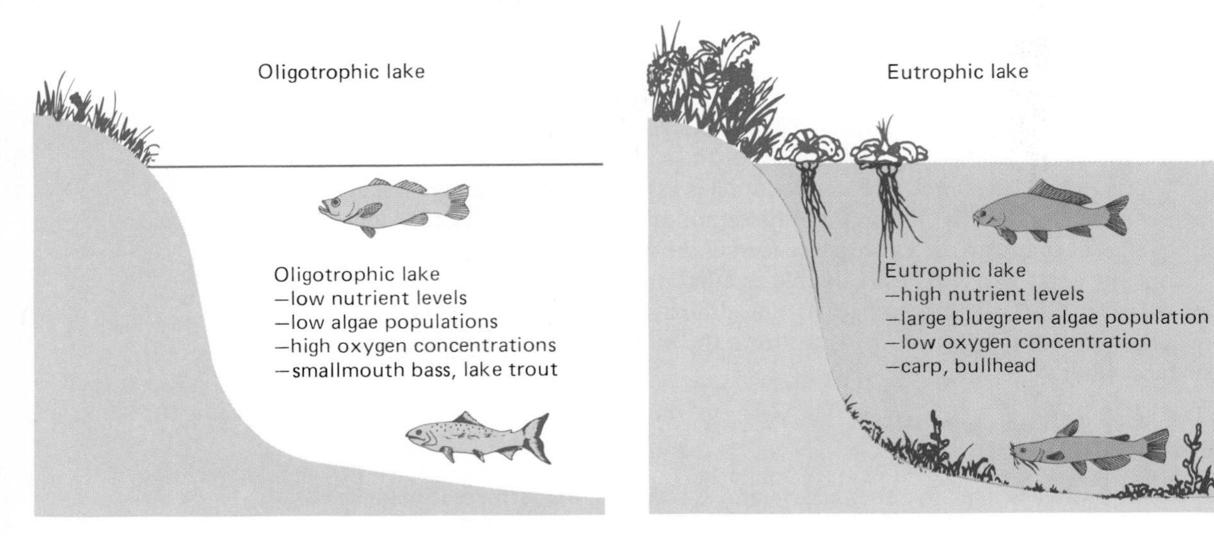


particulates) also increases, and the lake begins to fill with sediment. The lake has changed over time into a productive eutrophic lake. This series of changes called succession can theoretically occur to any body of water. However, in some cases, the natural process is so slow that we cannot detect it. On the other hand, it is possible to see accelerated aging of lakes when humans discharge nutrients into the water. Adding small amounts of nutrients can actually benefit some aquatic populations, but an excess of nutrients leads to a very large algae and plant population. When these organisms die, the decay process uses up the available oxygen, and many other forms of aquatic life die. Certain bacteria multiply rapidly under these anaerobic conditions. A lake in the advanced stages of eutrophication exhibits a decline in populations of many fish and insects and an increase in populations of anaerobic bacteria, sludge worms, and trash fish such as carp. (See fig. 20.5.)

The prevention of accelerated eutrophication requires proper waste-water treatment to reduce the amount of nutrients entering the water. It also requires a reduction in the amount of nutrients entering the water from agricultural runoff. The larger the lake and the slower the flow of water through the lake, the longer the time required to repair the damage done by human-accelerated eutrophication.

## Box 20.2
## International Joint Commission

Because water is a public resource used by many different interests, it is very difficult to prevent misuse. Everyone who uses the water sees his or her specific use as only a small modification of a large resource and, therefore, feels that their small activity is justifiable.

It is easy to dump our wastes and look the other way or to place blame for the pollution on someone else. While it is true that if one's community, family, or industry were the only polluter, there might be no problem, the use of a stream should not interfere with its use by others downstream.

Because water does not usually follow politically determined boundaries, its use is frequently the cause of disputes between various units of government. When one state blames another for polluting a waterway, the problems of local control and states' rights interferes with ensuring high-quality water. If we elevate this conflict to an international scale, the problems could be astronomical. However, the United States and Canada have a long history of cooperative planning to insure the quality of the water they share. In 1909, Canada and the United States signed The Boundary Waters Treaty. This treaty applies to all waters that form the boundaries between the two countries and to any water that flows across the boundary. The treaty specifically states that one country cannot pollute to the detriment of the other.

An International Joint Commission was created by the treaty to deal with boundary water problems. There are three commissioners from Canada and three from the United States. In 1964, the commission began an indepth study of the Lower Great Lakes. As a result of the study, the two governments entered into the Great Lakes Water Quality Agreement of 1972. This agreement provided funds to identify and stop point source discharges of pollutants into the Great Lakes. A second agreement signed in 1978 is more comprehensive than the first. It includes the watershed and drainage system of the lakes, as well as the lakes themselves.

This second agreement deals with such diverse topics as toxic substances management, air-borne pollutants, shipping, and dredging. The cooperative efforts of the two governments, their commissioners, several advisory boards, local committees, and the general public is evidence that natural resource management can be approached in a nonaccusatory, problem-solving way. The efforts of the commission and the various boards and committees have been rewarded by measurable improvement in water quality in Lake Erie, Lake Ontario, and the St. Lawrence River. This was possible because both countries supported the recommendations of the International Joint Commission by enacting necessary regulatory legislation, prosecuting those who continued to pollute and providing funding for research, surveillance, and administration. Because of the commission's success, both national governments have continued to support the commission's activities.

## Consider This Case Study
### Lake Erie

In the 1940s, Lake Erie was considered to be a very productive body of water. The lake contained walleye, small mouth bass, blue pike, whitefish, and cisco, among other species. The fish harvest from the lake was the highest of any freshwater lake in the United States.

Municipal sewage, agricultural runoff, and industrial wastes were discharged into the lake. Nutrients such as nitrates and phosphates stimulated algae and aquatic plant growth in the shallow water of the lake.

Therefore, the normal eutrophication process in Lake Erie was accelerated by the activities of the people around its shoreline. This led to many changes in the quality of the lake and changed fish populations. Currently, perch, smelt, and gizzard shad are the dominant fish species. These are not considered as valuable as the species that originally inhabited the lake. In 1970, the United States shoreline was almost always unsuitable for contact use, such as swimming, because of the high bacterial content in shore waters. Plant populations had significantly changed, and the appearance of the lake was aesthetically less pleasing.

Two major population centers polluted Lake Erie—Windsor-Detroit and Lorain-Akron-Cleveland. Should it have been their responsibility to construct new waste-water treatment facilities when this would create additional taxes for already overburdened city residents? In 1970, about 10 percent of the communities on the lake released their untreated sewage directly into the lake. A small community of only a few residents can hardly afford a treatment facility.

Fortunately, steps were taken to stop the pollution of Lake Erie, and as a result, Lake Erie is now an example of successful rejuvenation. Virtually all industrial pollution has been stopped. Swimming is once again allowed where it had been prohibited for a generation. Nearly all municipalities are meeting standards, so municipal sewage is a minor problem. The fish populations have also improved. Walleye are again being caught in Lake Erie. One of the reasons this change was so rapid is that Lake Erie is a shallow lake that can be flushed of its pollutants very rapidly. International efforts have also been involved in cleaning up the lake. (See box 20.2, International Joint Commission.)

## Summary

Water can accept large amounts of use without permanent damage. However, adding material to water may make it unfit for some uses but not for others. Water quality is related to the use we intend to make of the water.

Major sources of water pollution are municipal sewage, industrial wastes, and agricultural runoff. Organic matter in water requires oxygen for its decomposition and therefore has a large biochemical oxygen demand (BOD). Oxygen depletion can result in fish death and changes in the normal algae community, which leads to visual and odor problems. Nutrients, such as nitrates and phosphates from detergents and agricultural runoff, enrich water and stimulate algae and aquatic plant growth.

A point source of pollution is easy to identify and control, and a nonpoint source of pollution is more difficult to identify and control.

Thermal pollution occurs when an industry returns heated water to its source. Temperature changes in water can change the kinds and numbers of plants and animals that live in that area. The methods of controlling thermal pollution include cooling ponds, cooling towers, and dry cooling towers.

Marine oil pollution results from oil drilling and oil tanker accidents, runoff from streets and disposal of lubricating oil from machines and car crankcases, and intentional discharges from oil tankers during loading or unloading.

Waste-water treatment consists of primary treatment, a physical settling process; secondary treatment, biological degradation of the wastes; and tertiary treatment, chemical treatment to remove specific components. Two major types of secondary waste-water treatments are the trickling filter and the activated sludge methods.

Nonpoint sources of pollution, such as agricultural runoff and mine drainage, are more difficult to detect and control than those from municipalities or industries.

Eutrophication is the enrichment of water by natural or cultural processes. Eutrophic lakes are highly productive lakes, and oligotrophic lakes have lower levels of productivity. Human activity can accelerate the eutrophication process.

## Review Questions

1. Differentiate between point and nonpoint sources of water pollution.
2. How is water pollution related to agricultural practices?
3. Is the definition of water pollution related to its intended use? Give some examples to support your answer.
4. What is biochemical oxygen demand? How is it related to water quality?
5. How can marine oil pollution be reduced?
6. Describe primary, secondary, and tertiary sewage treatment.
7. How are most industrial wastes disposed of?
8. What is thermal pollution? How can it be controlled?
9. Describe eutrophication of a pond. How do people influence it?
10. What are the types of wastes associated with agriculture?

## Suggested Readings

Behrman, A. S. *Water Is Everybody's Business*. New York: Doubleday & Co., 1968. An introductory level text that includes all of the basic principles of water purification. As a starting point, this is a good book.

D'Itri, Patricia A., and D'Itri, Frank. *Mercury Contaminations Human Tragedy*. New York: John Wiley & Sons, 1977. A specific book that details the situation of mercury as a water pollutant. Covers this restricted topic very well.

Falkenmark, Malin, and Lindh, Gunnar. *Water for a Starving World.* Boulder, Colo.: Westview Press, 1976. A short book, translated from the original Swedish. A detailed and accurate account of the human need for water. Helpful for the general reader. Focuses on the real water issues for our growing population.

Goldstein, Jerome. *Sensible Sludge.* Emmaus, Pa.: Rodale Press, 1977. An interesting look at some creative ways of dealing with the problem of accumulated waste from waste-water treatment facilities.

Miller, Stanton. "Water Reuse: Learning about the European Experience." *Environmental Science and Technology* 15: 499–501. Interesting insight into European water reuse.

Morton, Stephen D. *Water Pollution—Causes and Cures.* Madison, Wisc.: Mimir Publishing, 1976. Details the problems concerned with recreational, domestic, and municipal, as well as industrial, uses of water.

U.S. Geological Survey. *National Water Summary 1983—Hydrologic Events and Issues.* U.S. Government Printing Office, Washington, D.C. Excellent overview of the national water situation.

# Part 6

## Where environmental decisions are made

Decisions concerning environmental issues are made at several different levels and in a variety of different ways. Often, government must provide the leadership and legal clout to force certain environmental values on the general public. It is important to remember that government responds to the will of the people in general, although it may be difficult to determine the true voice of the public amid the clamor of various vocal pressure groups who attempt to institutionalize their ideals through the governmental and judicial process. Chapter 21 explores the nature of government and its methods of operation.

Everyone makes dozens of environmental decisions daily. Individual purchases, voting decisions, recreational choices, and personal values ultimately affect the environment. Chapter 22 focuses on this level of decision making and illustrates some of the conflicts that occur between different people who support environmental causes.

Chapter 23 discusses the differences that can exist between individuals in a society and the different behaviors exhibited depending on whether the person is acting as an individual, as a part of a corporation, or as a part of government.

Finally, chapter 24 discusses the various schools of thought concerning the nature of our future environment and the nature of the forces that will act to bring about change.

## Chapter Outline

I. The democratic republic
  A. The legislative branch
  B. Pressures on the legislative branch
  C. The judicial branch
  D. The executive branch
II. Bureaucracy: The fourth branch of the government
III. The Environmental Protection Agency
  Box 21.1 Profile of an environmental organization
IV. Environmental policy
V. Consider this case study: The Reserve Mining case

## Objectives

Differentiate between a true democracy and a representative form of government.

Describe federalism.

Describe the legislative branch of government in terms of its members and their primary functions.

Describe the significance and powers of lobbies.

Describe the responsibility of the judicial and executive branches of government.

Explain why a bureaucracy is necessary and how it attained its power.

Explain how the executive, judicial, and legislative branches of government interact in forming policy.

List the forces that led to the creation of the EPA.

## Key Terms

administrative law

bureaucracy

conference committee

constitutional law

democracy

Environmental Protection Agency

executive branch

federalism

judicial branch

judicial review

legislative branch

policy

representative (republican) form of government

separation of powers

statutory law

# Chapter 21 Interactions between people and government

## The democratic republic

Government affects our lives in many ways. Food, air, and water quality are regulated by government. Individually, each of us supports the government by paying a variety of taxes. Unlike many countries, governmental policies in the United States are initiated and influenced by the people. Even though not all governmental policies have been perceived as being wise, many governmental policies have helped to solve complex environmental problems. We need a basic knowledge of the political process and institutional structure of our society before we can effectively influence public policies.

**Figure 21.1 True democracy vs. representative democracy.**
In a true democracy, each individual represents his or her own position on issues. Many small units of government, such as small towns, may operate in this manner. In most units of government, however, representatives are elected to act on behalf of large portions of the population, such as in the House of Representatives.

Unlike many other forms of government, the United States system is based upon the consent and will of the people. "Government is as good or as bad as we make it." This saying, while somewhat simplistic, has an element of truth when applied to our government. Historically, this concept was known as a **democracy.** Although examples of a true democracy can still be found in New England town meetings or in other small communities, a **representative** or a **republican form of government** has developed in most areas. (See fig. 21.1.)

The size of the population, the time required to deal with complex issues, and the size of the country make a true democracy impractical. In a representative democracy, we elect representatives to office to speak for us at the national, state, and local levels.

The Constitution is the foundation of the United States government. This document, which is nearly two hundred years old, established our form of government and gave it power but also imposed limitations on that power. Although the Constitution outlined the structure of our national government, it also distributed the power between the national government and the states. This form of government, in which power is shared between the national government and the state governments, is called **federalism.**

The Constitution is also specific in stating the rights of individuals. The will of the majority is expressed by either a direct vote of all citizens or by a vote of their representatives. A law cannot be passed if it conflicts with the Constitution; therefore, individual freedoms are protected. For example, no law can be passed that prevents you from acquiring an education.

The writers of the Constitution carefully divided the powers of the government between three separate branches—the legislative, judicial, and executive. This concept is called the **separation of powers.** (See fig. 21.2.) When powers are separate, it becomes very difficult for one branch to seize complete power over the government. Historically, the functions of the branches have been described as follows: the legislative branch makes laws; the judicial branch interprets the laws; and the executive branch implements laws and sets policy. Today, however, distinct functional divisions do not exist. There is substantial overlapping of functions, and policy is the result of the interactions of all three divisions of government.

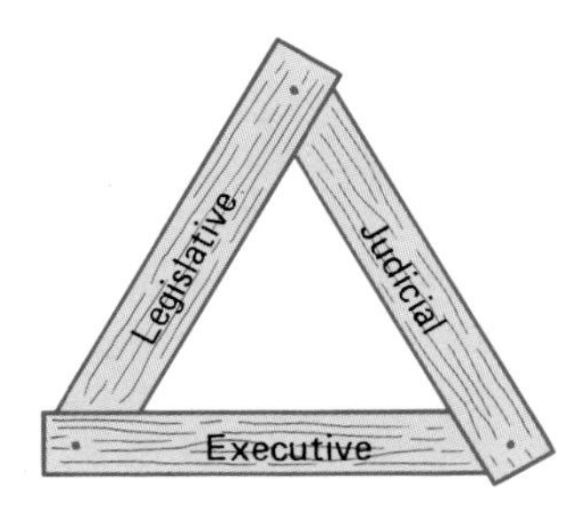


**Figure 21.2 The three branches of government.** Governmental power is divided among the legislative, judicial, and executive branches of government.

## *The legislative branch*

The legislature, known as the Congress of the United States, is divided into two bodies—the House of Representatives and the Senate. The members of the House of Representatives are elected from districts with approximately equal populations. Thus, the number of representatives for each state depends on the size of the state's population. The membership of the Senate consists of two senators per state.

The Constitution is precise in delegating powers to the **legislative branch:** Congress has the power to print money, declare war, and levy taxes. Congressional power, however, is not limited to the powers granted by the Constitution. A great deal of the actual power depends on the members of Congress, who they are, and what they actually want to do. (See table 21.1.)

The functions of Congress are varied. They include determining policy, providing funds, and overseeing how programs are carried out. A primary function is to develop and approve **policy.** Policy is a planned course of action on a question or a topic. Congress generally supports policy by passing bills or acts that establish a government agency or instruct an existing agency to take on new tasks or programs. (See table 21.2.)

**Table 21.1**
Powers of Congress

| Constitutional powers | Assumed powers |
|---|---|
| Print money | Determine policy |
| Declare war | Adopt budget |
| Levy taxes | Oversee programs |
| Create courts | Authorize spending |

**Table 21.2**
Major United States environmental legislation

| Topic | Legislation |
|---|---|
| *General* | National Environmental Policy Act of 1970 (NEPA) |
| *Air quality* | Clean Air Act of 1965 |
| | Clean Air Act of 1970 |
| | Clean Air Act of 1977 |
| *Water control* | Federal Water Pollution Control Act of 1972 |
| | Ocean Dumping Act of 1972 |
| | Safe Drinking Water Act of 1974 |
| | Toxic Substance Control Act of 1976 |
| | Clean Water Act of 1977 |
| | Municipal Wastewater Treatment Construction Grants of 1981 |
| *Land use* | Multiple Use Sustained Yield Act of 1960 |
| | Wilderness Act of 1964 |
| | Wild and Scenic River Act of 1968 |
| | National Coastal Zone Management Act of 1972 |
| | Forest Reserves Management Act of 1974 |
| | Forest Reserves Management Act of 1976 |
| | National Forest Management Act of 1976 |
| | National Strip Mining Control Act of 1977 |
| | Endangered American Wilderness Act of 1978 |
| | Alaska National Interest Lands Conservation Act of 1980 |
| *Solid waste management* | Solid Waste Disposal Act of 1965 |
| | Resource Recovery Act of 1970 |
| | Resource Conservation and Recovery Act of 1976 |
| | Comprehensive Environmental Response, Compensation and Liability Act of 1980 |
| *Wildlife* | Species Conservation Act of 1966 |
| | Marine Mammal Protection Act of 1972 |
| | Environmental Noise Control Act of 1972 |
| | Endangered Species Act of 1973 |
| | Quiet Communities Act of 1978 |
| | Aviation Safety and Noise Abatement Act of 1979 |
| *Pesticides* | Federal Pesticide Act of 1978 |
| *hazardous waste* | Comprehensive Environmental Response, Compensation and Liability Act of 1980 (Superfund) |

Once an agency or program is established, it may operate for many years without further long-term congressional policy making. The National Environmental Policy Act (NEPA) was passed by Congress in 1970. It was ultimately responsible for the establishment of the Environmental Protection Agency (EPA).

But enactment of policies into law may be only the beginning. Relatively few statutes are "self-executing" in the sense that they can be put into effect immediately by existing agencies. Many require funding, if only in the form of new personnel to perform the investigation, services, or enforcement called for in the statute. In such cases, the appropriations process following submission of the next year's budget serves as the second consideration of the issues addressed by the statute; the arena is not the policy-area committees but rather the subcommittees of the appropriations committees. When a budget has been passed by each house, differences are settled by an appropriations conference committee.

The allied processes of logrolling (helping push policies through government) and coalition building are basic to legislative policy making. Only a rare issue (such as a declaration of war or a strong civil rights bill) arouses a fairly strong interest, pro or con, in virtually all members of the legislative body. When this happens, it is because there is widespread concern or controversy over the issue in the country at large. Only a relatively small proportion of the members takes an active interest in most issues that come before a legislative body. Most have either a mild concern or none at all, largely reflecting the degree of interest among their constituents. A proposal to raise the tariff on foreign coal imports will certainly evoke strong interest in the coal-producing areas of Pennsylvania, West Virginia, and Illinois and thus in legislators representing these areas. But most other members of Congress are not likely to be strongly interested, even though some of their constituents use coal to heat their homes or factories. A proposal to appropriate federal funds to redirect the flow of the Colorado River will certainly awaken hopes or anxieties in legislators from the states bordering the river, but it will probably not deeply stir legislators from Georgia, Alaska, or New Jersey.

When there is a combination of strong interest among a few legislators and apathy from the rest, conditions are ripe for bargaining, logrolling, and coalition building. Political colleagues who are unconcerned about a particular issue will offer their support of that issue in return for reciprocal support of issues that are important to them. For example, if a Pennsylvania representative supports higher prices for cotton (an issue that Pennsylvania constituents have little interest in), he or she may then expect representatives from cotton-growing states to support a higher tariff on coal (an issue of little concern to constituents of cotton-growing states). Usually, there is no explicit agreement about vote trading; instead, there is the recognition that a favor granted today deserves a return favor in the future.

Table 21.3
Senate and House Committees

| Senate | House |
|---|---|
| Agriculture, Nutrition, and Forestry | Agriculture |
| Appropriations | Appropriations |
| Armed Services | Armed Services |
| Banking, Housing, and Urban Affairs | Banking, Finance, and Urban Affairs |
| Budget | Budget |
| Commerce, Science, and Transportation | District of Columbia |
| Energy and Natural Resources | Education and Labor |
| Environment and Public Works | Governmental Operations |
| Ethics | House Administration |
| Finance | Interior and Insular Affairs |
| Foreign Relations | International Relations |
| Governmental Affairs | Interstate and Foreign Commerce |
| Human Resources | Judiciary |
| Judiciary | Merchant Marine and Fisheries |
| Rules and Administration | Post Office and Civil Service |
| Select Indian Affairs | Public Works and Transportation |
| Select Intelligence | Rules |
| Select Nutrition and Human Needs | Science and Technology |
| Select Small Business | Small Business |
| Special Aging | Standards of Official Conduct |
| Veteran Affairs | Veterans Affairs |
| | Ways and Means |

Once money has been both authorized and appropriated (and the president has signed both bills), the agency or program can spend the money. The agency that is authorized by Congress is a part of the executive branch and implements policy.

A third major role of Congress is to oversee the various agencies of the executive branch. Congress looks at how well the various parts of the executive branch are carrying out the laws and how effectively they are spending their appropriations. This function is generally called congressional oversight.

Policy making, funding, and oversight are very large responsibilities. Given the range of issues faced by Congress and the amount of time that they must operate within, members of Congress inevitably tend to specialize in one or two areas. Because members of Congress are extremely busy and congressional agendas are so long, a great deal of the responsibility of formulating legislation is done in the committee and subcommittee system. Without the committees, Congress would not be able to properly investigate all the potential laws. Table 21.3 shows all the committees of the ninety-eighth Congress, but the actual number and type of committees will change with the need for expertise in certain areas. After a bill is introduced in either the House or Senate, it is then referred to the appropriate committee. If a proposed bill makes it through the committee, it stands an excellent chance of becoming law.

Because most bills do not make it out of committee, it is obvious that the committee system is a primary target for lobbyists.

## *Pressures on the legislative branch*

Because all federal legislation, programs, or agencies begin in Congress, members of Congress receive considerable pressure from individuals and groups. Although representatives to Congress are in theory supposed to represent both their constituents and their conscience, in practice this is not always the case. Some members of Congress have found the arguments from specific groups so persuasive that they have become spokespersons for these special interest groups. That is why we must carefully examine all the candidates in an election.

Lobbying efforts often have a tremendous influence on members of Congress. Members of congressional committees are extremely powerful in helping to pass or kill legislation, so they are prime targets for special interest groups. Special interest groups do not have the same impact on the judicial branch of government, because the courts are charged to assume a neutral position when interpreting legislation and because federal judges are appointed and not elected. Because they are not elected, they are removed from the subtle offers of "help" from interest groups during elections.

## *The judicial branch*

The Constitution provides "that judicial power of the United States shall be vested in one supreme court and in such inferior courts as the Congress may from time to time ordain and establish." The **judicial branch** of the United States government is a complex and layered series of courts ranging from local traffic courts to the United States Supreme Court. (See fig. 21.3.) The role of the courts is to interpret the law. There are, however, several different categories of laws, such as constitutional law, statutory law, and administrative law.

The foundation for all law is the Constitution, which is the highest law of the land. **Constitutional law** is involved in the interpretation and application of the Constitution and is concerned primarily with describing the limits of governmental power and the rights of individuals. Recent examples of questions of constitutional law involve civil rights and freedom of speech and press.

Most court cases do not concern questions of constitutional law. Rather, they are involved with **statutory law.** These are laws passed by legislation and cover a wide range of subjects. At the federal level, laws are made by Congress; at the state level, by state legislatures; and at the local level, by county, city, and township legislative boards. An example of a local law would be an ordinance to prohibit the burning of trash.

Laws with the greatest impact on improving our environment are **administrative laws.** Federal agencies, such as the Environmental Protection Agency, have developed their own methods of establishing regulations that are as binding as the laws passed by Congress. These regulations have the force of law because Congress has granted the agency the power to make its own regulations. Congress also allows or instructs agencies to set up their own internal

**Figure 21.3 The court systems of the United States.**
Although local court systems vary throughout the United States, the federal system was established by the Constitution and is, therefore, fixed. The numbers beside some of the courts on the chart indicate the number of the courts in the country. The arrows indicate how a case could get to the Supreme Court.

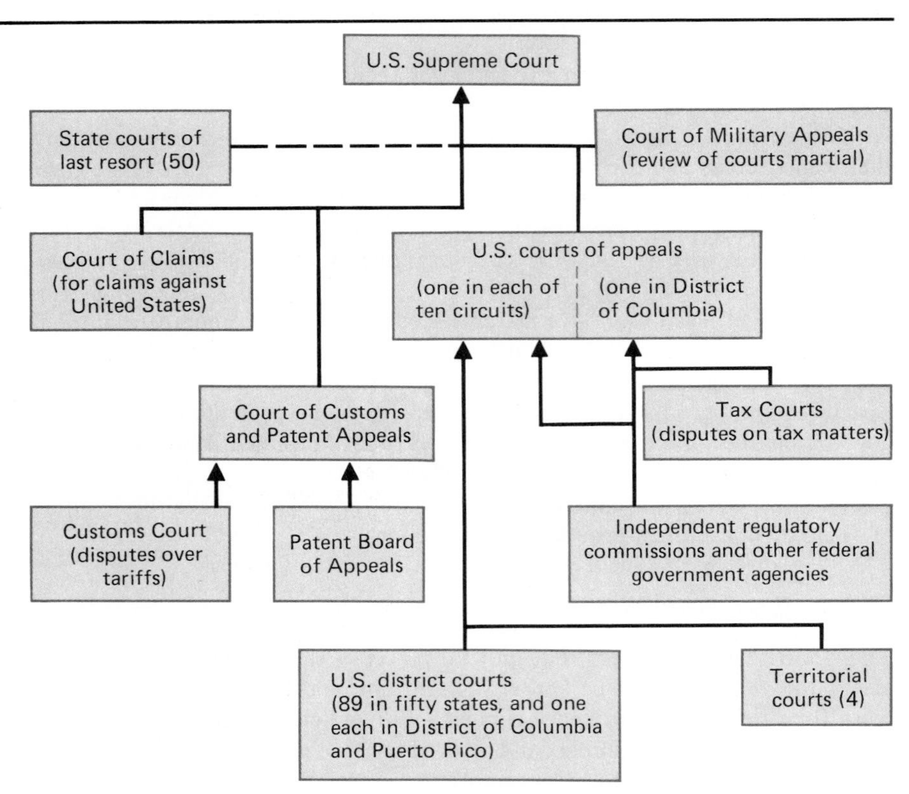


court system to enforce the regulations. However, if a citizen or business is unsatisfied with an agency decision, they still have the right to appeal outside the agency to the United States Court of Appeals. (See fig. 21.3.)

In addition to interpreting the meaning of laws, the judicial branch has another powerful role to play. This role is to decide whether or not laws are in conflict with the Constitution. This is referred to as **judicial review.** In this respect, the courts are more powerful than the legislative and executive branches of government. This power to "review the actions of individuals and agencies or the laws of Congress, of states, or of localities" may be exercised by a judge or a court at any level of the federal system. The final decision is made by the Supreme Court, because there is no further legal appeal. An example of judicial review was the 1974 decision by the Supreme Court that President Nixon's tape recordings of phone conversations were not his private property. In this example, the judicial branch was more powerful than the executive branch. However, the president is still the most powerful person in government.

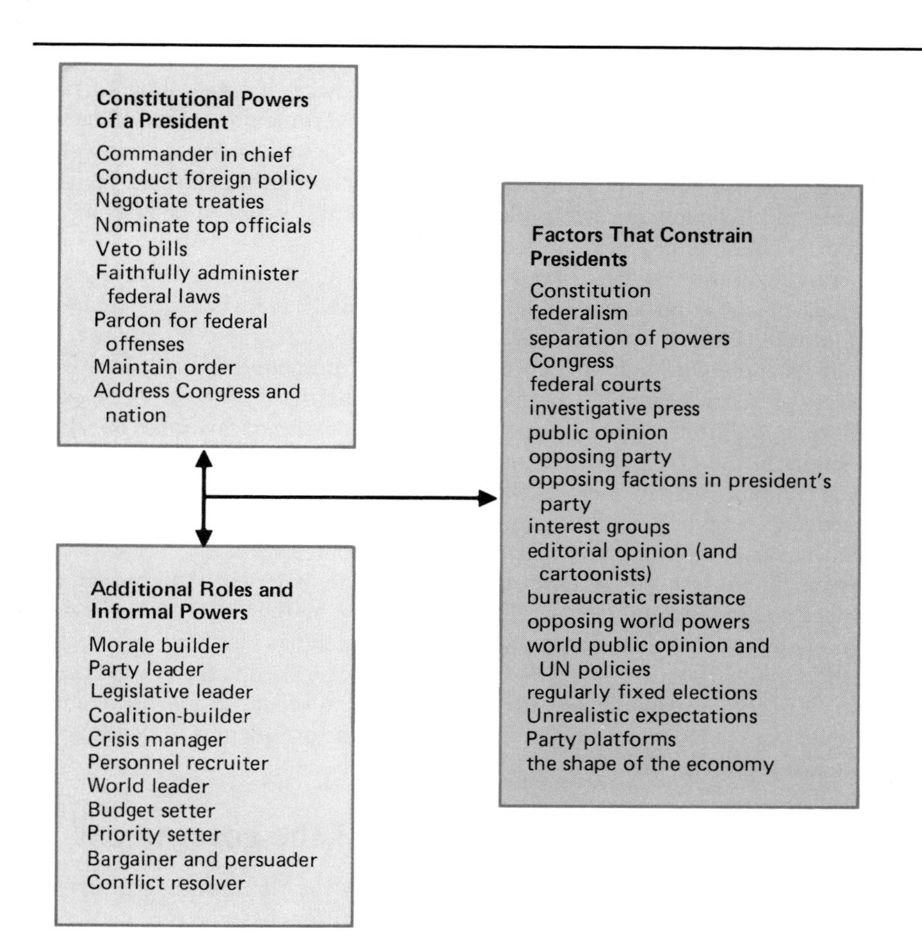


**Figure 21.4 Powers of the executive branch.** The Office of the president is granted both Constitutional powers and powers that are assumed by virtue of the office. There are, however, both legal and other factors that provide constraints to presidential powers.

## *The executive branch*

In general, the powers of the courts are less than that of the president. The powers of the **executive branch** (presidential power) come from a variety of sources. Some power is conferred by the Constitution. However, most executive power has been acquired over the years. (See fig. 21.4.) In addition to these written and acquired powers, the individual personality of the president also plays a major role in determining presidential power. Presidential powers are really permissions given to the office. When a president wants to gain support for programs or legislation, personality can be either an important asset or a serious liability. Examples of presidential "powers" are nominating and appointing people to important jobs (including judges to the Supreme Court), including specific items in budgets, determining who has access to the presidential office or presidential information, and providing individuals with political support.

Although the president is only one individual, presidential support can be crucial in deciding the fate of a program or a piece of legislation. In fact, presidential approval is essential for almost all governmental activity. The importance a president places on an issue will often dictate the issue's success or failure. This has been especially true with environmental regulations. Neither Presidents Nixon nor Ford considered environmental issues to be of vital importance. Nixon impounded or blocked money that had already been approved for water-pollution control. Ford twice vetoed legislation to control strip-mining, and he supported various bills that would have weakened clean-air regulations. (In fact, when the Ford Administration left office, at least twenty-one of its top environmental protection officials immediately took jobs with the very industries they had supposedly been policing.)

President Carter's environmental record was considered favorable by many, especially in the areas of water and air pollution control and land use. However, President Reagan is not of the same attitude, and his policies have been regarded as having a negative impact on environmental concerns, especially his cutbacks of environmental regulations and controls.

Even though the power that the president wields is vast, this power has limitations. Presidential power is limited formally by the Constitution and informally by the practicalities of time. Public opinion has also been an effective means of limiting or altering presidential action. The president is the executive officer of the government, but it is still an elected office. When presidential actions alienate too many legislators or citizens, the president is forced to moderate a stand on a position if reelection is sought.

## Bureaucracy: The fourth branch of the government

The United States government is not limited to the three branches just described. The legislative, judicial, and executive branches are the ones mentioned in the Constitution. A fourth branch of government has emerged during the last fifty years, and it continues to grow. This fourth branch of government is called the **bureaucracy.** A bureaucracy is a method of administration in which appointed officials manage departments by sets of inflexible and routine procedures. It is a familiar, yet frustrating, experience for many of us. As frustrated as we might become with the actions of a bureaucracy, few of us would want to return to the days when there was only a small governmental bureaucracy. As was mentioned earlier, government responds to the wishes and needs of its citizens. Over the years, the bureaucracy has grown in order to respond to these needs. Examples of this growth include government regulation of food and drug standards, social security programs, protection against false advertising, and environmental protection. (See fig. 21.5.) Today, one in every six United States workers is employed by federal, state, or local government. We willingly accept the benefits of government programs, but we continue to complain about the cost and method of administration.

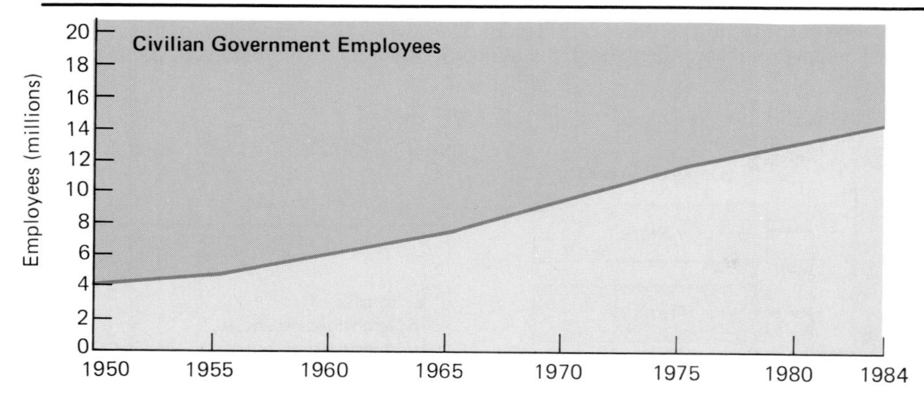


**Figure 21.5 The growth of governmental bureaucracy.** Federal and state governments have grown considerably over the years. It must be remembered, however, that what we expect and demand of government has grown just as fast. The size of the government has grown in order to keep up with the demands we place upon it. (Source: United States Department of Labor, Bureau of Labor Statistics.)

Presently, there are about 2.9 million federal bureaucrats who work in eighty-three major government bureaus, administering over a thousand different domestic programs and various foreign programs as well. (See fig. 21.6.)

As you can see, the federal bureaucracy is not about to disappear overnight. In fact, it has become a self-perpetuating, very powerful branch of government. Bureaucratic politics can have an important role to play in the kind of legislation proposed and the kinds of decisions made by the executive branch of government. Bureaucrats are often asked to give advice because they are the only ones who know how their department really works. This places them in a perfect position to build and protect their empires.

To understand how the United States political system really works, we must understand the nature of bureaucratic power. To better illustrate this point, we will analyze in depth the birth and development of one governmental regulatory agency—the Environmental Protection Agency.

## The Environmental Protection Agency

Even though environmental concerns have been raised throughout much of the twentieth century, it was not until the 1960s that concern over the environment became a popular social and political issue. (See fig. 21.7.) By the end of the 1960s, people in various parts of the world had been awakened to environmental problems. This was a result of a series of environmental disasters, such as air pollution alerts, pesticide misuse, and severe water pollution. Increasing awareness of environmental issues and popularization by media coverage was also important.

This generation of people began to see environmental concerns as one of the foremost problems facing the United States and the world. Growing concern began to show itself in surveys and opinion polls throughout the country. For example, in 1965 only 28 percent of the people surveyed thought air pollution was "very serious" or "somewhat serious," and 35 percent had the same

**Figure 21.6 The United States government.** This chart provides an overview of the organization of the federal government. Note where the Environmental Protection Agency is in the structure. (Source: United States Government Manual, 1979.)

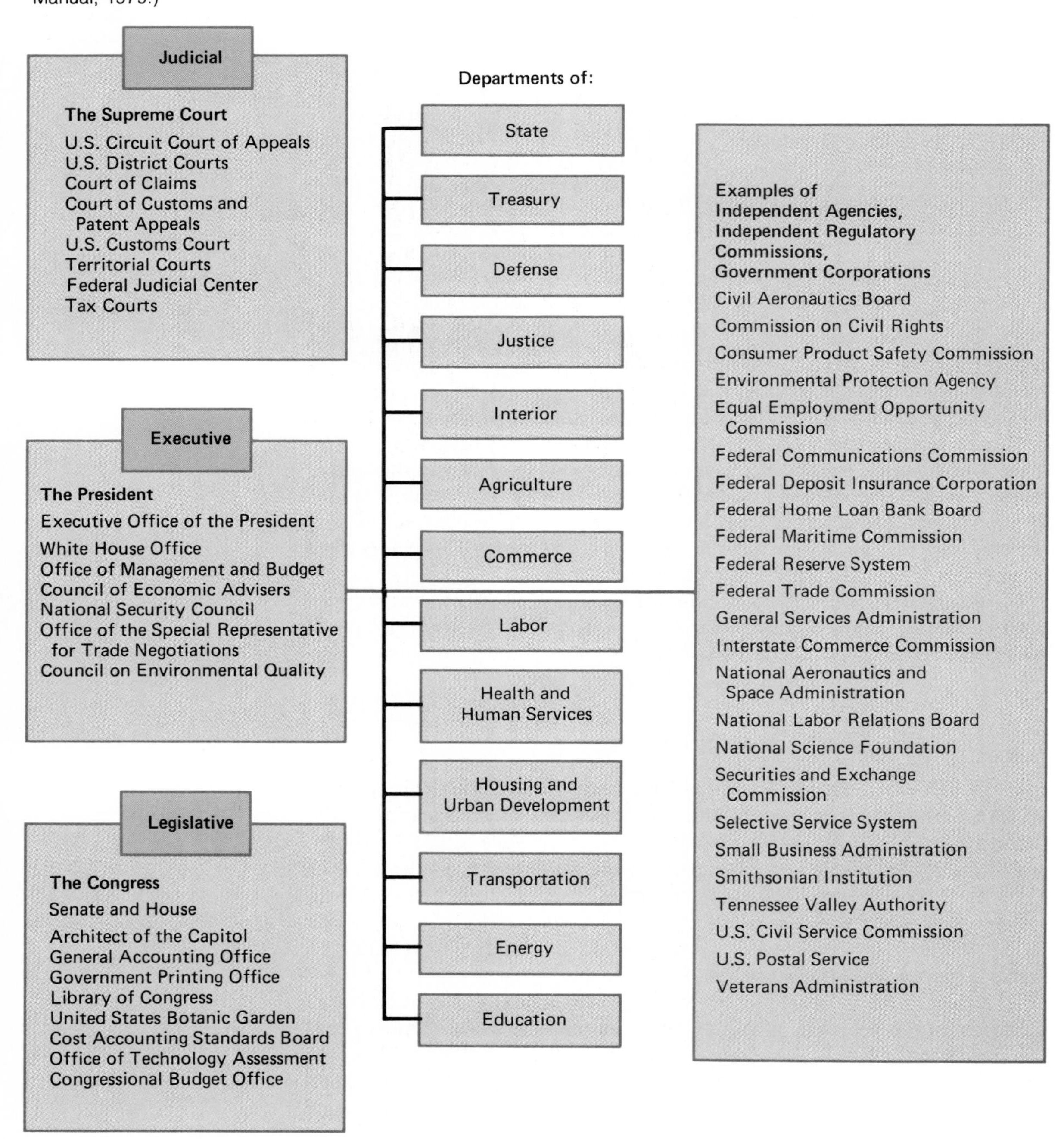

**Figure 21.7**
**Environmental activism.** Beginning in the 1960s, people became aware that groups of people demonstrating could influence governmental and industrial policy decisions.

feeling about water pollution. By 1970, however, 70 percent thought air pollution was serious and 74 percent thought water pollution was serious. People were not just concerned about their environment, they wanted something to be done to clean up existing problems and prevent new problems from developing.

The sudden rise of citizen awareness and concern over environmental issues was soon felt by decision makers in government. The federal government had been involved in water pollution legislation since 1948 and air pollution legislation since 1955. This pollution legislation assumed that prevention and control of pollution was the responsibility of state and local governments and not the federal government.

For two important reasons, states and cities had not acted to enforce this legislation. First, local polluting industries and related economic interests were so strong that they could usually prevent adoption of serious pollution-control programs. Second, these special interests were usually able to get their own people appointed to pollution-control boards, which meant the special interest groups had little to fear from "regulation." Therefore, the people had ample reason to be concerned about the lack of effective pollution control.

By 1969, the people were demanding comprehensive enforceable environmental protection legislation. In response to this pressure, Congress began to introduce new environmental legislation. In the House of Representatives, Congressman John Dingell (D., Michigan) introduced a bill creating a Council on Environmental Quality (CEQ), which was intended to advise the president and report to the public on environmental concerns. Dingell had long been a proponent of conservation and was also the chairman of the Fisheries and Wildlife Conservation subcommittee of the Merchant Marine and Fisheries Committee.

## Box 21.1
## Profile of an environmental organization

There are hundreds of environmental organizations in the United States. Some of these organizations have a small membership with a limited scope and purpose, but other organizations have millions of members with a diverse and constantly expanding scope and purpose. One of the largest environmental organizations in the United States is the National Wildlife Federation (NWF).

The NWF was organized in 1936 as a publicly supported, nonprofit, citizens' conservation education organization. Today the NWF is a federation of fifty-three state and territorial conservation organizations with over 4,500,000 members. The NWF is active in dealing with natural resource problems on regional, state, and local levels. This is accomplished by numerous educational programs, activities, and services. Examples of the services offered by the NWF include conservation publications, weekly reports on conservation legislation, environmental educational materials, press releases, public service announcements, legal involvement, and fellowships for graduate study. In addition, the NWF operates a conservation education center and conducts environmental education classes and conferences.

The NWF has been active over the past several decades in lobbying at every level of government. Areas of concern have included land use, acid rain, outdoor ethics, energy alternatives, and solid waste.

The organization maintains a large Washington, D.C., office and thirteen regional offices. The national office staff is divided into several divisions, each of which keeps a close watch on pertinent legislation. These divisions include affiliate services, conservation, creative services, education, information, resources defense, and wildlife heritage. In addition to nearly four hundred employees, a staff of the NWF maintains thirteen conservation consultants for areas of concern ranging from fisheries to military lands.

As you can readily see, the NWF is a diverse, powerful, and growing organization. Due to its solid reputation and its skills, the NWF has been instrumental in helping to pass important environmental legislation. Perhaps the NWF's own creed summarizes best the philosophy and direction of the organization.

THE
NATIONAL WILDLIFE
FEDERATION CREED

I pledge myself, as a responsible human, to assume my share of man's stewardship of our natural resources.

I will use my share with gratitude, without greed or waste.

I will respect the rights of others and abide by the law.

I will support the sound management of the resources we use, the restoration of the resources we have despoiled, and the safe-keeping of significant resources for posterity.

I will never forget that life and beauty, wealth and progress, depend on how wisely man uses these gifts . . . the soil, the water, the air, the minerals, the plant life, and the wildlife. This is my pledge!

In the Senate, two somewhat similar bills were introduced. Senator Henry Jackson (D., Washington) introduced a bill calling for the creation of a Council on Environmental Quality. Jackson's bill had two major provisions. It created a CEQ, and it required that all government agencies prepare an environmental impact statement (EIS). (See box 1.1.) However, Senator Edmund Muskie (D., Maine) feared that the bill introduced by Jackson would not be effective without an agency to enforce it. On June 12, 1969, Muskie introduced a bill that would create an Office of Environmental Quality, which would have enforcement powers. What evolved were two separate but strong bills in the Senate. Jackson's bill, numbered S. (for Senate) 1075, became the National Environmental Policy Act (NEPA) and created the Council of Environmental Quality in the Executive Office of the President. Muskie's bill, S. 2391, became known as the Environmental Quality Improvement Act. It established a new agency called the Office of Environmental Quality to be located in the Department of Housing and Urban Development.

Although the Senate passed two strong bills, the House bill was rather weak. Amendments to the original bill proposed by Congressman Dingell weakened it considerably. Dingell was willing to settle for a weak bill rather than no bill at all because he hoped that something could be worked out in a **conference committee** with the Senate to resolve differences between the Senate and House bills. A final version of the bill strongly resembled the original Jackson bill. It passed Congress with strong support. President Nixon signed the National Environmental Policy Act on January 1, 1970. (See box 21.1.)

Before NEPA was passed, responsibility for environmental affairs was scattered throughout various governmental departments and agencies. To have an efficient and effective environmental protection program, it was necessary to consolidate all the programs dealing with environmental protection into one agency. President Nixon favored this idea and issued an executive order combining fifteen different units into the **Environmental Protection Agency.** The president's order became effective on December 2, 1970.

The EPA is divided into ten regional offices. (See fig. 21.8.) Even though the responsibilities of the agency are varied, it is basically divided into six major areas—water and hazardous materials, toxic substances, air and waste management, research and development, enforcement, and planning and management. Within the agency, there are more than eleven thousand employees, more than any other regulatory agency.

The history of EPA reveals both successes and failures. The agency and the necessary watchdog mission it plays, however, would never have materialized had it not been for concerned people who realized a change was needed and petitioned the government for that change. Such citizen involvement is crucial to any attempt directed at bringing about change. The future of the agency is not, however, totally secure. The attitudes of Congress and the president toward the agency, in addition to the mood of the country, will be the final determinant of whether the EPA will continue as a strong regulatory agency or be reduced to a small ineffective government office. (See fig. 21.9.)

**Figure 21.8 Organizational chart of the Environmental Protection Agency.** As the chart indicates, the EPA is a diverse agency faced with a wide and growing range of responsibilities.

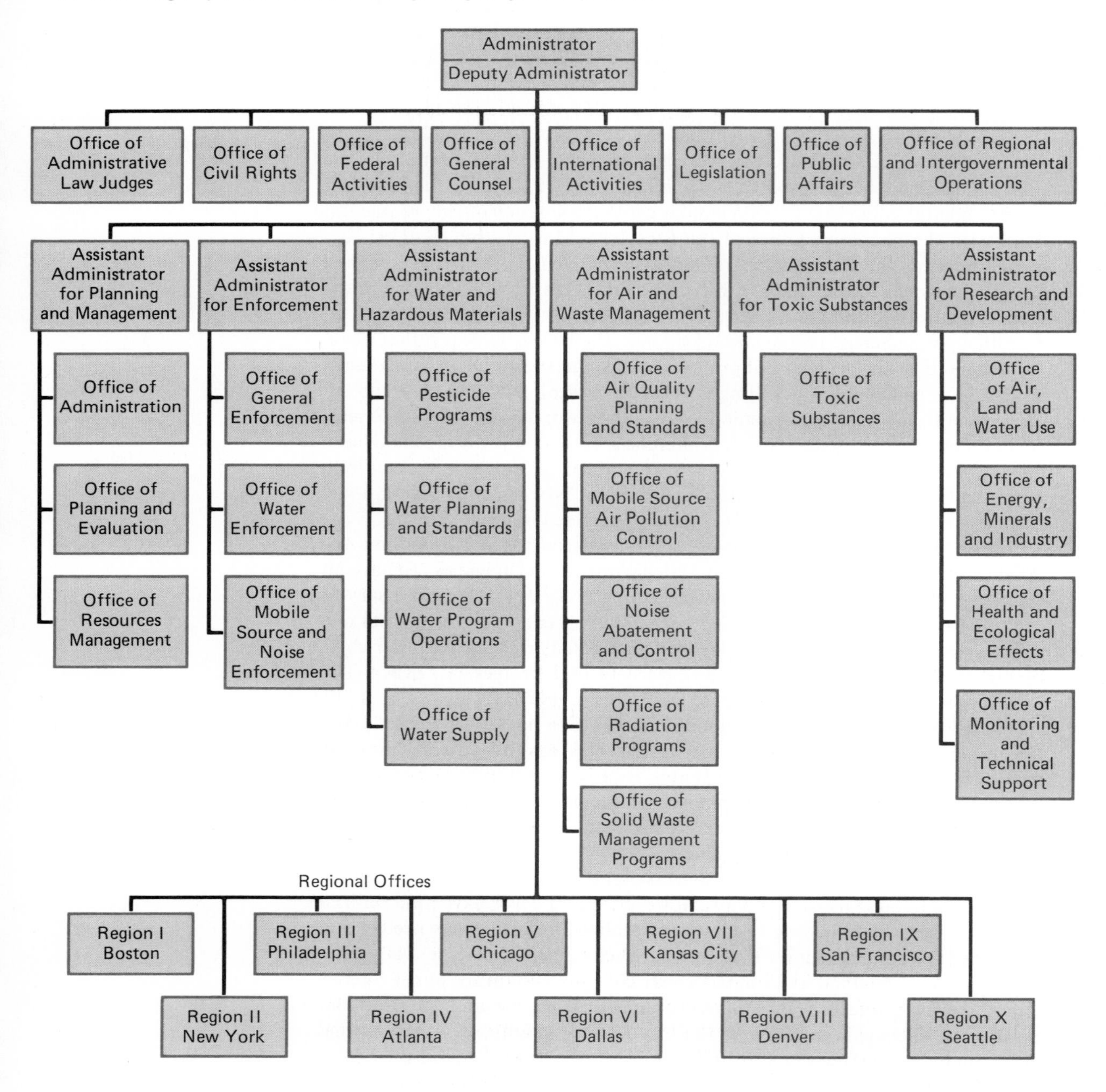

**Figure 21.9 Environmental Protection Agency operating program budget.** From 1971 to 1981, the amount of money allocated to the EPA increased each year. It has decreased every year since then. The change from the Carter to Reagan Administration had an impact on the budget of the EPA. (Source: Environmental Protection Agency.)

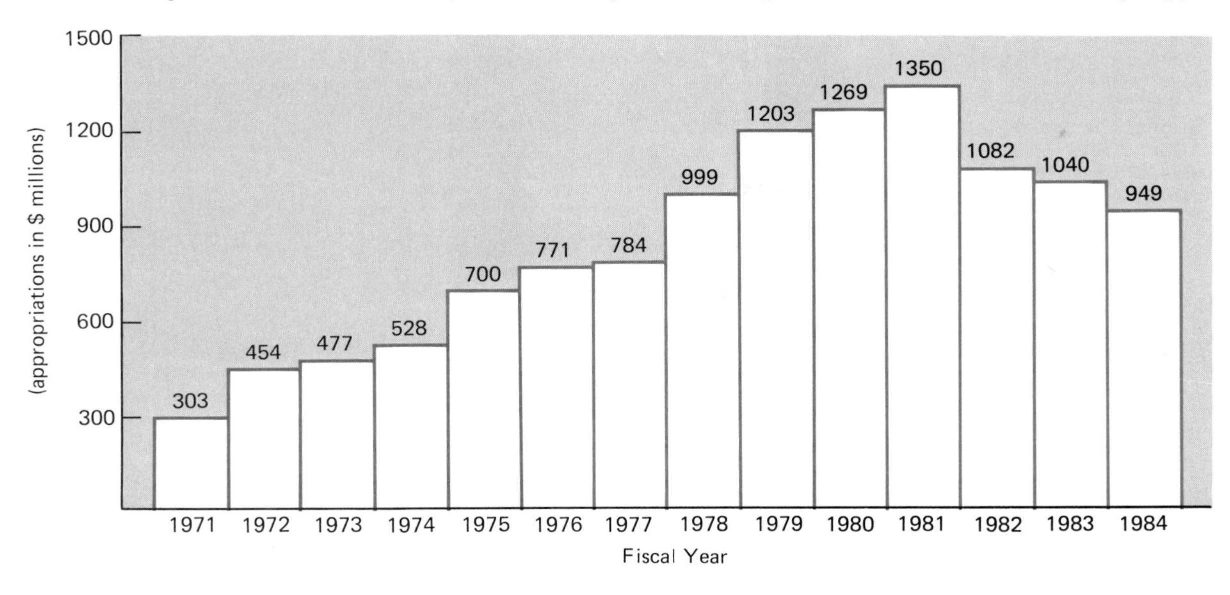


Perhaps one of the greatest challenges faced by the EPA was the controversy surrounding Anne McGill Burford (Gorsuch), who was appointed by President Reagan as administrator of the agency in 1981. On December 16, 1982, Gorsuch had become one of the highest executive officials ever cited for contempt of Congress by the U.S. House of Representatives. The citation was prompted by her refusal, on presidential orders, to deliver subpoenaed documents on alleged mismanagement of the "Superfund" program (a program for cleaning up the nation's chemical waste dumps). More than twenty other top EPA officials were also fired. On March 21, 1983, William D. Ruckelshaus, who had served as the first EPA administrator (1970–1973), was designated as her successor. One of the major problems that evolved from this scandal was the substantial loss of public confidence and trust in the agency; such a loss cannot be regained overnight.

## Environmental policy

Environmental policy is set up through a complex process—legislatures set basic quality standards, administrative agencies interpret and apply the law, and the courts review the administration of the law. The process is complicated further by conflict between the states and the federal government about their legal and proper roles. In general, environmental legislation has tended to centralize authority with the federal government, until the Reagan administration shifted a great deal of power to the states.

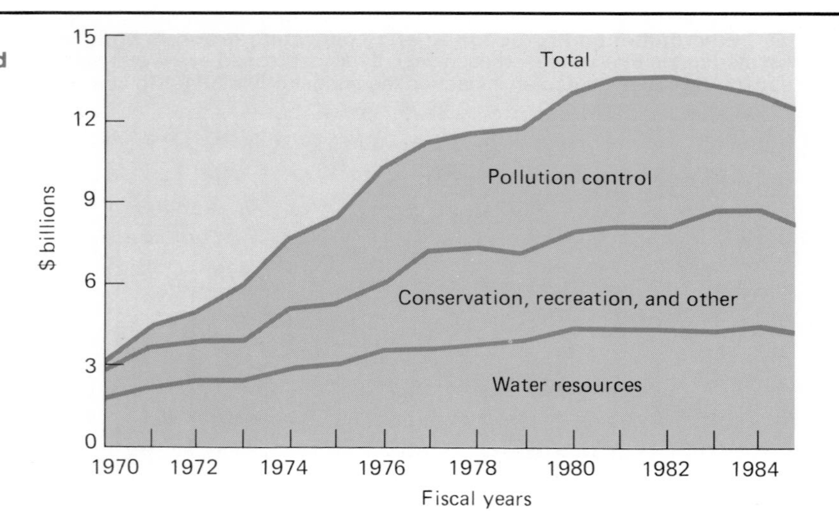


**Figure 21.10 Outlays for natural resources and environment.** Since 1970 the amount of federal monies for pollution control, conservation, and water resources has continued to increase. (Source: *The United States Budget in Brief, Fiscal Year 1984.* Washington, D.C.: United States Government Printing Office.)

In 1984 the federal government spent over $12 billion for environmental programs. (See fig. 21.10.) Even after taking into account inflation, the amount spent in 1984 was considerably greater than the $3 billion spent in 1970 for similar programs.

Beginning in 1981, however, there was noticeable pressure coming from President Reagan and his advisors. They felt that many environmental regulations were either unnecessary or were too restrictive. The direction President Reagan's environmental policy was to take became focused on his appointment of James Watt as the Secretary of the Interior. Before his appointment, Watt had served as president and chief legal advisor of a public interest law firm that represented business interests in environmental disputes over governmental regulation.

The appointment of Watt alarmed environmental organizations such as the Sierra Club, the Audubon Society, and the National Wildlife Federation. These organizations, along with numerous others, believed that many of the environmental regulations that were passed during the 1970s would be severely weakened by Secretary Watt. By October 1981, the Sierra Club had collected over one million signatures on petitions that asked for the resignation of Secretary Watt. Other organizations intensified their lobbying pressure in Washington to help protect legislation (such as the Clean Air Act and the Strip Mining Control Act) from being weakened.

On October 9, 1983, amid a series of controversies, James Watt resigned as Interior Secretary. Six months prior to the Watt resignation, the EPA Chief and more than twenty other top officials in the agency resigned or were fired. The problems resulting in the Interior Department and EPA were indicators that President Reagan perceived environmental matters somewhat differently than did his predecessors. In a March 1983 speech, President Reagan made the following remark about environmentalists: "I do not think they will be happy until the White House looks like a bird's nest."

To many, the President's "bird's nest" comment and his defense of Watt symbolized an attitude profoundly antagonistic to the environmental community. Reagan had entered office with a political agenda that was more conservative than any Presidential agenda in a half century, and he used his executive powers to the maximum in an attempt to carry it out during the first two years of his presidency. His deregulation program called for major changes in energy, health, consumer, and environmental policies that would reduce the government's scope and promote economic recovery. The policies of Watt, Gorsuch, and other agency officials accurately reflected these priorities, resulting in massive changes at the administrative level.

Although Reagan praised both Watt and Gorsuch for their outstanding performance in office and defended his environmental record in other public statements, it was obvious by the third year of Reagan's term that his environmental and natural resource policies had run into a wall of opposition. Rarely before had Congress exercised to such an extent its authority to check the powers of the executive branch as it did when driving these two agency heads from office. Nor did these remarkable events reflect only the competence and personal demeanor of the incumbents: underlying the dismissals were fundamental differences over the environmental policies of the Reagan administration.

In fact, the Reagan administration had embarked on a new policy course that has altered radically the environmental agenda of the 1980s. The guiding principles, set out by the president's Council on Environmental Quality, are (1) regulatory reform, including extensive use of cost-benefit analysis in determining the value of environmental regulations and programs; (2) reliance, as much as possible, on the free market to allocate resources; and (3) decentralization—shifting responsibilities for environmental protection to state and local governments whenever feasible. If carried out literally, these goals would reverse much of the policy development of the 1970s and return environmental administration to an earlier era in which private interests largely had their way. The federal government no longer would have a major role in controlling the social costs of pollution and economic development. Many public lands and facilities would be turned over to private ownership or management.

A 1979 Louis Harris poll found that 69 percent of the respondents favored cutbacks in federal spending. In the same poll, the public was asked to respond to this trade-off question: "Would you oppose a major cutback in federal government spending if it meant cutting back on spending for environmental protection?" A majority (57 percent) said they would oppose such a cutback.

In 1978, Resources for the Future (an environmental organization based in Washington, D.C.) conducted the first national environmental public opinion survey to assess trends in public attitudes toward environmental protection. In 1980, the President's Council on Environmental Quality and several other agencies decided to follow up with another survey to understand trends in environmental opinion. Final results showed that the public still supports environmental quality and is willing to pay the price for environmental quality.

A Gallup poll conducted for *Newsweek Magazine* in 1981 also found the public supportive of environmental policies. (See fig. 21.11.) The public, however, is not as supportive of environmental issues as it was only several years

**Figure 21.11 Public opinion and the polls.** The public still supports environmental quality, but the level of support is not as strong as it was only a few years ago.

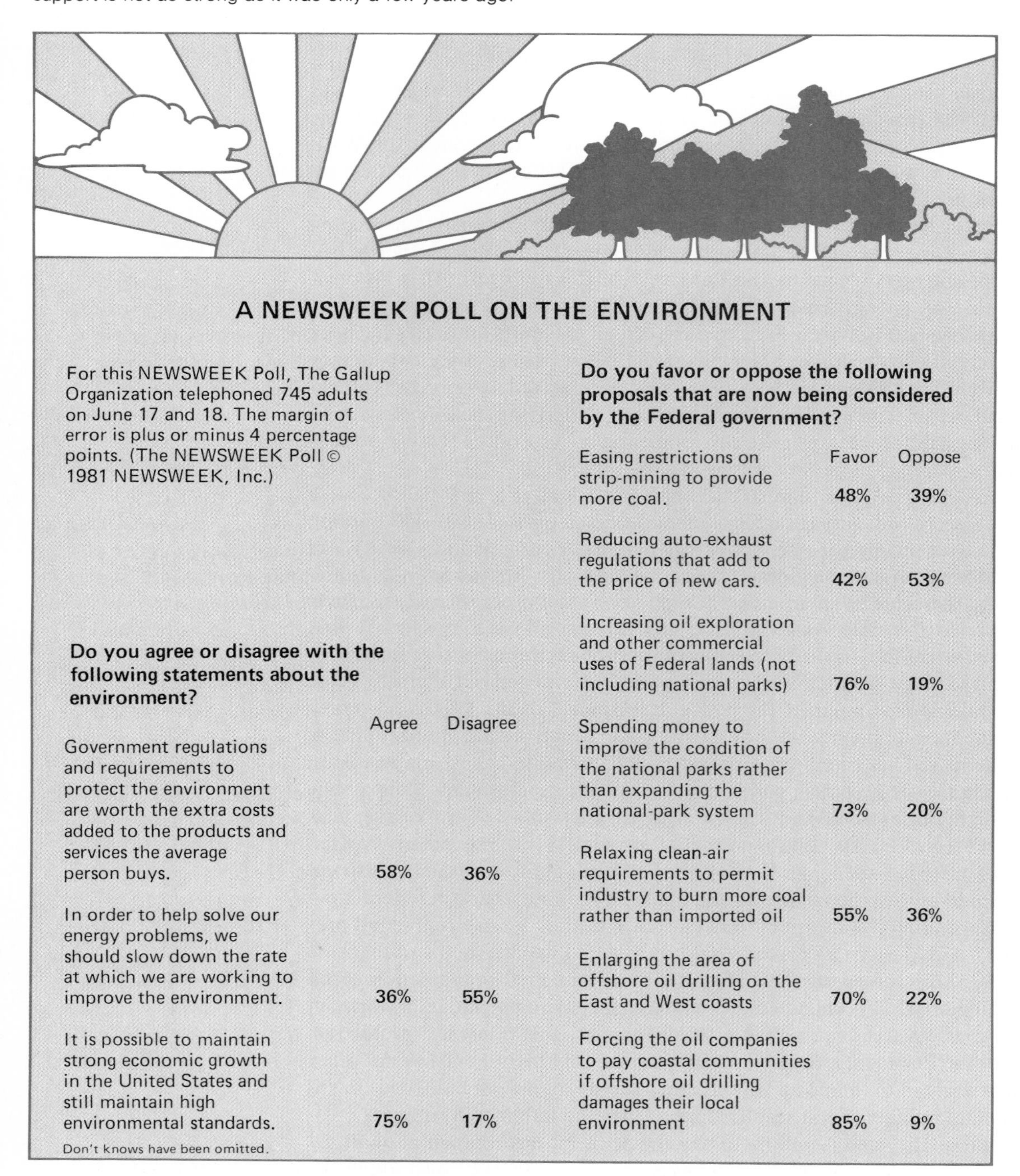

A NEWSWEEK POLL ON THE ENVIRONMENT

For this NEWSWEEK Poll, The Gallup Organization telephoned 745 adults on June 17 and 18. The margin of error is plus or minus 4 percentage points. (The NEWSWEEK Poll © 1981 NEWSWEEK, Inc.)

**Do you agree or disagree with the following statements about the environment?**

| | Agree | Disagree |
|---|---|---|
| Government regulations and requirements to protect the environment are worth the extra costs added to the products and services the average person buys. | 58% | 36% |
| In order to help solve our energy problems, we should slow down the rate at which we are working to improve the environment. | 36% | 55% |
| It is possible to maintain strong economic growth in the United States and still maintain high environmental standards. | 75% | 17% |

Don't knows have been omitted.

**Do you favor or oppose the following proposals that are now being considered by the Federal government?**

| | Favor | Oppose |
|---|---|---|
| Easing restrictions on strip-mining to provide more coal. | 48% | 39% |
| Reducing auto-exhaust regulations that add to the price of new cars. | 42% | 53% |
| Increasing oil exploration and other commercial uses of Federal lands (not including national parks) | 76% | 19% |
| Spending money to improve the condition of the national parks rather than expanding the national-park system | 73% | 20% |
| Relaxing clean-air requirements to permit industry to burn more coal rather than imported oil | 55% | 36% |
| Enlarging the area of offshore oil drilling on the East and West coasts | 70% | 22% |
| Forcing the oil companies to pay coastal communities if offshore oil drilling damages their local environment | 85% | 9% |

## Consider This Case Study
### The Reserve Mining case

One of the most controversial and costly environmental court battles was the Reserve Mining case. For more than a decade, the small Minnesota community of Silver Bay (population—3,500) on the northwest shore of Lake Superior was the focus of an epic environmental legal confrontation. At issue was the Reserve Mining Company's mammoth iron ore processing plant, whose construction in 1952 had created the town. Nearly 80 percent of the local labor force is employed at the facility, which processes about 15 percent of the United States iron ore supply.

The legal and environmental problems revolving around Silver Bay and Reserve Mining center on the waste products that are a result of processing ore. These waste products are called tailings. The tailings were dumped directly into Lake Superior—67,000 tons per day. Concern grew dramatically when it was discovered that the tailings contained tiny fibers similar to asbestos, a known carcinogen. The fibers became so abundant that on certain days they could be seen from the air flowing along the shore. The situation posed immediate concerns for cities that obtained their drinking water from Lake Superior. Duluth, Minnesota, had to spend $7 million for a new water filtration plant when the city's drinking water was found to contain the asbestos-like fibers.

Following the discovery of fibers in Duluth's water supply, the United States District Court ordered Reserve Mining to cease operations. The United States Court of Appeals reversed that order, however, and allowed the company to operate if it agreed to shift to on-land disposal of its tailings. The fact that two courts disagreed on the handling of the Reserve Mining case is not a surprise. Neither decision offered an easy choice. Forcing the plant to shut down would have eliminated more than three thousand jobs, virtually the entire working force in the town of Silver Bay. On the other hand, the fact that Lake Superior, the largest freshwater lake in North America, was being contaminated with carcinogenic fibers was also intolerable.

In late May of 1977 in United States District Court in Minneapolis, the case between Reserve Mining Company and the United States Environmental Protection Agency finally ended. Judge Edward Devitt ruled that in three years the company must stop all dumping of tailings into Lake Superior. The new dumping site would be a specially constructed basin with an area of about fifteen square kilometers, which was located eighteen kilometers west of the plant. The wastes would be transported by rail and pipeline to the basin. There the tailings would be deposited and kept covered under water so the fiber dust could not escape. The total cost of the alternative method of waste disposal is $370 million, more than the initial cost of the plant itself.

The Reserve Mining case points up the frustration of environmental litigation. After years, the case was finally decided. Neither side was totally satisfied. Reserve Mining continued to maintain that the new basin was not necessary and that it was built only because they were forced to do so. The EPA and the states involved felt that the case dragged on for far too long and that a solution should have been agreed upon before 1980, when the basin was completed. A spokesperson for the state of Minnesota summed up the feelings of many at the conclusion of the case: "Perhaps the whole problem might have been avoided if the lake had never been viewed as a convenient dump in the first place."

Why did this case take so long to decide?
Who paid for the construction of the on-land disposal site?
Should Reserve Mining pay for the new water treatment facility for Duluth?

ago. The Gallup poll showed that a solid majority of Americans are willing to pay the added price of environmental safeguards, but many issues advocated by Secretary Watt, such as increased oil exploration, were also favored by large majorities.

These public opinion polls show strong public support for environmental protection, but the need for environmental protection is no longer viewed as a crisis situation, as it was during the 1970s. Whether federal and state governments continue to enact and enforce environmental regulations, however, depends on how active citizens are in keeping pressure on decision makers.

## Summary

The United States has a representative form of government made up of elected officials. The Constitution is the foundation of our government. The Constitution distributed the power between the states and the national government. Our government is divided into three separate branches—the legislative, the judicial, and the executive.

The legislative branch is divided into two bodies—the House of Representatives and the Senate. The primary functions of this branch are policy making, providing funds, and overseeing programs. Lobbying can be very effective in educating and persuading members of Congress.

The judicial branch is responsible for interpreting laws and for reviewing their constitutionality.

The executive branch is responsible for implementing policy.

The bureaucracy is often referred to as the fourth branch of government. It is a method of administration in which appointed officials manage departments by sets of inflexible and routine procedures.

All branches of the government interact with one another in the development and enforcement of legislation.

The Environmental Protection Agency was developed as a result of public concern and has become the largest regulatory agency because it consolidated most environmental units of the government.

## Review Questions

1. What are the three branches of government? What are the major responsibilities of each?
2. Why is a committee system used in Congress?
3. List three kinds of law, and tell how they originated.
4. What is the federal bureaucracy? Why has it developed?
5. Describe three separate governmental activities that were involved in the creation of the Environmental Protection Agency.
6. Why do we have a representative democracy?
7. What is federalism?
8. In the government, where would it be most effective to apply pressure?
9. What is judicial review?
10. Why is the concept of separation of powers so clearly stated in the Constitution?

## Suggested Readings

Edwards, David V. *The American Political Experience.* Englewood Cliff, N.J.: Prentice-Hall, 1979. A good introductory text on United States government.

Ophuls, William. *Ecology and the Politics of Scarcity.* San Francisco, Calif.: W. H. Freeman & Company, 1977. A probing analysis of the political and social implications of the environmental problems facing us today.

Paulsen, David F., and Denhart, Robert B. *Pollution and Public Policy.* New York: Dodd, Mead & Company, 1973. A readable, interesting book of readings examining the policy process. Special attention to public policy and the problems of air and water pollution.

Quarles, John. *Cleaning up America.* Boston: Houghton Mifflin Company, 1976. A fascinating inside look at the development and responsibilities of the EPA by one of the chief administrators of the agency.

Rosenbaum, Walter A. *The Politics of Environmental Concern,* 2d ed. New York: Holt, Rinehart & Winston, 1977. A factual and interesting account of the political maneuvering and strategies behind much of our current environmental legislation.

## Chapter Outline

- I. History of United States economic development
- II. Outgrowths of economic development
  - A. Less work for more pay
  - B. Advertising
  - C. Easy credit
  - D. Materialism
  - E. Recreation and economics
    - Box 22.1 Tourism
- III. Outdoor recreation
  - A. Urban recreation
  - B. Conflict over recreational use of land
- IV. Consider this case study: Big Cypress National Preserve

## Objectives

List the conditions that allowed rapid and continuous economic development in the United States.

Explain how social changes were made.

Explain how advertising and easy credit increase consumption of goods and services.

Understand that the United States and other developed countries of the world have developed materialistic societies.

Describe the economic impact of recreation.

Recognize that people desire outdoor recreation.

Explain why recreational areas are needed in urban locations.

Explain why some land must be designated for particular recreational uses, such as wilderness areas and why that decision sometimes invites conflict from those who do not desire to use the land in the designated way.

## Key Terms

advertising

commercial recreation

materialistic society

outdoor recreation

productivity

wilderness areas

# Chapter 22
# Recreation and free time: Outgrowth of economic development

## History of United States economic development

Few countries have equalled the rapid economic growth of the United States. The United States began with a few poor colonies huddled on the eastern coast of the continent, and in only two hundred years it became one of the wealthiest and most technologically advanced nations in the world. What combination of factors led to this economic explosion? The most important factor was the extremely rich natural resources found in North America. Timber, coal, and oil were available to serve as energy sources for the production of material goods. Mineral resources and rich soil were also abundant. These were certainly not all used or even recognized as

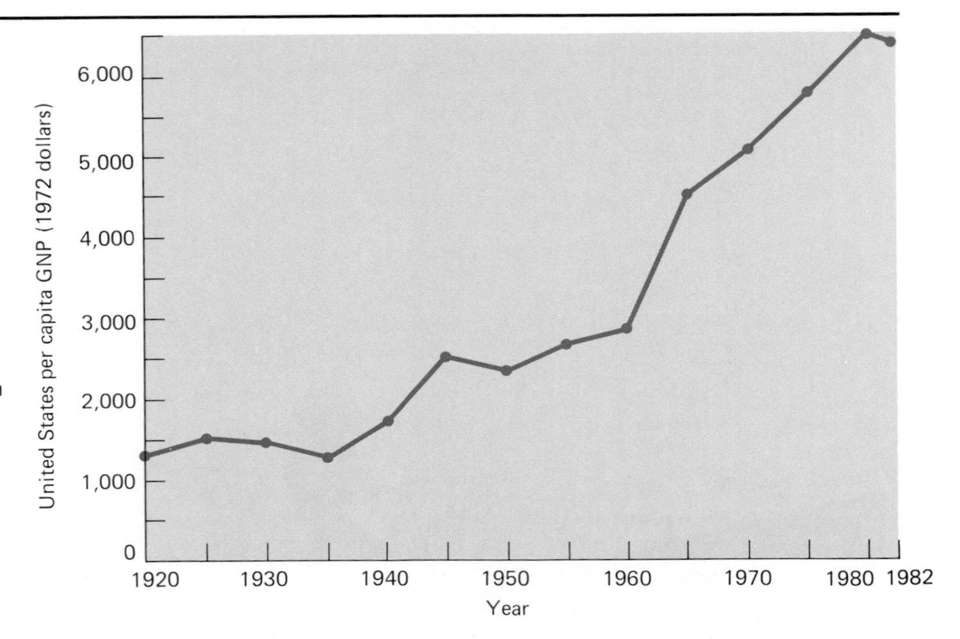


**Figure 22.1 United States per capita GNP (constant 1972 dollars).** The gross national product (GNP) per person is a good indicator of the economic status of a country. In this case, inflation has been considered so you can see the real growth. There has been a steady increase in the per capita GNP in the United States since the late 1800s, but in the last twenty years it has increased extremely rapidly. (Data from the United States Bureau of the Census, *Historical Statistics of the United States,* and *Statistical Abstracts.*)

valuable resources immediately. However, when the need was felt, the resource was available to satisfy the need; economic growth was the result.

Two other factors that determine economic growth are workers and machines. In the United States, the growth of technology and the growth of the population coincided. Natural resources were in such great supply that the increase in people could be assimilated with no difficulty. The Industrial Revolution that began in Europe was transferred to the United States, where it provided jobs for the growing population. There seemed to be no limits to growth, because there were plenty of jobs and abundant resources. This growth was based on an insensitive exploitation of resources and labor. Resources were used with no thought for the future. Often children worked many hours in mines and factories. Technological changes constantly increased productivity. These conditions were ideal for economic growth. This pattern of growth has continued to the present. It should be pointed out that the United States did not grow steadily. There were periods of rapid growth interspersed with economic declines, but looking at the overall picture of the last one hundred years, the upward trend in economic growth is unmistakable. (See fig. 22.1.)

However, today we are starting to feel the constraints that many other less richly endowed countries have experienced for generations. These countries have been unable to provide for the basic needs of their people because their populations were not in balance with the other things needed for economic growth. They had abundant labor but lacked the machines, energy resources, or raw materials for production.

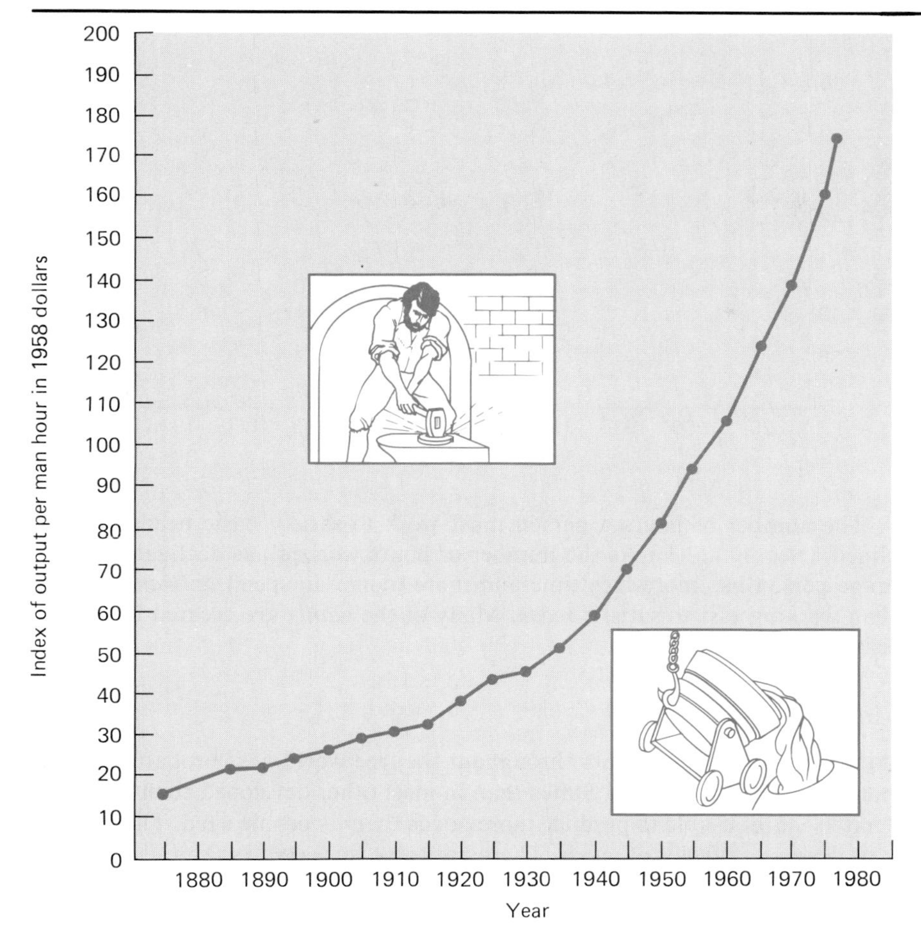


**Figure 22.2 Output per hour worked per person.** Increased technology and extensive use of fossil fuels has allowed a steady increase in the output achieved for each hour that a person works. Because energy has been cheap, this has allowed employers to pay high wages and substitute fossil fuel energy for human muscle energy. (Data from *Historical Statistics of the United States.*)

## Outgrowths of economic development

The history of continuous economic growth in the United States has contributed to the development of a particular set of attitudes.

### *Less work for more pay*

One attitude seems to be "less work for more pay." The ability to pay has been an issue in many labor contracts. If a company is making profits, then it has the ability to increase wages, and labor feels that the company should look after the welfare of its employees. When **productivity** (which is the amount of goods produced per worker) is high, business is able to maintain high wages and require less work for the wages paid. (See fig. 22.2.)

**Figure 22.3**
**Workweek.**
The workweek has been relatively stable at forty hours per week for the last forty years. The drop from nearly sixty hours per week in 1890 to forty hours per week in 1935 coincides with the change from human labor to machines. (Data from *Historical Statistics of the United States* and *Statistical Abstracts.*)

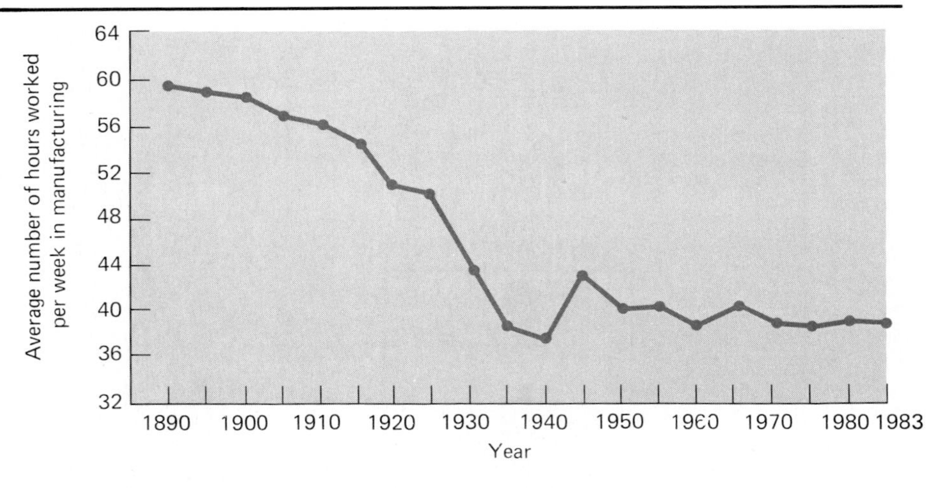


The number of hours a person must work to satisfy basic needs has declined. (See fig. 22.3.) As the number of hours worked has declined, the average person has more free time and more money to spend on *wants* rather than working just to satisfy *needs.* Many of the *wants* are created by advertising.

## *Advertising*

Although **advertising** occurs throughout the free world, it is probably a more potent force in the United States than in most other developed countries. The United States is able to produce more goods than its people need. This creates a problem—a surplus of goods. There are only two ways to solve this problem. One is to export the products to other countries. The other is to create a desire for goods where there is no need or to convince people that their *wants* are *needs* and must be satisfied. This is the aim of advertising. Advertising seeks to compel people to buy goods they don't need. When competition exists, advertising tries to convince people that one product is better than the alternatives. The costs of advertising are sometimes larger than the costs of raw materials. Advertising takes many forms, including newspaper advertising, direct mailings, radio and TV commercials, and outdoor signs with flashing lights. (See fig. 22.4.)

## *Easy credit*

Another function of advertising is to convince people that they need something *now.* This has been coupled with the ease of borrowing money and the increased use of credit cards in the United States today. The idea of "buy now, pay later" does make sense when the value of money is declining. Why not buy something that costs a dollar today and pay for it a month from now when the dollar is worth ninety-eight cents? One major drawback is the handling

**Figure 22.4 Advertising.** Advertising creates a demand for a product or service. Outdoor advertising of this type reminds people that a product or service is available at this site.

charges imposed by the credit-giving organizations. In most cases, these charges more than eliminate any "savings" the consumer would make. This is particularly true with short-term loans when a charge of 18 percent per year (1½ percent per month) or more is made for the money that is borrowed. However, this mechanism of easy, short-term loans makes it convenient for people to buy things on the "spur of the moment." They are able to borrow the money instantly.

High pay, advertising, and easy credit have developed along with our unprecedented economic growth and have produced a population with many more *wants* than *needs*. Our society has free time and the money to spend enjoying that free time.

## *Materialism*

The United States, Canada, Western Europe, Japan, and Australia are generally thought of as the economically developed regions of the world. They constitute about 14 percent of the world's population but consume a major portion of the world's goods. The United States is the dominant nation in this regard. In 1983, United States citizens owned 37 percent of the world's motor vehicles and 35 percent of the world's telephones; they flew 54 percent of the entire distance flown as commercial passengers and consumed 30 percent of the world's energy. In addition, there are 1.9 radios per person and 0.6 TV sets per person in the United States, compared to one radio for every forty-two people and one television set for every two thousand people in India. Obviously, there is a great difference between the material possessions of people in the United States and people in less-developed countries. The kind of society that places a high value on possessions is called a **materialistic society.**

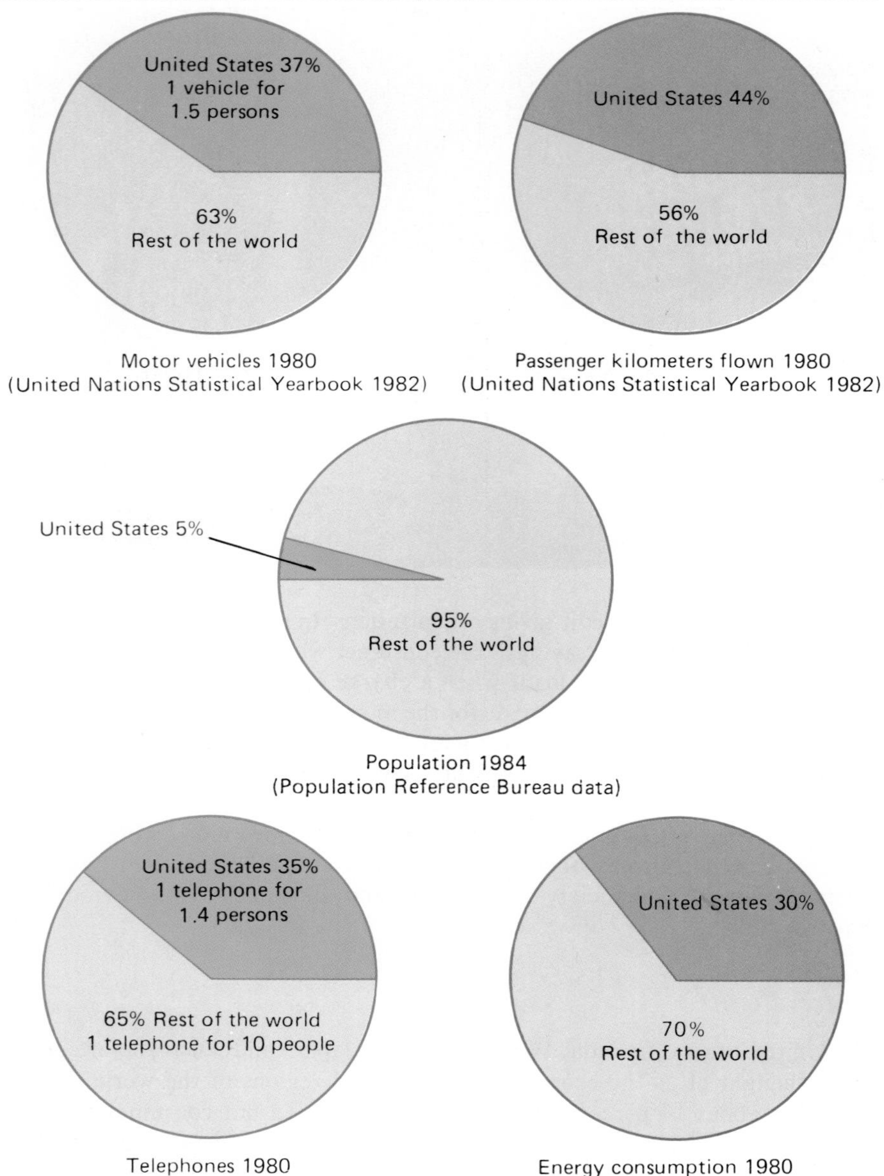


**Figure 22.5 United States material consumption.** When the consumption patterns of the United States are contrasted with other parts of the world, it becomes obvious that the United States consumes a large proportion of the world's resources. We have been fortunate to have abundant natural resources and a population growth pattern that coincided with our technological growth.

There is no doubt that the United States society is materialistic. (See fig. 22.5.) This society is one in which people buy things because they want them rather than because they need them. We buy new clothing because we want something new, not because we need it to keep warm. Indeed, if people bought only what they needed, the economy would decline and people would be forced out of work. Our economy is based on increasing consumption. The American dream is that of economic improvement for everyone, and it has come true for

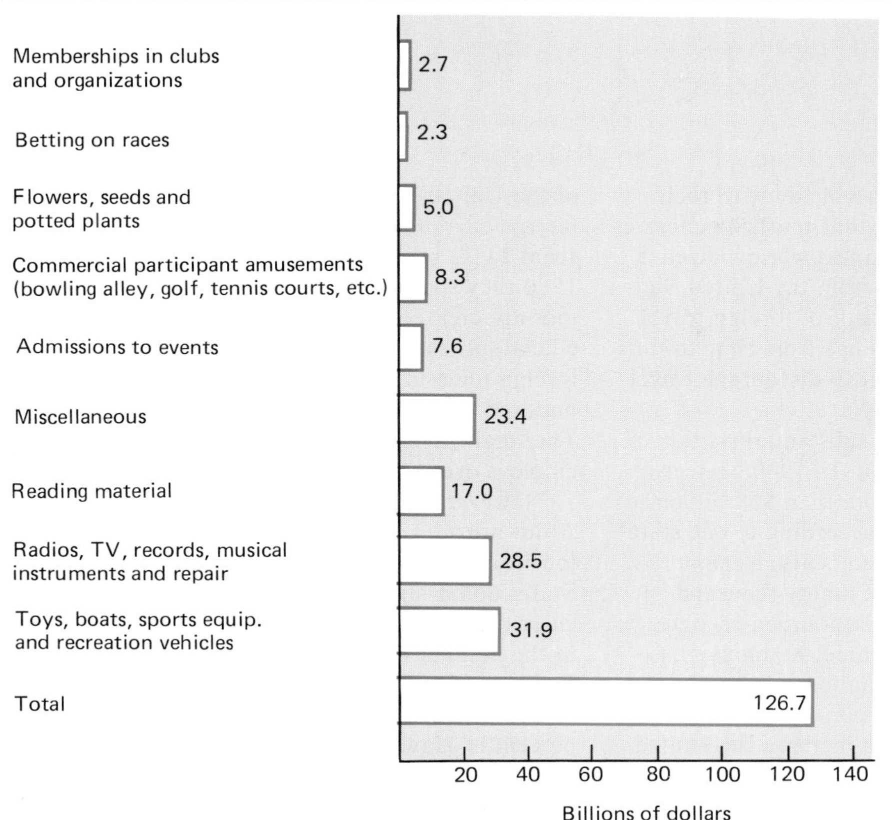


**Figure 22.6 Personal consumption expenditures for recreation.**
This graph shows that we spend billions of dollars on equipment and services used for recreation. The actual recreation expenditures related are much higher when we consider the expenditures associated with recreation, such as driving to a recreational location, lodging, and the purchase of public lands for recreational purposes. (Data from *Statistical Abstracts of the United States,* 1984.)

us. This kind of consumption requires exploitation of natural resources and tremendous inputs of energy to transform raw materials into the products we want.

## *Recreation and economics*

We spend money not only on services and material possessions but on leisure-time activities in a variety of kinds of recreational pursuit. As figure 22.3 shows, the workweek has been at the forty-hour level since World War II. However, what this does not show is the amount of free time that people have gained through increased vacation benefits. The amount of time now available for recreational pursuits is greater than the time the average person works. Consequently, leisure-time activities and tourism have become a very important part of the United States society and its economy. **Commercial recreation** ranks third behind manufacturing and agriculture in its impact on the economy. In 1982, United States citizens spent $126.7 billion on recreational goods and services ranging from memberships in clubs to toys and sports equipment. (See fig. 22.6.) In 1981, about sixty-seven million people paid over $445 million for

## Box 22.1
### Tourism

Travel is one thing that people do with their free time. It is estimated that tourism generates about $155 billion of business worldwide each year. The flow of dollars from the United States to other countries as a result of foreign travel by United States citizens has from time to time prompted our government to discourage travel abroad.

Many states receive a substantial portion of their income from tourism. In 1980, sixteen million tourists added more than $18 billion to the economy of Florida, according to one state official. In Hawaii, tourism is also a major source of jobs. More than ninety thousand Hawaiians are employed in tourism-related jobs. Hawaii's Senator Daniel K. Inouye has strongly supported the establishment of a nonprofit government corporation to develop a national tourism policy to increase the United States share of the tourism dollar.

From 1970 to 1980, foreign tourists visiting the United States increased from about 2.2 million to 7.7 million because of the low value of the United States dollar in relation to many foreign currencies. In 1982, foreign travelers spent $11.2 billion in the United States; in 1980 they spent around $10 billion. Additional jobs are created by increased tourism. A travel official has estimated that the 7.7 million foreign tourists in 1982 supported about three hundred thousand jobs in related industries. The total travel industry in the United States employs over six million people.

However, the tourism bubble is very fragile. A downturn in the economy drastically affects tourism business. In 1985 when the United States dollar strengthened against other currencies, the United States saw a sharp drop in the number of foreign travelers. Furthermore, energy costs are very closely related to the amount of travel. Thirty-six percent of Hawaii's energy consumption is in the form of jet fuel. If jet fuel becomes unavailable or increases in cost significantly, the Hawaiian economy would be severely injured.

hunting and fishing licenses. In addition, the federal government spent more than $355 million to acquire land and provide outdoor recreational opportunities. Obviously these dollar amounts show that recreation is a big business. But some recreational activities, such as walking and jogging, involve little or no cost. Nearly 70 percent of the United States population participates in walking and jogging for recreational purposes.

## Outdoor recreation

Not all people enjoy doing the same things with their free time. For some people, reading or watching television may consume most of their available free time. Commercial recreational activities, such as golf, tennis, bowling, amusement areas, race tracks, and skiing, are also very important uses of free time. Another major area of recreation, usually classified as **outdoor recreation,** uses the natural out-of-doors for hiking, camping, canoeing, and so forth.

**Table 22.1**
Outdoor activities

| Number of people that participated in activity in 1980 | |
|---|---|
| Fishing | 26 million |
| General exercise | 26 million |
| Bicycling | 19 million |
| Camping | 17 million |
| Running | 16 million |
| Hunting | 14 million |
| Hiking | 12 million |
| Golf | 10 million |
| Tennis | 10 million |
| Skiing | 8 million |
| Boating | 8 million |

Source: Data from *Statistical Abstracts of U.S.*, 1984.

Millions of people want to use public lands for these activities. However, some of these activities cannot occur in the same place at the same time without having conflicts develop among the various users. For example, wilderness camping and backpacking often conflict with off-road vehicles.

Most of our recreational activities require the consumption of natural resources. Resources, such as minerals, fuels, and timber, must be allocated for the building and operation of recreational devices. Some activities require more resources than others, but even the backpacker, who has traditionally been considered an ecologically frugal individual, requires considerable equipment to participate in that activity. Other activities, such as the use of off-road vehicles, require more resources in the form of equipment and fuel. Table 22.1 lists various outdoor activities and the number of people who participate in each activity. As energy and other resources become less available, we may be forced to abandon some of our more extravagant recreational activities for those that are more ecologically conservative.

Remember, recreation is not a basic biological need. It is, however, a very important psychological need and one that people will satisfy.

## *Urban recreation*

Recreation seems to be a basic human need. The most primitive tribes and cultures all have games or recreational activities. New forms of recreation are continually being developed.

A major problem with urban recreation is the development of recreational activities and facilities that are conveniently located near residential areas. The hundreds of thousands of square kilometers of national parks in Alaska will be visited by only a very small proportion of the United States population. Large urban centers are discovering that they must provide adequate, low-cost recreational opportunities within their jurisdictions. Some of these opportunities are provided in the form of commercial establishments such as

**Figure 22.7 Urban recreation.**
In urban areas, recreation often takes the form of sports programs, playgrounds, and walking. Most cities recognize the need for such activities and develop extensive recreation programs for their citizens.

bowling establishments, amusement parks, and theaters. Others must be subsidized by the community. (See fig. 22.7.) Playgrounds, organized recreational activities, and open space have usually been combined into an arm of the municipal government known as the Park and Recreation Department. Cities spend millions of dollars to develop and maintain recreation programs. Often there is conflict over the allocation of financial and land resources. These are very closely tied because open land is scarce in urban areas, and it is expensive. Riverfront property is ideal for park and recreational use, but it is also prime land for the development of industrial, commercial, or high-rise residential buildings. Therefore, a conflict is inevitable. Many metropolitan areas are beginning to see that recreational resources may be as important as economic growth for maintaining a healthy community.

An outgrowth of the trend toward urbanization is the development of nature centers. In many urban areas, there is so little natural area left that the people who live there need to be given an opportunity to learn about nature. Nature centers are basically teaching institutions that provide a variety of methods for people to learn about and appreciate the natural world. Zoos, botanical gardens, and some urban parks, combined with interpretative centers, also provide recreational experiences. Nature centers are usually located near urban centers, open to the public, where some appreciation of natural processes and phenomena can be developed. In some cases, they are operated by municipal governments. In other cases, they may be run by school systems or other nonprofit organizations.

**Figure 22.8 Conflict over recreational use of land.**
This land may be used for both hiking and jeep trails. The people who participate in these two kinds of recreation are often antagonists over the allocation of land for recreational use.

## *Conflict over recreational use of land*

Many people have a desire to use the natural world for recreational purposes because nature can provide a challenge that may be lacking in the workplace. Whether the challenge is hiking in the wilderness, underwater exploration, climbing mountains, or driving a vehicle through an area that has no roads, people experience a sense of adventure from these activities. Table 22.1 lists various outdoor activities and the number of people who participate in each kind of activity. All of these activities use the out-of-doors but not in the same way. Conflicts develop because some of these activities cannot occur in the same place at the same time.

There is a basic conflict between those who prefer to use motorized vehicles and those who prefer to use muscle power in their recreational pursuits. (See fig. 22.8.) This conflict is particularly strong because both groups would like to use the same publicly owned land for their activities. Both have paid taxes. Both "own" the land, and both feel that it should be available for them to use as they wish. When everyone "owns" something, there is often little desire to regulate activities.

For example, much of the range land in the West is publicly owned. Based on "Animal Use Months" established by the Bureau of Land Management or the Forest Service, ranchers are allowed to graze cattle on certain public lands. Technically, failure to comply can mean a loss of grazing rights. However, since the bureaucracy of setting and enforcing the regulation is politically motivated and understaffed, there is no incentive for ranchers to limit the number of cattle or sheep on this land unless all ranchers agree to do so. The usual result is that a few individuals exploit the land in hopes of a short-term profit. However, this exploitation could cause permanent damage to the range land.

An obvious solution to this problem is to allocate land to specific uses and to regulate the use once allocations have been made. Several United States governmental agencies, such as the National Park Service, the Bureau of Land

**Figure 22.9 Federal recreational lands in 1983.**
Of the approximately 108 million hectares of federal recreational lands in the United States, approximately 10 percent is designated as wilderness, primitive, or natural. (Data from *Statistical Abstracts of the United States,* 1984.)

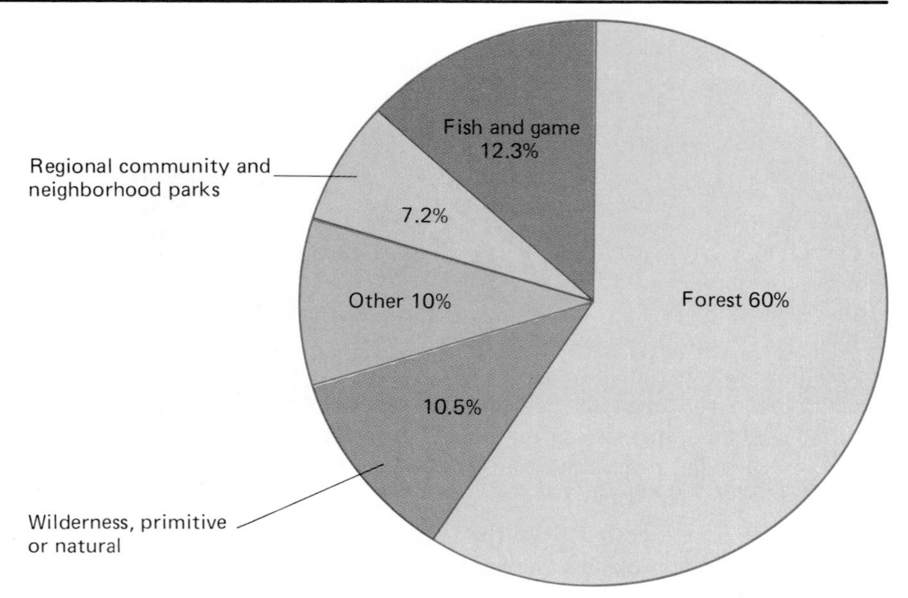


Management, the United States Forest Service, and the United States Fish and Wildlife Service, are engaged in allocating and regulating the use of the lands they control.

A particularly sensitive issue is the designation of certain areas as **wilderness areas.** As defined in the Wilderness Act of 1964, wilderness is "an area where the earth and its community of life are untrampled by man, where man himself is a visitor who does not remain." Obviously, if an area is to be wilderness, human activity must be severely restricted. This means that the vast majority of the people of the United States will never see or make use of these areas. Many people argue that this is unfair because they are paying taxes to provide for the recreation of a select few. Others argue that, if everyone were to use these areas, their charm and unique character would be destroyed; therefore, the cost of preserving wilderness is justifiable.

Areas designated as wilderness make up a very small proportion of the total public land available for recreation. (See fig. 22.9.) The fact that there are relatively few wilderness areas has resulted in a further problem. The areas are being loved to death. People pressure on these resources has become so great that in some cases the wilderness quality is being tarnished. The designation of additional wilderness would relieve some of this pressure.

## Consider This Case Study
### Big Cypress National Preserve

Big Cypress National Preserve was established by Congress in 1974. This area of approximately 2,300 square kilometers is located between Miami and Naples in the southern tip of Florida, where it adjoins Everglades National Park. Until the 1960s, this area had been largely undeveloped swampland, which was used primarily for hunting. Land-development companies sold thousands of small tracts within the swamp at up to $2,500 per hectare. There has been logging and some grazing in the Big Cypress area for years, and oil exploration is continuing.

Pressure to establish the Big Cypress National Preserve grew during the 1960s when a giant airport was planned east of the area and several land-development schemes began modifying the western portions of the Big Cypress area. The major concern was the amount of water being drained from the area. Drainage is necessary to develop the land. Development would also encourage drilling more wells to draw water from the underlying porous layers of limestone. Most of the water in the Big Cypress National Preserve flows slowly southward into the Everglades National Park. This water is particularly important during the dry season. A reduced water flow from the Big Cypress area would endanger the Everglades as well.

Once the Big Cypress National Preserve was established, the federal government began a program of buying the privately owned land within the preserve's boundaries. This amounted to 90 percent of the preserve. Many of the fifty thousand people who owned the land were not eager to sell, because they had paid up to three times what the government was offering them for their land. Therefore, the government exercised its power of eminent domain to force the sale.

Converting the land from private to public ownership was aided by the fact that the private owners would still be able to use their land as they had prior to the government purchase. In addition, hunting would still be permitted in the preserve. There are between four hundred and five hundred hunting camps in the area, most of which can be reached only with the use of swamp buggies, airboats, or all-terrain vehicles. These vehicles have created about one thousand kilometers of trails, which are readily visible from the air. The Park Service, which administers the preserve, recognizes that it will be necessary to regulate these vehicles to prevent further damage to the area. But the park service must also provide reasonable access to the people who want to use the area for hunting. Many environmental groups would like to see the area returned to wilderness. This would preserve habitats for certain endangered species, like the Florida panther. Some portions of the preserve have been set aside as wilderness, but many people would like to see the entire preserve revert to wilderness.

In reality, this entire scenario is a study of conflicting recreational interests. Southern Florida is a major vacation spot. The development of Miami, the proposed airport, and the land developments all serve to meet the needs of vacationists. Many who originally bought land in the Big Cypress National Preserve area were duped by land speculators. Others wanted the land for the recreation it would provide them in the form of hunting. Still others want to see the area preserved as wilderness.

What were the most critical issues in establishing Big Cypress National Preserve?

What impact has the relative wealth of the people of the United States had on the area?

## Summary

The United States has experienced continuous economic growth. It had abundant resources that were available for exploitation when the need arose. The growth of the United States population coincided with the growth in technology. This resulted in a nation with a continuous history of economic growth and expansion. As productivity rose, primarily due to technological advancements, wages increased, and the hours required of a worker decreased. The United States is able to produce more goods than its people need. This results in a surplus of goods. Advertising convinces people that they need goods, and easy credit helps them satisfy their wants.

The United States is one of the most materialistic nations in the world. We spend large amounts of money on free-time activities, services, and material possessions. Recreation is important socially and psychologically. People seem to need recreation as a change from their normal working life.

Conflicts sometimes arise over the use of resources for recreational purposes, because some recreational activities cannot occur at the same time and place. These conflicts are usually resolved by allocation and regulation of resources.

## Review Questions

1. What natural factors in the United States have contributed to its rapid technological growth?
2. In addition to resources, what other conditions are required for economic growth?
3. In a materialistic society, what motivates people to purchase items?
4. List some conflicts that arise when an area is designated strictly as recreational.
5. Define productivity.
6. Name three factors that have produced an attitude of wanting more than we need.
7. Where does recreation rank as a segment of the United States economy?
8. Define advertising.
9. If a country can overproduce, how does it solve the problem of overproduction?
10. How can recreational activities damage the environment?

## Suggested Readings

Bannon, Joseph J. *Leisure Resources: Its Comprehensive Planning.* Englewood Cliffs, N.J.: Prentice-Hall, 1976. Deals primarily with the planning and administration of urban recreational areas. Includes samples of survey questions useful in assessing public opinion.

Brockman, C. Frank, and Merriam, Lawrence C., Jr. *Recreational Use of Wild Lands.* 3d ed. New York: McGraw-Hill Book Company, 1979. A thorough treatment of the many aspects of recreational use of wildlands. Discusses the economics, visitor motivation, politics, and the kinds of lands used for recreation and the ways in which they are used.

Hales, Linda. "Who Is the Best Steward of America's Public Land?" *National Wildlife* 21: 5–10. Explores conflicting uses of public recreational lands.

Scheiber, Harry N.; Vatter, Harold G.; and Faulkner, Harold Underwood. *American Economic History.* 9th ed. New York: Harper & Row, Publishers, 1976. Chapter one has a section on the mechanisms of economic development, which is good background for this chapter. The remainder of the book gives a detailed history of United States economic growth.

Wigginton, Eliot, ed. *The Foxfire Book* series. San Francisco: The Anchor Society, 1972, 1973, 1975, 1977, 1979. This series of books about life in Appalachia gives a glimpse of what life was like in a simpler time. They also give a feeling for life that is less materialistic.

## Chapter Outline



## Objectives

Differentiate between ethics and morals.

Define personal ethics.

Explain the connection between material wealth and resource exploitation.

Describe how industry must exploit resources and consume energy in order to produce goods.

Explain how corporate behavior is determined.

Describe the tremendous power that corporations wield because of their size.

Explain how the courts have been used by both corporations and environmentalists as a delaying tactic.

Explain why governmental action was necessary to force all companies to meet environmental standards.

Describe what has been the general attitude of society toward the environment.

Explain the relationship between economic growth and environmental degradation.

List three conflicting attitudes toward nature.

## Key Terms

corporation
development ethic
economic growth
equilibrium ethic
ethics
morals
preservation ethic
profitability
resource exploitation

# Chapter 23
# Environmental ethics

## What is ethical?

**Ethics** is one branch of philosophy. Ethics seeks to define what is fundamentally right and what is wrong, regardless of cultural differences. All cultures have a reverence for life. All individuals have a right to live. It is unethical to deprive an individual of life.

**Morals** differ from ethics because morals reflect the predominant feelings of a culture about ethical issues. For example, it is certainly unethical to kill someone. However, when a country declares war, most of the people in the country accept the necessity for killing the enemy. Therefore, it is moral. (See fig. 23.1.) No nation has ever declared an immoral war.

**Figure 23.1 Ethical and moral behavior.** Ethics are clear-cut statements of right and wrong. Morals are much more difficult to define because they are reflections of a culture's view of how ethics should be applied to daily living. This set of moral statements would be quite different in other cultures.

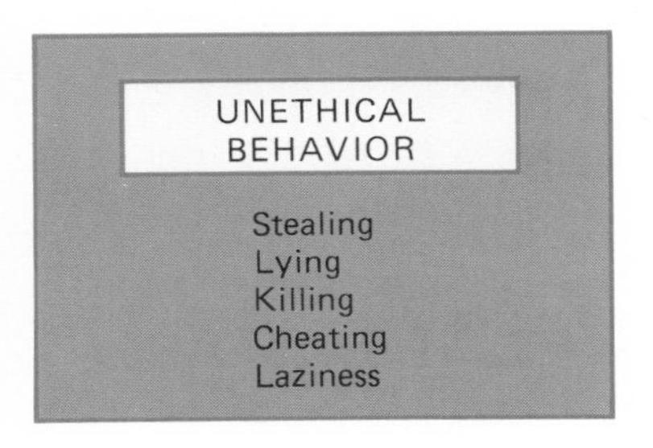


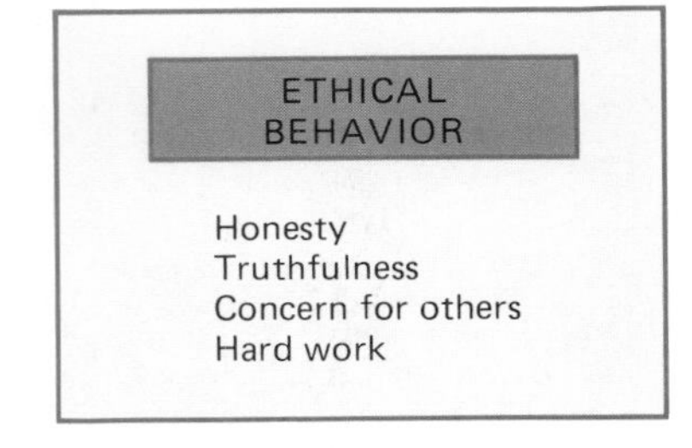


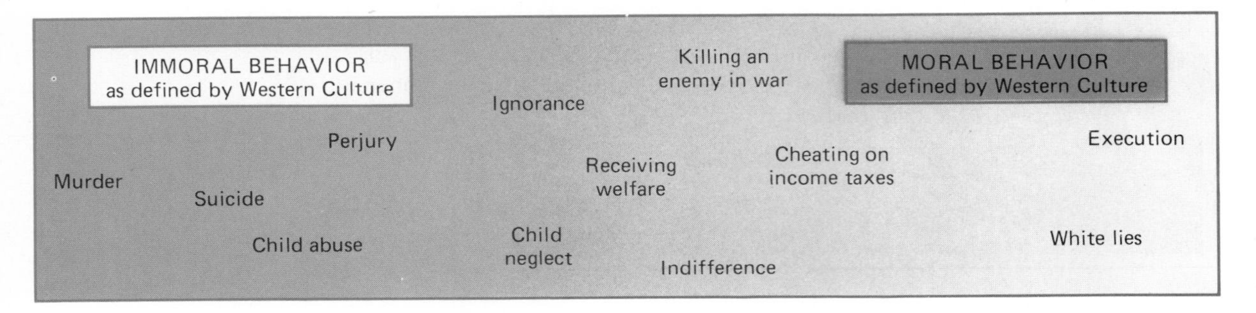


Because there is currently enough food in the world to feed everyone adequately, it is unethical to allow people to starve while others have more than enough food. However, the predominant mood of the developed world is one of indifference. In reality, this is a statement that it is all right to allow people to starve. This moral stand is not consistent with an ethical one.

Ethics and morals are not always the same; thus it is often difficult to clearly define what is right and what is wrong. Some individuals view the world's energy situation as serious and have reduced their own consumption. Others do not believe there is a problem, and, therefore, have not modified their energy use. Still others do not care what the situation is. They will use energy as long as it is available. Another similar issue is the population problem. Is it ethical to have more than two children when the world faces overpopulation? Should an industry try to persuade the public to vote "no" on a particular issue because it might reduce their profits, even though its passage would improve the environment? The stand taken on such issues often revolves around the position of the person making the statement. An industrial leader, for example, would probably not look upon pollution as negatively as someone who is active outdoors. In fact, many business leaders view the behavior of active preservationists as somewhat immoral because it restricts growth and, in some cases, causes unemployment.

Most ethical questions are very complex. Ethical issues dealing with the environment are no different. Therefore, it is important to explore environmental issues from several points of view before taking a stand. One point to consider is the difference between short-term and long-term effects of a course of action.

When you take an ethical stand, you become open to attack from those who disagree with your stand. Often, individuals are portrayed as villains for pursuing a course of action they consider to be righteous.

## Individual ethics

What are your "rights"? As a citizen of the United States, you are guaranteed certain rights or freedoms by the Constitution. These include freedom of the press, freedom of worship, freedom of speech, and the right to assemble peacefully. You are also guaranteed the right to a fair trial and the right to vote, as well as the right to own private property. Does this right extend to the point of owning goods at the expense of others in the world and at the expense of future generations? Is this behavior ethical? Nowhere is it written that you have a "right" to a car, color TV, clothes dryer, snowmobile, air conditioner, or even a stereo. In fact, one important ethical problem facing the industrialized western world is that we consider it our right to accumulate luxury items as if they were necessities. How often have you said that you "needed" something? Do you really need that new record or that new hair dryer? Such items as these are really not needs at all; they are luxuries. As a nation, we consume more and more luxury items each year, and we have begun to look upon them as absolute necessities of life. Few of us consider the natural resources used to produce such items. This situation then becomes an individual ethical question. Is it ethical to continue to consume our natural resources for such luxuries? Take a few moments and count the "necessities" you own, and ask yourself which of these items you *really* need. (See fig. 23.2.)

None of us would want to turn the clock back to a time when it was necessary to work long hours just to meet the bare necessities of life. Science, technology, and industry have advanced the standard of living for many. However, what price are you willing to pay for the continued convenience of owning a vast array of luxuries? Are you willing to see your environment fouled and other people deprived of their daily needs so that you can have a continuous supply of personal luxuries and conveniences? In some places, water or air pollution was not considered a problem until people began to die from the pollution. The ethical question is, How much of this world's resources is an individual entitled to use?

Take a quick glance at a billboard or television commercial. It is easy to understand why we continue to buy regardless of the economic and social costs. Although individuals are actually doing the consuming, incentives to do so come from industries and corporations.

## Corporate ethics

Many facets of industry, such as procuring raw materials, manufacturing and marketing, and disposing of wastes, are in large part responsible for pollution. This is not because any industry or company has adopted pollution as a cor-

**Figure 23.2 Necessities and luxuries.** These two photographs represent typical "well-equipped" kitchens in the economically developed world. Both are used to put good, attractive meals on the table. What are needs and what are luxuries?

porate policy. Industry is naturally dirty because all industries consume energy and resources. In the process of manufacturing, waste heat must be produced because of the second law of thermodynamics. (See chap. 2.) In addition, when raw materials are processed, some waste is inevitable. It is usually not possible to completely control the dispersal of all the by-products of a manufacturing process. Also, some of the waste materials produced may simply be useless.

For example, the food service industry uses energy in the preparation of meals. Much of this energy is lost as waste heat. In addition, smoke and odors are produced and not contained. Finally, bone, fat, and discolored food items become wastes (useless materials) that must be discarded.

The cost of controlling waste can be very important in determining the profit margin of a company. **Corporations** are designed to operate at a profit, which is not in itself harmful. Ethics are involved, however, when a corporation cuts corners in production quality or waste disposal in order to maximize profit. The cheaper it is to produce an item, the greater the possible profit. It is cheaper to dump wastes in a river than to install a waste-water treatment facility, and it is cheaper to release wastes into the air than it is to remove them. Many would consider such behavior unethical and immoral, but to the corporation it is just one of the factors that determines **profitability.** (See fig. 23.3.)

The amount of profit a corporation realizes determines the amount of expansion that is possible. In order to expand continually, industry creates an increased demand for its products through advertising. The more it expands, the more power it attains. The more power it has, the greater its influence over

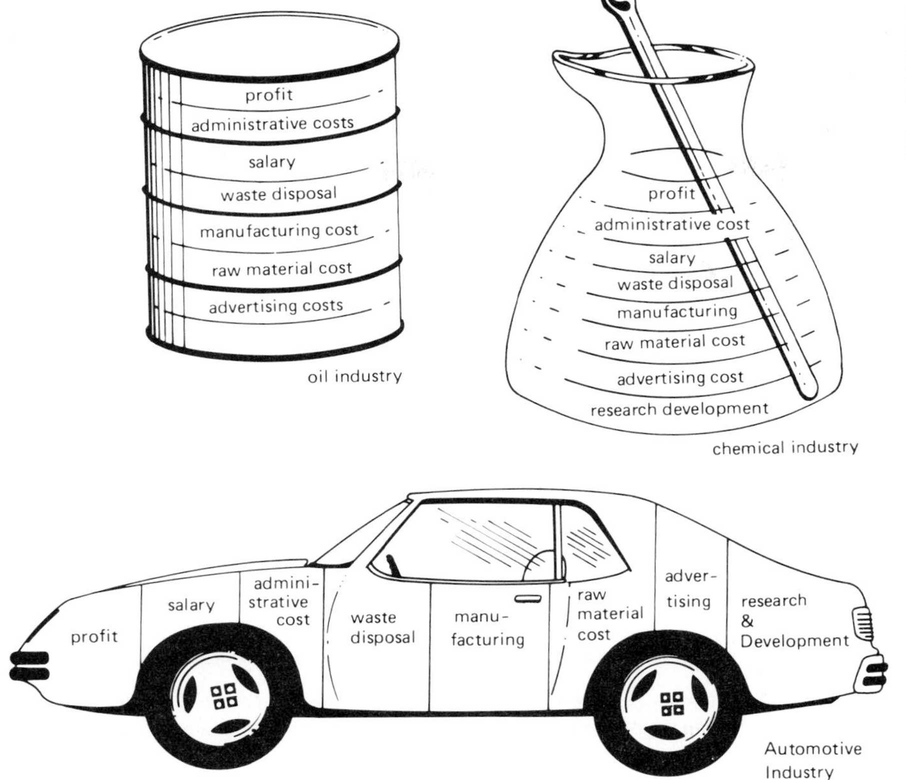


**Figure 23.3 Corporate decision making.** Corporations must make a profit. When they look at pollution control, they view its cost like any other cost. Any reductions in cost increase profits.

decision makers. These decision makers can create conditions favorable to corporations. Therefore, it becomes a seemingly never-ending spiral.

Nations of the world must confront the problem of corporate irresponsibility towards the environment. In business, incorporation allows for the organization and concentration of wealth and power far surpassing the capacity of individuals or partnerships. Some of the most important decisions affecting our environment are made by individuals who wield massive corporate power rather than by governments or the public. Often, these corporate decisions involve only minimal concessions to the public interest, while every effort is exerted to maximize profits.

Business decisions and technological developments have increased the exploitation of natural resources. In addition, our political and legal institutions have generally supported the development of private enterprise. They have also promoted and defended private property rights rather than social and environmental concerns. It is typical for businesses and individuals to use loopholes, political pressure, and the time-consuming nature of court action to circumvent or delay compliance with established criteria for social or environmental improvement. (See fig. 23.4, which details the time line of events concerning such a case.)

**Figure 23.4 Environmental litigation.** This illustrates the major events in a case determining a company's right to discharge materials into public waters. Notice that the case started almost ten years before it was brought to conclusion.

| | |
|---|---|
| 1967 | Verna Mize, a resident of Maryland who vacations in Minnesota, first finds out about the Reserve Mining Company's massive disposal of taconite tailings in Lake Superior. She begins a one woman crusade that will be responsible for much of the subsequent legal action. |
| 1972 | The U.S. Justice Department files suit against the Reserve Mining Company in U.S. District Court. |
| 1973 | The trial of United States vs Reserve Mining Company begins. |
| 1974 | Federal Judge Ford orders Reserve Mining Company to cease all discharges into Lake Superior. |
| 1974 | The U.S. Court of Appeals grants Reserve Mining Company a 70-day delay on the order to cease all discharges, conditional upon preparation of an acceptable abatement plan. |
| 1975 | The U.S. Court of Appeals requires that Reserve Mining Company be given a reasonable time to stop its discharge or to phase out its plant. |
| 1976 | The U.S. Court of Appeals removes Judge Ford from the case, citing bias against Reserve Mining Company. |
| 1976 | Judge Devitt, Judge Ford's replacement, finds Reserve Mining Company guilty of violating its discharge permits and fines the company $837,500. Reserve Mining Company is also found guilty of bad faith for improperly withholding evidence during the federal trial and is fined $200,000. |
| 1976 | On a motion by federal attorneys, Judge Devitt orders Reserve Mining Company to cease discharge of taconite tailings into Lake Superior in exactly one year, at midnight on 7 July 1977. |
| 1976 | The Minnesota Department of Natural Resources and the Minnesota Pollution Control Agency reject Reserve Mining Company's application for permits to construct an on-land disposal site. |
| 1976 | Reserve Mining Company appeals the state agencies' decisions in state court. |
| 1976 | November—The trial in Reserve Mining Company's appeal of the decisions denying permits for on-land disposal gets under way before three state district court judges. |
| 1977 | Minnesota Supreme Court orders state agencies to issue on-land disposal permits. |
| 1977 | Federal Judge Devitt stays his order for reserve Mining Company to close by 7 July 1977, provided that Reserve Mining Company begins construction of the on-land disposal project by 1 June and ceases tailing discharges into Lake Superior by 15 April 1980. |
| 1978 | Officials of Reserve Mining Company announce that, given reasonable interpretation of the permits for the on-land disposal system, Reserve Mining Company would be able to meet the conditions and be out of Lake Superior by the deadline of 15 April 1980. |
| 1981 | The on-land disposal of the taconite tailings is under operation with environmental monitoring conducted regularly by both the Minnesota Pollution Control Agency and Department of Natural Resources. |

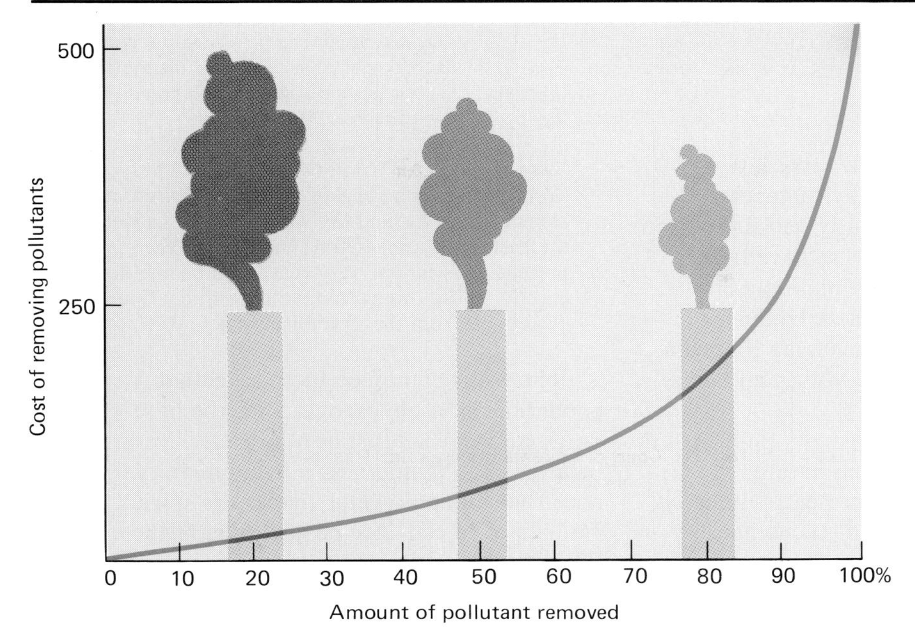


**Figure 23.5 Pollution control costs.** These costs increase very rapidly as companies are asked to remove higher and higher percentages of the pollutant. It often costs a company as much to remove the last 10 percent of the pollutant as it does to remove the first 90 percent.

Only in recent times have laws been enacted to prohibit industry from dumping toxic wastes into public waters and air. The most vocal opponents and critics of governmental health standards and environmental regulations have been corporations. Many felt that it was their right to decide how they should dispose of their wastes. This was a particularly pressing issue because the costs associated with waste disposal can be very large. This is especially true when a company is forced to improve the systems already installed. (See fig. 23.5.) Many of the industries that have practiced pollution control over the past ten years admit that they would not have done so if they had not been forced to by the government. Governmental action is often necessary to enforce compliance with government regulations because companies who are cleaning up their operations are forced to compete with those who do not. Governmental action requires all companies to meet the same standards. Some companies still rebel. The courts are full of cases in which corporations have failed to abide by federal or state regulations. It seems that this will be true for some time to come.

Environmentalists frequently think that companies initiate court action only to avoid compliance. Although this is sometimes true, it is also true that there are many legitimate cases before the courts to determine what the standards should be. It would be useless to impose standards that were so stringent that a company could not meet them. The company would not be profitable and could not stay in business. Workers would lose jobs, and consumers would lose the product. So it is important to remember that costly litigation is used frequently to determine the answer to a question and not just to circumvent regulations. On the other hand, some individuals take advantage of existing

## Box 23.1
## Naturalist philosophers

The philosophy behind the environmental movement of the 1970s actually had its roots in the last century. Although there have been many notable conservationist philosophers, several stand out. Some of these landmark naturalist thinkers are Ralph Waldo Emerson, Henry David Thoreau, John Muir, and Aldo Leopold.

In Emerson's first essay *Nature,* published in 1836, he claimed that "behind nature, throughout nature, spirit is present." Emerson was an early critic of rampant economic development, and he sought to correct what he considered to be the social and spiritual errors of his time. In his *Journals,* published in 1840, Emerson stated that "a question which well deserves examination now is the Dangers of Commerce. This invasion of Nature by Trade with its Money, its Credit, its Steam, its Railroads, threatens to upset the balance of Man and Nature."

Another naturalist who held beliefs similar to Emerson's was Henry David Thoreau. Thoreau's bias fell on the side of "Truth in nature and wilderness over the deceits of urban civilization." The countryside around Concord, Massachusetts, fascinated and exhilarated him as much as the commercialism of the city depressed him. It was near Concord that Thoreau wrote his classic *Walden Pond,* which describes a year in which he lived in the country in order to have direct contact with nature's "essential facts of Life." In his later writings and journals (1861), Thoreau summarized his feelings toward nature with prophetic vision.

> But most men, it seems to me, do not care for Nature and would sell their share in all her beauty, as long as they may live, for a stated sum—many for a glass of rum. Thank God, men cannot as yet fly, and lay waste the sky as well as the earth! We are safe on that side for the present. It is for the very reason that some do not care for those things that we need to continue to protect all from the vandalism of a few.

John Muir combined the intellectual ponderings of a philosopher with the hard-core, pragmatic characteristics of a leader. Muir believed that "wilderness mirrors divinity, nourishes humanity, and vivifies the spirit." Muir tried to convince people to leave the cities for a while to enjoy the wilderness. However, Muir felt that the wilderness was threatened. In the 1876 article entitled "God's First Temples: How Shall We Preserve Our Forests," published in the Sacramento *Record Union,* Muir argued that only enforced government control could save California's finest sequoia groves from the "ravages of fools." In the early 1890s, Muir organized the Sierra Club to "explore, enjoy, and render accessible the mountain regions of the Pacific Coast" and to enlist the support of the government in preserving these areas. Muir's actions in the West convinced the federal government to restrict development in the Yosemite Valley, which preserved its beauty for generations to come.

Another thinker as well as a doer in the early conservation field was Aldo Leopold. As a philosopher, Leopold summed up his feelings in *A Sand County Almanac.*

> Wilderness is the raw material out of which man has hammered the artifact called civilization. No living man will see again the long grass prairie, where a sea of prairie flowers lapped at the stirrups of the pioneer. No living man will see again the virgin pineries of the Lake States, or

Ralph Waldo Emerson

Henry David Thoreau

John Muir

Aldo Leopold

the flatwoods of the coastal plain, or the giant hardwoods.

Leopold was the founder of the field of game management. In the 1920s, while serving in the Forest Service, he worked for the development of a wilderness policy and pioneered his concepts of game management. He wrote extensively in the *Bulletin* of the American Game Association and stated that the amount of space and the type of forage of a wildlife habitat determine the number of animals that can be supported in an area. Furthermore, he said that regulated hunting can maintain a proper balance of wildlife in a habitat.

legislation to bring frivolous suits against companies in an effort to interfere with or delay the action of a company. This action is just as unethical as that of companies who use similar delaying tactics.

Both private industry and consumers must share the burden in repairing environmental damage and maintaining environmental quality. To ensure the quality of the environment, industry must recognize the impact of its operation on society as a whole and the implications for the future of human welfare. Industry's responsibility is to find ways of producing without excessive pollution and waste.

## Societal ethics

So far, we have examined some of the ethical stances taken by individuals and corporations with respect to the environment. The collective attitudes of society can also be analyzed from an ethical point of view. Society has long believed that the earth has unlimited reserves of natural resources, an unlimited ability to assimilate wastes, and an ability to accommodate unchecked growth.

The economic direction and rationale of the United States has been one of continual growth. Unfortunately, this growth has not always been carefully planned or even desirable. This "growthmania" has caused us to use our nonrenewable resources for comfortable homes, well-equipped hospitals, convenient transportation, as well as electric toothbrushes, fast-food outlets, and battery-operated toys. In the statistics of the economic index, such "growth" measures out as "productivity," and all looks rosy. The question then arises, What is enough? There are many poor societies that have too little, but you never hear a rich society say, "Halt! We have enough." As the Indian philosopher and statesman Mahatma Gandhi said, "The earth provides enough to satisfy every person's need but not for every person's greed."

There can be no human right that is more basic than a safe life-support system, and there can be no individual right that is more fundamental than the opportunity to breathe, drink water, eat, and move about with safety. Although these rights have long been taken for granted, they are not free. They are paid for daily by the work of an artificial life-support system processing society's wastes and by-products.

Growth, expansion, and domination remain the central sociocultural objectives of most advanced societies. **Economic growth** and **resource exploitation** are attitudes shared by developing societies. As a society, we continue to consume natural resources as if the supplies were never ending. All of this is reflected in our increasingly unstable relationship with the environment, which grows out of our tendency to take from the "common good" without regard for the future. The idea is deeply embedded in the fabric of our society. Since the first settlers arrived in the United States, nature has been considered an enemy. Frequently the pioneers expressed their relation to the wilderness in military terms. They viewed nature as an enemy to be "conquered," "subdued," or "vanquished" by a pioneer "army." (See fig. 23.6.) Any qualms the

**Figure 23.6 Early pioneer attitudes.** Early pioneers considered the forests as the enemy to be conquered in order to advance civilization.

pioneers may have felt about invading and exploiting the wilderness were justified in *Genesis* 1:28, "Be fruitful, and multiply, and replenish the earth, and subdue it: and have dominion over . . . every living thing that moveth upon the earth." This passage has been interpreted by many as being a God-given sanction to exploit wilderness resources for human benefit no matter what the consequences. This attitude toward nature is still extremely popular today. Many view wilderness solely as undeveloped land and see value in land only if it is farmed, built upon, or in some way developed. The notion that land and wilderness should be preserved is incomprehensible to some. The thought of purposely opting to not develop a resource is considered almost a sin by many.

## Environmental ethics

To better understand environmental values and attitudes, we will analyze three conflicting and differing environmental positions. Although there are many different personal attitudes pertaining to the environment (see fig. 23.7), most of these would fall under one of three headings. These three positions can be labeled the development ethic, the preservation ethic, and the equilibrium ethic. Each of these ethical positions has its own appropriate code of conduct against which ecological morality may be measured.

The **development ethic** is based on action. It assumes that the human race is and should be the master of nature and that the earth and its resources exist for our benefit and pleasure. This view is reinforced by the work ethic, which dictates that humans should be busy creating continual change and that things that are bigger, better, and faster represent "progress," which is of itself good. This philosophy is strengthened by the idea that, "if it can be done, it should be done," or that our actions and energies are best harnessed in creative work.

**Figure 23.7 Three views of nature.**
Each of these individuals envisions the same resources used differently.

Examples of the development ethic abound. The notion that bigger is better is certainly not new to us, nor is the belief that if something can be done or built, it should be. The American dream of upward mobility is embodied in this developmental ethic. In some circles, questioning growth is considered almost unpatriotic. Only recently have the by-products and waste associated with development been analyzed.

The **preservation ethic** considers nature to be special in itself. The supporters of this philosophy differ in their attitudes toward the reasons for preserving nature. Some preservationists have an almost religious belief regarding nature. These individuals hold a reverence for life and respect the right of all creatures to live no matter what the social and economic costs. Preservationists also include those whose interest in nature is primarily aesthetic or recreational. They believe that nature is beautiful and refreshing and should be available for picnics, hiking, camping, fishing, or just peace and quiet. In addition to the semireligious and recreational preservationists, there are also preservationists whose reasons are essentially scientific. These individuals would argue that the human species depends upon and has much to learn from nature. Rare and endangered species and ecosystems, as well as the more common ones, must be preserved because of their known or assumed long-range practical utility. In this view, natural diversity, variety, complexity, and wildness is thought to be superior to humanized uniformity, sameness, simplicity, and domesticity.

The third environmental ethic is referred to as the **equilibrium ethic.** The equilibrium ethic is related to the scientific preservationist view just mentioned but extends its rational consideration to the entire earth and for all time. It recognizes the desirability of decent living standards, but it works toward a balance of resource use and resource availability. The equilibrium ethic stresses a balance between total development and absolute preservation. It further stresses that rapid and uncontrolled growth in population and economics is self-defeating in the long run. The goal of the equilibrium ethic is one people living together in one world indefinitely.

## Summary

Ethics seeks to define what is right and what is wrong. Morals differ from ethics because they reflect the predominant feelings in a culture about what is right and what is wrong. Most ethical questions are complex. One important ethical problem facing us is that we consider it our right to accumulate luxury items as if they were necessities.

By the nature of their activities, industries are polluters. Corporations view waste control as only one item in determining profitability. Because waste control is expensive, it has been necessary for government to require all corporations to meet the same standards. This prevents polluting companies from gaining an unfair advantage over nonpolluting ones. However, large corporations have the power to influence policy makers in government. The courts have been used by both environmentalists and industries to circumvent or delay the actions of the other.

It is the responsibility of government to set realistic standards that corporations can achieve while maintaining their profitability. Industry's responsibility is to find ways of producing without excessive pollution and waste.

We need to recognize that natural resources are finite. However, our society has been based on economic growth and resource exploitation.

## Consider This Case Study
### The fur seal

Alaska's Pribilof Islands lie 320 kilometers north of the Aleutian Islands in the Bering Sea. In 1867, these islands were acquired by the United States from Russia. St. Paul, the main island, is only 20 kilometers long and 13 kilometers wide. Even though extremely harsh and bleak, the islands are home to seven hundred Aleut natives. The islands are also the breeding grounds for over a million fur seals.

The fur seal has been the center of attention on the Pribilofs for nearly two centuries. Commerical exploitation of the fur seals for their valuable skins began in 1786 and continued until 1911, when the International Fur Seal Treaty was signed. During this period, the population of seals had dropped from about three million to under two hundred thousand. Since the treaty was signed, the population has risen to over one million. Although much has changed over the years with regard to limiting the number, age, and sex of fur seals killed, one thing remains the same. The method of killing continues to be striking a blow to the seal's head with a long hickory club. Currently, about thirty thousand bachelor seals are killed for their skins each summer.

Opposition to the Pribilof seal kill has increased in recent years, prompted in part by news coverage. Many people remain staunchly opposed to the destruction of wild animals for luxury items, such as fur coats. Others are against killing wild animals for any reason. Several organizations have called for a complete halt to all seal killing. Caught in the

There are basically three types of environmental ethics. The developmental ethic espouses extensive use of resources, with minimal concern for environmental degradation. The preservationist ethic considers nature to be special and worthy of protection for a variety of reasons. An equilibrium ethic recognizes the practical need of people to use resources and the importance of using them wisely.

## Review Questions

1. Give an example of the difference between ethical and moral statements.
2. How does personal wealth relate to ethics?
3. Why do industries pollute?
4. Why would normal economic forces work against pollution control?
5. Is it reasonable to expect a totally unpolluted environment? Why or why not?
6. What was the prominent societal attitude toward resource use?
7. Describe the differences between the developmental, preservation, and equilibrium ethics.
8. What is the major motivating force for a corporation?
9. In what ways have environmental actions been affected by action in the courts?

middle of the seal killing controversy are the seven hundred Aleut natives.

The Aleuts depend upon sealing for a major portion of their incomes. They argue that if the sealing is ended, they will be forced to rely on public assistance. In addition to the income per family, proceeds from the sale of the skins go into the Pribilof Islands Fund. The fund is used to help cover the costs of harvesting the skins, but it also helps to pay for the health, welfare, education, and general support of the Aleut people. These obligations would have to be paid out of general tax funds if the harvest stopped.

What do you think the future of the seal hunt should be?

Should the seals be spared at the possible expense of the Aleuts?

What do you think would happen if the seal population were left alone with no human intervention?

Do you think there would be any controversy at all if the mammals in question were not fur seals but rats? Why?

10. Why do decision makers view the actions of corporations differently than they view the actions of individuals?

## Suggested Readings


Barbour, Sam G., ed. *Western Man and Environmental Ethics*. Reading, Mass.: Addison-Wesley Publishing Co., 1973. A very interesting and stimulating book that brings together statements of great value in understanding American attitudes on nature and technology.

Emerson, Ralph W. *Society and Solitude*. Boston: Houghton Mifflin Company, 1883. A great nature philosopher's thoughts on the human perspective in nature.

Leopold, Aldo. *A Sand County Almanac*. New York: Sierra Club/ Ballantine Books, 1949. An established environmental classic.

Strong, Douglas H. *The Conservationists*. Reading, Mass.: Addison-Wesley Publishing Co., 1971. Short, readable sketches of the leading conservationist thinkers and doers since the beginning of the United States.

Thoreau, Henry D. *Walden*. New York: W. W. Norton & Co., 1966. A classic on nature and philosophy.

Udall, Stewart. *The Quiet Crisis*. New York: Holt, Rinehart & Winston. 1963. With an emphasis on the use of conservation, Udall (who was Secretary of the Interior under Kennedy and Johnson) reviews the conduct of Americans toward the environment since the arrival of the first settlers.

## Chapter Outline

I. Predictions and predictors
   A. Thomas Malthus
   B. The Club of Rome
   C. Technological optimists
      Box 24.1 Environmental concerns
II. Forces for change
   A. Economics
   B. Governmental action
   C. Public attitudes

## Key Terms

arithmetic rate

Club of Rome

computer modeling

geometric rate

simulations

subsidy

# Chapter 24
# Our future

## Predictions and predictors

We have dealt with major principles that guide the natural world and have an understanding of the facts that relate to these principles. We are able to evaluate predictions about the future relationship between people and their environment. Some of the forces that will shape the future are influenced directly by the actions of individuals in society, and others are totally beyond human intervention. This book will have served its purpose if you now have a better understanding of the facts and forces that will shape the future.

We do not lack predictions or predictors about the future. The wide range of predictions can be put on a continuum. Some people see no

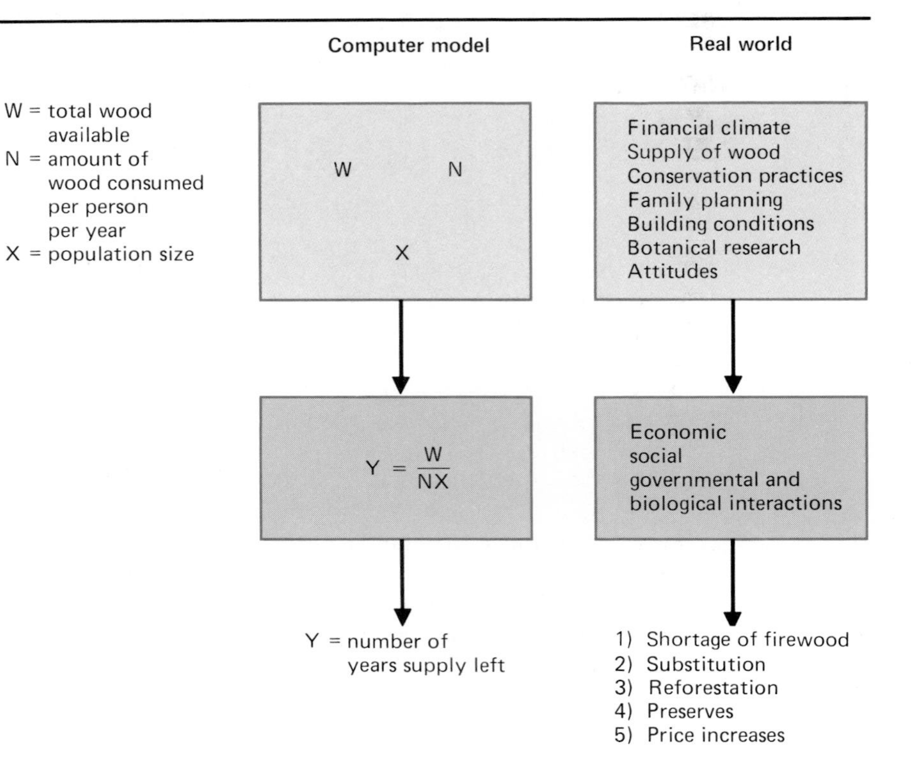


**Figure 24.1 Computer modeling.** The example on the left is a computer simulation, a model of how long a supply of firewood will last. With the simple data fed into the computer and the relationships the computer uses, the outcome is reasonable. If the data is wrong or the relationships are not accurate, the prediction will be faulty. The real-world example shows how complex the issue really is. Many more parameters influence the supply of firewood than those used in the computer model.

problems for the future; others see no hope. Because of the wide variety of predictions, it is helpful to be able to evaluate their validity. When evaluating the validity of predictions, the following should be taken into consideration: the purposes of the people making the prediction, the kinds of data used by the predictors, and whether or not the prediction is a logical extension of the facts. Some predictions can be dismissed because they are mere speculation and cannot be backed up by any logic or fact. Other predictions are generated directly from facts.

Various techniques have been used to generate predictions. A technique commonly used is **computer modeling** and **simulation.** A computer model attempts to establish relationships between facts in ways that are similar to those in the real world. Once these relationships are described, the computer can extend these relationships into the future, and statements can be made about the future.

Although computer modeling is used frequently, most people do not really understand the process of computer modeling. Unless the basic assumptions used to generate the computer model are well understood, critical evaluation of predictions is difficult. This doesn't mean that computer simulations are useless. What it does mean is that you must understand the parameters put into the computer in order to evaluate the significance of the data coming out in the form of predictions. (See fig. 24.1.) The real benefit of using the computer is the speed with which it can generate data based on the relationships

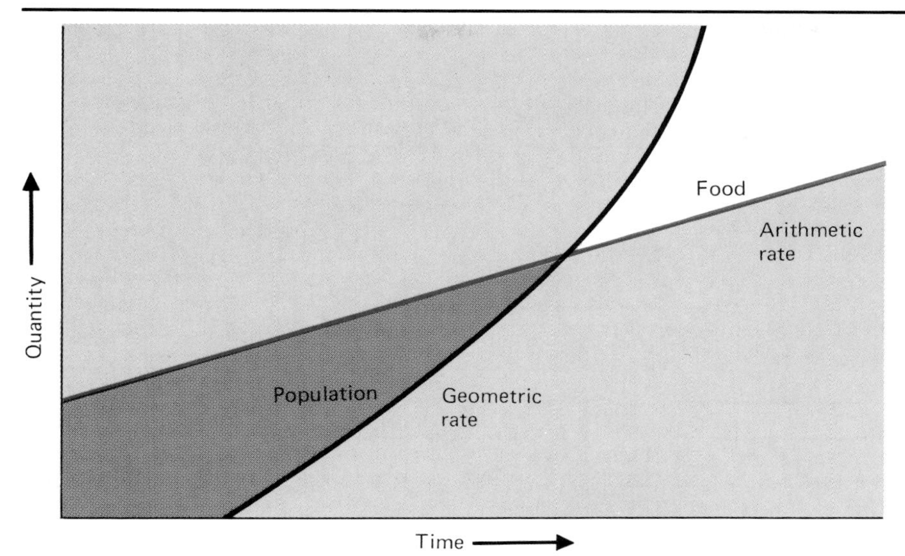


**Figure 24.2 Malthus's prediction.**
Malthus predicted that the world would run out of food, because it looked to him that the rate at which food production increased was a simple arithmetic one, while the growth of the human population was geometric. If the population grows faster than food production, the point will eventually be reached where there will not be enough food to support the human population.

that are fed into it. However, a computer prediction is only as good as the information fed into it. The variables can be manipulated to reflect the range of numerical values considered reasonable by the simulators. Although the computer is a valuable tool, it is not necessary; individuals have made predictions for centuries without the aid of computers.

## *Thomas Malthus*

One of the earliest predictions related to human welfare was made by Thomas Malthus in 1797. He predicted that populations would grow faster than the ability to feed them. He based his prediction on two basic assumptions: that population grew at a **geometric rate,** while food production increased at an **arithmetic rate.** (See fig. 24.2.) Malthus concluded that if these two assumptions were correct, the ultimate result of these two interacting forces would be starvation, disease, and war. His predictions were hotly debated by the intellectual community of his day. His assumptions and conclusions were attacked as being erroneous and against the best interest of society.

When we look at his predictions two centuries later, we find that they have some validity. People are overpopulating the world, and food is scarce for many. Starvation is a real problem. However, he predicted that the problems would occur much earlier than they actually did. His assumptions did not include the possibility of changing birthrates, changing agricultural technology, and the colonization of new lands in the Americas. His purpose in making the predictions was to alert the general public to the dangers of overpopulation. It seems that he used all of the facts at his disposal and related them to one another in a logical fashion. Other predictors have reached similar conclusions with more elaborate models and more data. The Club of Rome is one of these.

**Figure 24.3 Club of Rome predictions.** The predictions of the Club of Rome are based on a computer simulation that integrates five major parameters.

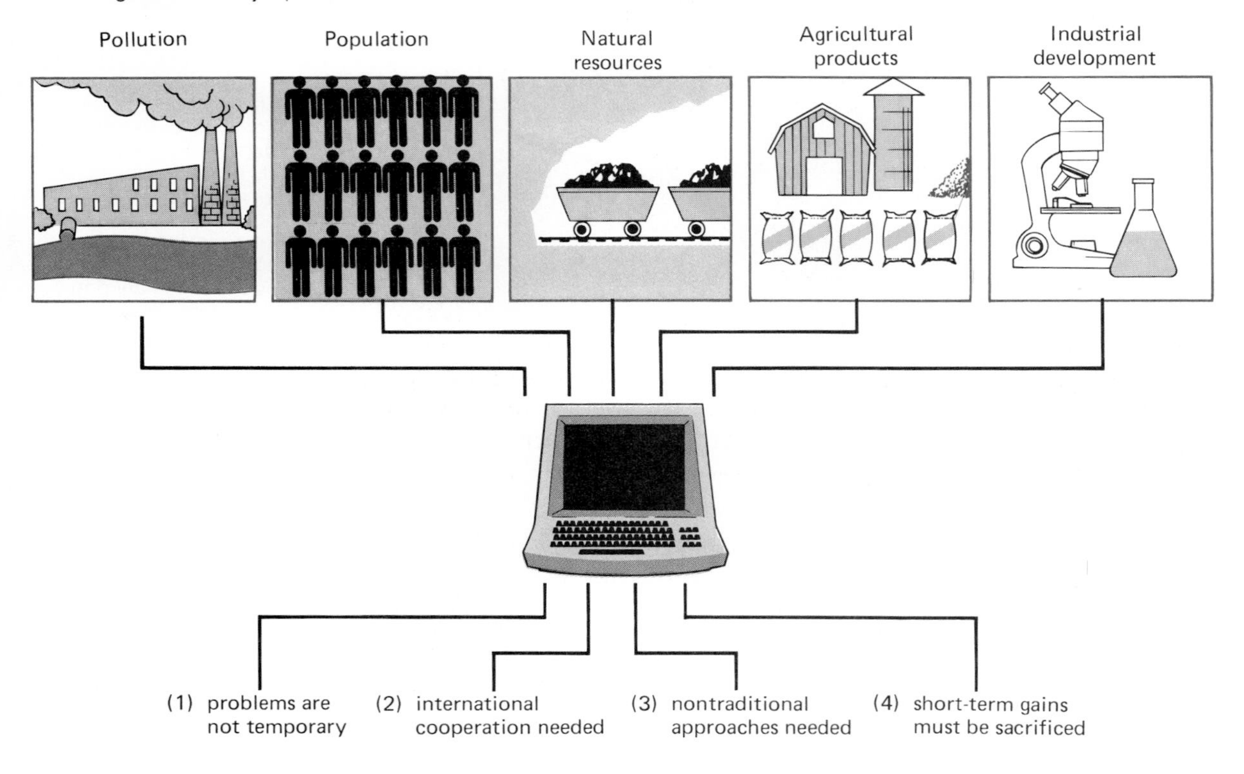


## *The Club of Rome*

The **Club of Rome** is an informal assemblage of renowned individuals from all parts of the world and many fields of study. Their initial meeting in 1970 was held to discuss the "predicament of mankind." A group from the Club of Rome decided there are five critical parameters that influence social and economic growth. These are population size, agricultural production, natural resource availability, industrial production, and environmental pollution. *At this point, you should be able to appreciate why these five items were considered to be of primary importance.* Following the identification of these parameters, computer simulation was used to manipulate relative values of each of these to see how the parameters could affect the welfare of people throughout the world. (See fig. 24.3)

Four interrelated conclusions were reached after analyzing the information generated by the computer. Significant changes in the way people and countries act and interact are needed to prevent disaster. There are four critical areas of attitude change:

1. We must understand that our current problems are not temporary and seem to be getting worse.
2. We must understand that it will take international cooperation on social, scientific, and economic levels to initiate a change for the better.
3. We must understand that the solutions will require nontraditional approaches integrating the knowledge from all aspects of human endeavor.
4. We must understand that short-term gains must be sacrificed for the long-term betterment of all.

### *Technological optimists*

Just as Malthus was criticized by those who felt that problems were temporary and easily solved, the Club of Rome report and other similar predictions have been criticized by technological optimists. These people look at history and maintain that technological changes have always solved our problems. They also maintain that science will continue to bring solutions to our current and future problems. This position simply avoids looking at problems. To have blind faith that technology will always be able to provide solutions to the changes humans inflict on the world overlooks the fact that we can change the world in ways that were not possible a century ago. Nuclear weapons can destroy the world. We are using all of our best agricultural land. The consequences of modern pollutants are much different from the smoke of a cooking fire. In addition, the technological optimists usually take a short-sighted approach to problem solving. A common attitude is "if we take care of the little problems as they come up, the future will take care of itself."

Since 1940, nuclear power has been hailed as the answer to our problems of energy, transportation, national defense, pollution, and economic well-being. Today this enthusiasm has been dulled by the realities of nuclear accidents, decreasing availability of suitable fuels, the costs of construction and operation, and the problems associated with the containment of nuclear wastes.

A current panacea is "space-age technology." Outer space has been suggested as a cure for our environmental problems. We can shoot our dangerous wastes into an endless universe, where they can be forgotten. We can trap sunlight in space and return it to earth to solve our energy shortage. We can send excess people to space stations. We can even mine the moon and grow food in space colonies. This line of thinking makes many people uncomfortable because the solutions are based on nonexistent technology, which may or may not come into being.

It should be obvious that there is a wide range of opinions relating to our future. Two extreme viewpoints are called the Doomsday Philosophy and the Pie in the Sky attitude. Many people feel that these two extreme viewpoints are only helpful in stating the problems, not in solving them.

## Box 24.1
## Environmental concerns

The Conservation Foundation evaluated six recent international studies that attempt to rank the major environmental threats facing the world today. The foundation placed these forty-seven issues into four broad categories:

1. War, accident and natural disasters,
2. Population growth and distribution,
3. Contaminants, including those that are physical, chemical, or biological in nature, and
4. Natural resource depletion.

The criteria used to establish the rank of these threats relate to the severity and controllability of the threat and the irreversibility of its effects. The value of the study and the ranking system is that it allows the reader to gauge the importance of local as well as national and international issues. It is interesting that the less significant issues seem to occupy more of our time than those of much greater significance. However, you must realize that individual action on the smaller problems is more fruitful than attempts to change the course of international events. Individual action on any of these issues is necessary and has been proven effective. The study ranks these forty-seven issues as follows.

*Forty-seven environmental issues*

War
Nuclear accidents, terrorism
Chemical plant explosions
Failure of aging infrastructure
Intentional weather modification (unintentional effects)
Drought
Floods
Earthquakes, volcanoes, and other natural disasters
Population growth
Crowding and impacts of urbanization
Sprawl problems
Mass migration, immigration
Radioactive waste disposal
Debris from space
Microwave radiation
Electronic pollution
Solid waste disposal
Noise
Pathogens from human waste
Proliferation of biological organisms, bioengineering wastes and mistakes
Mutagens
Carbon dioxide accumulation in the atmosphere
Acid deposition
Depletion of the ozone layer
Hazardous waste management
Conventional pollutants, ambient air
Toxic pollutants in air
Indoor air pollution
Conventional pollutants in water, from point sources
Nonpoint source water pollution
Toxic pollutants in surface water
Groundwater, drinking water contamination
Pesticides
Chemical fertilizers
Chemicals in food chains
Water scarcity
Loss of agricultural land due to salinization, desertification, or urbanization
Soil erosion and overexploitation of agricultural soils
Ocean fisheries depletion
Plant and animal species loss
Energy scarcity

Critical materials scarcity
Damage to the marine environment
Loss of tropical forests
Coastal area degradation
Loss of wetlands
Wilderness and wild and scenic rivers degradation

*Criteria for evaluating the gravity of environmental problems*

Severity of the effects associated with the threat

A. *Severity of effects.* The seriousness of the effects to people/environment exposed to them, assuming a worst-case situation.
B. *Extent of population/environment likely to be affected.* The number of people or amount of area likely to be affected if threat occurs.
C. *Geographic extent.* The amount of area likely to be threatened by several distinct manifestations or occurrences of the threat.
D. *Special populations or areas affected.* Specific population subgroups or unique natural areas at particularly high risk.
E. *Nature of effects.* Types of effects that the threat might create—for example, damage to amenities, ecological stability, human economic and social welfare, or human health.
F. *Certainty of effects.* The extent of scientific evidence that effects of concern will in fact occur if threat is manifested.
G. *Indirect effects.* Whether the threat and its effects are likely to generate indirect additional threats and effects that are of significant concern.
H. *Benefits associated with threat.* The extent to which society benefits from the activities that may create the threat.

*Possibilities of avoiding the threat*

I. *Immediacy of threat.* The extent to which this is a problem that is being faced now or that may only appear in the long term.
J. *Probability of threat occurring.* The likelihood that the threat which would cause the effects of concern will actually occur, assuming the current situation or trend or if no new action is taken.
K. *Controllability of threat—technological.* The extent to which physical or technological methods for controlling the threat are currently known and can be technologically adopted.
L. *Controllability of threat—political, social, economic.* The extent to which high costs and social or political constraints may interfere with adoption of controls if the technology is available.

*Possibilities of reversing or mitigating the effects*

M. *Irreversibility of effects—physical.* The ability to physically reverse or undo the effects if they occur.
N. *Irreversibility of effects—economic, political, social.* The feasibility of reversing effects if there are technological means available.
O. *Adequacy of existing institutions.* The ability of existing institutions to effectively implement any programs needed to control or respond to the threat.

**Figure 24.4 City farmers.**
Many cities have encouraged residents to make use of vacant lots as a place to grow some of their food. Not only do these people get food for their effort, but they derive pleasure from the activity. This picture is of the garden display at the Second Annual City of Detroit Harvest Festival.

## Forces for change

The severity and effects of current and future environmental problems will be determined by several variables, including economics, governmental action and public attitudes.

### *Economics*

If there is a demand for a product or service, the price will reflect the strength of the demand and the availability of the commodity. When demand exceeds supply, the price rises. The cost increase causes people to seek alternatives or to decide not to use the product or service. This results in a lower demand. Currently the prices for petroleum products are rising because the supply is nearly fixed, but the demand remains high. As a result, people are making adjustments by buying more energy-efficient automobiles, insulating their homes, lowering their thermostats, and using public transportation more often.

Food production depends on petroleum for the energy to plant, harvest, and transport food crops. In addition, petrochemicals are used for fertilizers and the production of chemical pest control agents. The price of food must rise as the price of petroleum rises. We will use substitutes for expensive sources of food, and it will become profitable to grow our own food in small garden plots. Seed companies are doing a very good business, and many cities have made vacant lots available to local residents to raise gardens. (See fig. 24.4.) For many, gardening is both a method of saving money on food and a source of recreation.

| | Initial outlay in $ | Cost of use/hour |
|---|---|---|
| 1. Jogging | $50-100 | $0 |
| 2. Skiing, cross country | 200-300 | 0 |
| 3. Golfing | 400-500 | 5-25 |
| 4. Hunting | 100-500 | 1-5 |
| 5. Snowmobiling | 1000-5000 | 1.50-2.00 |

**Figure 24.5 Recreational costs.** The cost of recreation varies greatly. Some sports include both a high initial cost and a high operating cost; others have a low initial cost and almost no operation cost.

Because of economic pressure, recreation is becoming more locally oriented. Cities are developing parks and recreation facilities to be used by their inhabitants. This is necessary because people are less willing to leave the city to satisfy their recreational needs. Jogging, biking, community sports programs, reading, and home entertainment are all becoming more popular. This is partly because recreational costs are becoming higher, and people are seeking lower cost alternatives. (See fig. 24.5.) As inflation decreases purchasing power, less costly alternatives will be considered for all activities.

## *Governmental action*

Government is often blamed for inflation, and government rules and regulations are blamed for adding to the cost of production. Pollution controls on industries as well as on their products have resulted in somewhat higher prices. Coal miners have much better working conditions today, but their productivity has declined significantly. This means that coal has become more expensive as a direct result of governmental intervention.

Another way in which government influences our future is through governmental subsidies. Agriculture, transportation, space technology, and communication have all been subsidized for years. A **subsidy** is a gift from government to a private enterprise that is important to the public but is having temporary economic difficulty. The money that government grants as a subsidy is collected from the public through taxes. Subsidies increase costs of products on the short term but are considered important because of the long-term benefits.

Subsidies are costly in two ways. The bureaucracy necessary to administer a subsidy costs money. In addition, the subsidy is an indirect way of keeping the market price of a product low. (The actual cost is higher because government gifts must be added to the market price.) In many cases, subsidies are indirect gifts rather than direct grants and are not called subsidies. For example, the monies that are granted to states to build new highways are really subsidies for the automobile industry.

An area in which we should expect increased government subsidy is in the area of public transportation. Current subsidies are rather small, but as more people begin to understand that public transportation is more energy efficient than individual automobile transportation, we will begin to encourage mass transportation industries.

## *Public attitudes*

Attitudes toward government intervention have been changing. Today we accept such government intervention as a matter of course. For instance, prior to the fifty-five miles per hour speed limit mandated by executive order in 1973, legal speed limits were much higher. Today, we might not always drive at exactly fifty-five miles per hour, but we drive considerably slower than the seventy or eighty miles per hour common in the early 1970s. Our attitudes have changed.

There are several other areas of life in which attitudes have changed significantly in the last few years. Many people now exhibit a willingness to accept smaller cars, a preference for smaller families, greater acceptance of all forms of family planning (including abortion), and an increasing awareness and appreciation of natural areas and open space. In addition, the majority of society is demanding that industry control its pollution.

Information is necessary to change attitudes. We receive information formally and informally. Informal information from newspapers, magazines, TV, and radio can change attitudes. In fact, some media productions are specifically designed to do so. Advertising and editorial comments shape people's feelings about issues. Today many environmental organizations publish their points of view in an attempt to shape public opinion. A few years ago, most publications were biased in favor of business and industry, but now both points of view are readily available. This gives people the full spectrum of information concerning environmental issues, which facilitates attitude changes.

Formal education can also be a factor that changes attitudes. Since 1970, environmental education has been incorporated into the curricula of many school systems.

One of the most effective forces for changing attitudes (or reinforcing them) is peer pressure. Few want to be left out or be nonconforming. Pressure from peers was partly responsible for the initiation of the Environmental Protection Agency. People were encouraged to become politically active and to participate in marches, write letters, and demonstrate. Without such public participation, politicians probably would not have responded.

Informed citizen opinion will be more important in shaping our future than the action of courts, business, industry, or politicians. The single most important action you can take is to become informed so you can become effectively involved in influencing decisions.

## Suggested Readings


Birch, L. *Confronting the Future.* New York: Penguin Books, 1975. One of the more optimistic views of the future.

Kahn, Herman, et al. *The Next 200 Years: A Scenario for America and the World.* New York: William Morrow & Co., 1976. An interesting but somewhat oversimplified model of our technological future.

Meadows, Donella H., et al. *The Limits to Growth.* New York: Universe Books, 1972. A controversial and popular description of the second world model sponsored by the Club of Rome.

Oltmans, William L. *On Growth.* New York: G. P. Putnam's Sons, 1974. Collection of interviews of seventy of the world's great thinkers in many disciplines on the debate over the *Limits to Growth.*

Stavrianos, L. S. *The Promise of the Coming Dark Age.* San Francisco: W. H. Freeman Company, 1976. An optimistic view of the future in which the author describes "a vision of the transcendence of homo sapiens to homo humanus—of humankind at last realizing its humanity."

U.S. Executive Office of the President, Council on Environmental Quality. *Annual Environmental Quality Report.* An annually published report that presents the current status of all major environmental problems.

# Appendix 1
# Active environmental organizations

Alliance for Environmental Education, Inc.
1619 Massachusetts Avenue, N.W.
Washington, DC 20036
(Works to further formal and informal education activities at all levels.)

Common Cause
2030 M Street, N.W.
Washington, DC 20036
(An active citizens lobby that covers a broad range of political issues.)

The Conservation Foundation
1717 Massachusetts Avenue, N.W.
Washington, DC 20036
(Nonprofit conservation research and educational organization.)

Ducks Unlimited
P.O. Box 66300
Chicago, IL 60666
(Primarily dedicated to protection of migrating waterfowl through acquisition of vital breeding habitats.)

Environmental Defense Fund, Inc.
475 Park Avenue, S.
New York, NY 10016
(Works to link laws and science in defense of the environment before courts and regulatory agencies.)

Environmental Policy Center
317 Pennsylvania Avenue, S.E.
Washington, DC 20003
(Lobbying for all aspects of energy development.)

Friends of the Earth
124 Spear Street
San Francisco, CA 94105
(Active organization committed to the preservation, restoration, and wise use of the earth.)

The Izaak Walton League of America
1800 North Kent Street
Suite 806
Arlington, VA 22209
(Promotes conservation of renewable natural resources and development of outdoor recreation.)

League of Women Voters of the United States
1730 M Street, N.W.
Washington, DC 20003
(Works for political responsibility through an informed and active citizenry.)

National Audubon Society
950 Third Avenue
New York, NY 10011
(Operates over forty wildlife sanctuaries across the United States and provides a wide array of environmental education services.)

National Parks and Conservation Association
1701 Eighteenth Street, N.W.
Washington, DC 20009
(Works for acquisition and protection of public parklands.)

National Wildlife Federation
1412 Sixteenth Street, N.W.
Washington, DC 20036
(Promotes citizen and governmental action for conservation.)

The Nature Conservancy
Suite 800
1800 North Kent Street
Arlington, VA 22209
(A membership organization dedicated to protecting natural areas.)

Population Reference Bureau
1337 Connecticut Avenue, N.W.
Washington, DC 20036
(Clearinghouse for data concerning the effects of worldwide population explosion.)

Sierra Club
530 Bush Street
San Francisco, CA 94108
(Works to protect and conserve the world's natural resources.)

Smithsonian Institution
1000 Jefferson Drive, S.W.
Washington, DC 20560
(Promotes environmental education through a wide variety of programs.)

Soil Conservation Society of America
7515 N.E. Ankeny Road
Ankeny, IA 50021
(Professional society dedicated to soil and water conservation.)

Student Conservation Association, Inc.
Box 550
Charlestown, NH 03603
(Helps provide opportunities for conservation-minded students to work and learn during summer vacations.)

The Wilderness Society
1901 Pennsylvania Avenue, N.W.
Washington, DC 20006
(An organization dedicated to protection of wild lands and acquisition of additional wilderness by the federal government.)

Wildlife Society
Suite 611
7101 Wisconsin Avenue, N.W.
Washington, DC 20014
(Professional society for the preservation of wildlife.)

Worldwatch Institute
1776 Massachusetts Avenue, N.W.
Washington, DC 20036
(Research and education on major worldwide environmental problems.)

World Wildlife Fund
1319 Eighteenth Street, N.W.
Washington, DC 20036
(Research and education on endangered species.)

Zero Population Growth
1346 Connecticut Avenue, N.W.
Washington, DC 20036
(Primary emphasis is on family planning, population stabilization, and voluntary sterilization.)

For a more detailed list of national, state, and local environmental organizations, see *Conservation Directory,* published annually by the National Wildlife Federation, 1412 Sixteenth Street, N.W., Washington, DC 20036.

For international organizations, see Thaddeus C. Trzyna and Eugene V. Coan, eds., *World Directory of Environmental Organizations,* San Francisco: Sequoia Institute, 1976.

# Appendix 2
# Metric unit conversion table

Metric unit conversion table

| | | |
|---|---|---|
| Kilocalorie | 1,000 calories | 3.97 BTU |
| calorie | | 0.00397 BTU |
| Kilometer | 1,000 meters | 0.621 miles |
| Meter | | 39.4 inches |
| Centimeter | 0.01 meter | 0.394 inches |
| Millimeter | 0.001 meter | 0.0394 inches |
| Hectare | 1,000 ares | 2.471 acres |
| Liter | | 1.06 quart |
| Milliliter | 0.001 liter | 0.00106 quart |
| Metric ton | 1,000 kilograms | 2,200 pounds |
| Kilogram | 1,000 grams | 2.205 pounds |
| Gram | | 0.035 ounces |
| Milligram | 0.001 gram | 0.000035 ounces |

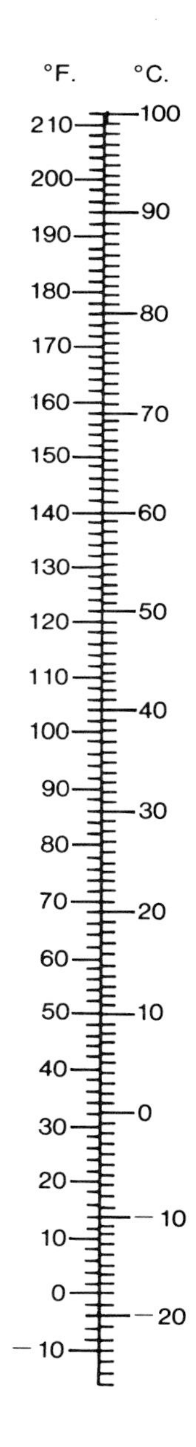

# GLOSSARY

**abiotic** Nonliving.

**acid deposition** The accumulation of potential acid forming particles on a surface.

**acid mine drainage** A kind of pollution, associated with coal mines, in which bacteria form sulfuric acid that enters local streams.

**activated sludge** Material removed from sewage that is reintroduced into the waste-water stream to promote decay of organic matter.

**activation energy** The initial energy input required to start a reaction.

**active solar heat system** Traps sunlight energy in one place and with the use of mechanical devices moves it to a desired location.

**administrative law** Regulations established by governmental agencies.

**advertising** A method of calling attention to a product in an effort to encourage its purchase.

**age distribution** The comparative percentages of different age groups within a population.

**agricultural runoff** Surface water that drains from a field and contains both agricultural chemicals and soil particles.

**algal bloom** The rapid accumulation of microscopic green organisms (algae).

**applied science** Deals with the use of information gained through science to influence the human condition.

**aquifer** A porous, subsurface layer that accumulates water.

**arithmetic rate** Growth by continual adding.

**atom** Basic subunit of elements.

**atomic fission** The disintegration of the nucleus of an atom that releases atomic particles and large amounts of energy.

**auxin** A plant growth regulator.

**benefit-cost analysis** A study of the benefits of a project weighed against the costs of the proposed project.

**biochemical oxygen demand (BOD)** The amount of oxygen required to decay organic molecules in aquatic ecosystems.

**biocide** A chemical control agent that kills many forms of life.

**biodegradable** Able to be broken down by natural biological processes.

**biological amplification** Concentration of a chemical as it passes through a food chain.

**biological control** A method of pest control that uses natural diseases, predators of the pest, or particular pest species characteristics.

**biomass** Weight of material comprising organisms.

**biome** Major regional climax community.

**biotic environment** The living surroundings of an organism that help to determine its survival.

**biotic factors** Living portions of the environment.

**biotic potential** The inherent reproductive capacity.

**birthrate** The number of individuals added to the population through reproduction per thousand individuals per year.

**black lung disease** A respiratory condition resulting from the accumulation of large amounts of fine coal dust particles in miners' lungs.

**bureaucracy** A method of administration in which appointed officials manage departments according to inflexible and routine regulations and procedures.

**carbamates** Nerve poisons used as pesticide.

**carbon cycle** The cyclic flow of carbon from the atmosphere to living organisms and back to the atmospheric reservoir.

**carbon dioxide** A compound consisting of carbon and oxygen, which is a normal component of the atmosphere.

**carbon monoxide (CO)** A primary air pollutant, containing carbon and oxygen, that results from incomplete combustion of organic materials.

**carcinogenic** Cancer causing.

**carnivores** Animals that eat other animals.

**carrying capacity** The optimum number of individuals of a species that can be supported in an area over an extended period of time.

**catalyst** A substance that alters the rate of a reaction but is not itself changed.

**cellular respiration** The process that organisms use to release chemical bond energy from food.

**chain reaction** A nuclear reaction in which the products of the disintegration of one nucleus cause the disintegration of other nuclei, which in turn cause further disintegrations.

**chemical bond** The physical attraction between atoms that results from the interaction of their electrons.

**chlorinated hydrocarbon** A group of widely used pesticides containing carbon, hydrogen, and chlorine.

**clear-cutting** A forest harvest practice by which all of the trees in an area are removed.

**climax community** A relatively stable, long-lasting interrelated group of plants and animals in an area.

**Club of Rome** An informal assemblage of renowned individuals from all parts of the world and many fields of study who have used computer simulations to predict the future.

**coal** Solid fossil organic material, which is used as a source of energy.

**combustion** The process of releasing chemical bond energy from fuel.

**community** The interacting groups of different species.

**competition** An interaction between two organisms in which both require the same limited resource.

**compost system** A waste disposal system whereby organic matter is allowed to decay to a usable product.

**computer modeling** Using mathematical statements in an attempt to establish relationships between facts in ways that are similar to those in the real world.

**conference committee** Congressional committee designed to resolve differences between a senate and house version of a bill.

**conservation** The wise use of a commodity so it maintains a maximum sustained yield or a continuous supply of a resource.

**consumers** Organisms that rely on other organisms for food.

**contour farming** Tilling across a slope, thus preventing fast water runoff.

**controlled experiment** Compares two situations that differ in only one fundamental way.

**corporation** A business structure that has a particular legal status.

**cottage industry** A decentralized system of manufacturing that relies on human labor performed in the home.

**cover** A place where an organism can go to escape predators or to be sheltered from the elements.

**crop rotation** Farming practice that involves planting the same field with a series of different crops.

**death phase** The portion of the population growth curve that shows the population declining.

**deathrate** The number of deaths per thousand individuals per year.

**deciduous trees** Trees that lose their leaves each fall.

**decommissioning costs** The cost to stop using a facility and clean up the site.

**decomposer** Microconsumers that cause decay of dead organic matter and recycle nutrients.

**decreasing energy growth pattern** Trend toward using less energy, which necessitates changes in life-style.

**democracy** Governmental system based on the consent and will of the people.

**demographic transition** The hypothesis that economics proceed through a series of stages resulting in stable populations and high economic development.

**demography** The study of human populations, their characteristics, and changes.

**denitrifying bacteria** Bacteria that convert soil nitrates into nitrogen gas.

**desert** Biome characterized by very low rainfall and sparse vegetation.

**detritus** Organic material that results from the decomposition of plants, animals, or fecal waste material.

**developed countries** The countries with high GNPs—economically rich countries.

**development ethic** Philosophy that states that the human race should be the master of nature and that the earth and its resources exist for human benefit and pleasure.

**dispersal** Migration of organisms from a concentrated population into areas with lower population densities.

**domestic water** Water diverted for household uses.

**dominant species** A key organism in a food chain. One that has significant impact on the area in which it lives.

**ecological niche** The total role an organism plays in a habitat.

**ecological succession** The process of one natural community replacing another.

**ecology** Branch of science that deals with the interrelationship between organisms and their environment.

**economic costs** Monetary expenditures required to exploit a natural resource.

**economic growth** The perceived increase in monetary growth within a society.

**ecosystem** Groups of interacting species combined with the physical environment.

**ectoparasite** A parasite that is adapted to live on the outside of its host.

**electrons** The lightweight, negatively charged particles that move around the nucleus of an atom.

**element** A form of matter consisting of a specific kind of atom.

**emigration** Movement out of an area that was one's place of residence.

**endoparasite** A parasite that is adapted to live within a host.

**energy costs** Expenditures of energy that are required to exploit a natural resource.

**environment** Everything that affects an organism during its lifetime.

**environmental costs** Perceived degradation of the environment resulting from the exploitation of a natural resource.

**environmental impact statements** Required assessment of the potential effects of changing the use of an area.

**Environmental Protection Agency** United States government organization responsible for the establishment and enforcement of regulations concerning the environment.

**environmental quality** Perceived status of natural surroundings.

**environmental resistance** The combination of all environmental influences that tend to keep populations stable.

**environmental science** An interdisciplinary area of study that includes both applied and theoretical aspects of human impact on the world.

**equilibrium ethic** Philosophy that espouses a balance between total development and absolute preservation.

**erosion** The wearing away and transportation of soil by water or wind.

**ethics** A discipline that seeks to define what is fundamentally right and wrong.

**eutrophication** The enrichment of water with nutrients.

**executive branch** The office of the presidency of the United States.

**experiment** An artificial situation designed to test the validity of a hypothesis.

**exploitive use** Philosophy that supports the belief that enhancing human comfort should be the sole consideration when determining how we should use nature.

**exponential growth** The period during population growth when the population increases at an ever-increasing rate.

**extinction** The disappearance of an entire interbreeding species.

**factory system** A centralized method of manufacturing.

**federalism** A governmental system in which power is shared between national and regional units.

**first law of thermodynamics** A statement about energy that says that under normal physical conditions, energy is neither created nor destroyed.

**fission** The decomposition of the nucleus of an atom.

**floodplain** Lowland on either side of a river that periodically floods.

**floodplain zoning ordinances** Governmental regulations designed to restrict the use of areas that are periodically covered with water.

**fly ash** Particles of partially burned materials that are present in smoke.

**food chain** The series of organisms involved in the passage of energy from one trophic level to the next.

**food web** Intersecting and overlapping food chains.

**fossil fuels** Organic compounds derived from chemical modification of ancient plant or animal remains.

**friability** The ability of a soil to crumble, due to its texture, structure, and moisture content.

**fungicide** Pesticide that controls undesirable fungus (mold, mildew, rust, and smut).

**fusion** The union of smaller atomic nuclei to form a heavier atomic nucleus.

**gasohol** An automotive fuel that is a mixture of 10 percent alcohol and 90 percent gasoline.

**geometric rate** Growth by continual doubling.

**geothermal energy** The heat energy from the molten core of the earth that converts water to steam.

**government regulation** Rules, or standards of behavior, established by branches of government.

**gradual extinction** The replacement of one group of organisms by their evolutionary descendants.

**grassland** A relatively dry, treeless biome.

**grass-roots politics** Organized citizen action at the local level.

**greenhouse effect** The accumulation of heat because certain substances such as glass and atmospheric $CO_2$ are transparent to light but less transparent to heat.

**gross national product (GNP)** An index that measures the total goods and services generated annually within a country.

**groundwater** Water that infiltrates the soil and may be stored in ground reservoirs.

**groundwater mining** Removal of water from an aquifer that is in excess of the water's natural replacement.

**habitat** An identifiable region in which a particular kind of organism lives.

**habitat management** Modification of an area to enhance its ability to support particular species of wildlife.

**hard pesticides** A persistent pesticide.

**hazardous waste** Substances that could threaten life if they are released into the environment.

**heavy oil** A type of fossil fuel with the consistency of asphalt.

**hectare** Metric unit of a thousand ares (an are is a hundred square meters), equal to 2.471 acres.

**herbicide** A pesticide that controls undesirable vegetation.

**herbivores** Primary consumers. Animals that eat plants.

**historical energy growth pattern** Trend in energy growth that suggests continual increases in energy demand.

**host** The organism a parasite uses for its source of food.

**humus** The soil component that consists of dead organic material.

**hydrocarbons** A group of organic compounds consisting of carbon and hydrogen atoms.

**hydrologic cycle** The continual recycling of water from surface water and oceans to atmospheric water vapor to precipitation, powered by energy from the sun.

**hypothesis** A logical guess that explains an event or answers a question.

**immigration** Movement into an area in which one has not previously resided.

**Industrial Revolution** A period of time during which machinery replaced human labor.

**industrial uses** Employment of water for manufacturing purposes.

**infiltration** Water absorbed into the soil.

**in-migration** The introduction of new individuals into an area.

**insecticide** Pesticides that control undesirable insects.

**instream uses** Use of water without removing it from its natural channel or basin.

**integrated pest management** A form of control of unwanted organisms that is harmonious with natural ecosystems because control methods are specifically selected based on the characteristics of the target organisms.

**interspecific competition** Competition between members of different species for a limited resource.

**intraspecific competition** The competition among members of the same species for a limited resource.

**irrigation** Water diverted for agricultural production.

**isotope** Atoms of the same element that have different numbers of neutrons.

**judicial branch** That portion of the government that includes the court system.

**judicial review** Power of the United States supreme court to review the acts and actions of the United States government.

**kerogen** The tarlike organic substance found dispersed in tar sands or oil shales.

**kinetic energy** Energy contained by moving objects.

**lag phase** The initial stage of population growth during which growth occurs very slowly.

**land** The part of the world that is not covered by the oceans.

**land-use planning** The construction of an orderly list of priorities for the use of land.

**leaching** The process of transporting soluble materials downward through the soil.

**legislative branch** That portion of the government that is responsible for the development of laws.

**less-developed countries** The countries with low GNPs—economically poor countries.

**limiting factors** The one condition of the environment that determines the success of an organism.

**liquid metal fast breeder reactor** A nuclear fission reactor that uses liquid metal as a primary heat transfer material.

**liquified natural gas** Natural gas that has been converted to a liquid by cooling to −162°C.

**loam** A kind of soil with excellent nutrient retention, good drainage, and aeration properties because of its mixture of soil particles.

**lobbying** Attempts to influence action of government.

**materialistic society** A society that places high value on possessions.

**matter** Substance with measurable mass and volume.

**megalopolis** Large regional urban centers.

**migratory** Organisms that travel seasonally, usually along traditional routes.

**mine tailings** Crushed rock and other wastes from the mining process.

**moderator** Water that absorbs energy in nuclear fission reactors.

**molecule** Two or more atoms combined to form a stable unit.

**monoculture** A practice of planting large areas in a single crop.

**morals** Predominant feeling of a culture about ethical issues.

**mortality** The number of deaths per thousand individuals per year.

**multiple land use** The use of an area for two or more compatible purposes.

**mutualism** The association between organisms in which both benefit.

**natality** The number of individuals added to the population through reproduction per thousand individuals per year.

**natural gas** Volatile hydrocarbon formed by the decomposition of ancient organisms.

**natural resources** Those structures and processes that can be used by humans but cannot be created by humans.

**natural selection** A process that determines which individuals within a species will reproduce most effectively.

**natural sinks** Areas that serve as storage reservoirs of unwanted materials.

**neutron** Neutrally charged particle located in the nucleus of an atom.

**niche** The total role an organism plays in a habitat.

**nitrogen cycle** The series of stages in the flow of nitrogen in ecosystems.

**nonpersistent pollutant** Those pollutants that do not remain in the environment for long periods or are biodegradable.

**nonpoint source** Pollutants that come from a broad locale, which makes the specific polluter difficult to identify.

**nonrenewable resources** Those resources that are not replaced by natural processes within a reasonable time span.

**northern coniferous forest** The biome characterized by low temperature. Forests with evergreens as the dominant plants.

**nuclear breeder reactor** Nuclear fission reactor designed to produce radioactive fuel from nonradioactive uranium and at the same time release energy to use in the generation of electricity.

**nuclear reactor** Site where energy is released by fission of radioactive isotopes.

**nucleus** The central region of an atom that contains protons and neutrons.

**oil** Liquid hydrocarbon formed by the decomposition of ancient organisms.

**oil shale** Deposits of shale rock containing dispersed oil droplets.

**omnivore** An organism that eats both plants and animals.

**organophosphates** Chemical compound that disrupts the nervous system of an organism.

**outdoor recreation** The use of natural out-of-doors for leisure time activities.

**out-migration** Movement of organisms from an area.

**overburden** The layer of soil and rock that covers deposits of desirable minerals.

**oxides of nitrogen** Compounds (such as $NO_2$ and $NO_3$) composed of nitrogen and oxygen.

**ozone** A molecule consisting of three atoms of oxygen.

**parasite** An organism adapted to survival by using another living organism (host) for nourishment.

**parent material** Ancient layer of rock or more recent geological deposits that can be modified to form soil.

**particulates** Small pieces of solids that are dispersed into the environment.

**passive solar heat system** The design that allows for the entrapment and transfer of heat from the sun to a building without the use of moving parts or machinery.

**patchwork clear-cutting** A method of harvesting one section of timber by clear-cutting while leaving contiguous patches of forest uncut.

**persistence** The long-term retention of toxic substances in the environment due to the stability of their complex organic structures.

**persistent pesticide** A chemical that is retained in the environment in its lethal form over a long period of time.

**persistent pollutant** Pollutants that remain in the environment for many years in an unchanged condition.

**pest** Unwanted organism.

**pesticide** A chemical agent that kills pests.

**petrochemicals** Compounds from crude oil that are used in the manufacture of synthetic organic materials such as herbicides or plastics.

**photochemical smog** The secondary pollutant that is a mixture of complex organic compounds and appears as a yellowish-brown haze produced by the interaction of HC, $NO_x$, and sunlight.

**photosynthesis** The process by which plants manufacture food. Light energy is used to convert carbon dioxide and water to sugar and oxygen.

**photovoltaic cell** Bimetallic sandwich that allows the direct conversion of sunlight to electricity.

**physical environment** The nonliving surroundings of an organism that help to determine its survival.

**phytoplankton** Free-floating microscopic chlorophyll-containing organisms.

**pioneer community** The first group of organisms established in an area.

**plutonium–239** A radioactive isotope produced in a breeder reactor and used as a nuclear fuel.

**point source** Pollution that comes from an obvious localized place.

**policy** A planned course of action on a question or topic.

**pollution** Waste material that people produce in such large quantities that it interferes with our health or well-being.

**polygamous** A mating pattern that is characterized by one male mating with many females.

**population** A group of organisms that is capable of interbreeding and producing offspring.

**post-war "baby boom"** A large increase in the birth rate immediately following World War II.

**potable water** Unpolluted fresh water suitable for human consumption.

**potential energy** The energy of position.

**predator** An organism that kills and eats another organism.

**preservation** To keep from harm or damage; to maintain.

**preservation ethic** Philosophy that considers nature to be so special that it should remain intact.

**pressurized water reactor** A type of nuclear reactor in which water is used to moderate the fission.

**prey** The organism that is killed and eaten by a predator.

**primary consumers** Organisms that eat plants (producers) directly.

**primary pollutant** A material released into the environment that can interfere with human well-being.

**primary sewage treatment** The first stage of removal of impurities from water, generally by simple physical methods such as filtering and settling.

**primary succession** The sequence of events that starts with bare mineral substrate and progresses from a pioneer community to a climax community.

**producers** Organisms that can manufacture food from inorganic compounds and light energy.

**productivity** The amount of goods produced per worker (economics). The amount of biomass produced per unit time or unit area (biological).

**profitability** The extent to which economic benefits exceed the economic costs of doing business.

**protons** The positively charged particles located in the nucleus of an atom.

**public resources** That part of the environment that is owned by everyone.

**radiation** Energy that travels through space in the form of waves or particles (i.e., heat and light).

**radioactive** Unstable nuclei that release particles and energy as they disintegrate.

**radioactive half-life** The time it takes for one half of a sample of a radioactive isotope to spontaneously decompose.

**radioactive waste** Materials contaminated by the use of radioactive isotopes.

**range of tolerance** The ability that organisms have to succeed under a variety of environmental conditions. The breadth of this tolerance is an important ecological characteristic of a population.

**recycling** Reprocessing a material from obsolete or contaminated sources.

**reforestation** The process of establishing new trees on a site.

**renewable resources** Those resources that can be regenerated by natural processes.

**replacement fertility** The number of children per woman needed to just replace the parents.

**representative (republican) form of government** Pattern of government whereby elected individuals represent segments of society.

**reserves** The known deposits from which materials can be extracted profitably with existing technology under present economic conditions.

**resource** A naturally occurring substance that is potentially feasible to extract under prevailing conditions.

**resource exploitation** The use of natural materials by society.

**ribbon sprawl** Urban development of commercial and industrial sites along transportation routes at the edge of a city.

**rodenticides** Pesticides that control rodents, such as rats and mice.

**runoff** Water that flows over the surface of the earth and enters a river system rather than infiltrating the soil.

**rutting season** The time of the year when mating occurs.

**sanitary landfill** An area used for the containment of solid wastes.

**savanna** Warm, tropical grassland with scattered trees maintained by periodic fires.

**scavengers** Animals that feed on animals they did not kill.

**science** A method for gathering and organizing information that involves observation, hypothesis formation, and experimentation.

**secondary consumers** Organisms that eat animals that have eaten plants.

**secondary pollutant** A pollutant that results from the interaction of primary pollutants in the presence of an appropriate energy source.

**secondary recovery** Advanced technological methods used to obtain additional yields from oil pools previously considered exhausted.

**secondary sewage treatment** The removal of impurities from water by the digestive action of various small organisms in the presence of oxygen.

**secondary succession** The sequence of events that progresses from an altered area to mature ecosystem.

**second law of thermodynamics** A statement about energy conversion that says that whenever energy is converted from one form to another, some of the useful energy is lost.

**selective harvest** Cutting only trees of certain ages or species and maintaining other nonharvested organisms.

**separation of powers** Division of governmental responsibilities in the United States among the executive, judicial, and legislative branches.

**sex ratio** Comparison between the number of males and females in a population.

**shaft mining** Mining that uses deep shafts to reach buried deposits.

**simulations** Creation of a model that mimics real situations.

**soft pesticide** One that is more readily biodegradable.

**soil** Organized mixture of minerals, organic compounds, living organisms, air, and water that supports plant life.

**soil horizons** Recognizably different layers within the soil.

**soil profile** Composed of various soil layers or horizons.

**soil structure** The way the various soil particles clump together.

**soil texture** Property of soil determined by the size of the rock fragments.

**solid waste** Unusable or unwanted solid products that result from human activity.

**standard of living** The necessities and luxuries that are essential to a level of existence that is customary within a society.

**stationary growth** That portion of population growth curve in which the population has stopped increasing.

**statutory law** Regulations passed by legislation.

**storm-water runoff** Surface water flow from streets and buildings caused by precipitation.

**strip farming** The alternation of strips of row crops and closely sown crops.

**strip mining** A type of mining in which soil and rock above a mineral deposit is removed to procure the underlying deposit.

**subsidy** A gift from government to private enterprise. It is thought to be prudent when the enterprise is important to the public but is having temporary economic difficulty.

**sulfur dioxide** The chemical compound of sulfur and oxygen that results from the oxidation of a sulfur-containing fossil fuel.

**symbiosis** A close, long-lasting physical relationship between members of two different species.

**symbiotic nitrogen-fixing bacteria** Bacteria that grow within a plant's root system that can convert nitrogen gas from the atmosphere to organic nitrogen compounds the plant can use.

**synergism** The interaction of materials or energy that increases the potential for harm.

**taiga** Northern coniferous forest biome.

**target organism** The specific organism that should be controlled by a pesticide or another control measure.

**tar sands** Deposits of sand that contain dispersed droplets of tarlike organic molecules.

**technological advances** Increasing use of machines to replace human labor.

**temperate deciduous forest** A biome characterized by a seasonal climate in which trees lose their leaves.

**terracing** Construction of flat, horizontal steps on steep slopes, which are then farmed.

**tertiary sewage treatment** A variety of techniques used to remove dissolved pollutants remaining after primary and secondary treatment.

**theoretical science** Deals with increasing the base of scientific knowledge.

**theory** A unifying principle that binds together large areas of scientific knowledge.

**thermal inversion** The condition in which warm air is sandwiched between two layers of cold air.

**thermal pollution** Waste heat that industries release to the environment.

**threshold levels** The minimum amount of something that is required to cause measurable effects.

**tract development** The building of many similar houses on blocks of land near cities.

**transpiration** The evaporation of water from the surfaces of plants.

**trophic levels** A stage in the energy flow through ecosystems.

**tropical rain forest** A biome characterized by high rainfall, extensive plant and animal diversity, and lack of frost.

**tundra** A biome characterized by permanently frozen subsoil and the absence of trees.

**uranium–235** A naturally occurring radioactive isotope of uranium used as fuel in nuclear reactors.

**uranium–238** A stable isotope of uranium that can be converted to radioactive plutonium–239 with a nuclear breeder reactor.

**urban sprawl** Unplanned growth of cities outward from their center.

**water diversion** The physical process of transferring water from one area to another for societal uses.

**watershed** The area drained by a body of water.

**watershed management** Economic, social, and political management of a watershed.

**water table** The top of an aquifer.

**waterways** Depressions on sloping land that allow water to flow off the land.

**wetlands** Land permanently or periodically inundated by water.

**wilderness** Areas that are designated as such cannot be used by humans for any disruptive activities.

**wilderness areas** Naturally occurring areas that have restricted human use to activities that do not change the nature of the area.

**wildlife** All of the undomesticated animals in an area. (Often refers to game species of fish, birds, and mammals.)

**wildlife management** A branch of biology that seeks to maintain populations of game animals through habitat modification and predator control.

**windbreaks** Any structure or planting that reduces the velocity of the wind.

**wood** Central portion of the stem of trees, which is most frequently used for construction or fuel.

**zero energy growth pattern** Trend in energy use in which the amount of energy used remains constant.

**zero population growth** The stabilized growth stage of human population during which births equal deaths and equilibrium is reached.

**zoning** Legal designation of land for specific uses.

**zooplankton** Swimming microscopic animals.

# Credits

## PHOTOGRAPHS

PART OPENERS—**Part One:** © John D. Cunningham; **Part Two:** © David C. Fritts/*Animals Animals;* **Part Three:** © Bob Coyle; **Part Four:** U.S. Department of the Interior; **Part Five:** © Bob Coyle; **Part Six:** © UPI-Bettmann Archives, Inc.

CHAPTER 1 **opener:** © Bob Coyle; **1.1 (left):** © John D. Cunningham, **(right):** © Bob Coyle; **1.3:** © John D. Cunningham; **1.5:** USDA-Soil Conservation Office; **1.8 (left):** Author, **(right):** © Larry Stepanowicz; **1.9:** © Michigan Department of Natural Resources.

CHAPTER 2 **opener:** © Bob Coyle; **2.2:** © H. Armstrong Roberts.

CHAPTER 3 **opener:** © Carolina Biological Supply; **3.6:** © David C. Fritts/*Animals Animals;* **3.7:** © Michael DiSpezio; **3.8:** © John D. Cunningham; **3.9:** © The Nitragin Co., Inc.

CHAPTER 4 **opener:** © Buffalo Bill Historical Center, Cody, Wyoming; **4.2:** © Bob Coyle; **4.3:** © Edward S. Ross; **4.4:** © *The Still Hunt* by J. H. Moser from National Park Service, Jefferson National Expansion Memorial; **4.6:** © *The Buffalo Hunt* by Frederic Remington from Buffalo Bill Historical Center, Cody, Wyoming; **4.7 (top):** *Taking the Robe* by Frederic Remington from Remington Art Museum, Ogdensburg, New York, **(bottom):** photo by Western History Collections, University of Oklahoma; **page 79 (upper left and right):** © John D. Cunningham, **(lower left):** © Edward S. Ross, **(lower right):** © Jerg Kroener; **4.10:** © Russ Kinne/Photo Researchers, Inc.

CHAPTER 5 **opener:** © Bob Coyle; **5.4:** © Edward S. Ross; **5.5:** © Field Museum of Natural History; **5.7:** Department of Natural Resources, Puerto Rico—Douglas J. Pool; **5.8:** © Bob Coyle; **5.9:** © Chris Grajczyk; **5.10:** Bureau of Land Management; **5.11:** © John D. Cunningham; **5.12:** © H. Armstrong Roberts; **5.13 (top):** © Bob Coyle, **(bottom):** © Josephus Daniels/Photo Researchers, Inc.; **5.15:** © Edward S. Ross.

CHAPTER 6 **opener:** © UPI-Bettmann Archives, Inc.; **6.4 (top left):** © Bob Coyle, **(top right):** © UPI-Bettmann Archives, Inc.; **(bottom):** © John D. Cunningham; **page 120 (top left and right):** © The Pennsylvania State University, **(middle and lower left and right):** © Bob Coyle.

CHAPTER 7 **opener:** © Jean-Claude Lejeune; **page 135:** © Owen Franken/Sygma.

CHAPTER 8 **opener:** © UPI-Bettmann Archives, Inc.; **8.1:** © Irven DeVore/Anthro-Photo; **8.2–8.8:** © UPI-Bettmann Archives, Inc.; **8.9:** Department of Energy; **8.10:** © James L. Shaffer; **8.13:** © Japan National Tourist Organization; **page 160:** © Department of Community Development, Davis, CA.

CHAPTER 9 **opener:** © Tennessee Valley Authority; **9.1:** © Field Museum of National History; **9.7:** © National Coal Association; **9.9(a):** U.S. Department of Surface Mining Reclamation and Enforcement, **(b):** U.S. Department of Interior, Office of Surface Mining; **9.13:** © American Petroleum Institute; **9.14:** © John D. Cunningham; **9.15:** © Tennessee Valley Authority.

CHAPTER 10 **opener:** © Combustion Engineering, Inc./Frost Publishing Group; **10.4:** © Consumers Power Company; **10.5:** © UPI-Bettman Archives, Inc.; **10.10:** © TASS from SOVFOTO.

CHAPTER 11 **opener:** © Bob Coyle; **11.2:** U.S. Department of Energy, by Westcott; **11.5 (left):** © UPI-Bettman Archives, Inc., **(right):** © Bob Coyle; **11.6:** © Pacific Gas and Electric Company; **page 221:** © Courtesy of IBM Corporation; **11.7:** © French Embassy Press and Information Division, Michel Brigaud; **11.8:** © Bob Coyle; **page 228:** © Henningson, Durham, and Richardson.

CHAPTER 12 **opener:** © E. R. Degginger/*Animals Animals;* **12.1–12.3:** USDA-Soil Conservation Service; **12.5:** © John D. Cunningham; **12.6:** © Michigan Department of Natural Resources; **12.7:** © Culver Pictures, Inc.; **12.9:** USDA-Soil Conservation Service.

CHAPTER 13 **opener:** USDA-Soil Conservation Service; **13.2:** © Arizona Office of Tourism; **13.8:** © John D. Cunningham; **13.9–13.12 (both), page 276, 13.14–13.15:** USDA-Soil Conservation Service; **13.16(a):** © Fred Mayer/Woodfin Camp and Associates, **(b):** USDA-Soil Conservation Service; **13.17 (a):** © James L. Shaffer, **(b):** USDA-Soil Conservation Service; **13.18 (both):** USDA-Soil Conservation Service; **13.20:** © Bureau of Land Management; **13.21:** © Edward S. Ross.

CHAPTER 14 **opener:** © Bob Coyle; **14.1:** © Historical Pictures Service, Chicago; **14.2:** © UPI-Bettmann Archives, Inc.; **14.3:** © From the Chicago Historical Society; **14.4 (both):** © Culver Pictures, Inc.; **14.6:** USDA-Soil Conservation Service; **14.7:** © Maine Department of Natural History; **14.11:** © Bob Coyle; **14.12:** © Michigan State University Cooperative Extension Service; **14.13:** © Larry Stepanowicz; **page 309:** © UPI-Bettmann Archives, Inc.; **14.14:** © Jacques Jangoux/Peter Arnold, Inc.

CHAPTER 15 **opener:** © Fred Forbes; **15.4:** © Ewing Galloway; **15.5 (left):** © AP/Wide World Photos, **(middle):** USDA-Soil Conservation Service, **(right):** © James L. Shaffer; **15.8:** © AP/Wide World Photos; **15.9:** Author; **15.11:** USDA-Soil Conservation Service.

CHAPTER 16 **opener, 16.6:** © Valmont Industries, Inc.; **16.7:** © John D. Cunningham; **16.8 (left):** © Bob Coyle, **(right):** Author; **16.9 (left):** © Michael DiSpezio, **(right):** © John D. Cunningham; **page 351:** © Clivus Multrum, Inc.; **16.11:** © UPI-Bettmann/George Remaine; **16.13:** © California Department of Water Resources; **16.15:** Author.

CHAPTER 17 **opener:** © John D. Cunningham; **17.1 (top left):** © Bob Coyle, **(top right):** © John D. Cunningham, **(second from top left):** Author, **(second from top right):** © John D. Cunningham, **(third from top left):** EPA-DOCUAMERICA, **(third from top right):** © Bob Coyle, **(bottom left):** © Las Vegas News Bureau, **(bottom right):** © Bob Coyle; **17.2:** © UPI-Bettmann Archives, Inc.; **17.4:** EPA-DOCUAMERICA/Gary Miller; **17.5:** © John D. Cunningham; **17.7:** National Library of Medicine; **17.8:** © Courtesy of the American Lung Association; **17.11:** © AP/Wide World Photos.

CHAPTER 18 **opener:** © H. Armstrong Roberts; **18.1:** © Bob Coyle; **18.3:** © John D. Cunningham; **18.5 (all):** © Bob Coyle; **18.6:** © John D. Cunningham; **18.9:** EPA-DOCUAMERICA; **18.10:** © Michigan Department of Natural Resources; **page 398:** © Dave Dieter.

CHAPTER 19 **opener:** © Bob Coyle; **19.4** © Peter Arnold, Inc.; **19.6 (both):** © Air Pollution Control District; **19.8 (both):** © John D. Cunningham; **19.11:** © Canapress Photo Service, Inc.; **19.13:** © John Maher/EKM-Nepenthe.

CHAPTER 20 **opener:** U.S. Department of Interior; **20.4 (top):** © James L. Shaffer, **(bottom):** © General Filter Company, Ames, Iowa.

CHAPTER 21 **opener:** © James L. Shaffer; **21.1 (top):** © AP/Wide World Photos, Inc., **(bottom):** © James L. Shaffer; **21.7:** © UPI-Bettmann Archives, Inc.

CHAPTER 22 **opener:** © Bob Coyle; **22.4:** © Las Vegas News Bureau; **22.7:** © David Strickler; **22.8 (both):** © Bob Coyle.

CHAPTER 23 **opener:** EPA-DOCUAMERICA; **23.2 (left):** © Michael Putnam/Peter Arnold, Inc., **(right):** © James L. Shaffer; **page 491 (top left, top right, bottom left):** © *Dictionary of American Portraits,* Dover 1967, **(bottom right):** © State Historical Society of Wisconsin; **23.6:**

© UPI-Bettmann Archives, Inc.; **23.7:** © H. Armstrong Roberts; **page 497:** © E. R. Degginger/*Animals Animals.*

CHAPTER 24 **opener:** © Bob Coyle; **24.4:** City of Detroit; **24.5:** © James L. Shaffer.

## ILLUSTRATIONS/FIGURES

CHAPTER 2—**2.4** and **2.12:** From Enger, Eldon D., et al., *Essentials of Allied Health Science.* © 1978 Wm. C. Brown Publishers, Dubuque, Iowa. All Rights Reserved. Reprinted by permission. **2.6:** From Enger, Eldon D., et al., *Concepts in Biology,* 3d. ed. © 1976, 1979, 1982 Wm. C. Brown Publishers, Dubuque, Iowa. All Rights Reserved. Reprinted by permission.

CHAPTER 5—**5.14:** Reprinted with permission of Macmillan Publishing Company, from *Too Many: A Study of Biological Limitations* by George Borgstrom. Copyright © 1969 by George Borgstrom.

CHAPTER 6—**6.1:** Birthrates and deathrates from the 1984 *World Population Data Sheet* of the Population Reference Bureau. **6.2:** Data from "Mortality rates of cottontail rabbits," by Rexford D. Lord, Jr., appearing in *Journal of Wildlife Management,* 25 (1): 33–40, 1961. Copyright 1961, The Wildlife Society, Inc. Washington, D.C. **6.3:** From Arthur Haupt and Thomas T. Kane, *Population Handbook,* Washington, D.C.: Population Reference Bureau, 1978, p. 14. **6.9:** From Jean Van der Tak, Carl Hub, and Elaine Murphy, "Our Population Predicament: A New Look," *Population Bulletin,* Vol. 34, No. 5, Population Reference Bureau, Inc., December, 1979. **6.10:** Data from the 1981 *World Population Data Sheet* of the Population Reference Bureau, Inc. **Illustration on p. 120:** From Enger, Eldon D., et al., *Concepts in Biology,* 3d. ed. © 1976, 1979, 1982 Wm. C. Brown Publishers, Dubuque, Iowa. All Rights Reserved. Reprinted by permission.

CHAPTER 7—**7.3** and **7.4:** Data from the 1984 *World Population Data Sheet* of the Population Reference Bureau, Inc. **7.7:** Data from the 1980 *World's Children Data Sheet* of the Population Reference Bureau, Inc.

CHAPTER 8—**8.11:** Source: *1983 United Nations Statistical Yearbook* and *1984 World Population Data Sheet* of the Population Reference Bureau, Inc. **8.12:** Modified from *Exploring Energy Choices: A Preliminary Report,* Energy Policy Project, copyright the Ford Foundation, 1974.

CHAPTER 9—**9.2:** From *The Earth Sciences,* 2d. ed., by Arthur N. Strahler. Copyright © 1963, 1971 by Arthur N. Strahler. Fig. 22.17 on p. 382. Reprinted by permission of Harper and Row, Publishers, Inc. **9.4:** Adapted with permission from A. N. Strahler, *Planet Earth,* Fig. 8.23 on p. 182. Copyright © 1972 by Arthur N. Strahler. **9.5:** Courtesy of the Exxon Corporation © 1973. **9.12:** From *Man, Energy, Society,* by Earl Cook. W. H. Freeman and Company. Copyright © 1976. Reprinted by permission.

CHAPTER 10—**10.3:** Source: Environmental Protection Agency, in the June–July 1980 issue of *National Wildlife Magazine.*

CHAPTER 11—**11.2:** Diagram from *Solar Energy: A Biased Guide,* in the International Library of Ecology Series by Domus Books, 1977. **11.3:** Adapted from Richie, James, *Successful Alternate Energy Methods.* Edited by Peggy Frohn. Copyright © 1980 by Ideals Publishing Company. Reprinted by permission.

CHAPTER 12—**12.10:** Aldo Leopold, excerpted from *Game Management.* Copyright © 1933 Charles Scribner's Sons; copyright renewed © 1961 Estella B. Leopold. Reprinted by permission of Charles Scribner's Sons.

CHAPTER 13—**13.5:** From *Ecology and Field Biology,* 3d. ed. by Robert Leo Smith. Copyright © 1980 by Robert Leo Smith. Fig. 9.5 on p. 281. Reprinted by permission of Harper and Row, Publishers, Inc. **13.7** and **13.11:** From Dasmann, R. F., *Environmental Conservation.* New York: John Wiley and Sons, 1968.

CHAPTER 14—**14.8:** From Enger, Eldon D., et al., *Concepts in Biology,* 3d. ed. © 1976, 1979, 1982 Wm. C. Brown Publishers, Dubuque, Iowa. All Rights Reserved. Reprinted by permission. **14.10:** From *Pesticides and Human Welfare,* by D. C. Gunn and J. G. H. Stevens, eds. Oxford, England: Oxford University Press, 1976. *Science* 226: 1293, December 14, 1984. **14.15:** Reprinted with permission of Macmillan Publishing Company from *Man and the Environment,* Second Edition, by Arthur S. Boughey. Copyright © 1975 by Arthur S. Boughey.

CHAPTER 16—**16.2:** From *Goode's World Atlas,* 1980. Copyright © by Rand McNally and Company, R. L. 82–5–56. **16.5:** Copyright © 1984 by the National Wildlife Federation. Reprinted from the February/March issue of *National Wildlife Magazine.* **16.14:** From Audubon Action Special Report on the Garrison Diversion Unit. Reprinted by permission.

CHAPTER 17—**17.6:** Copyright © 1971 by the National Wildlife Federation. Reprinted from the August–September issue of *National Wildlife Magazine.* **17.10:** Reprinted with permission of the Idaho Statesman.

CHAPTER 19—**19.9:** From Rhodes, Steven L., and Middleton, Paulete, "Public Pressures, Technical Options: The Complex Challenge of Controlling Acid Rain," *Environment* 25, No. 4 (1983): 6–9. © Helen Dwight Reid Educational Foundation. Reprinted by permission of Heldref Publications.

CHAPTER 21—**12.11:** Courtesy of Bob Conrad.

CHAPTER 24—**pp. 504, 505:** From the December 1983 issue of the Conservation Foundation Letter, 1717 Massachusetts Ave., N.W., Washington, D.C. Reprinted by permission.

# INDEX

Numbers in italics indicate pages on which tables, charts, or maps appear.